DATE DUE

LIGHT SCATTERING BY LIQUID SURFACES AND COMPLEMENTARY TECHNIQUES

SURFACTANT SCIENCE SERIES

Volume 1: NONIONIC SURFACTANTS, edited by Martin J. Schick (*see also Volumes 19 and 23*)

Volume 2: SOLVENT PROPERTIES OF SURFACTANT SOLUTIONS, edited by Kozo Shinoda (*out of print*)

Volume 3: SURFACTANT BIODEGRADATION, by R. D. Swisher (*See Volume 18*)

Volume 4: CATIONIC SURFACTANTS, edited by Eric Jungermann (*see also Volumes 34 and 37*)

Volume 5: DETERGENCY: THEORY AND TEST METHODS (*in three parts*), edited by W. G. Cutler and R. C. Davis (*see also Volume 20*)

Volume 6: EMULSIONS AND EMULSION TECHNOLOGY (*in three parts*), edited by Kenneth J. Lissant

Volume 7: ANIONIC SURFACTANTS (*in two parts*), edited by Warner M. Linfield (*out of print*)

Volume 8: ANIONIC SURFACTANTS–CHEMICAL ANALYSIS, edited by John Cross

Volume 9: STABILIZATION OF COLLOIDAL DISPERSIONS BY POLYMER ADSORPTION, by Tatsuo Sato and Richard Ruch

Volume 10: ANIONIC SURFACTANTS–BIOCHEMISTRY, TOXICOLOGY, DERMATOLOGY, edited by Christian Gloxhuber (*out of print*)

Volume 11: ANIONIC SURFACTANTS–PHYSICAL CHEMISTRY OF SURFACTANT ACTION, edited by E. H. Lucassen-Reynders (*out of print*)

Volume 12: AMPHOTERIC SURFACTANTS, edited by B. R. Bluestein and Clifford L. Hilton

Volume 13: DEMULSIFICATION: INDUSTRIAL APPLICATIONS, by Kenneth J. Lissant

Volume 14: SURFACTANTS IN TEXTILE PROCESSING, by Arved Datyner

Volume 15: ELECTRICAL PHENOMENA AT INTERFACES: FUNDAMENTALS, MEASUREMENTS, AND APPLICATIONS, edited by Ayao Kitahara and Akira Watanabe

Volume 16: SURFACTANTS IN COSMETICS, edited by Martin M. Rieger

Volume 17: INTERFACIAL PHENOMENA: EQUILIBRIUM AND DYNAMIC EFFECTS, by Clarence A. Miller and P. Neogi

Volume 18: SURFACTANT BIODEGRADATION, Second Edition, Revised and Expanded, by R. D. Swisher

Volume 19: NONIONIC SURFACTANTS: CHEMICAL ANALYSIS, edited by John Cross

Volume 20: DETERGENCY: THEORY AND TECHNOLOGY, edited by W. Gale Cutler and Erik Kissa

Volume 21: INTERFACIAL PHENOMENA IN APOLAR MEDIA, edited by Hans-Friedrich Eicke and Geoffrey D. Parfitt

Volume 22: SURFACTANT SOLUTIONS: NEW METHODS OF INVESTIGATION, edited by Raoul Zana

Volume 23: NONIONIC SURFACTANTS: PHYSICAL CHEMISTRY, edited by Martin J. Schick

Volume 24: MICROEMULSION SYSTEMS, edited by Henri L. Rosano and Marc Clausse

Volume 25: BIOSURFACTANTS AND BIOTECHNOLOGY, edited by Naim Kosaric, W. L. Cairns, and Neil C. C. Gray

Volume 26: SURFACTANTS IN EMERGING TECHNOLOGIES, edited by Milton J. Rosen

Volume 27: REAGENTS IN MINERAL TECHNOLOGY, edited by P. Somasundaran and Brij M. Moudgil

Volume 28: SURFACTANTS IN CHEMICAL/PROCESS ENGINEERING, edited by Darsh T. Wasan, Martin E. Ginn, and Dinesh O. Shah

Volume 29: THIN LIQUID FILMS, edited by I. B. Ivanov

Volume 30: MICROEMULSIONS AND RELATED SYSTEMS: FORMULATION, SOLVENCY, AND PHYSICAL PROPERTIES, edited by Robert S. Schecter and Maurice Bourrel

Volume 31: CRYSTALLIZATION AND POLYMORPHISM OF FATS AND FATTY ACIDS, edited by Nissim Garti and Kiyotaka Sato

Volume 32: INTERFACIAL PHENOMENA IN COAL TECHNOLOGY, edited by Gregory D. Botsaris and Yuli M. Glazman

Volume 33: SURFACTANT-BASED SEPARATION PROCESSES, edited by John F. Scamehorn and Jeffrey H. Harwell

Volume 34: CATIONIC SURFACTANTS: ORGANIC CHEMISTRY, edited by James M. Richmond

Volume 35: ALKYLENE OXIDES AND THEIR POLYMERS, by F. E. Bailey, Jr. and Joseph V. Koleske

Volume 36: INTERFACIAL PHENOMENA IN PETROLEUM RECOVERY, edited by Norman R. Morrow

Volume 37: CATIONIC SURFACTANTS: PHYSICAL CHEMISTRY, edited by Donn N. Rubingh and Paul M. Holland

Volume 38: KINETICS AND CATALYSIS IN MICROHETEROGENEOUS SYSTEMS, edited by M. Grätzel and K. Kalyanasundaram

Volume 39: INTERFACIAL PHENOMENA IN BIOLOGICAL SYSTEMS, edited by Max Bender

Volume 40: ANALYSIS OF SURFACTANTS, by Thomas M. Schmitt

Volume 41: LIGHT SCATTERING BY LIQUID SURFACES AND COMPLEMENTARY TECHNIQUES, edited by Dominique Langevin

OTHER VOLUMES IN PREPARATION

LIGHT SCATTERING BY LIQUID SURFACES AND COMPLEMENTARY TECHNIQUES

edited by
Dominique Langevin
Laboratoire de Physique Statistique
Ecole Normale Supérieure
Paris, France

Marcel Dekker, Inc. **New York • Basel • Hong Kong**

Library of Congress Cataloging-in-Publication Data

Light scattering by liquid surfaces and complementary techniques / edited by Dominique Langevin.
p. cm. -- (Surfactant science series; v. 41)
Includes index.
ISBN 0-8247-8607-6
1. Liquids--Surfaces. 2. Light--Scattering. I. Langevin, D. (Dominique). II. Series.
QC176.8.S8L54 1992
530.4'27--dc20 91-41702
CIP

This book is printed on acid-free paper.

Marcel Dekker, Inc.
270 Madison Avenue, New York, New York 10016

Current printing (last digit):
10 9 8 7 6 5 4 3 2 1

PRINTED IN THE UNITED STATES OF AMERICA

Preface

Light scattering is a powerful nonpertubative optical technique that can be applied to the study of liquid surfaces. Although elastic light scattering can in principle be performed, in practice the scattering is frequently dominated by the bulk phases, particularly at large scattering angles, and the contribution of the surface is difficult to extract. However, inelastic small angle scattering by the surface can be accurately studied and in this way, information about the dynamic behavior of the surface can be obtained. The typical length scale of the measurements is in the range 10–100 μ. This means that the technique does not give information about molecular processes, but rather about collective surface excitations, which in this range of length scales are capillary waves. These capillary waves are thermally excited, and their temporal evolution is governed as for all types or capillary waves by surface tension and by surface and bulk viscoelasticity. The time analysis is made as in conventional inelastic light scattering experiments, by studying the frequency dependence of the scattering light. This allows the determination of the value of the surface tension and of the surface and bulk viscoelasticity.

There are many other less sophisticated methods for the measurement of the interfacial tension. The interest of the surface light scattering (SLS) method is the absence of perturbation of the surface. This can be interesting, for instance, when working with anisotropic liquids which can be aligned by the external devices used in the conventional methods. The SLS method can be used in extreme temperature and pressure conditions. The only requirement is that the liquid cell have an optical window to allow the passage of a light beam. The range of surface tension that can be measured with the method is very large and practically unlimited; from above 300 mN/m (mercury) to below 0.0001 mN/m (microemulsion systems). In the low tension range, SLS becomes relatively easy to operate, because the amplitude of the thermally excited capillary waves is large and the scattered light intensity is also large. The only competitive method in this tension range is the spinning drop method.

SLS also gives information about the surface viscoelasticity of adsorbed surface layers. Contrary to most surface rheometers, which allow the measurement of shear viscoelasticity, SLS allows the measurement of compression surface viscoelasticity. The corresponding coefficients are usually much larger than the shear coefficients, especially in the case of soluble surfactant monolayers (typically two orders of magnitude larger). They therefore play an important role in surface motion. They need to be accurately determined to gain more insight into such problems as emulsion and foam stability, spreading of thin liquid films on solid substrates. SLS, being a nonpertubative technique, is particularly well suited for this type of determination.

SLS also gives information about the viscoelasticity of the coexisting bulk phases. Because the frequency range of the technique is larger than for mechanical devices but smaller than for ultrasonic devices, it can be interesting to use it to explore a new frequency range in viscoelastic behavior. This can be helpful, for instance, in the case of concentrated polymer solutions.

The technique is described in detail in Part I of the book; Part II illustrates with many examples, systems in which the method has already proved useful. These include: simple fluids, critical fluids, liquid crystals, polymer solutions, monolayers, lipid bilayers, microemulsions, soap films, and the crystal-melt interface. In each corresponding chapter, the data obtained with SLS are related to the physiochemical properties of the system under study. In particular, the problems of surface viscoelasticity and low interfacial tensions in surfactant systems are discussed in great detail. Several complementary techniques—excited surface waves, light reflectivity and ellipsometry, X-ray and neutron reflectivity, and scattering and fluorescence methods—are presented in Part III. The complementary character of these techniques with SLS is emphasized. The scattering of light by solid surfaces is discussed in Part IV.

One of the aims of the book is to help the reader interested in developing the SLS method, with descriptions of the difficulties involved in the technical realization and many illustrations of the potentials of the technique. The reader not interested in the technique itself will also find extensive up-to-date information about the properties of liquid interfaces.

I am grateful to Elizabeth K. Mann for a critical reading of the manuscript and for her numerous suggestions, including linguistic improvements. I am also greatly indebted to Sandra Beberman and Joseph Stubenrauch from Marcel Dekker, Inc., for their invaluable help, and to my husband and daughters for their almost unlimited patience during the preparation of the book.

Dominique Langevin

Contributors

M. A. Bouchiat Laboratoire de Spectroscopie Hertzienne, Ecole Normale Supérieure, Paris, France

Herman Z. Cummins Department of Physics, The City College, The City University of New York, New York, New York

J. C. Earnshaw Department of Pure and Applied Physics, The Queen's University of Belfast, Belfast, Northern Ireland

Dominique Langevin Laboratoire de Physique Statistique, Ecole Normale Supérieure, Paris, France

Daniel Maystre Laboratoire d'Optique Electromagnétique, University of Aix-Marseille 3, Marseille, France

Jacques Meunier Laboratoire de Physique Statistique, Ecole Normale Supérieure, Paris, France

Kenjiro Miyano Department of Applied Physics, University of Tokyo, Tokyo, Japan

Peter S. Pershan Division of Applied Sciences and Department of Physics, Harvard University, Cambridge, Massachusetts

F. Rondelez Laboratoire de Structure et Reactivité aux Interfaces, Université Pierre et Marie Curie, Paris, France

Mark L. Schlossman* Division of Applied Sciences and Department of Physics, Harvard University, Cambridge, Massachusetts

Lloyd M. Williams Department of Physics, The City College, The City University of New York, New York, New York

*Current affiliation: James Franck Institute, Center for Advanced Radiation Sources, University of Chicago, Chicago, Illinois.

Contents

Part II: Applications

Part III: Complementary Techniques

Part IV: Solid Surfaces

I

General Principles

1

Historical Development

DOMINIQUE LANGEVIN Laboratoire de Physique Statistique, Ecole Normale Supérieure, Paris, France

I. THE EARLY WORK

The scattering of light from liquid surfaces was first predicted by Smoluchowski in 1908 [1]. He made the remark that thermal motion should constantly distort the surface and that the surface should be rough on average. In those special circumstances where the surface energy becomes vanishingly small, as in the vicinity of a critical point, he suggested that surface opalescence should be observed, in addition to the familiar bulk opalescence. The first attempt to check this prediction was made in 1913 by Mandelstam [2] who reported qualitative observations on the critical mixture of carbon disulfide and methyl alcohol. He noted that there was indeed some scattered light in directions not greatly removed from that of regular reflection. In order to interpret his observations, he derived the mean square amplitude of the surface roughness by using the equipartition theorem established by Einstein for thermal fluctuations. By applying a theory of Rayleigh for the diffuse reflection by a rough surface, Mandelstam calculated the intensity of the light scattered in the plane of incidence. He could then account for the observed surface opalescence phenomena. In 1926, A. A. Andronov and M. A. Leontovich [3] and independently R. Gans [4] derived formulas for the scattered light intensity and polarization in all space directions.

A new field of investigations on liquid surfaces was open. At that time, there were many unsolved questions relative to liquid surfaces. In 1925, Raman and Ramdas [5] wrote: "Should not the molecular layer forming the boundary of a liquid be doubly refractive? How is the configuration of the molecules at the boundary of a liquid surface influenced by the thermal agitation? Is the surface layer capable of producing (by reason of the thermal agitation or its optical anisotropy or other cause) the observed elliptic polarization of the reflected light at the Brewster angle?" With these questions in mind, Raman and Ramdas undertook a systematic study of the light scattered by the surface of a large number of liquids. They made visual measures using a pencil of sunlight and reported the following observations. (1) Surface scattering is easier to observe with liquids that produce little bulk scattering, such as water, alcohols, ether, and paraffins; (2) surface scattering increases when the surface tension decreases; (3) surface scattering is the greatest when the angle of observation is close to the critical angle where the light is totally reflected by the interface.

They also studied the surface scattering as a function of temperature for carbon dioxide and of carbon disulfide-methanol mixtures and extended the earlier observations by Mandelstam close to a critical point. They reported a number of observations relative to the polarization of the scattered light. The detailed comparison with the theory was made by Andronov and Leontovich in their paper, who found a remarkable concordance between predicted and measured polarizations; the intensity determinations were not precise enough to obtain more than a good qualitative agreement.

New experiments on methanol surface, using photographic photometry to determine scattered intensities, were undertaken by Hariharan in 1942 [6]. The agreement with the theory was substantially improved, but some discrepancies remained unresolved.

II. THE ROLE OF LASERS

In the 1960s, helium-neon lasers became available. The study of thermal fluctuations in fluids, much easier with these new light sources, underwent a considerable development. The laser sources have greater intensities than spectral lamps, and they are easier to use than sunlight. Laser beams are also better defined in direction, improving the accuracy on the scattering angle and allowing the study of the very small angles where the surface scattering is much larger than bulk scattering. But the main advantage is the good frequency definition, making possible the frequency analysis of the scattered light. There is indeed, as we shall see later, much more and better quality information contained in the spectrum than in the intensity of the scattered light.

When different scattering angles are successively selected, the Fourier image of the surface is constructed. One angle is related to one Fourier component of the surface deformation. This Fourier component behaves like a surface wave of wavelength λ. For the observable Fourier components, λ is smaller than the capillary length, and the waves are capillary waves. Levich has shown that if the liquid viscosity is small, the waves propagate; otherwise they are overdamped [7]. Each Fourier component behaves like a moving diffraction grating: It does not scatter light elastically but changes its frequency by the Doppler effect. In the case of a small-viscosity liquid, one should observe a Brillouin-like frequency doublet; in the case of a large viscosity, spectral broadening only. This was noted by Papoular [8], who associated the Mandelstam results with classical hydrodynamics. The first experiments were made by Katyl and Ingard under very difficult conditions [9]. They detected the light scattered by the free surface of methanol with a Fabry-Perot, whose resolution is intrinsically too small for the phenomena under study; they were unable to resolve the frequency doublet splitting. Since their scattering angles were large, the bulk scattering contributed approximately half of the observed intensity. They later reported results obtained by using light beating spectroscopy; these results were in good agreement with the hydrodynamic theory. At the same time and independently, Bouchiat and Meunier also used light beating spectroscopy to study the free surfaces of ethanol and of glycerol [10]. They confirmed the predictions by Papoular in the two extreme viscosity regimes. The method of light beating spectroscopy is very well suited to the study of surface light scattering by fluids, because of its practically unlimited frequency resolution. It was adopted in all the further studies.

When testing the method with simple fluids, discrepancies between the Papoular theory and the experiments were found for liquids of intermediate viscosities, i.e., in the transition region between propagating and overdamped capillary waves. The use of the power spectrum of a thermally excited harmonic oscillator, having the same characteristic frequencies as the surface, proved to be satisfactory [11]. Later, when studying liquids of large refractive index that scatter more light, new discrepancies were found [12]. Bouchiat and Meunier pointed out that the surface motion was coupled to the bulk, and that extra energy dissipation by the bulk was at the origin of the deviations between the experiment and the harmonic oscillator spectrum. A new theoretical expression was proposed for the spectrum, which properly satisfies the fluctuation dissipation theorem and proved since to be able to describe accurately all the experimental features [13].

It was then established that the study of the spectrum of the light scattered by surfaces of simple fluids allows the determination of the surface tension γ and of the bulk viscosity η. Many other, often more precise methods can be used to measure these quantities. The advantage of the light scattering technique is that the system is studied at thermal equilibrium: The only perturbation to the surface is the light beam, whose intensity can be lowered if necessary. The technique can therefore be used when the simpler methods fail, for instance close to critical points, where such methods introduce substantial perturbations.

III. DEVELOPMENTS OF THE 1970s AND AFTER

In the early 1970s, extensive work was done using the surface light scattering technique to study surface critical opalescence by Meunier and coworkers on carbon dioxide [14], Webb and coworkers on sulfur hexafluoride [15] and cyclohexane-methanol mixtures [16], and Benedek and coworkers on Xenon [17]. Bouchiat and Langevin investigated the free surface of nematic liquid crystals [18] and the nematic-isotropic interface [19]. A new theoretical spectrum had to be derived to take into account the anisotropy of the bulk viscosity of the nematic liquid crystals. The technique was particularly interesting in this case, because more conventional methods to measure bulk viscosity perturb the alignment of the molecules.

In the late 1970s, Hard and Lofgren improved the light beating procedure by adding a diffraction grating as local oscillator [20]. This substantially decreases the sensitivity of the method to mechanical vibrations and allows the study of liquid surfaces of very small scattering power without extremely stringent mechanical stability requirements. Since then, the number of existing experimental devices in the world has increased regularly with time.

The technique has been also applied to monolayers adsorbed at the surface. Monolayers not only change the surface tension; they possess surface viscoelasticity that changes both the peak frequency and the width of the spectrum. Surface light scattering is a unique method to measure surface viscoelasticity. The corresponding coefficients are indeed strongly frequency dependent, and light scattering offers a frequency range inaccessible to other existing techniques. Moreover, these other techniques are frequently perturbative: Differences of one to two orders of magnitude between measurements on the same monolayer using different techniques are currently found in the literature.

The light scattering technique has also been applied to multiphase microemulsion systems where the interfacial tensions are ultralow and where the other existing techniques are not applicable, excepted the spinning drop method developed for the study of these particular systems. However, the light scattering method is the only one applicable under high pressure and eventually high temperature conditions. For this reason, devices have been developed within several oil companies: Exxon, Sohio, and Chevron. The light scattering method is particularly well suited for the measurement of low tensions, because the amplitude of the thermal fluctuations is large in this case and the scattered intensity is easier to detect.

Other systems relevant to the method are thin liquid films—soap films and lipid bilayers, where not only surface tension but also surface viscoelasticity and interactions between limiting surfaces (playing a role in thickness fluctuations) can be studied. The method has also been applied to study bulk viscoelasticity of polymer solutions; the frequency range is interesting because it is intermediate between mechanical and ultrasonic frequencies.

Details on the measurements done in these different types of systems will be found in Part II. They will illustrate the considerable potentials of the technique.

REFERENCES

1. M. von Schmoluchowski, *Ann Physik 25*: 225 (1908).
2. L. Mandelstam, *Ann Physik 41*: 609 (1913).
3. A. A. Andronov and M. A. Leontovich, *Z. Phys. 38*: 485 (1926).
4. R. Gans, *Ann. Physik 79*: 204 (1926).
5. C. V. Raman and L. A. Ramdas, *Proc. Roy. Soc. A108*: 561 (1925) and *A109*: 150, 272 (1925).
6. P. S. Hariharan, *Proc. Indian Acad. Sci. A16*: 290 (1942).
7. V. G. Levich, *Physicochemical Hydrodynamics,* Prentice Hall, Englewood Cliffs, New Jersey, 1962.
8. M. Papoular, *J. de Phys. 29*: 81 (1968).
9. R. H. Katyl and U. Ingard, *Phys. Rev. Lett. 19*: 64 (1967); *20*: 248 (1968).
10. M.-A. Bouchiat, J. Meunier, and J. Brossel, *C. R. Acad. Sci. 266B*: 255 (1968); M.-A. Bouchiat and J. Meunier, *C. R. Acad. Sci. 266B*: 301 (1968); M.-A. Bouchiat and J. Meunier in *Polarization, Matter and Radiation,* Presses Universitaires, Paris, 1969.
11. J. Meunier, D. Cruchon and M.-A. Bouchiat, *C. R. Acad. Sci. 268B*: 92, 422 (1969).
12. D. Langevin, M.-A. Bouchiat, *C. R. Acad. Sci. 272B*: 1422 (1971).
13. M.-A. Bouchiat and J. Meunier, *J. de Phys. 32*: 561 (1971).
14. J. Meunier, *J. de Phys. 30*: 933 (1969); M.-A. Bouchiat and J. Meunier, *Phys. Rev. Lett. 23*: 752 (1969); J.-C. Herpin and J. Meunier, *J. de Phys. 35*: 847 (1974). *N. B.*: In Eq. 1, the expression for (S) must be corrected: In the S^2 term, $m'(m-1)$ and $m(m'-1)$ should be replaced by $m'(m+1)$ and $m(m'+1)$.
15. E. S. Wu and W. W. Webb, *J. de Phys. 33*: C1–149 (1972); E. S. Wu and W. W. Webb, *Phys. Rev. A8*: 2070 (1973).
16. J. S. Huang and W. W. Webb, *Phys. Rev. Lett. 23*: 160 (1969).
17. J. Zollweg, G. Hawkins and G. Benedek, *Phys. Rev. Lett. 27*: 1182 (1971).
18. D. Langevin and M. A. Bouchiat, *J. de Phys. 33*: C1–77 (1972); D. Langevin, *J. de Phys. 33*: 249 (1972); *36*: 745 *(1975); 37:* 901 (1976).
19. D. Langevin and M.-A. Bouchiat, *Mol. Cryst. Liq. Cryst. 22*: 317 (1973).
20. S. Hard, Y. Hamnerius, O. Nilsson, *J. Appl. Phys. 47,* 2433 (1976).

2

Principle of the Technique

DOMINIQUE LANGEVIN Laboratoire de Physique Statistique, Ecole Normale Supérieure, Paris, France

In this chapter, we shall present the basic theoretical principles of the technique in their simplest version. The complete treatments are somewhat lengthy and will be presented in Chapters 3 and 4.

I. MEAN SQUARE VALUE OF THE SURFACE VERTICAL DISPLACEMENT

The vertical displacement of a given point of the surface **r** at a given instant t can be written as a sum of Fourier components:

$$\zeta(\mathbf{r}, t) = \sum_{\mathbf{q}} \zeta_{\mathbf{q}}(\mathrm{t}) e^{iqr} \tag{1}$$

In order to calculate the mean square value of the displacement ζ due to thermal motion, one needs to evaluate the increase in free energy F. Adding the contributions of capillarity and gravity, we obtain

$$F = \int \left\{ \gamma \left[\sqrt{1 + \left(\frac{\partial \zeta}{\partial x}\right)^2 + \left(\frac{\partial \zeta}{\partial y}\right)^2} - 1 \right] + \frac{\rho}{2} g \, \zeta^2 \right\} d^2 r \tag{2}$$

where γ is the surface tension and ρ the liquid density (or the difference between densities of lower and upper fluids $\rho_2 - \rho_1$). Assuming that the derivatives $\partial\zeta/\partial x$ and $\partial\zeta/\partial y$ are small compared to 1 (i.e., that the surface slope is much less than 1):

$$F = \frac{1}{2} \sum_{\mathbf{q}} (\gamma q^2 + \rho g)\, \zeta_{\mathbf{q}}^2 \tag{3}$$

The equipartion theorem then implies for a surface of area unity:

$$\langle \zeta_{\mathbf{q}}^2 \rangle = \frac{k_B T}{\gamma q^2 + \rho g} \tag{4}$$

The total mean square amplitude is $\langle \zeta^2 \rangle = \sum_{\mathbf{q}} \langle \zeta_{\mathbf{q}}^2 \rangle$. By replacing the sum by an integral and introducing an upper cut-off $q_{max} = 1/a$ where a is a molecular size:

$$\langle \zeta^2 \rangle = \int_0^{q_{max}} \frac{k_B T}{\gamma q^2 + \rho g} \frac{q\,dq}{2\pi} = \frac{k_B T}{4\pi\gamma} \ln \frac{\gamma q_{max}^2}{\rho g} = \frac{k_B T}{2\pi\gamma} \ln \frac{l_c}{a} \tag{5}$$

where l_c is the capillary length $(\gamma/\rho g)^{1/2}$. Typically $\gamma \sim 50$ mN/m; $l_c \sim 2$ mm; $a \sim 2$ Å; and $\langle \zeta^2 \rangle^{1/2} \sim 5$ Å. The vertical displacement of the surface is thus extremely small. It increases only as $\gamma^{-1/2}$ when γ decreases, so that it only reaches large values near critical points or in multiphase microemulsion systems. For instance, when $\gamma \sim 10^{-3}$ mN/m, $\langle \zeta^2 \rangle^{1/2} \sim 1000$ Å.

II. SCATTERED INTENSITY PER UNIT SOLID ANGLE

A given Fourier component $\zeta_{\mathbf{q}}$ behaves optically as a diffraction grating and scatters light in equally spaced directions around both the reflected and the refracted beam. Let us consider the case of the reflected beam (Fig. 1) (the

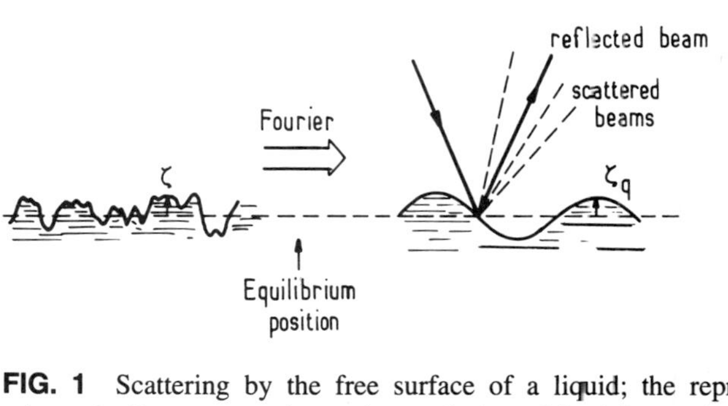

FIG. 1 Scattering by the free surface of a liquid; the represented scattered beams schematically correspond to orders −1, 1, and 2 of the grating equivalent to $\zeta_{\mathbf{q}}$.

refracted beam case is similar). The scattered intensities will be important in only two of these directions: the direction of the regular reflection (order 0) and the adjacent direction (order 1). As the amplitude $\zeta_{\mathbf{q}}$ is much smaller than the wavelength, the scattering in the higher-order directions is negligible. The scattered light is therefore concentrated in a well-defined direction simply related to $\mathbf{q}$ by expressing momentum conservation in the surface plane:

$$\mathbf{q} = \mathbf{k}_{\mathrm{d}_\Sigma} - \mathbf{k}_{\mathrm{r}_\Sigma} \tag{6}$$

where $\mathbf{k}_{\mathrm{d}_\Sigma}$ and $\mathbf{k}_{\mathrm{r}_\Sigma}$ are respectively the projections of the wave vectors of the scattered and reflected beams in the surface plane.

The scattered intensity per unit solid angle around the reflected beam in the limit *of small scattering angles* is given by

$$\frac{dI}{d\Omega} = I_{\mathrm{R}} \frac{k_0^4}{\pi^2} \langle \zeta_q^2 \rangle \cos^3 \theta_{\mathrm{i}} \tag{7}$$

where I_{R} is the reflected intensity, k_0 the modulus of the wave vector of the electromagnetic radiation ($k_0 = 2\pi/\lambda$), and θ_{i} the angle of incidence.

Let us estimate the intensity, considering, for instance, the case of an incident beam normal to the surface of water. If I_{o} is the intensity of the incident beam and n is the refractive index, then

$$\frac{dI}{d\Omega} = I_{\mathrm{o}} \frac{k_0^4}{\pi^2} \left(\frac{n - 1}{n + 1}\right)^2 \frac{k_{\mathrm{B}}T}{\gamma q^2 + \rho g} \tag{8}$$

For a scattering angle of 10 min, using a He-Ne laser ($\lambda = 6328$ Å),

$$\frac{1}{I_{\mathrm{o}}} \frac{dI}{d\Omega} \sim 10^{-3}$$

This is large compared to the scattering by density fluctuations in the bulk of the liquid (which is 2 orders of magnitude lower for a scattering volume of thickness 1 cm). But the surface scattering decreases rapidly as the scattering angle increases, whereas the bulk scattering remains constant. In practice, the scattering angles $\Delta\theta$ never exceed a few degrees. They cannot be less than a few minutes, because of the finite size of the light beam. Eq. 6 yields

$$q = k_0 \, \Delta\theta \tag{9}$$

and the q range is typically (for $k_0 \sim 10^5$ cm^{-1})

$$200 < q < 2000 \text{ cm}^{-1}$$

For these wave vectors, the gravity contribution is negligible compared to the surface tension contribution, ρg and γq^2 respectively in Eqs. 4 and 8. This is because the wavelengths $\lambda_q = 2\pi/q$ of the investigated surface deformations are much smaller than l_c: $30 < \lambda_q < 300\ \mu$.

The scattered intensity depends on the reflection angle through the factor $R \cos^3 \theta_i$, R being the reflection coefficient. Let n be the relative refraction index between the two media, $n = n_t/n_i$, n_i and n_t being the indices for the incident and transmitted beams respectively. When $n > 1$, $\frac{1}{I_o}\frac{dI}{d\Omega}$ varies smoothly θ_i; for $n > 1.2$, the maximum corresponds to normal incidence. When $n < 1$, there is a pronounced maximum when the angle of incidence corresponds to total internal reflection. The scattered intensity at this maximum is much larger than for normal incidence when n is close to 1.

III. SPECTRUM OF THE SCATTERED LIGHT

The angular study of the scattered intensity reflects the spatial distribution of the fluctuations of the surface position. The spectral analysis of the scattered light allows us to gain information on their temporal evolution. It can be shown generally that the intensity of light scattered with a wave vector $\mathbf{k}_d$ and frequency ω_d is such that [1]

$$I(\mathbf{k}_d, \omega_d) \propto P(\mathbf{q}, \omega = \omega_d - \omega_0)$$

ω_0 being the frequency of the incident beam. The proportionality constant depends only on the properties of the incident beam, P only on those of the scattering system: P is the space-time Fourier transform of the correlation function of ζ, and the time Fourier transform of the correlation function of $\zeta_{\mathbf{q}}$: $G(\mathbf{q}, \tau) = \langle \zeta_{\mathbf{q}} (t + \tau)\, \zeta^*_{\mathbf{q}}(t)\rangle$. It can be also shown that the time evolution of the correlation function is the same as the evolution of a well-defined surface deformation mode $\zeta_{\mathbf{q}}$. P can therefore be derived from the hydrodynamic equations subject to appropriate limit conditions at the surface.

After its creation, a surface mode or *ripplon* will tend to disappear under the influence of restoring forces (capillary + gravity) and damping forces due to bulk viscosity in the case of simple fluids. Depending on whether the restoring force is larger or smaller than the damping force, the ripplon will either propagate as a dampled oscillation or be damped exponentially without propagation. It behaves optically as a moving diffraction grating, and the scattered light is frequency shifted by the Doppler effect. If the fluid viscosities η_1 and η_2 are small, the frequency of propagation ω_q and the lifetime of ripplons τ_q are (propagating case)

$$\omega_q = \left(\frac{\gamma}{\rho_1 + \rho_2}\right)^{1/2} q^{3/2} \qquad \frac{1}{\tau_q} = \Delta\omega_q = 2\,\frac{(\eta_1 + \eta_2)}{\rho_1 + \rho_2}\, q^2 \tag{10}$$

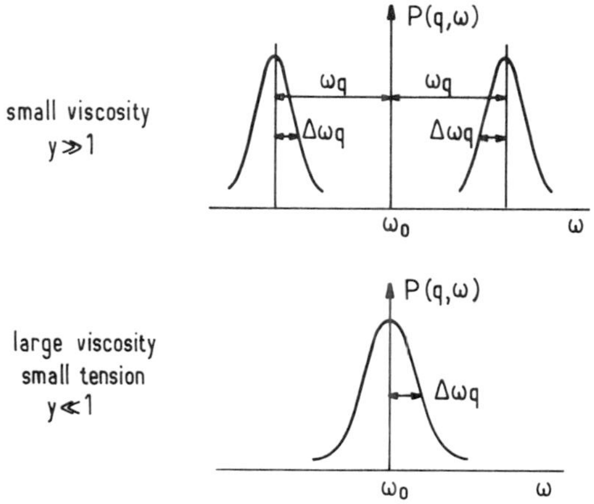

FIG. 2 Schematic aspect of the spectrum of the scattered light in the two extreme cases of liquids of small viscosity ($y \gg 1$) and of large viscosity or small surface tension ($y \ll 1$).

and if the fluid viscosities are large, or if the surface tension γ is very small, their lifetime is (overdamped case)

$$\frac{1}{\tau_q} = \Delta\omega_q = \frac{\gamma q}{2(\eta_1 + \eta_2)} \tag{11}$$

As a consequence, the spectrum of the scattered light has two Lorentzian components symmetric with respect to ω_0 in the first case. They are centered at $\omega_0 \pm \omega_q$, and their half width is $\Delta\omega_q$ (Eq. 10). In the second case, the spectrum is a single Lorentzian curve centered at ω_0, of half width $\Delta\omega_q$ (Eq. 11) (see Fig. 2).

In the extreme limits of large or small bulk viscosities, the analysis of the spectrum is therefore very simple; its components have Lorentzian shapes whose characteristics (frequency, widths) are deduced from the roots of the dispersion equation of surface phonons. But in the intermediate viscosity range, which is encountered very often experimentally, the spectrum is much more complex and will be derived in Chapter 4.

When complex fluids are used, or when monolayers are adsorbed at the surface, other properties like surface and bulk viscoelasticity come into play and can therefore be investigated.

REFERENCE

1. R. Pecora, *J. Chem. Phys. 40:* 1604 (1963).

3

Theoretical Background—Scattered Intensity

DOMINIQUE LANGEVIN Laboratoire de Physique Statistique, Ecole Normale Supérieure, Paris, France

There are several methods allowing the calculation of the intensity scattered by a surface. The method used by Rayleigh and Mandelstam [1] and later by Andronov, Leontovich, and Gans [2, 3] is an extension of the Fresnel calculations of the reflection and refraction coefficients to the case of an undulated surface. We reproduce below this calculation for the simple case of normal incidence, with an incident polarization normal to the incidence plane and a small scattering angle (Fig. 1). We shall assume that the surface roughness is small enough so that only the first-order scattering is important. We have seen in Chapter 2 that typically the amplitude of the roughness integrated over all wavelengths is about 5 Å. The amplitude of a given Fourier component is obviously much smaller.

Grating theories predict that the scattered intensities decrease from one scattering order to the next by a factor of the order of $(\zeta_q/\lambda)^2$. Since $\lambda \sim 5000$ Å, the scattering orders larger than 1 can thus be safely neglected. In the case of solid surfaces, this approximation will not always be valid. This problem will be addressed in Chapter 20.

I. SCATTERED ELECTRIC FIELDS

Electromagnetic theory predicts that when an electromagnetic radiation hits a surface separating two different media, the tangential component of the electric field **E** and the normal component of the magnetic field **B** are continuous. When the media are not magnetic, the tangential component of the magnetic field is also continuous. For planar electromagnetic radiation,

$$\mathbf{B} = \frac{\mathbf{k} \times \mathbf{E}}{\omega} \tag{1}$$

where $k = k_0 = \frac{\omega}{c}$ in vacuum. Let us consider an undulated surface: $z = \zeta_q e^{iqx}$, where **q** is perpendicular to **E** (Fig. 1). The continuity conditions are therefore

$$E_i + E_r = E_t \tag{2}$$

where i, r, and t stand for incident, reflected, and transmitted waves, and

$$B_i - B_r = B_t \tag{3}$$

Contrary to **E, B** is not exactly parallel to the surface. But, since we deal with a calculation to first order in ζ_q and θ, we have neglected the projection corrections

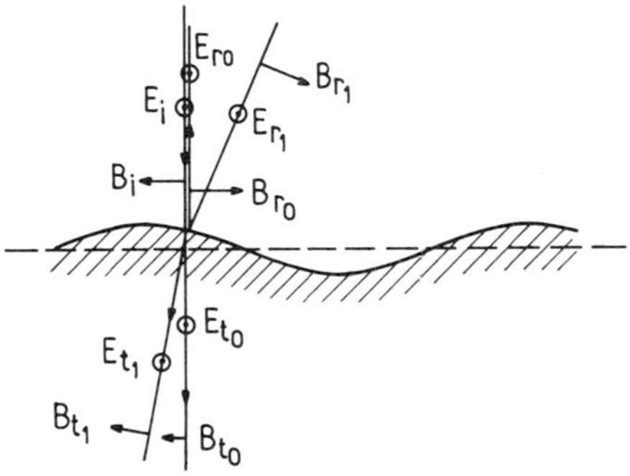

FIG. 1 Scattering by an undulated surface; case of normal incidence and of an electric field polarized perpendicular to the plane of incidence. The subscripts 0 correspond to reflection and refraction, the subscripts 1 to scattering.

on the undulated surface (leading to $q^2\ \zeta_q^2$ terms), and the normal components of the scattered fields (leading to θ^2 terms) (see Fig. 1). Using Eq. 1 and $k = n\omega/c$, where n is the refractive index, Eq. 3 becomes

$$E_i - E_r = nE_t \tag{4}$$

for the interface between vacuum and a medium of index n. Eqs. 2 and 4 remain valid for an interface between two fluids of refractive indices n_i for the incident beam and n_t for the transmitted beam, if one uses $n = n_t/n_i$.

Let us now make explicit the contributions to the first-order scattering in the electric fields

$$\begin{aligned} &\text{incident field} && E_i = e^{ikz} \\ &\text{reflected field} && E_r = A_0\, e^{-ikz} + A_1\, e^{ik(-z\cos\theta + x\sin\theta)} \\ &\text{refracted field} && E_t = B_0\, e^{inkz} + B_1\, e^{ink(z\cos\theta_d + x\sin\theta_d)} \end{aligned} \tag{5}$$

where θ and θ_d are the angles between the normal to the surface and the scattering directions around reflection and refraction respectively and $k = n_i\omega/c$. The continuity conditions must be fulfilled for $z = \zeta_q\, e^{iqx}$. Expanding to first order in ζ_q, this leads to

$$\begin{aligned} 1 + ik\,\zeta_q\, e^{iqx} + A_0(1 - ik\,\zeta_q\, e^{iqx}) + A_1\, e^{ikx\sin\theta} \\ = B_0\,(1 + ink\,\zeta_q\, e^{iqx}) + B_1\, e^{inkx\sin\theta_d} \end{aligned} \tag{6}$$

and

$$\begin{aligned} 1 + ik\,\zeta_q\, e^{iqx} - A_0\,(1 - ik\,\zeta_q\, e^{iqz}) - A_1\, e^{ikx\sin\theta} \\ = nB_0\,(1 + ink\,\zeta_q\, e^{iqx}) + nB_0\, e^{inkx\sin\theta_d} \end{aligned} \tag{7}$$

These equations must be satisfied whatever x is. This leads to

$$\begin{aligned} \sin\theta &= \frac{q}{k} \\ \sin\theta_d &= \frac{q}{nk} \end{aligned} \tag{8}$$

which are the usual grating equations for the first order scattering, and $1 + A_0 = B_0$, $1 - A_0 = nB_0$, that is, $A_0 = \dfrac{1-n}{1+n}$, $B_0 = \dfrac{2}{1+n}$, which are the usual Fresnel conditions, and

$$\begin{aligned} ik\,\zeta_q\,(1 - A_0) + A_1 &= ink\,\zeta_q\, B_0 + B_1 \\ ik\,\zeta_q\,(1 + A_0) - A_1 &= in^2\, k\,\zeta_q\, B_0 + nB_1 \end{aligned}$$

which give, finally,

$$A_1 = B_1 = 2ik\,\zeta_q\, A_0 \tag{9}$$

II. SCATTERED INTENSITY PER UNIT SOLID ANGLE

Let us now integrate the contributions of the scattered electric field in a solid angle $d\Omega$ around the scattering direction. For this purpose, we have to generalize slightly Eqs. 8 to take into account the wave vectors **q** that are not parallel to OX. If ϕ is the azimuth of the scattering plane (Fig. 2), one finds simply from Eq. 6 of Chapter 2:

$$\frac{q_x}{k} = \sin\theta\cos\phi = n\sin\theta_d\cos\phi$$
$$\frac{q_y}{k} = \sin\theta\sin\phi = n\sin\theta_d\sin\phi \tag{10}$$

The surface fluctuations of wave vectors varying in a small domain $dq_x\, dq_y$ around a given value **q** scatter light in the directions comprised in the solid angles $d\Omega$ and $d\Omega_d$ around θ and θ_d respectively such that $d\Omega = \sin\theta_d\, d\theta\, d\phi$ and $d\Omega_d = \sin\theta_d\, d\theta_d\, d\phi$. It follows from Eq. 10 that

$$dq_x\, dq_y = k^2\cos\theta\, d\Omega = n^2 k^2 \cos\theta_d\, d\Omega_d \sim k^2\, d\Omega \sim n^2 k^2\, d\Omega_d$$

The total surface deformation is the integral over **q** wave vectors of ζ_q contributions (Chapter 2). The field amplitude scattered by the surface deformations

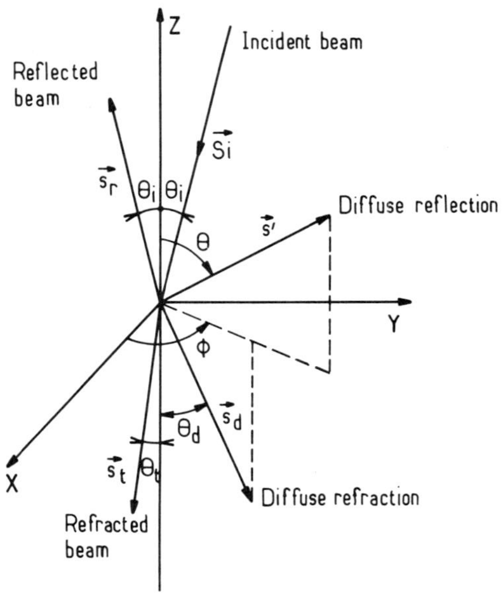

FIG. 2 Angles and unit wavevectors relative to the different directions: incidence, reflection, refraction, and scattering.

in the range $dq_x\, dq_y$ is then

$$dA = \frac{A_1\, dq_x\, dq_y}{(2\pi)^2}$$

the corresponding scattered intensity is

$$dI = \frac{1}{2}\, \varepsilon_0\, c\, \langle |A_1|^2 \rangle\, \frac{dq_x\, dq_y}{(2\pi)^2}$$

the reflected intensity being $I_R = \frac{1}{2}\, \varepsilon_0\, c\, A_0^2$, so that

$$dI = I_R\, \frac{\langle |A_1|^2 \rangle}{A_0}\, \frac{dq_x\, dq_y}{(2\pi)^2}$$

Similarly, the intensity scattered around the transmitted beam is

$$dI_d = I_t\, \frac{\langle |B_1|^2 \rangle}{B_0}\, \frac{dq_x\, dq_y}{(2\pi)^2}$$

The expressions for the intensities scattered per unit solid angle are finally

$$\begin{aligned} \frac{dI}{d\Omega} &= I_R\, \frac{k^2}{4\pi^2}\, \frac{\langle |A_1|^2 \rangle}{A_0^2} = I_R\, \frac{k^4}{\pi^2}\, \langle |\zeta_q|^2 \rangle \\ \frac{dI_d}{d\Omega_d} &= I_t\, \frac{k^2}{4\pi^2}\, \frac{\langle |B_1|^2 \rangle\, n^2}{B_0^2} = I_t\, \frac{k^4}{\pi^2}\, \langle |\zeta_q|^2 \rangle \end{aligned} \tag{11}$$

as quoted earlier in Chapter 2.

III. GENERAL CASE

The scattered electric fields can be calculated by the same method when the incidence is not normal and the scattering angle is not small (Fig. 2). One finds

$$\begin{aligned} \frac{dI}{d\Omega} &= I_R\, \frac{k^4}{4\pi^2}\, \langle |\zeta_q|^2 \rangle\, f(\theta_i, \theta, \phi, n) \\ \frac{dI_d}{d\Omega_d} &= I_t\, \frac{k^4}{4\pi^2}\, \langle |\zeta_q|^2 \rangle\, f_d\, (\theta_i, \theta, \phi, n) \end{aligned} \tag{12}$$

n being the ratio of the refractive indices for the refracted and incident beam respectively, $n = n_t/n_i$. The functions f and f_d are tabulated in Table 1. They simplify considerably when the scattering angles are small; in this case,

$$\begin{aligned} f(\theta_i, \theta, \phi, n) &= 4 \cos^3 \theta_i \\ f_d(\theta_i, \theta, \phi, n) &= \frac{\sin^2 (\theta_i - \theta_t)}{\sin^2 \theta_t}\, n^2 \cos \theta_t \end{aligned} \tag{13}$$

TABLE 1 Angular Factors in the Scattered Intensity for Polarizations Parallel (//) and Perpendicular (⊥) to the Plane of Incidence

$$f_d^{//} = n^2 (n^2 - 1)^2 \cos\theta_d \frac{\sin^2\theta_d}{\sin^2(\theta + \theta_d)} \left[\frac{(\cos\theta_t \cos\theta \cos\phi + \sin\theta_t \sin\theta)^2}{\cos^2(\theta - \theta_d)} + \cos^2\theta_t \sin^2\phi \right]$$

$$f_d^{\perp} = n^2 (n^2 - 1)^2 \cos\theta_d \frac{\sin^2\theta_d}{\sin^2(\theta + \theta_d)} \left[\cos^2\phi + \frac{\cos^2\theta \sin^2\phi}{\cos^2(\theta - \theta_d)} \right]$$

$$f^{//} = (n^2 - 1)^2 \frac{4\cos^2\theta_i \sin^2\theta_t}{\cos^2(\theta_i + \theta_t)\sin^2(\theta_i - \theta_t)} \cos\theta \frac{\sin^2\theta_d}{\sin^2(\theta + \theta_d)}$$

$$\times \left[\cos^2\theta_t \sin^2\phi + \frac{(\cos\theta_t \cos\theta_d \cos\phi - \sin\theta_i \sin\theta)^2}{\cos^2(\theta - \theta_d)} \right]$$

$$f^{\perp} = (n^2 - 1)^2 \frac{4\cos^2\theta_i \sin^2\theta_t}{\sin^2(\theta_i - \theta_t)} \cos\theta \frac{\sin^2\theta_d}{\sin^2(\theta + \theta_d)} \left[\cos^2\phi \frac{\sin^2\phi \cos^2\theta_d}{\cos^2(\theta - \theta_d)} \right]$$

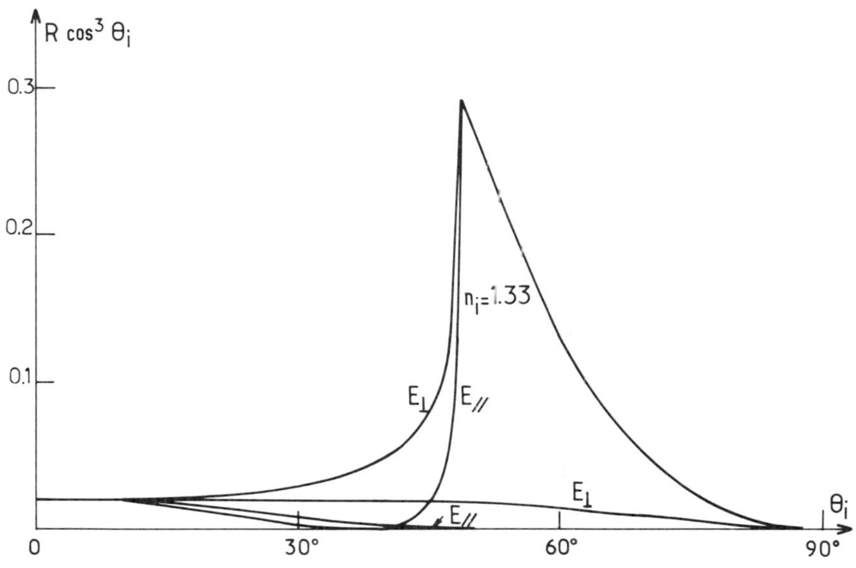

FIG. 3 Reflectivity coefficient times $\cos^3\theta_i$ for polarizations parallel (‖) and perpendicular (⊥) to the plane of incidence in the case of the free surface of water. The upper curves correspond to $n_i = 1.33$, $n_t = 1$, i.e., to an incident beam travelling in the water; the lower curves correspond to $n_t = 1.33$; $n_i = 1$, i.e., to an incident beam travelling in air.

We have represented in Fig. 3 the product of the reflectivity coefficient $R = I_R/I_0$ and $f = \cos^3 \theta_i$ for polarizations parallel (||) and perpendicular ($\perp$) to the plane of incidence, and for an incident beam travelling alternatively in the medium of smaller and larger refractive index. We know from Fresnel that

$$R_{\parallel} = \left|\frac{\tan(\theta_i - \theta_t)}{\tan(\theta_i + \theta_t)}\right|^2 \qquad R_{\perp} = \left|\frac{\sin(\theta_i - \theta_t)}{\sin(\theta_i + \theta_t)}\right|^2 \tag{14}$$

$R_{\parallel}$ and $R_{\perp} = 1$ if θ_i is larger than the limit angle θ_l such as $\sin \theta_l = n$; this is the total external reflection which is found when $n < 1$, i.e., when $n_i > n_t$: the incident beam should travel in the medium of larger refractive index.

It is seen in Fig. 3 that there is a pronounced maximum of the scattered intensity around the limit angle. It is therefore interesting to work around this particular angle when possible.

IV. ALTERNATIVE METHOD

The above method does not allow the calculation of the intensity scattered by other types of thermal fluctuations, for instance concentration fluctuations in a surfactant monolayer. It is then necessary to take better into account molecular details at the surface.

It is known that the electric field reflected by the interface between a vacuum and a medium arises from the radiation of the dipoles created in the medium. The existence of these dipoles also explains why light does not propagate in the medium as fast as in a vacuum: The dipoles excited by the electromagnetic incident waves radiate two electric fields; the first field cancels exactly the incident field, and the second field propagates with a smaller velocity. This is called the extinction theorem [4].

The first calculations of the scattered intensity were an extension of the Fresnel calculations for a flat surface. In the same way, the more complex calculations using dipolar radiation are an extension of the corresponding derivations of reflection and refraction coefficients for a planar surface due to Born and Wolf [4]. These calculations can be easily generalized to treat rough surfaces. More interestingly, the alternative method allows us to calculate the scattering by other types of surface fluctuations [5].

A. Reflected and Refracted Electric Fields

This section summarizes the Born and Wolf calculation [4], to which the reader can refer for further details.

We consider the propagation of an electromagnetic wave in a homogeneous, isotropic, nonmagnetic medium. Let Σ denote the boundary of the medium. The effective electric field $\mathbf{E}^*$ acting on the dipoles is the sum of the incident field $\mathbf{E}_i$ and the dipole field $\mathbf{E}_d$:

$$\mathbf{E}^* = \mathbf{E}_i + \mathbf{E}_d \tag{15}$$

If α is the molecular polarizability, and N the number of molecules per unit volume, then the electric dipole $\mathbf{P}$ per unit volume is given by

$$\mathbf{P} = N\alpha\mathbf{E}^* \tag{16}$$

The contribution from the dipoles is

$$\mathbf{E}_d\,(\mathbf{r}, t) = \int \textbf{rot rot}\,\frac{\mathbf{P}}{R}\left(t - \frac{R}{c}, \mathbf{r}'\right) d^3r' \qquad \text{with } R = |\mathbf{r} - \mathbf{r}'| \tag{17}$$

If the point of observation $\mathbf{r}$ is outside the medium, the integral is taken throughout the entire medium. If it is inside the medium, a small domain occupied by the atom must first be excluded; we consider a small sphere σ of radius a, and proceed to the limit $a \to 0$.

Combining Eqs. 15, 16, and 17 for $\mathbf{r}$ within the medium, we obtain an integral equation for $\mathbf{P}$:

$$\mathbf{P} = N\alpha\left\{\mathbf{E}_i + \int_\sigma^\Sigma \textbf{rot rot}\,\frac{\mathbf{P}}{R}\left(t - \frac{R}{c}, \mathbf{r}'\right) d^3r'\right\}$$

If the incident field is monochromatic, $\mathbf{E}_i(\mathbf{r}, t) = \mathbf{A}_i(\mathbf{r})e^{-i\omega t}$, and propagates with velocity c, we will take, as a trial solution for $\mathbf{P}$, a wave of the form $\mathbf{P}(\mathbf{r}, t) = \mathbf{Q}(\mathbf{r})e^{-i\omega t}$ which has a different velocity of propagation, say c/n. We also assume that there are no charges in the medium: div $\mathbf{P} = 0$.

Defining $G(R) = e^{ik_0R}/R$ with $k_0 = \omega/c$, we obtain

$$\left(1 - \frac{4\pi\,N\alpha}{3}\,\frac{n^2 + 2}{n^2 - 1}\right)\mathbf{Q}(\mathbf{r})$$
$$= N\alpha\left\{\mathbf{A}_i\,(\mathbf{r}) + \frac{\textbf{rot rot}}{k_0^2\,(n^2 - 1)}\int_\Sigma \left[\mathbf{Q}(\mathbf{r}')\,\frac{\partial G(R)}{\partial \nu'} - G(R)\,\frac{\partial \mathbf{Q}(r')}{\partial \nu'}\right] d^2r'\right\}$$

where $\partial/\partial\nu'$ denotes differentiation along the outward normal to the boundary Σ. In this equation, the term on the left is a wave propagated with the velocity c/n, whereas the term on the right represents a wave which, like G, is propagated with the vacuum velocity c. The equality is only possible if each term vanishes separately, so that

$$\frac{4\pi N\alpha}{3} = \frac{n^2 - 1}{n^2 + 2} \tag{18}$$

$$\mathbf{A}_i(\mathbf{r}) + \frac{\textbf{rot rot}}{k_0^2 (n^2 - 1)} \int_\Sigma \left\{ \mathbf{Q} \frac{\partial G}{\partial \nu'} - G \frac{\partial \mathbf{Q}}{\partial \nu'} \right\} d^2 r' = 0 \tag{19}$$

Relation 19 expresses the extinction of the incident wave $\mathbf{E}_i = \mathbf{A}_i\, e^{-i\omega t}$ at any point inside the medium, by interference with part of the dipole field. The incident wave is replaced by another wave, $\mathbf{E}^* = \mathbf{P}/N\alpha$, which is propagated inside the medium with velocity c/n. n is given by Eq. 18 in terms of the number density N and the polarizability α; this relation will be recognized as the Lorentz-Lorentz formula.

The electric field is $\mathbf{E} = 4\pi\mathbf{P}/(n^2 - 1)$. It is related to the effective field by

$$\mathbf{E} = \mathbf{E}^* - (4\pi/3)\mathbf{P}$$

Outside the medium, $\mathbf{P} = 0$, and consequently $\mathbf{E} = \mathbf{E}^* = \mathbf{E}_i + \mathbf{E}_r$. The reflected electric field is therefore the dipole field:

$$\mathbf{E}_r = \int \textbf{rot rot} \frac{\mathbf{P}}{R} \left(t - \frac{R}{c}, \mathbf{r}' \right) d^3 r'$$

which can be written in the form

$$\mathbf{E}_r = \frac{e^{-i\omega t}}{k_0^2 (n^2 - 1)} \textbf{rot rot} \int_\Sigma \left\{ \mathbf{Q} \frac{\partial G}{\partial \nu'} - G \frac{\partial \mathbf{Q}}{\partial \nu'} \right\} d^2 r' \tag{20}$$

We now consider the case of a medium that fills the half-space $z < 0$ and show that this leads to the laws of refraction and reflection. The incident wave is written in the form

$$\mathbf{E}_i = \mathbf{A}_i \exp(i[k_0 \mathbf{r} \cdot \mathbf{s}_i - \omega t])$$

$\mathbf{s}_i$ is the unit vector in the direction of propagation and has components $(-\sin \theta_i, 0, -\cos \theta_i)$ (see Fig. 4). We take as a trial solution for the transmitted wave

$$\mathbf{E} = \frac{4\pi\mathbf{P}}{n^2 - 1} = \mathbf{A}_t \exp(i[k_0\, n\, \mathbf{r} \cdot \mathbf{s}_t - \omega t])$$

$\mathbf{s}_t$ has the components $(-\sin \theta_t, 0, -\cos \theta_t)$. The solution of Eq. 19 can easily be obtained if we consider points of observation whose distance from the boundary $z = 0$ is large compared to the wavelength, i.e., $k_0 \gg 1/R$. We can then carry out the surface integral by the stationary phase method:

$$\mathbf{A}_i\, e^{ik_0\mathbf{r}\cdot\mathbf{s}_i} - \frac{1}{2k_0^2} \textbf{rot rot} \left\{ \mathbf{A}_t \frac{\sin(\phi + \theta_t)}{\cos\phi \sin\theta_t} e^{ik_0\mathbf{r}\cdot\mathbf{s}} \right\} = 0$$

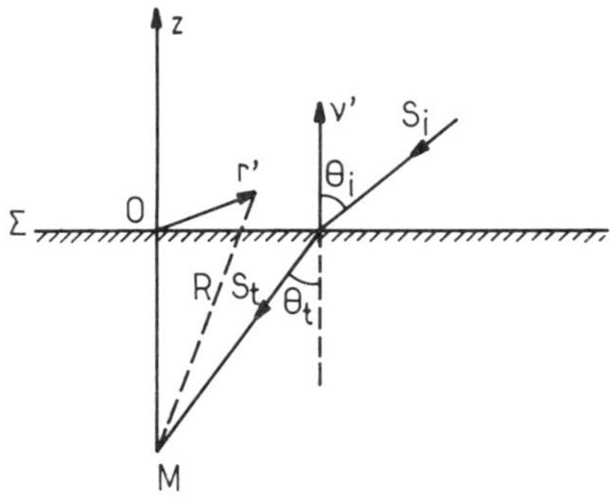

FIG. 4 Refraction at the interface between two transparent media.

where **s** is a vector whose components are $(-\sin\phi, 0, -\cos\phi)$ and ϕ is determined by $n \sin\theta_t = \sin\phi$. The above equality is possible only if $\mathbf{s} = \mathbf{s}_i$. We thus obtain $n \sin\theta_t = \sin\theta_i$, which expresses the law of refraction, and

$$\mathbf{A}_i = -\frac{1}{2}\frac{\sin(\theta_i + \theta_t)}{\sin\theta_t \cos\theta_i}\mathbf{s}_i \times (\mathbf{s}_i \times \mathbf{A}_i) \tag{21}$$

An analogous calculation from Eq. 20 leads to the reflected field:

$$\mathbf{E}_r = \mathbf{A}_r \exp(i[k_0 \mathbf{r} \cdot \mathbf{s}_r - \omega t])$$

$\mathbf{s}_r$ is the vector $(-\sin\theta_r, 0, -\cos\theta_r)$ with $\theta_r = \pi - \theta_i$, which expresses the law of reflection, and

$$\mathbf{A}_r = -\frac{1}{2}\frac{\sin(\theta_t - \theta_i)}{\sin\theta_t \cos\theta_i}\mathbf{s}_r \times [\mathbf{s}_r \times \mathbf{A}_t]. \tag{22}$$

The Fresnel formulas can be easily obtained from Eqs. 21 and 22.

Remark: We suppose here that the transition between the medium and the vacuum is abrupt, and that the value of **P** at the surface Σ does not differ from its value inside the medium: $\mathbf{P} = (n^2 - 1)\mathbf{E}/4\pi$. In a vacuum, $\mathbf{P} = 0$, and in fact the value of **P** at the surface may be intermediate between these two values. The exact calculation of **P** requires detailed knowledge of the nature of the molecular distribution at the surface.

We point out that the hypothesis of a transition region allows one to account for the small ellipticity of the reflected light observed near the Brewster angle (when the polarization of the incident light is rectilinear). One is lead to attribute a thickness of one to two molecular lengths to the transition region. This problem will be discussed in more detail in Chapter 17.

In this section, we retain the hypothesis of an abrupt transition at the surface.

B. Light Scattering by Surface Roughness

We take the plane $z = 0$ as the equilibrium position of the surface and call $\zeta(\mathbf{r}, t)$ the vertical displacement of the point on the surface whose projection on the horizontal plane has the coordinates $\mathbf{r}$: $(x, y, 0)$. We consider a surface fluctuation of wave vector $\mathbf{q}$: $\zeta(\mathbf{r}, t) = \zeta_q(t)\, e^{i\mathbf{q}\cdot\mathbf{r}}$, whose amplitude is very small compared to the wavelength: $k_0\zeta_q \ll 1$. We label the equilibrium values of the variables by an index 0 and their variations due to the fluctuation ζ by the symbol δ. We seek for the scattered field a solution of first order with respect to $k_0\zeta_q$. Finally, we consider observation points at distances from the surface large compared to the wavelength, as in the preceding section.

1. Scattering Around the Refracted Wave

Equations 15 and 16 can be written as

$$\mathbf{P} = \mathbf{P}_0 + \delta\mathbf{P} = N\alpha\, [\mathbf{E}_i + \mathbf{E}_{0d} + \delta\mathbf{E}_d]$$

from which we deduce

$$\delta\mathbf{P} = N\alpha\delta\mathbf{E}_d \tag{23}$$

and Eq. 17 becomes

$$\mathbf{E}_{0d} + \delta\mathbf{E}_d = \int_{\sigma}^{\Sigma'} \mathbf{rot\ rot} \left(\frac{\mathbf{P}_0 + \delta\mathbf{P}}{\mathbf{R}}\right) \left(t - \frac{R}{c}, \mathbf{r}'\right) d^3\,\mathbf{r}'$$

with Σ' the distorted surface. Dropping terms of order larger than $k_0\zeta$ (in the first term, we interchanged the integral symbol and the operator **rot rot** since the function $\mathbf{P}_0/\mathbf{R}$ does not present singularities):

$$\delta\mathbf{E}_d = \mathbf{rot\ rot} \int_{\Sigma} \zeta(\mathbf{r}', t)\, \frac{\mathbf{P}_0}{R} \left(t - \frac{R}{c}, \mathbf{r}'\right) d^2r' + \int_{\sigma}^{\Sigma} \mathbf{rot\ rot}\ \delta\mathbf{P} \left(t - \frac{R}{c}, \mathbf{r}'\right) d^3\,r' \tag{24}$$

with Σ the undistorted surface.

The fluctuation of the dipolar field δE_d is the sum of two terms. The first term is the field radiated by the dipoles in the vicinity of the surface (it has been evaluated neglecting effective field variations at the surface, which should lead to higher-order contributions in $k_0\zeta$); the second term is the variation of the field radiated by all the dipoles when they experience an effective field change due to the surface deformation.

Combining Eqs. 23 and 24, we obtain an integral equation for $\delta\mathbf{P}$ similar to that obtained for $\mathbf{P}_0$ in the last section

$$\delta\mathbf{P} = N\alpha \left\{ \mathbf{rot}\ \mathbf{rot} \int_{\Sigma} \zeta(\mathbf{r}', t) \frac{\mathbf{P}_0}{R} \left(t - \frac{R}{c}, \mathbf{r}'\right) d^2 r' + \int_{\sigma}^{\Sigma} \mathbf{rot}\ \mathbf{rot}\ \delta\mathbf{P} \left(t - \frac{R}{c}, \mathbf{r}'\right) d^3 r' \right\} \quad (25)$$

$\delta\mathbf{P}$ will correspond to a wave of frequency ω, propagating with velocity c/n: $\delta\mathbf{P} = \delta\mathbf{Q}\, e^{-i\omega t}$ with div $\delta\mathbf{P} = 0$ (in fact, Eq. 25 shows that $\delta\mathbf{P}$ varies with time as $\zeta_q\,(t)\, e^{-i\omega t}$. Since the characteristic frequencies for the time variation of ζ are much lower than ω, one can easily verify that this assumption does not noticeably affect the final result).

Transforming the second integral as in the preceding section and with the help of the Lorentz-Lorentz formula, we obtain

$$0 = \mathbf{rot}\ \mathbf{rot} \int_{\Sigma} \left\{ \zeta(\mathbf{r}')Q_0(\mathbf{r}')G(R) + \frac{1}{k_0^2(n^2 - 1)} \left[\delta\mathbf{Q}(\mathbf{r}') \frac{\partial G(R)}{\partial \nu'} - G(R) \frac{\partial \delta\mathbf{Q}(r')}{\partial \nu'} \right] \right\} d^2 r' \quad (26)$$

We take for the transmitted wave a trial solution of the form

$$\mathbf{E} = \mathbf{A}_{0t} \exp(i[k_0 n\mathbf{r} \cdot \mathbf{s}_t - \omega t]) + \delta\mathbf{A}_t \exp(i[k_0 n\mathbf{r} \cdot \mathbf{s}_d - \omega t])$$

where $\mathbf{s}_d$ is the unit vector parallel to the scattering direction with components $(-\sin\theta_d, 0, -\cos\theta_d)$. Let us point out that with this choice ZOX is no longer the incidence plane except when the scattering takes place in this plane.

The integral has been evaluated by the stationary phase method (see the appendix to this chapter). The result is

$$\mathbf{s}_d \cdot \mathbf{r}' = \left(\mathbf{s}_t + \frac{\mathbf{q}}{nk_0}\right) \cdot \mathbf{r}'$$

This equation expresses momentum conservation in the surface plane. $\mathbf{k}_t = nk_0\mathbf{s}_t$ is the wave vector of the refracted wave, $\mathbf{k}_d = nk_0\mathbf{s}_d$ that of the scattered wave. If $\mathbf{k}_{t_\Sigma}$ and $\mathbf{k}_{d_\Sigma}$ represent their projections in the surface plane, then

$$\mathbf{k}_{d_\Sigma} = \mathbf{k}_{t_\Sigma} + \mathbf{q} \quad (27)$$

Note that, since we neglected the time dependence of the fluctuations, $\mathbf{k}_d$ and $\mathbf{k}_t$ have the same modulus.

If we introduce the unit vector $\mathbf{s}\,(-\sin\theta, 0, -\cos\theta)$, where θ is the incidence angle related to the scattered wave inside the medium by $n \sin\theta_d = \sin\theta$, we have the further result

$$\delta\mathbf{A}_t = ik_0\,\zeta_q\,(n^2 - 1)\,\frac{\sin\theta_d}{\sin(\theta + \theta_d)}\left\{\mathbf{A}_{0t} - \frac{\mathbf{s}(\mathbf{s}_d \cdot \mathbf{A}_{0t})}{\cos(\theta - \theta_d)}\right\} \tag{28}$$

For small scattering angles, $q \ll k_0$, $\mathbf{s}_d$ is close to $\mathbf{s}_t$ ($\theta \approx \theta_i$, $\theta_d \approx \theta_t$); the above equation shows that $\delta\mathbf{A}_t$ is almost parallel to $\mathbf{A}_{0t}$, and therefore

$$\delta\mathbf{A}_t = ik_0\zeta_q(n^2 - 1)\,\frac{\sin\theta_t}{\sin(\theta_i + \theta_t)}\,\mathbf{A}_{0t} \tag{29}$$

2. Scattering Around the Reflected Wave

We have seen in Sec. I that the reflected wave arises from the radiation of the dipoles outside the medium:

$$\mathbf{E}_r = \mathbf{E}_{0r} + \delta\mathbf{E}_r = \mathbf{rot\ rot}\int \frac{\mathbf{P}_0 + \delta\mathbf{P}}{R}\left(t - \frac{R}{c}, \mathbf{r}'\right)d^3r'$$

Keeping again only the terms of order $k_0\zeta$:

$$\delta\mathbf{E}_r = \mathbf{rot\ rot}\int_\Sigma \zeta(r')\,\frac{\mathbf{P}_0}{R}\left(t - \frac{R}{c}, \mathbf{r}'\right)d^2r' + \mathbf{rot\ rot}\int \frac{\delta\mathbf{P}}{R}\left(t - \frac{R}{c}, \mathbf{r}'\right)d^3r'$$

After a transformation similar to that of Sec. I,

$$\delta\mathbf{E}_r = e^{-i\omega t}\,\mathbf{rot\ rot}\int_\Sigma\left\{\zeta_q\,e^{i\mathbf{q}\cdot\mathbf{r}'}Q_0(\mathbf{r}')G(R) + \frac{1}{k_0^2(n^2 - 1)}\left[\delta\mathbf{Q}\frac{\partial G}{\partial\nu'} - G\frac{\partial\delta\mathbf{Q}}{\partial\nu'}\right]\right\}d^2r' \tag{30}$$

These integrals have been carried out by the stationary phase method (cf. Appendix). $\delta\mathbf{E}_r$ can be expressed as $\delta\mathbf{E}_r = \delta\mathbf{A}_r \exp(i[k_0\mathbf{r}\cdot\mathbf{s}' - \omega t])$, where $\mathbf{s}'$ is the unit vector parallel to the diffuse reflection direction. It corresponds to an incident wave refracted along $\mathbf{s}_d$, where $\mathbf{s}_d$ is the unit vector for the diffuse refraction. The result of the integration gives

$$\mathbf{s}'\cdot\mathbf{r}' = \left(\mathbf{s}_r + \frac{\mathbf{q}}{k_0}\right)\cdot\mathbf{r}'$$

This equation expresses momentum conservation in the surface plane. $\mathbf{k}_r =$

$k_0\mathbf{s}_r$ is the wave vector of the regular reflected wave, $\mathbf{k} = k_0\ \mathbf{s}'$ that of the scattered wave outside the medium:

$$\mathbf{k}_\Sigma = \mathbf{k}_{r_\Sigma} + \mathbf{q} \tag{31}$$

We have also

$$\delta\mathbf{A}_r = -\mathbf{s}' \times \left\{\mathbf{s}' \times \left[ik_0\ \zeta_q\ (n^2 - 1)\frac{\sin\theta_d}{\sin(\theta + \theta_d)}\right]\left[\mathbf{A}_{0t} - \frac{tg\ (\theta_d - \theta)}{2\cos\theta\sin\theta_d}\mathbf{s}(\mathbf{s}_d \cdot \mathbf{A}_{0t})\right]\right\} \tag{32}$$

Within the approximation of small scattering angles, $q \ll k_0$ ($\mathbf{s}' \approx \mathbf{s}_r$; $\theta_d \approx \theta_t$; $\theta \approx \theta_i$), we obtain

$$\delta\mathbf{A}_r = -\mathbf{s}_r \times \left\{\mathbf{s}_r \times \left[ik_0\ \zeta_q\ (n^2 - 1)\frac{\sin\theta_t}{\sin(\theta_i + \theta_t)}\mathbf{A}_{0t}\right]\right\}$$

and, expressing $\mathbf{A}_{0t}$ in terms of $\mathbf{A}_{0r}$ from Eq. 22,

$$\delta\mathbf{A}_r = -2ik_0\ \zeta_q\ \cos\theta_i\ \mathbf{A}_{0r} \tag{33}$$

Note that $\delta\mathbf{A}_r$ can also be put in the form

$$\delta\mathbf{A}_r = -\mathbf{s}_r \times (\mathbf{s}_r \times \delta\mathbf{A}_t) \tag{34}$$

where we see that the scattering amplitudes $\delta\mathbf{A}_r$ and $\delta\mathbf{A}_t$ outside and inside the medium are very similar.

These results are in complete agreement with the Rayleigh-Mandelstam [1] and Leontovich-Andronov-Gans [2, 3] theory.

V. SCATTERING BY DENSITY FLUCTUATIONS IN A MONOLAYER

We shall now apply the preceding method to density fluctuations in a monolayer.

Let $\delta\Gamma$ be a surface concentration (or density) fluctuation of the film. It creates a variation of the dipole density in the film: $\delta\mathbf{P}_f = (\delta\Gamma/\Gamma_0)\ \mathbf{P}_{0f}$, with $\mathbf{P}_{0f}$ and Γ_0 the polarization and density in the film at equilibrium. If we suppose the film insoluble and if $\mathbf{v}$ is the hydrodynamic velocity, we can write a conservation equation for the mass of the film: $\partial\Gamma/\partial t + \text{div}(\Gamma\mathbf{v}) = 0$. For a fluctuation of wave vector $\mathbf{q}$ parallel to OX, $\delta\Gamma + iq\xi\Gamma_0 = 0$, where ξ is the horizontal displacement of the film along OX. Then

$$\delta\mathbf{P}_f = -iq\xi\mathbf{P}_{0f}$$

These fluctuations will create effective field variations in the liquid substrate and therefore fluctuations of $\mathbf{P}$: $\delta\mathbf{P} = N\alpha\delta\mathbf{E}_d$ (Eq. 23). Consequently, we obtain the following integral equation for $\delta\mathbf{P}$:

$$\delta\mathbf{P} = N\alpha \left\{ \int_{\sigma}^{\Sigma} \mathbf{rot\ rot}\, \frac{\delta\mathbf{P}}{R} \left(t - \frac{R}{c}, \mathbf{r}'\right) d^3 r' + \int_{\Sigma} h\, \mathbf{rot\ rot}\, \frac{\delta\mathbf{P}_f}{R} \left(t - \frac{R}{c}, \mathbf{r}'\right) d^2 r' \right\}$$

where h is the film thickness. This thickness, of the order of 10 Å for the usual monolayers (fatty acids on water, for instance), is much smaller than the wavelength. The effective field is therefore approximately the same in the film and in the underlying liquid very close to the surface. Then

$$\delta\mathbf{P}_f = -iq\xi\, N_f\, \alpha_f\, \mathbf{E}_0^* = -iq\zeta \frac{N_f\, \alpha_f}{N\alpha}\, \mathbf{P}_0$$

where α_f is the molecular polarizability in the film, and N_f is the number of molecules per unit volume in the film.

If we put $\xi(\mathbf{r}, t) = \xi_q(t)\, e^{iqx}$, the calculation of the scattered electric field becomes formally identical to those of Sec. 4, provided ζ_q is replaced by ζ_q $iqhN_f\, \alpha_f/N\alpha$. We have seen in Sec. 4 that the scattered field is proportional to ζ_q. Therefore, in transmission and in reflection, the ratio of the scattered intensities I_x due to density fluctuations and I_z due to amplitude fluctuations of the film are independent of the incidence and scattering angles:

$$\frac{I_x}{I_z} = \frac{\langle|\xi_q|^2\rangle}{\langle|\zeta_q|^2\rangle} (qh)^2 \left(\frac{N_f\, \alpha_f}{N\alpha}\right)^2$$

The mean square value of ξ_q can be derived from thermodynamic considerations (Part II, Chap. 11):

$$\langle|\xi_q|^2\rangle = k_B T/\varepsilon q^2$$

where ε is the compressibility modulus of the film. This modulus plays the same role in the horizontal restoring force opposed to the displacement ξ as the surface tension γ in the vertical restoring force against ζ (cf. Part II, Chap. 11). We deduce (with $\gamma q^2 \gg \rho g$):

$$\frac{I_x}{I_z} = \frac{\gamma}{\varepsilon} (qh)^2 \left(\frac{N_f \alpha_f}{N\alpha}\right)^2$$

qh is typically of the order of 10^{-4} ($q = 1000\ \text{cm}^{-1}$; h $= 10$ Å) and I_x/I_z is generally very small. It will therefore be extremely difficult to observe the scattering by density fluctuation of the monolayer except perhaps in the following cases: (1) when the monolayer polarizability is large (this may be the case, for instance, with a dye monolayer having an absorption band near the wavelength of the incident beam); (2) close to a two-dimensional critical point. Let us recall that a monolayer can be characterized by a surface pressure Π defined by $dF = -\Pi dA$, where dF is the free-energy variation due to surface area variation dA (for more details, see Chap. 11). Π is the two-dimensional analog of the pressure in a fluid, and varies with the surface area. The isotherms $\Pi(A)$ are analogous to the pressure versus volume isotherms of a fluid. The monolayer can exist in several phases: solid, liquid, gas, and the transitions between these different states appear as isotherm levels. Studies in the vicinity of a two-dimensional critical point have shown that the compressibility modulus $\varepsilon = A(d\Pi/dA)$ goes to zero. The scattered intensity by density fluctuations should therefore become very large.

The dipole method can also be applied to other types of surface fluctuations; for example, orientation fluctuations in a two-dimensional nematic, and, at the surface of a three-dimensional nematic, thickness fluctuations of bilayers and soap films [5].

APPENDIX

In this appendix, we compute the surface integrals appearing in Eqs. 26 and 30 with the help of the stationary phase method:

$$\int_\Sigma \zeta(\mathbf{r}')\, \mathbf{Q}_0\,(\mathbf{r}')\, G(R)\, d^2 r' = \left(\zeta_q\, \mathbf{A}_{0t}\, \frac{n^2 - 1}{4\pi}\right) I$$

$$\int_\Sigma \left(\delta\mathbf{Q}\, \frac{\partial G}{\partial \nu'} - G\, \frac{\partial \delta\mathbf{Q}}{\partial \nu'}\right) d^2 r' = \left(\frac{ik_0}{R}\, \delta\mathbf{A}_t\, \frac{n^2 - 1}{4\pi}\right) J$$

with $\mathbf{E} = 4\pi\mathbf{P}/(n^2 - 1)$ and with

$$I = \int_\Sigma \frac{\exp(i[k_0\{n\mathbf{r}' \cdot \mathbf{s}_t + R\} + \mathbf{q} \cdot \mathbf{r}'])}{R}\, d^2 r'$$

$$J = \int_\Sigma \exp(i[k_0 n\mathbf{s}_d \cdot \mathbf{r}' + R]) \left[\frac{\partial R}{\partial \nu'} - n\frac{\partial R}{\partial \nu'}\, (\mathbf{s}_d \cdot \mathbf{r}')\right] d^2 r'$$

We consider two cases: $\mathbf{r}$ inside the medium (half-space $z < 0$) or outside ($z > 0$).

Let us recall that $R = |\mathbf{r} - \mathbf{r}'|$, that Σ is the plane $z = 0$, that $\mathbf{s}_t$ and $\mathbf{s}_d$ are the unit vectors along the regular and diffuse refraction directions ($\mathbf{s}_d$ has components $-\sin\theta_d$, 0, $-\cos\theta_d$), and that $\partial/\partial\nu'$ denotes differentiation along the outward normal to Σ.

We use the stationary phase method in the following form [1]: consider the integral $I = \iint g(x', y') \exp[ik_0 f(x', y')]\, dx'\, dy'$; if x'_j and y'_j are the points at which F is stationary and

$$\alpha_j = \left(\frac{\partial^2 f}{\partial x'^2}\right)_{x'_j, y'_j} \qquad \beta_j = \left(\frac{\partial^2 f}{\partial y'^2}\right)_{x'_j, y'_j} \qquad \gamma_j = \left(\frac{\partial^2 f}{\partial x' \partial y'}\right)_{x'_j, y'_j}$$

then

$$I \approx \frac{2i\pi}{k_0} \sum_j \frac{\sigma_j}{|\alpha_j \beta_j - \gamma_j^2|^{1/2}} g(x'_j, y'_j) \exp[ik_0 f(x'_j, y'_j)] \tag{A1}$$

with

$$\begin{aligned} \sigma_j &= +1 && \text{when} && \alpha_j \beta_j > \gamma_j^2 \text{ and } \alpha_j > 0 \\ &= -1 && \text{when} && \alpha_j \beta_j > \gamma_j^2 \text{ and } \alpha_j < 0 \\ &= -i && \text{when} && \alpha_j \beta_j < \gamma_j^2 \end{aligned}$$

A. Case of an Observation Point Inside the Medium (z < 0)

In the integral J the function f has the form:

$$f(x', y') = n\mathbf{r}' \cdot \mathbf{s}_d + R = -nx' \sin \theta_d + R$$

The components of $\mathbf{r}$ and $\mathbf{r}'$ are (see Fig. 5) $\mathbf{r}(0, 0, -r)$, $\mathbf{r}'(x', y', 0)$, and $R = |\mathbf{r} - \mathbf{r}'| = (x'^2 + y'^2 + r^2)^{1/2}$. Consequently,

$$\frac{\partial f}{\partial x'} = -n \sin \theta_d + \frac{x'}{R} \qquad \frac{\partial f}{\partial y'} = \frac{y'}{R}$$

For f to be stationary, $\partial f/\partial x' = \partial f/\partial y' = 0$. Hence f is seen to be stationary only when $x'_1 = R_1 n \sin \theta_d$, $y'_1 = 0$. At this point, introducing the incidence angle θ relative to the refracted wave along $\mathbf{s}_d$, $n \sin \theta_d = \sin \theta$, we obtain

$$x'_1 = R_1 \sin \theta = r\, tg\theta \qquad y'_1 = 0$$

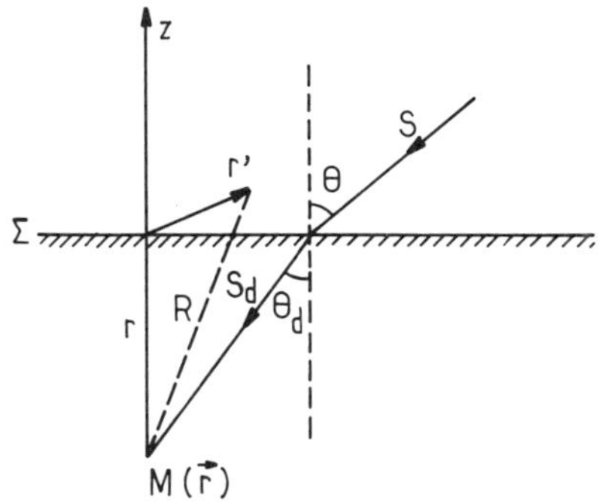

FIG. 5 Scattering at the interface between two transparent media.

If we write now

$$\mathbf{s}'_d \cdot \mathbf{r}' = \left(\mathbf{s}_t + \frac{\mathbf{q}}{nk_0}\right) \cdot \mathbf{r}'$$

the function f in the integral I will have the same form as in J provided $\mathbf{s}_d$ is replaced by $\mathbf{s}'_d$. As these functions are stationary only for one point, the integrals are, from Eq. A1, exponentials $\exp[ik_0 f(x'_1, y'_1)]$. Equality 26 will therefore only be possible, whatever the value of $\mathbf{r}$, if

$$\mathbf{s}'_d \cdot \mathbf{r}' = \mathbf{s}_d \cdot \mathbf{r}'$$

On the other hand,

$$\frac{\partial^2 f}{\partial x'^2} = \frac{1}{R} - \frac{x'^2}{R^3} \qquad \frac{\partial^2 f}{\partial y'^2} = \frac{1}{R} - \frac{y'^2}{R^3} \qquad \frac{\partial^2 f}{\partial x' \partial y'} = \frac{-x'y'}{R^3}$$

and

$$\alpha_1 = \frac{\cos^3 \theta}{r} \qquad \beta_1 = \frac{\cos \theta}{r} \qquad \gamma_1 = 0$$

Consequently,

$$I = \frac{2i\pi}{k_0} \frac{1}{\cos \theta} e^{ik_0 \mathbf{r} \cdot \mathbf{s}}$$

with $\mathbf{s}$ $(-\sin \theta, 0, -\cos \theta)$. Let us calculate the function g in J:

$$\frac{\partial R}{\partial v'} = \frac{r}{R} \qquad \frac{\partial \mathbf{r}' \cdot \mathbf{s}_d}{\partial v'} = s_{zd} = -\cos \theta_d$$

$$g(x', y') = \frac{1}{R}\left(\frac{r}{R} + n \cos \theta_d\right)$$

and

$$g(x_1, y_1) = \frac{\cos \theta}{r} (\cos \theta + n \cos \theta_d) = \frac{\cos \theta \sin(\theta + \theta_d)}{r \sin \theta_d}$$

Thus,

$$J = \frac{2i\pi}{k_0} \frac{\sin(\theta + \theta_d)}{\cos \theta \sin \theta_d} e^{ik_0 \mathbf{r} \cdot \mathbf{s}}$$

Equation 26 can therefore be written in the form

$$0 = \mathbf{rot}\ \mathbf{rot}\left\{(n^2 - 1)\mathbf{A}_{0t}\zeta_q \frac{1}{\cos \theta} + \frac{i\delta \mathbf{A}_t}{k_0} \frac{\sin(\theta + \theta_d)}{\sin \theta_d \cos \theta}\right\} e^{ik_0 \mathbf{r} \cdot \mathbf{s}}$$

so that

$$0 = \mathbf{s} \times \left\{ \mathbf{s} \times \left[\zeta_q(n^2 - 1)\, ik_0\, \mathbf{A}_{0t} - \frac{\sin(\theta + \theta_d)\, \delta\mathbf{A}_t}{\sin \theta_d} \right] \right\}$$

We therefore obtain the following expression for the scattered field amplitude:

$$\delta\mathbf{A}_t = ik_0\, \zeta_q\, (n^2 - 1) \frac{\sin \theta_d}{\sin(\theta + \theta_d)} \left[\mathbf{A}_{0t} - \frac{\mathbf{s}(\mathbf{s}_d \cdot \mathbf{A}_{0t})}{\cos(\theta - \theta_d)} \right]. \qquad \text{(A2)}$$

B. Case Where the Observation Point is Outside the Medium

The calculations are strictly analogous to those of the preceding section, provided the $\mathbf{r}$ components are taken to be (0, 0, r). This is equivalent to replacing s_z by $-s_z$, i.e., to replacing θ by $\theta' = \pi - \theta$; θ' is the reflection angle relative to the direction $\mathbf{s}$ of the incident wave, refracted along $\mathbf{s}_d$. The unit vector along the reflection direction $\mathbf{s}'$ has the components

$$s'_x = -\sin \theta' = -\sin \theta \qquad s'_y = 0 \qquad s'_x = -\cos \theta' = \cos \theta.$$

Taking into account that $\partial R/\partial \nu' = -r/R = -\cos \theta$, we finally obtain

$$I = \frac{2i\pi}{k_0} \frac{1}{\cos \theta} e^{ik_0 \mathbf{r}\cdot\mathbf{s}'}$$

$$J = \frac{2i\pi}{k_0} \frac{\sin(\theta - \theta_d)}{\sin \theta_d \cos \theta} e^{ik_0 \mathbf{r}\cdot\mathbf{s}'}$$

Equation 30 can therefore be written in the form

$$\delta E_r = \frac{1}{2} \textbf{rot rot} \left[\frac{i}{k_0} \frac{(n^2 - 1)}{\cos \theta} \zeta_q\, \mathbf{A}_{0t} + \frac{1}{k_0^2} \delta\mathbf{A}_t \frac{\sin(\theta_d - \theta)}{\sin \theta_d \cos \theta} \right] e^{i(k_0 \mathbf{r}.\mathbf{s}' - \omega t)}$$

If we express $\delta\mathbf{A}_t$ in terms of $\mathbf{A}_{0t}$ using Eq. A2, we obtain the following expression for the scattered field amplitude:

$$\delta\mathbf{A}_r = -\mathbf{s}' \times \left\{ \mathbf{s}' \times \left[ik_0\, (n^2 - 1) \frac{\sin \theta_d}{\sin(\theta + \theta_d)} \right] \left[\mathbf{A}_{0t} - \frac{tg(\theta_d - \theta)}{2 \cos \theta \sin \theta_d} \mathbf{s}(\mathbf{s}_d \cdot \mathbf{A}_{0t}) \right] \right\}$$

REFERENCES

1. L. Mandelstam, *Ann. Physik 41*: 609 (1913).
2. A. A. Andronov and M. A. Leontovich, *Z. Phys. 38*: 485 (1926).

3. R. Gans, *Ann. Physik 79*: 204 (1926).
4. M. Born and E. Wolf, *Principles of Optics,* Chap. 2, Sec. 4, Pergamon, 1959.
5. M.-A. Bouchiat and D. Langevin, *J. Colloid Interface Sci. 63:* 193 (1978).

4

The Spectral Shape Problem

M. A. BOUCHIAT Laboratoire de Spectroscopie Hertzienne*, Ecole Normale Supérieure, Paris, France

*Unité de Recherche de l'ENS et de l'Université Paris VI associeé au CNRS.

The free surface of a liquid is never perfectly planar, because of thermal excitation. As early as 1913–1925, Mandelshtam [1], then Raman and Ramdas [2] were able to study the thermally excited asperities by studying the intensity of scattered light, in particular its dependence on the scattering angle. Much later, in 1968 [3–5], owing to the monochromaticity and the directivity offered by newly developed laser sources, the time evolution of the asperities at last became accessible. This opened new perspectives for the study of dynamic processes occurring at a liquid interface. As mentioned in the introductory chapters of this book, it is somewhat easy to predict the general aspect of the spectrum on the basis of only qualitative arguments. The first observations actually confirmed the anticipated shape: a doublet for a liquid of low viscosity—associated with slowly damped propagating waves—and a unique component centered at the incident light frequency for a highly viscous liquid, where surface waves are highly damped and do not propagate. A potentiality offered by the spectral analysis of the scattered light is the determination of some of the liquid parameters without causing any disturbence of the equilibrium state. As an illustration of measurements, inaccessible to traditional methods, made possible, let us cite the measurements of the surface tension between the liquid and gas phases of a fluid very close to its critical point [6] or the determination of various pressure and viscosity parameters of a film spread on the surface of water [7].

We deal here with the problem of reliably connecting the relevant liquid parameters with the spectral shape of the scattered light. In the two extreme cases of low and high damping, the spectrum consists simply of Lorentzian lines. However, such a simple result no longer holds in intermediate damping conditions; in particular, it is not at all valid close to the critical damping situation (corresponding to the transition from an aperiodic to a damped periodic motion). It turns out that several interesting physical situations correspond precisely to situations of this kind, e.g., the liquid-gas interface of a pure fluid close to its critical point, or the free surface of light oils. Here, for the sake of simplicity, we shall explicitly discuss the case of the free surface of a simple liquid. The computation is easily adapted to other cases where the dispersion law is more complex. At the end of this chapter we shall quote the result relevant for a fluid interface close to the critical point.

The power spectrum of the vertical fluctuations thermally excited on the free surface of a simple liquid is a problem that can be solved by a method suggested by Landau and Placzek [8] and applied by Mountain [9] to density fluctuations in the bulk of a fluid. One seeks the solution of the hydrodynamic equations with well-defined initial values, specially chosen to obtain a correct description of the statistical system at thermal equilibrium. However, for a liquid surface, an additional subtilty occurs and deserves special attention [10]. At first sight one is inclined to consider only the statistical state of the free surface at thermal equilibrium and to deduce the time evolution of the fluctuations from a particular exponential-type solution of the hydrodynamic surface wave problem. From a physical point of view, this procedure is shown to be equivalent to likening the liquid surface to a thin membrane having the same vibration modes. It is also equivalent to considering thermal excitation of capillary waves as that of a free surface phonon gas. The spectrum obtained in this way is identical to the fluctuation spectrum of a thermally excited harmonic oscillator. In Sec. I below we recapitulate this naive approach. Then we discuss and criticize its result, noting that it actually implies an assumption on the statistical state of the bulk phase of the liquid. We can find conditions in which this assumption is unreasonable. For instance, it would imply the existence of a certain correlation between the vertical displacement at the surface and the velocity field in the fluid at the same time, the magnitude of this correlation being particularly large when the damping becomes close to the critical condition. For this reason, we are lead in Sec. II to reconsider the problem, taking into consideration the statistical state of both the surface and the bulk phase of the fluid underneath. The hydrodynamic problem is then solved in its complete generality. The initial values are prescribed so that the displacement at the surface and the simultaneously observed velocity field in the fluid are statistically independent variables. The new spectrum differs from the fluctuation spectrum of a thermally excited harmonic oscillator. We next compute the surface response to an external sinusoidal pressure. This gives us the opportunity of verifying that the modified spectrum satisfies the fluctuation-dissipation theorem. The origin of the spectral shape modification is physically interpreted as resulting from a coupling between the surface motion and the motion that it induces in the fluid underneath. In Sec. III we discuss the deviations of this modified spectrum from the fluctuation spectrum of a thermally excited harmonic oscillator. For damping conditions close to critical damping, these deviations cannot be neglected without causing serious alterations of the parameters, such as the viscosity and the surface tension extracted from the fit of experiment to theory. We show experimental spectra, which agree very well with theory. The inclusion of the surface-bulk phase coupling changes the spectrum by several percent in the wings.

I. NAIVE SIMPLIFIED APPROACH

A. Definitions and Properties of the Correlation Function

In surface light scattering experiments, the variable of interest is the vertical displacement of small surface elements having an arbitrarily specified spatial periodicity. To be specific we consider surface elements of dimensions large at the molecular scale and very small at the scale of the characteristic periodicity length. Let $\zeta(\mathbf{r}, t)$ denote the vertical displacement at time t of the surface element which projects to $\mathbf{r}$ on the horizontal plane. At thermal equilibrium the macroscopic variable $\zeta(\mathbf{r}, t)$ is random. It is defined so that $\overline{\zeta(\mathbf{r}, t)} = 0$. The space-time correlation function $\overline{\zeta(\mathbf{r}, t)\zeta(\mathbf{r} + \boldsymbol{\rho}, t + \tau)}$ satisfies invariance with respect to space and time translation. Consequently, it depends only on $\boldsymbol{\rho}$ and τ and is denoted $g(\boldsymbol{\rho}, \tau)$. It is useful to decompose $\zeta(\mathbf{r}, t)$ into spatial modes

$$\zeta_{\mathbf{q}}(t) = \mathcal{A}^{-1} \int d^2r\, \zeta(\mathbf{r}, t)\, exp(i\mathbf{q} \cdot \mathbf{r})$$

$\mathcal{A}$ being the liquid surface area. Using translational invariance we write

$$\overline{\zeta_{\mathbf{q}}'(t)\, \zeta_{\mathbf{q}}^*(t + \tau)} = \mathcal{A}^{-2} \int d^2\rho\, g(\boldsymbol{\rho}, \tau)\, exp \frac{i(\mathbf{q} + \mathbf{q}') \cdot \boldsymbol{\rho}}{2} \times \int d^2R\, exp\, i(\mathbf{q}' - \mathbf{q}) \cdot \mathbf{R} \tag{1}$$

with $\mathbf{R} = (\mathbf{r}' + \mathbf{r})/2$ and $\boldsymbol{\rho} = \mathbf{r}' - \mathbf{r}$. The above expression cancels unless $\mathbf{q} = \mathbf{q}'$, so that $g(\boldsymbol{\rho}, \tau)$ appears as a summation over the correlation functions $G_{\mathbf{q}}(\tau)$ relative to the various modes:

$$g(\boldsymbol{\rho}, \tau) = \mathcal{A}(2\pi)^{-2} \int d^2q\, \overline{\zeta_{\mathbf{q}}(t)\, \zeta_{\mathbf{q}}^*(t + \tau)}\, e^{i\mathbf{q}\cdot\boldsymbol{\rho}} \tag{2}$$

$$= (2\pi)^{-2} \int d^2q\, G_{\mathbf{q}}(\tau)\, e^{i\mathbf{q}\cdot\boldsymbol{\rho}} \tag{3}$$

For a simple liquid, rotational invariance makes $g(\boldsymbol{\rho}, \tau)$ dependent on $|\boldsymbol{\rho}|$ only, and consequently $G_{\mathbf{q}}(\tau)$ depends on $|\mathbf{q}|$ only.

The light scattering experiments give access either to $G_{\mathbf{q}}(\tau)$ or to its Fourier transform, the surface fluctuation power spectrum

$$P_{\mathbf{q}}(\omega) = (2\pi)^{-1} \int_{-\infty}^{+\infty} d\tau\, G_{\mathbf{q}}(\tau)\, e^{i\omega\tau} \tag{4}$$

for a well-specified wave number (cf. Chap. 2).

B. Statistical Model

The method used to construct the correlation function $\overline{X(0)\, X(\tau)}$ of a state variable X is first presented in quite a general way before it is applied to the case $X = \zeta_{\mathbf{q}}$.

We first introduce what we call a complete set of state variables $\{x_k\}$ for the system under consideration. When the system deviates macroscopically from thermal equilibrium, the instantaneous values of all the variables of this set are sufficient to specify totally at any later time the state of the system governed by the hydrodynamic equations. The index k can be either discrete or continuous. For instance, in our problem, the x_k will include $\zeta_{\mathbf{q}}$ and $\dot{\zeta}_{\mathbf{q}}$.

If we consider an ensemble of identical thermodynamic systems, the instantaneous values $X(t)$, $x_k(t)$ are randomly distributed around the value at equilibrium, supposed to be zero. At a given time $t = 0$, we can concentrate our attention on all the systems that are in the same state, i.e., all the systems with $\{x_k(0)\}$ identical. By convention from here on, this condition will specify a particular class of systems.

To predict the evolution of X for the systems of one class, we admit that the average evolution (i.e., $X(t)$ averaged over the systems of this class) obeys the macroscopic equations relevant to describe the fluid, here the hydrodynamic equations [8]. (This hypothesis was first inferred by Onsager.) Since the deviations with respect to equilibrium are very small, the equations can be linearized. Consequently, we can write $X(t)$ under the quite general form:

$$X(t) = \sum_k x_k(0)\, f_k(t) \qquad t > 0 \tag{5}$$

For a given class of systems we thus obtain the average value of the considered quantity:

$$X(0)\, X(t) = \sum_k X(0)\, x_k(0)\, f_k(\mathrm{t}) \qquad t > 0 \tag{6}$$

We now consider the different classes. From one class to another, $x_k(0)$ varies randomly. However, $f_k(t)$ does not change. Consequently, the right-hand side of Eq. 6 behaves as a well-specified function of time and a random function of $x_k(0)$ When the average is taken over all classes, we finally arrive at the correlation function in the form

$$\overline{X(0)\, X(t)} = \sum_k \overline{X(0)\, x_k(0)}\, f_k(t) \qquad t > 0 \tag{7}$$

Because the random process is stationary, this function must be even in time. Thus we find

$$\overline{X(0)\, X(t)} = \sum_k \overline{X(0)\, x_k(0)}\, f_k(|t|) \tag{8}$$

The average quantities that appear on the right-hand side must be evaluated by applying the principles of statistical mechanics to a system at equilibrium.

In summary, the calculation of the fluctuation spectrum consists first in the resolution of the hydrodynamic initial value problem and secondly in a proper prescription of the initial values to arrive at a correct statistical description of the system at thermal equilibrium. This type of procedure was initially suggested by Landau and Placzek [8]. It has been used by Mountain [9] to study density fluctuations in the bulk phase.

C. A Simple Useful Remark

In view of its practical importance in the following calculations, we shall here insist on a straightforward consequence of Eq. 8. If X is statistically independent of one or several state variables $x_1 \ldots x_j$ (i.e., if $\overline{X(0)\, x_j(0)} = 0$), then the correlation function of X can be built directly from the solution of the hydrodynamic problem that satisfies the initial conditions $x_1(0) = 0 \ldots x_j(0) = 0$.

D. Summary of the Surface Wave Problem and Its Resolution Proposed by Levich

We are dealing with liquid displacements that are transverse (i.e., orthogonal to **q**). In a medium without elastic rigidity, this is possible only in the vicinity of the interface that separates two fluids. The restoring force is essentially a capillary force. Gravity forces are generally involved at wavelengths much longer than those probed by light scattering. Because the restoring force is weak, the motion takes place at low frequencies, so that the fluid can be considered incompressible.

Linearized hydrodynamic equations expressing mass and momentum conservation in the liquid volume are written in the form

$$div\ \mathbf{v} = 0 \tag{9a}$$

$$\frac{\partial \mathbf{v}}{\partial t} = -\frac{1}{\rho}\,\mathbf{grad}\ p + \frac{\eta}{\rho}\,\Delta\mathbf{v} \tag{9b}$$

In addition, the motion must satisfy boundary conditions that express that normal and tangential stresses exerted on a small element of the interface are identical on the two sides of the surface:

$$\left.\begin{aligned} &-p + 2\eta\frac{\partial v_z}{\partial z} - \gamma\frac{\partial^2\zeta}{\partial x^2} = -p_0 && (10a)\\ &\eta\left(\frac{\partial v_x}{\partial z} + \frac{\partial v_z}{\partial x}\right) = 0 && (10b)\end{aligned}\right\}\ \text{at the surface}$$

For the sake of simplicity we consider the free surface of a liquid; p_0 is then the pressure in the gas phase above the liquid. The x axis is chosen along **q.** The z axis is vertical; the liquid occupies the half-space $z < 0$.

The liquid depth is supposed large compared to q^{-1} and the bottom of the container prevents motion in its proximity. Consequently, we can write

$$\mathbf{v} \to 0 \qquad \text{when} \qquad z \to -\infty \tag{10c}$$

As the amplitude of the displacement is small compared to the wavelength, we can make the approximation

$$v_z = \frac{d\zeta}{dt} = \frac{\partial\zeta}{\partial x}\frac{dx}{dt} + \frac{\partial\zeta}{\partial t} \approx \frac{\partial\zeta}{\partial t}$$

It is useful to introduce the two parameters

$$\tau_0 = \frac{\rho}{2\eta q^2} \tag{11a}$$

$$y = \frac{\gamma\rho}{4\eta^2 q} \tag{11b}$$

The method of resolution used by Levich [11] consists in seeking a solution of Eqs. 9 and 10 in the form of a periodic function of the x coordinate and an exponential function of time and of depth, i.e., $\propto e^{iqx}e^{\alpha t}e^{mz}$ (α complex). He thus obtained solutions expressed as the real parts of the functions

$$v_x = \sum_{\nu=1,2} - ie^{-iqx+S_\nu t/\tau_0}\,(A_\nu e^{qz} + B_\nu e^{m_\nu z}) \tag{12a}$$

$$v_z = \sum_{\nu=1,2} e^{-iqx+S_\nu t/\tau_0}(A_\nu e^{qz} + \frac{q}{m_\nu} B_\nu e^{m_\nu z}) \tag{12b}$$

$$\zeta = \sum_{\nu=1,2} e^{-iqx+S_\nu t/\tau_0} \cdot \frac{\tau_0}{S_\nu}\,(A_\nu + \frac{q}{m_\nu} B_\nu) \tag{12c}$$

$$p = p_0 - \sum_{\nu} \frac{S_\nu\rho}{q\tau_0} A_\nu\, e^{-iqx+S_\nu\tau/\tau_0+qz} \tag{12d}$$

with

$$m_\nu = q\sqrt{1 + 2S_\nu} \tag{13}$$

$$\frac{A_\nu}{B_\nu} = -\frac{m_\nu^2 + q^2}{2qm_\nu} \tag{14}$$

S_ν must satisfy simultaneously the dispersion equation

$$y + (S + 1)^2 - \sqrt{1 + 2S} = 0 \tag{15a}$$

and the condition

$$\operatorname{Re} S < 0 \tag{15b}$$

The symbol $\sqrt{}$ specifies the square root whose real part is positive.

For each value of y there are two values S_1 and S_2 that satisfy Eq. 15a and Ineq. 15b. They are real if $y < 0.145$ and complex imaginary conjugates if $y > 0.145$, the system evolution being respectively aperiodic and periodic with damping.

The above solution involves two arbitrary constants, for instance A_1 and A_2. It can be verified that they are totally specified if the initial conditions completely define $\zeta_{\mathbf{q}}$ and $\dot{\zeta}_{\mathbf{q}}$ at $t = 0$:

$$A_\nu = \frac{\zeta_{\mathbf{q}}(0)\, S_\mu - \dot{\zeta}_{\mathbf{q}}(0)\, \tau_0}{(S_\mu - S_\nu)\tau_0}(1 + S_\nu) \tag{16a}$$

$$B_\nu = -\frac{\zeta_{\mathbf{q}}(0)\, S_\mu - \dot{\zeta}_{\mathbf{q}}(0)\, \tau_0}{(S_\mu - S_\nu)\tau_0}\sqrt{1 + 2S_\nu} \tag{16b}$$

(indices ν and μ both hold for 1 or 2 with the convention $\nu \neq \mu$).

E. Calculation of the Correlation Function

Since we have at our disposal a solution of the hydrodynamic problem (Eqs. 12), it seems quite natural to use it and to construct the correlation function of $\zeta_{\mathbf{q}}(t)$. To specify this solution, all we have left to do is to prescribe $\zeta_{\mathbf{q}}(0)$ and $\dot{\zeta}_{\mathbf{q}}(0)$. As explained in Sec. I.B this requires the evaluation of $\overline{|\zeta_{\mathbf{q}}(0)|^2}$ and $\overline{\zeta_{\mathbf{q}}(0)\, \dot{\zeta}_{\mathbf{q}}(0)}$ at thermal equilibrium. The mean square of a thermodynamic fluctuation is evaluated from the minimum work necessary to create the fluctuation, here $R_{min} = \frac{1}{2}\gamma q^2 \mathcal{A}\, \overline{|\zeta_{\mathbf{q}}(0)|^2}$, and consequently

$$\overline{|\zeta_{\mathbf{q}}(0)|^2} = \frac{k_B T}{\gamma q^2 \mathcal{A}} \tag{17a}$$

and

$$\overline{\zeta_{\mathbf{q}}(0)\dot{\zeta}_{\mathbf{q}}(0)} = 0 \tag{17b}$$

This last property is actually valid for any state variable of a physical system obeying classical mechanics. Since $\zeta_{\mathbf{q}}$ and $\dot{\zeta}_{\mathbf{q}}$ considered at the same time are statistically independent, according to a preceeding remark (Sec. I.C) $G_{\mathbf{q}}(\tau)$ can be deduced from Eq. 12c expressed for $\dot{\zeta}_{\mathbf{q}}(0) = 0$, i.e.,

$$\zeta_{\mathbf{q}}(t) = \zeta_{\mathbf{q}}(0)\,(S_2 - S_1)^{-1}\,(S_2 e^{S_1 t/\tau_0} - S_1 e^{S_2 t/\tau_0}) \tag{18}$$

We thus obtain

$$G_{\mathbf{q}}(\tau) = \frac{k_B T}{\gamma q^2}(S_2 - S_1)^{-1}\,[S_2 e^{S_1|\tau/\tau_0|} - S_1 e^{S_2|\tau/\tau_0|}] \tag{19}$$

$$P_{\mathbf{q}}(\omega) = \frac{k_B T}{\pi\gamma q^2}\,\frac{S_2 S_1(S_2 + S_1)\tau_0}{(\omega^2\tau_0^2 - S_1 S_2)^2 + \omega^2\tau_0^2(S_1 + S_2)^2} \tag{20}$$

S_1 and S_2 being the two values of S that simultaneously verify Eq. 15a and in Eq. 15b.

II. THE PROBLEM REVISITED: A MORE RIGOROUS APPROACH

A. Criticism of the Above Calculation

We have recalled a simple calculation of the correlation function of $\zeta_{\mathbf{q}}$ making use of a particular mathematical solution of the hydrodynamic problem. To specify this solution completely, only parameters relative to the surface, as $\zeta_{\mathbf{q}}(0)$ and $\dot{\zeta}_{\mathbf{q}}(0)$, have been required. The correlation function obtained in this way involves the statistical state of the surface but ignores the statistical state of the fluid underneath. In this procedure, the liquid surface is physically likened to an elastic membrane having the same vibration modes. In these conditions, the spectrum is found to be identical to the fluctuation spectrum of the coordinate of a thermally excited harmonic oscillator.

However, if one inquires about the statistical state of the fluid in the bulk phase, it is unambiguously specified, and an important remark arises. The mathematical solution used above gives v_z according to Eq. 12b. If A_ν and B_ν are expressed explicitly in terms of $\zeta_{\mathbf{q}}(0)$ and $\dot{\zeta}_{\mathbf{q}}(0)$, and if we use Fourier transformation with respect to x,

$$\mathbf{v}_{\mathbf{q}}(z, t) = \mathcal{A}^{-1} \int d^2r \, e^{i\mathbf{q}\cdot\mathbf{r}} \mathbf{v}(\mathbf{r}, z, t)$$

then we obtain the vertical component of the velocity at $t = 0$ in the form

$$\begin{aligned} v_{z,\mathbf{q}}(z, 0) = \dot{\zeta}_{\mathbf{q}}(0) \, e^{qz} &+ \frac{\zeta_{\mathbf{q}}(0) \, S_2 - \dot{\zeta}_{\mathbf{q}}(0) \, \tau_0}{(S_2 - S_1)\tau_0} \left(e^{qz} - e^{qz\sqrt{1+2S_1}}\right) \\ &+ \frac{\zeta_{\mathbf{q}}(0) \, S_1 - \dot{\zeta}_{\mathbf{q}}(0) \, \tau_0}{(S_1 - S_2)\tau_0} \left(e^{qz} - e^{qz\sqrt{1+2S_2}}\right) \end{aligned} \tag{21}$$

With the help of Eqs. 17 we deduce that, once solution 12 is adopted, the statistical average of the quantity $\overline{v_{z,\mathbf{q}}(z, 0) \, \zeta^*_{\mathbf{q}}(0)}$ is prescribed exactly as

$$\begin{aligned} \overline{v_{z,\mathbf{q}}(z, 0)\zeta^*_{\mathbf{q}}(0)} = \frac{\overline{|\zeta_{\mathbf{q}}(0)|^2}}{\tau_0} \Bigg\{ e^{qz} &- \frac{S_2}{S_2 - S_1} \, e^{qz\sqrt{1+2S_1}} \\ &+ \frac{S_1}{S_2 - S_1} \, e^{qz\sqrt{1+2S_2}} \Bigg\} \end{aligned} \tag{22}$$

In general, this expression is nonzero, except of course at the surface ($z = 0$). Consequently, the solution used above implies that the fluid velocity in the bulk phase, and the displacement at the surface considered at the same time, are partially correlated. As proved later, such a statistical assumption is unreasonable and actually violates an important symmetry principle. At present, it is worth noting that numerically Eq. 22 strongly depends on the damping conditions of the system: it actually takes very small numerical values, whatever z, when the damping is very weak ($y \gg 1$, $\tau_0 \rightarrow \infty$) or also when it is very strong ($y \ll 1$, $S_1 \lesssim 0.5$, $S_2 \ll 1$). It is essentially for damping conditions close to critical damping ($y \approx 1$) that the numerical values obtained for $\overline{v_{z,\mathbf{q}}(z, 0)\, \zeta_{\mathbf{q}}^*(0)}$ are the largest. For instance, Fig. 1 indicates the dependence of this quantity on qz for a particular value close to critical damping. This will help us to understand why it is more particularly in these damping conditions that the approximation of the free surface to a membrane fails.

We shall now reconsider the problem of the displacement correlation function. Our main goal is to arrive at a solution of the hydrodynamic initial value problem with a fixed value of $\zeta_{\mathbf{q}}$ and an initial velocity field in the fluid that is zero everywhere. Then we shall know how to construct the correlation function so that the variables $\zeta_{\mathbf{q}}(t)$ and $v_{z,\mathbf{q}}(z, t)$, considered anywhere at the same time, shall be statistically independent. However, it is important first to justify rigorously why such a condition must be prescribed.

If the Hamiltonian that describes the microscopic state of the atoms is supposed to be invariant under time reversal in a zero magnetic field, then the probability density must be invariant under the transformation that changes all atomic velocities into their opposite while the coordinates are unchanged. Consequently, any physical quantity that is odd in such a transformation must have its statistical average vanish at equilibrium [12]. In particular, this leads to $\overline{v_{z,\mathbf{q}}(z, t)\, \zeta_{\mathbf{q}}^*(t)} = 0$.

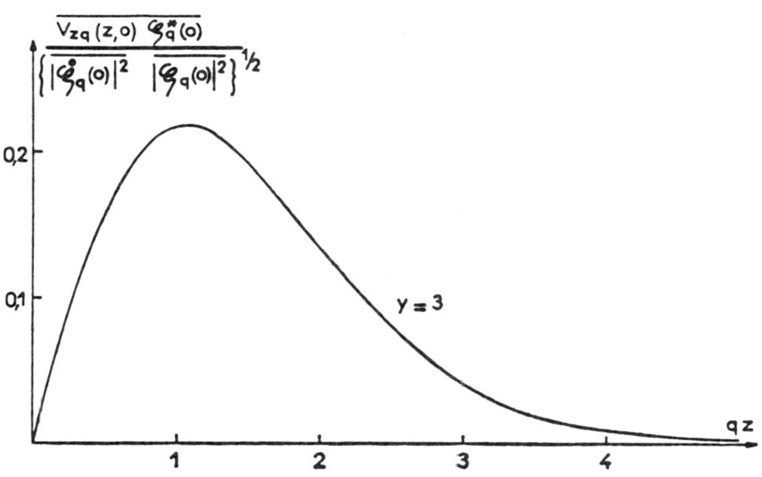

FIG. 1 The figure illustrates the criticism of the naive approach used in Sec. I.E by presenting, versus qz, (for $y = 3$) the values of the correlation between the displacement and the velocity field, while this correlation should vanish everywhere.

At this point it may look surprising that, in the preceeding calculation, the above condition was not automatically satisfied in the bulk phase although it was imposed at the surface. The reason is that $\zeta_{\mathbf{q}}(0)$ and $\dot{\zeta}_{\mathbf{q}}(0)$ are sufficient to specify the Levich solution, because it is a particular solution (of exponential type). On the contrary, as will appear below, $\zeta_{\mathbf{q}}(0)$ and $\dot{\zeta}_{\mathbf{q}}(0)$ are insufficient to specify the most general solution of the hydrodynamic problem. In other words, $\zeta_{\mathbf{q}}$ and $\dot{\zeta}_{\mathbf{q}}$ do not actually constitute a complete set of state variables. We shall demonstrate that this set, to be complete, must also include $\mathrm{v}_{z,\mathbf{q}}(z, 0)$ as a function of z-satisfying condition 10c. Next we shall determine the unique solution of the hydrodynamic problem that satisfies the complete set of initial conditions corresponding to $\zeta_{\mathbf{q}}(0)$ given and $\mathrm{v}_{z,\mathbf{q}}(z, 0) = 0$ everywhere. Using this solution and the remark of Sec. I.C, we shall easily obtain the fluctuation spectrum within the assumption of a statistical state corresponding to thermal equilibrium, i.e.,

$$\overline{|\zeta_{\mathbf{q}}(0)|^2} = \frac{k_B T}{\gamma q^2 \mathcal{A}} \tag{23a}$$

$$\overline{\mathrm{v}_{z,\mathbf{q}}(z, t)\, \zeta_{\mathbf{q}}^*(t)} \equiv 0 \tag{23b}$$

B. General Resolution of the Hydrodynamic Initial Value Problem

Introducing $\mathbf{v}_{\mathbf{q}}$ and $p_{\mathbf{q}}$, the Fourier transforms of $\mathbf{v}$ and $p - p_0$ with respect to $\mathbf{r}$, Eq. 9 relative to the bulk phase becomes

$$iq\mathrm{v}_{x,\mathbf{q}} + \frac{\partial}{\partial z}\mathrm{v}_{z,\mathbf{q}} = 0 \tag{24a}$$

$$\rho \frac{\partial}{\partial t}\mathrm{v}_{z,\mathbf{q}} = \eta\left(-q^2\,\mathrm{v}_{z,\mathbf{q}} + \frac{\partial^2}{\partial z^2}\mathrm{v}_{z,\mathbf{q}}\right) - \frac{\partial}{\partial z}p_{\mathbf{q}} \tag{24b}$$

$$\rho \frac{\partial}{\partial t}\mathrm{v}_{x,\mathbf{q}} = \eta\left(-q^2\,\mathrm{v}_{x,\mathbf{q}} + \frac{\partial^2}{\partial z^2}\mathrm{v}_{x,\mathbf{q}}\right) - iqp_{\mathbf{q}} \tag{24c}$$

We eliminate $\mathrm{v}_{x,\mathbf{q}}$ and $\mathrm{v}_{z,\mathbf{q}}$ and obtain

$$-q^2 p_{\mathbf{q}} + \frac{\partial^2}{\partial z^2}p_{\mathbf{q}} = 0 \tag{25}$$

The solution of this equation that satisfies the boundary conditions of Eqs. 10 is expressed in the form

$$p_{\mathbf{q}} = e^{qz}\left\{\gamma q^2\,\zeta_{\mathbf{q}} + 2\eta\left[\frac{\partial}{\partial z}\mathrm{v}_{z,\mathbf{q}}\right]_{z=0}\right\} \tag{26}$$

When this solution is taken back to Eq. 24b, this latter becomes a partial differential equation relative to $v_{z,\mathbf{q}}$:

$$\rho \frac{\partial}{\partial t} v_{z,\mathbf{q}} = \eta\left(-q^2 v_{z,\mathbf{q}} + \frac{\partial^2}{\partial z^2} v_{z,\mathbf{q}}\right) - qe^{qz}\left\{ \gamma q^2 \zeta_{\mathbf{q}} + 2\eta \left[\frac{\partial}{\partial z} v_{z,\mathbf{q}}\right]_{z=0}\right\} \tag{27}$$

With a set of boundary conditions reexpressed in terms of $v_{z,\mathbf{q}}$ alone, we have

$$\frac{\partial^2}{\partial z^2} v_{z,\mathbf{q}}(0, t) + q^2 v_{z,\mathbf{q}}(0, t) = 0 \tag{28b}$$

$$v_{z,\mathbf{q}}(z, t) \to 0 \qquad \text{when} \qquad z \to -\infty \tag{28c}$$

To take into account the initial conditions that concern $\zeta_{\mathbf{q}}$ and $v_{z,\mathbf{q}}$ at $t = 0$, it is convenient to introduce the Laplace transforms of $v_{z,\mathbf{q}}$ and $\zeta_{\mathbf{q}}$ with respect to t:

$$\tilde{v}_{z,\mathbf{q}}(z, s) = \int_0^\infty dt\, e^{-st}\, v_{z,\mathbf{q}}(z, t) \tag{29a}$$

$$\tilde{\zeta}_{\mathbf{q}}(s) = \int_0^\infty dt\, e^{-st}\, \zeta_{\mathbf{q}}(t) \tag{29b}$$

We shall use the relation

$$\tilde{v}_{z,\mathbf{q}}(z = 0, s) = -\zeta_{\mathbf{q}}(0) + s\tilde{\zeta}_{\mathbf{q}}(s) \tag{30}$$

easily derived by integrating by parts.

Taking the Laplace transform of both sides, Eq. 24 becomes

$$\rho s\tilde{v}_{z,\mathbf{q}}(z, s) - \eta\left[-q^2 \tilde{v}_{z,\mathbf{q}}(z, s) + \frac{\partial^2}{\partial z^2} \tilde{v}_{z,\mathbf{q}}(z, s)\right] + qe^{qz}\left\{ \gamma q^2 \zeta_{\mathbf{q}}(s) + 2\eta\left[\frac{\partial}{\partial z} \tilde{v}_{z,\mathbf{q}}(z, s)\right]_{z=0}\right\} = \rho v_{z,\mathbf{q}}(z, 0) \tag{31}$$

If we multiply both sides by e^{-qz} and differentiate them with respect to z, we then obtain a third-order differential equation with respect to z:

$$\frac{d}{dz}[e^{-qz}(au'' + bu)] = \frac{d}{dz}[e^{-qz}\varphi(z)] \tag{32}$$

or

$$au''' - aqu'' + bu' - bqu = \psi(z) \tag{33}$$

with

$$a = -\eta \qquad b = s\rho + \eta q^2 \tag{34a}$$

$$\varphi(z) = \rho v_{z,\mathbf{q}}(z, t = 0) \tag{34b}$$

$$\psi(z) = \frac{d}{dz}\varphi(z) - q\,\varphi(z) \tag{34c}$$

$$u(z) = \tilde{v}_{z,\mathbf{q}}(z, s) \tag{34d}$$

s being considered a parameter.

We are interested in the solution of Eq. 33 that satisfies the set of conditions

$$au''(0) + \alpha u'(0) + (b + \beta)\,u(0) = \delta \tag{35a}$$

$$u''(0) + q^2\,u(0) = 0 \tag{35b}$$

$$u(z) \rightarrow 0 \qquad \text{when} \qquad z \rightarrow -\infty \tag{35c}$$

The first condition is deduced from Eq. 31 rewritten for $z = 0$ with the help of Eq. 30. The new notations are defined as

$$\beta = \frac{\gamma q^3}{s} \qquad \alpha = 2\eta q \qquad \delta = \frac{-\gamma q^3\,\zeta_{\mathbf{q}}(0)}{s} + \rho\,\dot{\zeta}_{\mathbf{q}}(0) \tag{36}$$

Conditions 35b and 35c are equivalent to 28b and 28c.

The mathematical problem to be solved has been reduced to a particularly simple form. Its solution is unique. This is easily verified by noting that if two functions u_1 and u_2 both satisfy Eqs. 33 and 35, their difference is a solution of Eqs. 33 and 35 which corresponds to the particular case $\delta = 0$ and $\psi(z) = 0$. Only a function that vanishes identically can fulfill such conditions, so that $u_1 - u_2 \equiv 0$.

The important result established above is that $\zeta_{\mathbf{q}}(0)$ and $v_{z,\mathbf{q}}(z, 0)$ constitute a complete set of state variables. Using now the remark presented in Sec. I.C we need only write explicitly the solution that satisfies the initial conditions $v_{z,\mathbf{q}}(z, 0) \equiv 0$ and $\zeta_{\mathbf{q}}(0)$ given. This means that we seek the solution of Eqs. 33 and 35 in the particular case $\Psi(z) \equiv 0$, i.e.,

$$\tilde{v}_{z,\mathbf{q}}(z, s) = f(s)\,e^{qz} + g(s)\,e^{m(s)z} \tag{37a}$$

with

$$m(s) = q\sqrt{1 + 2s\tau_0} \tag{37b}$$

If f and g are written explicitly we obtain

$$\tilde{v}_{z,\mathbf{q}}(z = 0, s) = -\frac{y\,\zeta_{\mathbf{q}}(0)}{[y + (1 + S)^2 - \sqrt{1 + 2S}]_{S = s\tau_0}} \tag{38}$$

which leads to

$$\tilde{\zeta}_{\mathbf{q}}(s) = \frac{\tilde{v}_{z,\mathbf{q}}(z = 0, s)}{s} + \frac{\zeta_{\mathbf{q}}(0)}{s}$$
$$= \left\{\frac{(1 + S)^2 - \sqrt{1 + 2S}}{s[y + (1 + S)^2 - \sqrt{1 + 2S}]}\right\}_{S=s\tau_0} \zeta_{\mathbf{q}}(0) \quad (39)$$

(For the definitions of τ_0 and y see Eq. 11.)

As can be noted, the inverse Laplace transform of $\tilde{\zeta}_{\mathbf{q}}(s)$ is not simple. For the particular initial values of interest here, $\zeta_{\mathbf{q}}(t)$ is not an exponential function of time. We can see also, as announced previously, that $v_{z,\mathbf{q}}(z, t)$, deduced from Eq. 37*a* by inverse Laplace transformation, does not belong to the kind of solutions selected by Levich.

C. Calculation of the Surface Fluctuation Spectrum

We are now in position to write the Laplace transform of the correlation function:

$$\overline{\tilde{\zeta}_{\mathbf{q}}(s)\, \zeta^*_{\mathbf{q}}(0)} = \left\{\frac{(1 + S)^2 - \sqrt{1 + 2S}}{[y + (1 + S)^2 - \sqrt{1 + 2S}]s}\right\}_{S=s\tau_0} \overline{|\zeta_{\mathbf{q}}(0)|^2} \quad (40)$$

with the assumption that at thermal equilibrium

$$\overline{v_{z,\mathbf{q}}(z, t)\, \zeta^*_{\mathbf{q}}(t)} = 0 \qquad \text{for all } z$$

In view of our previous definitions (Eqs. 4 and 29b), the surface fluctuation spectrum is related to the above expression by

$$\{\overline{\tilde{\zeta}_{\mathbf{q}}(s)\, \zeta^*_{\mathbf{q}}(0)} + c.c.\}_{S=-i\omega} = \int_{-\infty}^{+\infty} G_{\mathbf{q}}(\tau)\, e^{i\omega\tau}\, d\tau = 2\pi\, P_{\mathbf{q}}(\omega) \quad (41)$$

We obtain

$$P_{\mathbf{q}}(\omega) = \frac{1}{\pi} \operatorname{Re}\{\overline{\zeta_{\mathbf{q}}(s)\, \zeta^*_{\mathbf{q}}(0)}\}_{s=-i\omega}$$
$$= -\frac{1}{\pi\omega}\, y \left\{\frac{Im[y + (1 + S)^2 - \sqrt{1 + 2S}]}{|y + (1 + S)^2 - \sqrt{1 + 2S}|^2}\right\}_{S=-i\omega\tau_0} \overline{|\zeta_{\mathbf{q}}(0)|^2} \quad (42)$$

Inserting the value of $\overline{|\zeta_{\mathbf{q}}(0)|^2}$ from Eq. 17a, and using the fact that $P_{\mathbf{q}}(\omega)$ is even in ω, we finally obtain

$$P_{\mathbf{q}}(\omega) = \frac{q}{\rho}\, \tau_0^2 \left\{\frac{Im[2S - \sqrt{1 + 2S}]}{|y + (1 + S)^2 - \sqrt{1 + 2S}|^2}\right\}_{S=i\omega\tau_0} \frac{k_B T}{\pi\omega} \quad (43)$$

In the denominator of this expression is recognized the left-hand side of the dispersion equation (Eq. 15a). When the latter has two complex roots, $S_{1,2} = S_r \pm iS_i$, the spectrum presents two maxima in the vicinity of $\omega = \pm S_i/\tau_0$ as the spectrum given by Eq. 20. A detailed numerical comparison (see Sec. III below) shows that the two different spectra (Eqs. 20 and 43) are practically identical in the strong and weak damping conditions ($y \gg 1$ and $y \ll 1$). On the other hand, there exists a small damping range in the vicinity of critical damping where the difference between the two is nonnegligible ($0.3 \lesssim y \lesssim 3$).

D. Inspection of the Result in the Light of the Fluctuation-Dissipation Theorem

As is well known, there is a connection between the fluctuations of a physical quantity characteristic of a system at thermal equilibrium and the dissipative part of the response of this system to an external force [13]. In the present case, this leads us to consider the free surface deformed under the effect of external forces. In the surface energy this corresponds to a perturbative contribution of the type $-\int d^2r\, \zeta(\mathbf{r})\, p_{ext}(\mathbf{r}, t)$, p_{ext} being a time dependent pressure exerted by an external source.

The most general surface displacement $\zeta(\mathbf{r},t)$ in response to this external perturbation can be written as

$$\zeta(\mathbf{r},t) = \int_A d^2r' \int_0^\infty \chi(\mathbf{r} - \mathbf{r}', \tau) p_{ext}(\mathbf{r}', t - \tau)\, d^2r'\, d\tau \qquad (44)$$

in the limit of linear effects.

For a well-defined surface mode, this leads to

$$\zeta_{\mathbf{q}}(t) = \int_0^\infty \chi_{\mathbf{q}}(\tau)\, p_{ext,\mathbf{q}}(t - \tau)\, d\tau \qquad (45)$$

If $p_{ext,\mathbf{q}}(t)$ is time varying like $p^{\omega}_{ext,\mathbf{q}}\, e^{-i\omega t}$, then $\zeta_{\mathbf{q}}(t)$ is also a periodic function of time and

$$\tilde{\tilde{\zeta}}_{\mathbf{q}}(\omega) = \tilde{\tilde{\chi}}_{\mathbf{q}}(\omega)\, p^{\omega}_{ext,\mathbf{q}} \qquad (46)$$

with

$$\tilde{\tilde{\chi}}_{\mathbf{q}}(\omega) = \int_0^\infty d\tau\, \chi_{\mathbf{q}}(\tau)\, e^{i\omega\tau} \qquad (47)$$

We wish to calculate explicitly the response function and to verify that it is related to the modified surface fluctuation spectrum as the fluctuation-dissipation spectrum predicts. In the classical limit ($\hbar\omega \ll k_BT$) the relation stated by this theorem is written as

$$P_{\mathbf{q}}(\omega) = \frac{k_B T}{\pi\omega} \operatorname{Im} \tilde{\tilde{\chi}}_{\mathbf{q}}(\omega). \tag{48}$$

In order to calculate $\tilde{\tilde{\chi}}_{\mathbf{q}}(\omega)$, we must solve the hydrodynamic problem described by the same equations as before, as far as the bulk phase is concerned (Eq. 9); the only difference is in the boundary conditions at the surface. In Eq. 10a, which expresses the equilibrium between the various stresses normal to the surface, now appears the new contribution coming from an external pressure:

$$-p + 2\eta \frac{\partial v_z}{\partial z} - \gamma \frac{\partial^2 \zeta}{\delta x^2} = -p_{ext} - p_0 \qquad \text{for} \qquad z = 0 \tag{49}$$

Conditions 10b and 10c remain unchanged.

This new problem can be solved in a way very similar to that previously used for the initial value problem, and we shall indicate only the essential steps.

Starting with Eqs. 24 and 25, we obtain the following solution for the pressure:

$$p_{\mathbf{q}} = e^{qz}\left\{\gamma q^2 \zeta_{\mathbf{q}} + 2\eta\left[\frac{\partial}{\partial z} v_{z,\mathbf{q}}\right]_{z=0} + p_{ext,\mathbf{q}}\right\} \tag{50}$$

This result is carried back into Eq. 24b, and both sides are time Fourier transformed with the notations

$$\tilde{\tilde{v}}_{z,\mathbf{q}}(z, \omega) = (2\pi)^{-1} \int_{-\infty}^{+\infty} dt\, v_{z,\mathbf{q}}(z, t)\, e^{i\omega t}$$

and

$$\tilde{\tilde{p}}^{\omega}_{ext,\mathbf{q}} = p^{\omega}_{ext,\mathbf{q}}\, \delta(\omega) \tag{51}$$

We obtain

$$-i\omega\rho \tilde{\tilde{v}}_{z,\mathbf{q}} - \eta\left\{-q^2 \tilde{\tilde{v}}_{z,\mathbf{q}} + \frac{\partial^2}{\partial z^2} \tilde{\tilde{v}}_{z,\mathbf{q}}\right\}$$

$$+ q e^{qz}\left\{\left[\gamma q^2 \frac{\tilde{\tilde{v}}_{z,\mathbf{q}}}{-i\omega} + 2\eta \frac{\partial}{\partial z} \tilde{\tilde{v}}_{z,\mathbf{q}}\right]_{z=0} + \tilde{\tilde{p}}^{\omega}_{ext,\mathbf{q}}\right\} = 0 \tag{52}$$

The boundary conditions are rewritten as

$$\left\{\frac{\partial^2}{\partial z^2} \tilde{\tilde{v}}_{z,\mathbf{q}} + q^2 \tilde{\tilde{v}}_{z,\mathbf{q}}\right\}_{z=0} = 0 \tag{53b}$$

$$\tilde{\tilde{v}}_{z,\mathbf{q}}(z, \omega) \to 0 \qquad \text{when} \qquad z \to -\infty \tag{53c}$$

The problem now becomes identical to the preceeding one: we seek that solution of Eq. 33 that satisfies the three conditions 35 with the new definitions

$$u = \tilde{\tilde{v}}_{z,\mathbf{q}}(z, \omega) \tag{54a}$$

$$a = -\eta \qquad b = -i\omega\rho + \eta q^2 \tag{54b}$$

$$\beta = -\frac{\gamma q^3}{i\omega} \qquad \alpha = 2\eta q \qquad \delta = q\tilde{\tilde{p}}^{\omega}_{ext,\mathbf{q}} \tag{54c}$$

The solution is written as

$$\tilde{\tilde{v}}_{z,\mathbf{q}}(z, \omega) = \{(1 - i\omega\tau_0)\, e^{qz} - e^{qz\sqrt{1-2i\omega\tau_0}}\}\, \tilde{\tilde{\zeta}}_{\mathbf{q}}(\omega)\tau_0^{-1} \tag{55}$$

leading to

$$\tilde{\tilde{\zeta}}_{\mathbf{q}}(\omega) = \left[\frac{\tilde{\tilde{v}}_{z,\mathbf{q}}(z, \omega)}{-i\omega}\right]_{z=0}$$

$$= -\tau_0^2 \frac{q}{\rho} \frac{\tilde{\tilde{p}}^{\omega}_{ext,\mathbf{q}}}{[y + (1 + S)^2 - \sqrt{1 + 2S}]_{S=-i\omega\tau_0}} \tag{56}$$

We thus obtain the dissipative part of the response function as

$$Im\, \tilde{\tilde{\chi}}_{\mathbf{q}}(\omega) = -\tau_0^2 \frac{q}{\rho} \left\{\frac{Im(2S - \sqrt{1 + 2S})}{[y + (1 + S)^2 - \sqrt{1 + 2S}]^2}\right\}_{S=-i\omega\tau_0} \tag{57}$$

Comparing $P_{\mathbf{q}}(\omega)$ (Eq. 43) and the above result, we arrive at the conclusion that the fluctuation-dissipation theorem (Eq. 48) is satisfied.

E. Properties of the Surface Response to an External Sinusoidal Pressure

The above calculation gives the system response

$$\tilde{\tilde{\chi}}_{\mathbf{q}}(\omega) = \frac{-\tau_0^2\, q/\rho}{y + (1 - i\omega\tau_0)^2 - \sqrt{1 - 2i\omega\tau_0}} \tag{58}$$

It can be verified that for complex values of the ω variable this is an analytic function all of whose singularities are located in the same half plane ($\text{Im}\ \omega < 0$)—a property general to all physical systems that obey the causality principle [13]. Among these singularities there are two poles corresponding to the two roots of the dispersion equation having physical signification, i.e., the eigenfrequencies of the system. However, there is another singularity that is a branch cut located on the imaginary axis at $\omega = -i/2\tau_0$.

Let us notice that the approximated spectrum (Eq. 20) obtained in the naive approach verifies the fluctuation-dissipation theorem only if the exact expression of the response function is replaced by the approximated expression

$$\tilde{\tilde{\chi}}_{\mathbf{q}}(\omega) = \frac{-S_1S_2/\gamma q^2}{-\omega^2\tau_0^2 - (S_1 + S_2)i\omega\tau_0 + S_1S_2} \tag{59}$$

The analytic function associated with Eq. 59 has the same two poles as Eq. 58 and differs only by the absence of the branch cut. This last singularity can be related to the dispersion process associated with the propagation of the motion from the surface to the fluid layers underneath. A transverse velocity field in the fluid builds up and then decays. More precisely, as a result of the previous calculation, a surface displacement $\zeta_{\mathbf{q}}(t)$ generates a vertical displacement of the fluid at the depth z, $u_{\mathbf{q}}(z, t)$, which is expressed as

$$u_{\mathbf{q}}(z, t) = \int_0^{\infty} d\tau\, K(z, \tau)\, \zeta_{\mathbf{q}}(t - \tau) \tag{60}$$

with

$$\tilde{\tilde{K}}(z, \omega) = (2\pi)^{-1} \int_0^{\infty} d\tau\, K(z, \tau)\, e^{i\omega\tau}$$

$$= e^{qz} + \frac{i}{\omega\tau_0}\left(e^{qz\sqrt{1-2i\omega\tau_0}} - e^{qz}\right) \tag{61}$$

The first term is ω independent. It describes an instantaneous excitation, while the second term is characterized by a certain delay. The delay function is nonexponential, but its order of magnitude is roughly τ_0.

The two terms in $\tilde{\tilde{K}}$ correspond to the building-up in the bulk phase of two distinct velocity modes, both associated with the same surface mode. The first one corresponds to a potential flow (**rot v** $= 0$), and the second to a rotational flow (**rot v** $\neq 0$) [11]. For an ideal fluid ($\eta = 0$), only the first mode exists, and the motion at the surface generates a pressure change inside the fluid that is immediately transmitted to the layers underneath. The liquid, supposedly incompressible, is put into motion as a whole. In a viscous fluid the rotational mode establishes itself in a different way: The surface deformation induces vortex lines of axis orthogonal to **q** that diffuse in the fluid layers underneath with energy dissipation. The characteristic time needed for the motion to set up is roughly $\tau_0 = \rho/2\eta q^2$. By feedback action the rotational motion inside the volume alters the evolution of the surface waves after a certain delay of order τ_0. For this effect to be sizeable, the viscosity must not be too low, otherwise the rotational motion amplitude would be negligible (e.g., no effect for $y \gg 1$). However, in addition, the characteristic evolution time of the surface waves and of the rotational motion excited in the fluid must be roughly of the same size. This last condition excludes an appreciable effect for the highly damped surface waves ($y \ll 1$), so that the dominant damping time $\tau_q = 2\eta/\gamma q$ is much longer than $\tau_0 = \rho/2\eta q^2$. Both conditions happen to be simultaneously fulfilled at the vicinity of critical damping ($y \approx 0.145$) and in fact only there.

In conclusion, the branch cut singularity is associated with a coupling of the surface motion with motions in the bulk fluid. Its best chance to show up as a deformation of the surface fluctuation spectrum is close to critical damping. Experimental confirmations of the effect have actually been obtained in such damping conditions (see Sec. III).

As has been emphasized by Nelkin [14], there is an interesting physical similarity that exists between the thermally excited displacement of a free surface and the Brownian motion of a sphere in a fluid. In this latter case also the motion of the boundary excites transverse velocity modes in the fluid that decay because of viscosity. However, these modes contribute a delayed reaction back on the boundary, which must be included in a proper solution for the boundary motion. In the Brownian motion of a sphere this is responsible for the long-time behavior of the velocity correlation function (asymptotically proportional to $t^{-3/2}$). This differs from the delayed effects of the rotational velocity field excited in the bulk phase by the surface motion, which behave asymptotically as $(t/\tau_0)^{-3/2}\ e^{-t/2\tau_0}$. The additional exponential factor arises from the periodic deformation of the surface with wavenumber $\mathbf{q}$. The boundary condition does not lead to excitation of transverse velocity components of wavenumber smaller than q. In the Brownian motion problem, the boundary condition allows excitation of all wavenumbers in the fluid down to zero. It is the summation over all excited modes that leads to the $t^{-3/2}$ asymptotic behavior [14]. (For experimental results consistent with the existence of a $t^{-3/2}$ term in the velocity autocorrelation function for spherical Brownian particles, see Ref. [15].)

III. DETAILED COMPARISON BETWEEN SPECTRAL SHAPES

A. Comparison Between Theoretical Spectra

Figure 2 shows the modified surface fluctuation spectrum (curve a corresponding to Eq. 43) at different y values and, for comparison, the fluctuation spectrum of a one-dimensional harmonic oscillator (curve b, Eq. 20). Outside the damping range considered here ($0.145 \lesssim y \lesssim 3$), the differences between curves a and b are so small that they would not be apparent in the figure. We note that the two spectra present the largest difference in the strongly damped periodic case $y = 3$, at low frequencies, i.e., in a range where precise measurements are delicate. For more damped regimes ($y = 0.3$ and $y = 0.145$), where the spectra are centered at zero frequency, we see that the differences between curves a and b are small and that actually both spectra present the same ω^{-4} decrease at high frequencies. For each value of y the figure shows also, in dash, the sum of two

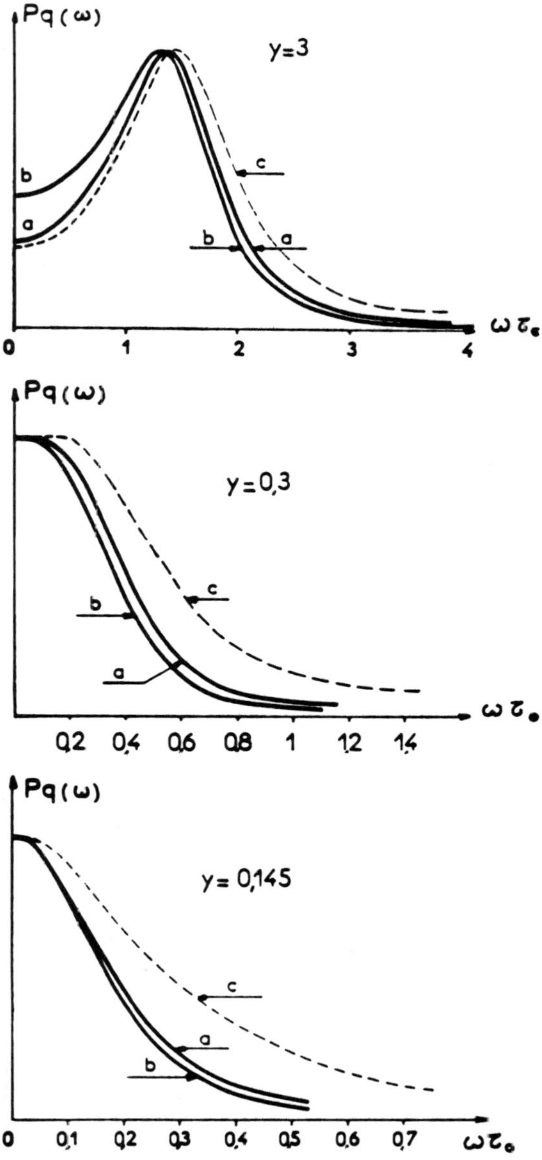

FIG. 2 Comparison between different theoretical spectral shapes at different values of $y = \gamma\rho/4\eta^2 q$, using reduced units of pulsations ($\tau_0 = \rho/2\eta q^2$). Curve a: the modified surface fluctuation spectrum (Eq. 43); curve b:the fluctuation spectrum of a one-dimensional harmonic oscillator (Eq. 20); curve c: superposition of two Lorentzian curves. At each y the three curves correspond to the same eigenfrequencies.

Lorentzian curves centered at $\omega = \pm S_i\tau_0^{-1}$ having a width $S_r\tau_0^{-1}$ (curve c). As can be seen, curve c deviates much more from curves a and b than these two from one another.

B. Comparison Between Theoretical and Experimental Spectra

At first, during the course of a detailed experimental study [16], no systematic disagreement was discovered between experimental spectra and the harmonic oscillator fluctuation spectrum then used to fit the data. In fact, at that time the available experimental accuracy was insufficient to settle the question of the line shape unambiguously. We shall now present the results of two different experiments that soon after provided clear evidence for a difference between the experimental spectrum and the harmonic oscillator fluctuation spectrum attributable to the surface-bulk coupling. These two experiments benefit from an improved signal-to-noise ratio due to a large amount of scattered light. Both also are performed in the strongly damped periodic regime where the predicted effect is the largest.

1. Surface Waves at the Free Surface of MBBA in Its Isotropic Phase ($T - T_c = 7°$) [17]

The scattered intensity proportional to the reflexion coefficient $[(n - 1)/(n + 1)]^2$ is large because of the large index of refraction of the liquid (methoxy benzylidene butylaniline: $n = 1.6$).

The value of $\mathbf{q}$ is selected to obtain a strongly damped periodic regime. Figure 3 shows an experimental spectrum that has been fitted by the two different theoretical shapes. The value of y that leads to the best fit is different in the two cases. Using the surface fluctuation spectrum (Eq. 43), one arrives at $y = 1.07$, leading to $\gamma = 34.1$ dyn/cm and $\eta = 0.178$ P in excellent agreement with a direct determination with a viscosimeter. On the other hand, the best fit with the harmonic oscillator fluctuation spectrum (Eq. 20) gives $y = 1.32$ (leading to $\gamma = 36.4$ dyn/cm and $\eta = 0.165$ P). Obviously the first fit leads to a better agreement with the experimental curve, especially on the low-frequency wing of the peak. We remark that sizeable errors in the determination of η and γ would result from the use of the harmonic oscillator fluctuation spectrum. The surface-bulk coupling is here a fairly important effect.

2. Surface Waves at the Liquid-Gas Interface of CO_2 Close to the Critical Point ($T_c - T = 0.7°$) [6]

Close to the critical point, the amplitude of the surface fluctuations tends to become very large because the surface tension becomes quite low (see Eq. 17), and consequently the scattered intensity increases considerably. The lowering

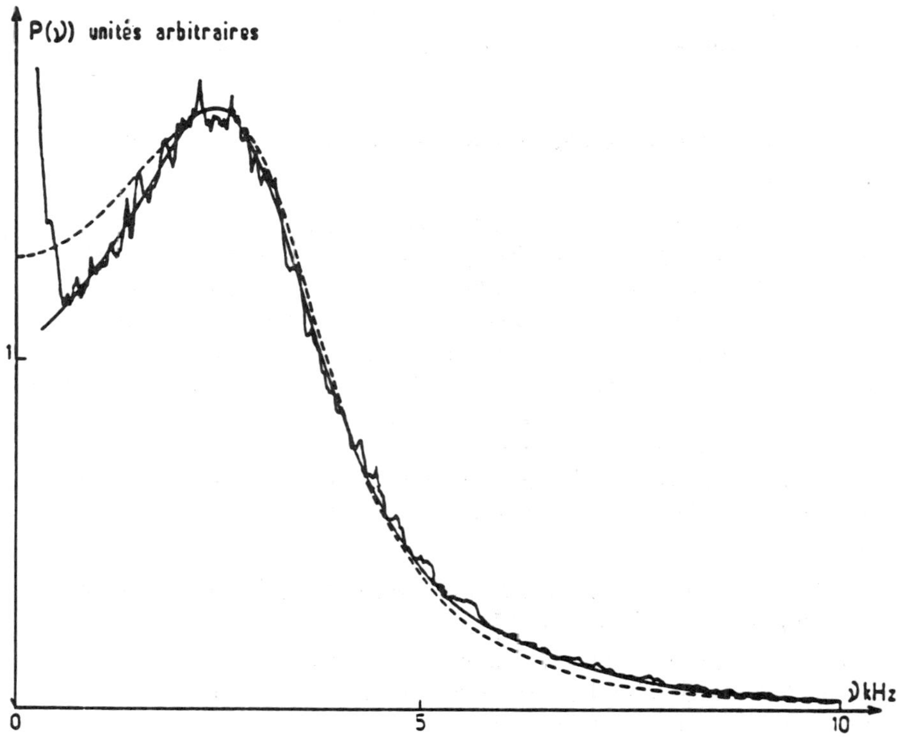

FIG. 3 Spectrum of the light scattered from the surface of MBBA in the isotropic phase. In solid line, best fit obtained with the modified surface fluctuation spectrum (Eq. 43 and $y = 1.07$); in dash, best fit obtained with the fluctuation spectrum of a one-dimensional harmonic oscillator (Eq. 20 and $y = 1.32$).

of the surface tension also favors a strong damping of the waves in spite of the low viscosity of the fluid.

Figure 4 shows an experimental spectrum in the interesting critical damping range and again the best fit obtained with either theoretical spectrum. The best result is clearly obtained with the modified surface fluctuation spectrum and particularly at low frequencies where the difference between the two theoretical curves is the largest.

Actually the spectrum presented in Fig. 4 is not exactly compared to $P_{\mathbf{q}}(\omega)$ but to

$$P_{\mathbf{q}}(\omega) + \lambda \frac{\int d\omega' \, P_{\mathbf{q}}(\omega') \, P_{\mathbf{q}}(\omega - \omega')}{\int d\omega' \, P_{\mathbf{q}}(\omega')} \tag{62}$$

λ being an adjustable parameter that represents the ratio between the homodyne

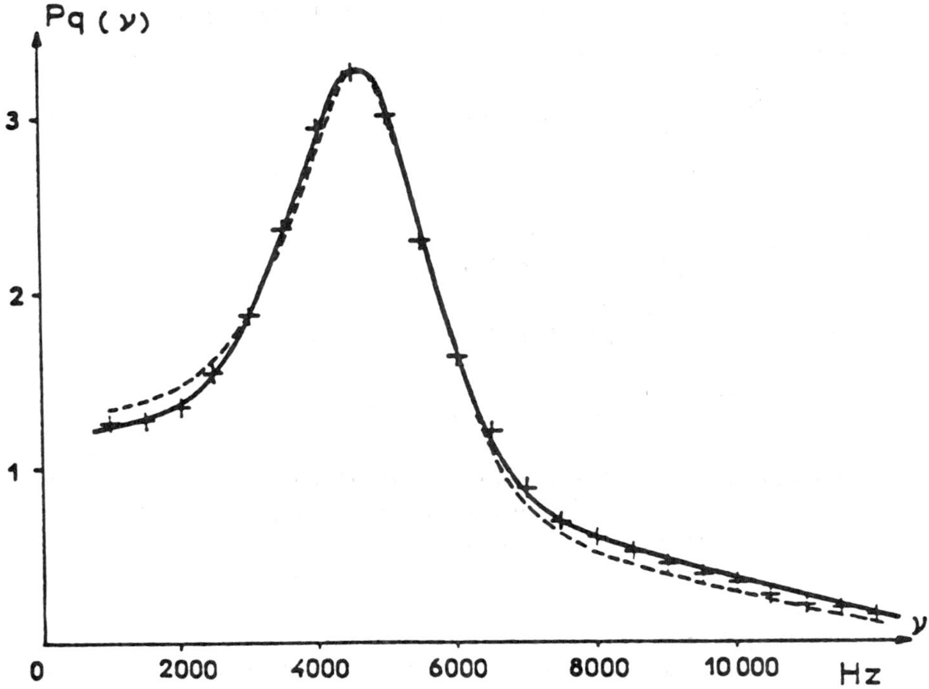

FIG. 4 Spectrum of the light scattered from the CO_2 liquid vapor interface at t = 30.438°C for q = 3325 cm^{-1}. Experimental points and theoretical curves: in solid line, best fit obtained with the modified surface fluctuation spectrum (Eq. 62 and 64, y = 4.27); in dash, best fit obtained with the fluctuation spectrum of a one-dimensional harmonic oscillator (Eqs. 62, 20, and 67 and y = 4.00).

and heterodyne detection efficiencies in the experiment (see Chap. 5). For the two spectral shapes, the best fit is obtained with $\lambda \simeq 0.3$.

IV. GENERALIZATION TO MORE COMPLEX INTERFACES

The generalization of the calculations presented in Sec. II presents in principle no difficulties. We shall only quote here the result corresponding to the case of the liquid-gas interface of a liquid close to the critical point [6], since this case has lead to an interesting comparison with experiment. Denoting ρ_1,ρ_2 and η_1,η_2 the densities and viscosities of the two fluid phases, we can use the simplifying relation

$$\frac{\rho_1}{\rho_2} = \frac{\eta_1}{\eta_2} \tag{63}$$

which is valid to a very good approximation close to T_c. The surface fluctuation spectrum generalized to this case is

$$P_{\mathbf{q}}(\omega) = \frac{k_B T}{\pi\omega} \cdot \frac{q}{\rho + \rho'} \tau_0^2 \operatorname{Im}\{\Delta^{-1}(i\omega\tau_0)\} \tag{64}$$

with

$$\tau_0 = \frac{\rho + \rho'}{2(\eta + \eta')q^2} \qquad y = \frac{\gamma(\rho + \rho')}{4(\eta + \eta')^2 q} \tag{65}$$

$$\begin{aligned}\Delta(S) = {} & y + \frac{2\rho\rho'}{(\rho + \rho')^2} S\sqrt{1 + 2S}\,[1 + \sqrt{1 + 2S}] \\ & + \frac{(\rho - \rho')^2}{(\rho + \rho')^2}[(1 + S)^2 - \sqrt{1 + 2S}]\end{aligned} \tag{66}$$

The theoretical spectrum represented as a solid line in Fig. 4 makes use of the above expression, while the dashed curve is given by Eq. 20, S_1 and S_2 being the two roots of the dispersion equation

$$\Delta(S) = 0 \tag{67}$$

having a negative real part.

A derivation of the spectral shape for the interface between two liquids in the general case can be found in chapter 7.

A derivation of the spectral shape for thermal fluctuations of a thin film at the surface of a liquid [17] can be found in chapter 11. The case of a liquid crystal is also considered in chapter 9.

REFERENCES

1. L. Mandelshtam, *Ann. Physik 41*: 609 (1913).
2. C. V. Raman and L. A. Ramdas, *Proc. Roy. Soc. A108*: 561 (1925); *A109*: 150, 272 (1925).
3. M. Papoular, *J. de Phys. 29*: 81 (1968).
4. R. H. Katyl and U. Ingard, *Phys. Rev. Lett 19*: 64 (1967); *20*: 248 (1968).
5. M.-A. Bouchiat, J. Meunier, and J. Brossel, *C.R. Acad. Sci. 266B*: 255 (1968); M.-A. Bouchiat and J. Meunier, *C.R. Acad. Sci. 266B*: 301 (1968); M.-A. Bouchiat and J. Meunier, in *Polarization, Matter and Radiation,* Presses Universitarires, Paris, 1969.
6. J. Meunier, *J. de Phys. 30*: 933 (1969); M.-A. Bouchiat and J. Meunier, *Phys. Rev. Lett. 23*: 752 (1969); J.-C. Herpin and J. Meunier, *J. de Phys. 35*: 847 (1974). *N.B.*: In Eq. 1, the expression for (S) must be corrected. In the S^2 term, $m'(m - 1)$ and $m(m' - 1)$ should be replaced by $m'(m + 1)$ and $m(m' + 1)$.
7. D. Langevin and C. Griesmar, *J. Phys. D 13:* 1189 (1980).

8. L. D. Landau and G. Placzek, *Physik Z. Sowjetunion 5*: 172 (1934).
9. R. D. Mountain, *Rev. Mod. Phys. 38*: 205 (1966).
10. M.-A. Bouchiat and J. Meunier, *J. de Phys. 32*: 561 (1971).
11. V. G. Levich, *Physicochemical Hydrodynamics,* Prentice Hall, 1962, chap XI.
12. C. Bouchiat, private communication.
13. L. D. Landau and E. M. Lifshitz, *Statistical Physics,* 3rd ed., Pergamon, 1980.
14. M. Nelkin, *Phys. Fluids 15*: 1685 (1972).
15. A. Bouiller, J. P. Boon, P. Deguent, *J. de Physique 39*: 159 (1978); G. L. Paul and P. N. Pusey, *J. Phys. A 14*: 3301 (1981).
16. J. Meunier, D. Langevin and M.-A. Bouchiat, C.R.A.S. *268B*: 92 and 422 (1969).
17. D. Langevin and M.-A. Bouchiat, *C.R.A.S. 272B*: 1422 (1971).

5

Experimental Realization

DOMINIQUE LANGEVIN Laboratoire de Physique Statistique, Ecole Normale Supérieure, Paris, France

The principle of the technique is simple: a laser beam of frequency ω_0 is sent to the liquid surface. This beam is reflected and refracted; in addition, because of surface thermal roughness, a small part of the incident light is scattered around both reflected and refracted beams. In existing devices, the surface scattering around reflection has been selected. For this purpose, the scattered beams are focused onto a small diaphragm to select the contributions of the different surface modes ζ_q. The scattered light of frequency $\omega_0 + \omega$ is received onto a photomultiplier where it is mixed with light of frequency ω_0, called "local oscillator," scattered by optical defects or by an external grating. The photomultiplier gives an output current proportional to the square of the total electric field and contains components at the frequency ω. These components are typically in the kilohertz range or below, and can be analyzed with wave analyzers or correlators.

I. MECHANICAL STABILITY

One of the main difficulties of the experiment is to reduce the mechanical vibrations of the liquid surface. If the surface moves, the beams also move and the scattering direction is no longer defined. Because the scattered intensity decreases rapidly when the scattering angle increases, in practice only small scattering angles (typically between 1° and 0.1°) can be studied. This means that the angular deflections of the beam due to mechanical vibrations must be much less, at most about one minute (300 μrad).

This is particularly difficult to achieve for the free surfaces of liquids, which are very sensitive to vibrations. Low-frequency vibrations have long wavelengths and are damped when the liquid height is small, because of the boundary conditions at the solid bottom of the liquid cell. It is therefore easy to suppress some of the vibrations by reducing the liquid depth typically to 1 mm. At smaller depths, the studied capillary waves also become damped by the cell bottom. Other methods that have been used to reduce vibrations involve the use of cells suspended in air or of cells floating on glycerin [1]. Liquid-liquid interfaces are less well coupled to mechanical vibrations and easier to study.

An obvious way to reduce mechanical vibrations is to build the equipment in a room isolated from vibrations, thus preferentially in a basement. The first experiments by Meunier and coworkers were made in a basement where the floor was isolated from the rest of the building and in direct contact with the ground. Another obvious way to reduce vibrations is to mount the system on a heavy table (typically 1 ton) whose eigenfrequencies are very small, so that they strongly damp mechanical vibrations of larger frequencies. The tables can be mounted on springs, rubber tubes, or air cylinders. In the two last cases, air servosystems can allow one to keep the table top horizontal and at a fixed position. This method also was used by Meunier and coworkers. Still better mechanical isolation can be obtained using actively stabilized platforms on top of the first table. In the Newport electronic vibration isolation system used by Mazur and Chung [2], stabilization is achieved with passive springs and an electromechanical servosystem made of eight accelerometers and eight force transducers (one vertical and one horizontal in each of the four corners). This device produces an angular beam stability better than 60 μrad although not operating in a basement.

II. LIQUID CELL

The first requirement relative to the liquid container is the planarity of all the optical windows of the cell crossed by the light beams. The windows should be "optically polished" (planarity at least of order $\lambda/4$). This requirement also applies to all other optical elements of the system: $\lambda/4$ for lenses, $\lambda/8$ for mirrors. The second requirement deals with the diameter of the cell, which should be sufficiently large that there is a flat portion (within $\lambda/4$) of liquid surface in the center: indeed, the surface is curved close to the cell vertical borders, where the typical height of the capillary meniscus is the capillary length $l_c = [\gamma/(\rho_2 - \rho_1)g]^{1/2}$. For the free surface of water ($\gamma \sim 70$ mN/m), the minimum cell diameter is about 70 mm; for liquids of smaller surface tension like alcohols or silicon oils ($\gamma \sim 20$ mN/m) it can be smaller, about 40 mm. For liquid-liquid interfaces of very low interfacial tensions, still smaller cells can be used. For ordinary liquid-liquid interfaces, however, cells larger than for free surfaces are needed, because $\rho_2 - \rho_1$ is typically 10 times smaller than ρ, so that l_c is three times larger than for a free surface.

When the measurements are made at high temperatures, good thermal isolation must be provided, in order to avoid surface distortions due to surface tension gradients. When free surfaces are studied, it is useful to warm the upper cell window a few degrees above the liquid temperature, in order to avoid liquid condensation on that window. Another advantage of the procedure is to suppress the convection currents.

III. LASER SOURCE

Helium-neon lasers are the best suited lasers for this type of experiment. Their noise level is very low, and the frequency dependence of this noise is small. These lasers are air cooled and therefore do not produce mechanical vibrations in the system. However, when the intensity scattered by the surface is very small, as in the case of water, the signal-to-noise ratio can become very bad at large scattering angles. In this case, argon-ion lasers allow the measurements at larger scattering angles because of their higher power. Similarly, lipid bilayers, which scatter very little light, can barely be studied with the most powerful He-Ne lasers. One disadvantage of argon-ion lasers is their larger noise level. This noise contains sharp low frequency components that are multiples of 100 Hz and a broad component around 50 kHz (gas plasma frequency); these components could cause problems if they fall in the frequency range of the scattered light. These lasers must be mounted on a separate table, in order to avoid the communication of mechanical vibrations, due to the laser water cooling system, to the liquid surface.

In the case of ultralow interfacial tension systems, very-low-power lasers can be used. The only requirement is power stability, which should be very good: the power should not fluctuate at time scales comparable to the inverse characteristic frequencies under study. For this purpose, stabilized power supplies could be used instead of the laser standard power supply.

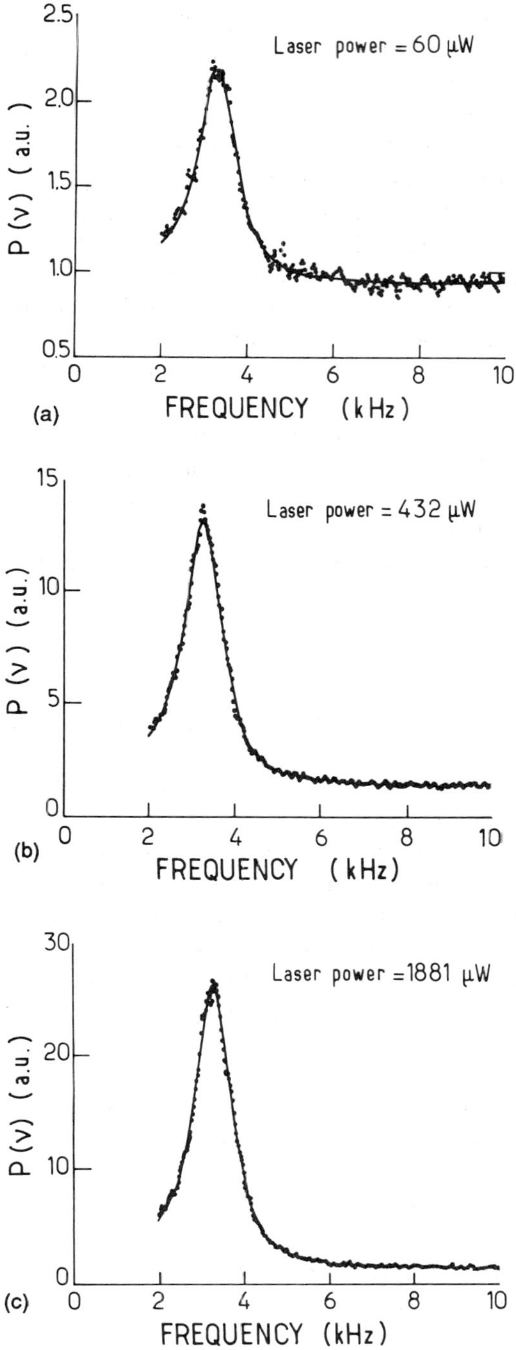
Laser power = 60 μW
P (v) (a.u.)
2.5
2.0
1.5
1.0
0.5
0
2
4
6
8
10
(a)
FREQUENCY (kHz)
Laser power = 432 μW
P (v) (a.u.)
15
10
5
0
(b)
FREQUENCY (kHz)
Laser power = 1881 μW
P (v) (a.u.)
30
20
10
(c)
FREQUENCY (kHz)

When one of the liquid phases absorbs light, the laser can produce surface heating. This generally results in surface deformations and defocusing of the laser beam. To ensure that residual heating has been completely suppressed, measurements of the scattering spectrum must be made as a function of laser power: convenient power levels correspond to the range in which the spectral characteristics are not changing. For instance, Dorshow and Swofford studied the scattering of argon-ion laser light by crude oil surfaces [3]. They found that photothermal "blooming" of the surface occurred above 10 mW, and that the scattering spectra remain the same in the range between 60 μW and 5 mW (Fig. 1). Obviously, when the liquid phase does not absorb equally over the visible spectrum, the laser should be chosen with a frequency as far as possible from the liquid absorption frequencies.

IV. LOCAL OSCILLATOR

The local oscillator is the source of an electric field E_{l_0} of frequency ω_0 equal to that of the incident electric field and scattered in one or more space directions around the reflected or transmitted beams. The field E_{l_0} scattered in a given direction can interfere with the field E_s scattered by the surface in the same direction; the interference conditions will be discussed in Sec. V. If the scattering direction is selected with a pinhole placed in front of the photomultiplier, the photocurrent is

$$i \sim (E_{l_0} + E_s)^2 = |A_{l_0} e^{i\omega_0 t} + A_s e^{i(\omega_0 \pm \omega)t} + \text{c.c}|^2$$

c.c. standing for complex conjugate. The photocurrent low frequency contribution analyzed in the experiment is

$$A_{l_0} A_s^* e^{i\omega t} + |A_s|^2 e^{2i\omega t} + \text{c.c.}$$

Because of the intrinsic laser noise, A_{l_0} is not constant, and its time variation reflects the time variation of the laser electric field

$$A_{l_0} = A_{l_0}^0 + \delta A_{l_0} e^{i\delta\omega t}$$

Several cases can be distinguished.

$A_s \gg A_{l_0}$	The analyzed signal is $|A_s|^2 e^{2i\omega t}$ + c.c.; this is the "homodyne detection."

FIG. 1 Spectrum of the light scattered by the free surface of a crude oil for different laser powers (a) 60 μW; (b) 432 μW; (c) 1.881 mW. (After Ref. 3.)

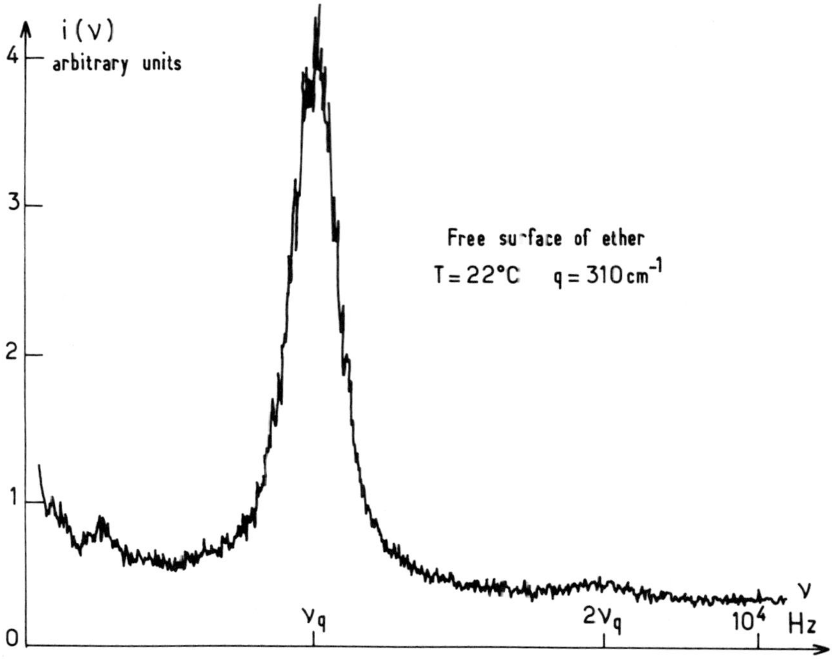

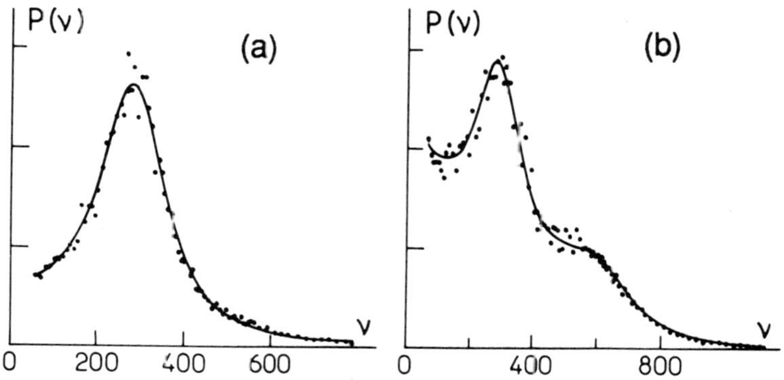

FIG. 2 (A) Frequency dependence of the photocurrent for the scattering by the free surface of ethyl-ether; a small bump is observed at twice the frequency of the main peak and corresponds to homodyne detection. When $i(\nu)$ is squared, this bump is no longer visible; let us recall that $P(\nu) \sim |i(\nu)|^2$. (B) Spectrum of the light scattered by the liquid-vapor interface of CO_2 close to its critical point for different homodyne-to-heterodyne ratio. (a) 4%; (b) 40%; the variation of this ratio has been achieved by coating the windows of the liquid cell with thin layers of silicon oil. (After Ref. 9.)

$A_{l_0} \gg A_s \gg \delta A_{l_0}$ The analyzed signal is $A_{l_0} A_s^* e^{i\omega t}$ + c.c.; this it the "heterodyne detection."

$\delta A_{l_0} \gg A_s$ The analyzed signal is $A_{l_0}\, \delta A_{l_0}\, e^{i\delta\omega t}$ + c.c., i.e., the laser intrinsic noise.

Even when no external grating is used, there are always local oscillator sources, such as the optical defects of the system elements or dust deposited on these elements. The corresponding stray light is scattered mainly in the forward directions, i.e., the directions in which surface scattering is studied. Homodyne contributions are observed only in cases where the surface scattering is strong, such as close to critical points or for oil-water interfaces of ultralow tensions (Fig. 2). In these cases, it can be difficult to separate homodyne and heterodyne contributions. Indeed, the surface waves are frequently overdamped because γ is small, and the spectrum is a superposition of two Lorentzians of half widths $\Delta\omega_q$ and $2\Delta\omega_q$. If this spectrum is used to measure γ, the accuracy is very poor: 100%. The problem can be solved by working close to the limit angle around which the scattered intensity varies strongly. When varying the incidence angle

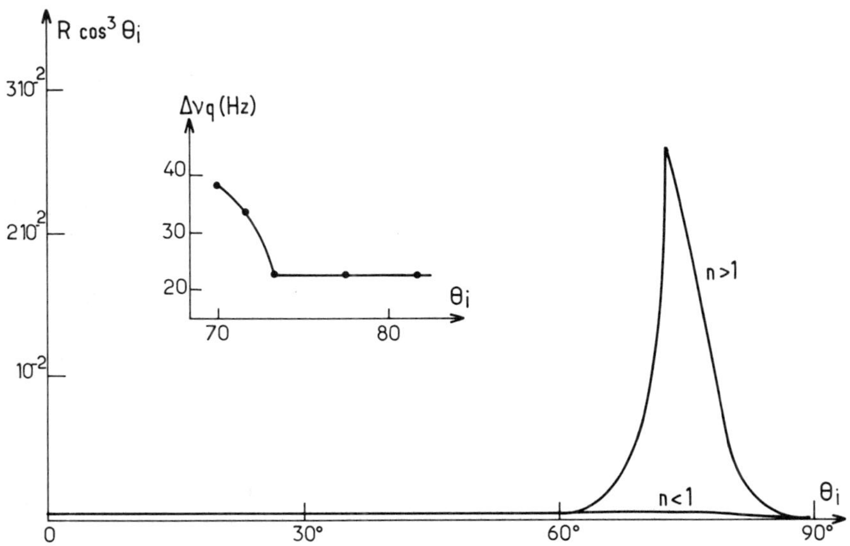

FIG. 3 Variation of R cos θ_i with θ_i for the interface between toluene (n = 1.430) and water (n = 1.352) (in the presence of salt, alcohol, and surfactant). Insert: variation of the width of the spectrum of the light scattered by this interface versus θ_i. The points are experimental; the line is a guide for the eye. For $\theta_i \sim \theta_l$, the detection is homodyne; for larger θ_i, it is heterodyne. (After Ref. 10.)

θ_i around the limit angle θ_l, the ratio A_s/A_{l_0} varies, as well as the spectral width, if the detection is intermediate between homodyne and heterodyne. An example is shown on Fig. 3. The width of the spectrum close to the limit angle is twice (homodyne detection) its value at larger angles (heterodyne detection). An alternative solution to the problem is to use an external local oscillator such as a diffraction grating, and to vary A_{l_0} with a variable attenuator; in this way the ratio A_s/A_{l_0} is also varied.

The ratio of heterodyne to homodyne signals is A_{l_0}/A_s and should be as large as possible. But the ratio $\delta A_{l_0}/A_{l_0}$ is fixed and typically of the order of 1%. The currently accepted optimum value for A_{l_0}/A_s is therefore about 10. This corresponds to a signal-to-noise ratio of about 10. When a grating is used, its scattering power should be adjusted with an attenuator to this optimal condition, ideally each time the scattering angle is changed (because A_s is angle dependent).

V. ADVANTAGES OF AN EXTERNAL GRATING

An external grating allows the good control of the level of the local oscillator A_{l_0}. The optimal ratio A_{l_0}/A_s being about 10 and A_s being very small, special gratings of small scattering power must be used. Usually, they are made of black stripes a few μm wide, separated by 100 μm intervals. They can be obtained by reproduction of computer drawings on photographic plates [4], or by deposition of metals on optically flat glass (microelectronics technology). The local oscillator strength can be easily adjusted with a variable rotating attenuator.

In order to be able to interfere, the wavefronts associated with the fields E_s and E_{l_0} must be parallel, and these fields must be coherent; otherwise the products $A_s A^*_{l_0}$ will average to zero. The second requirement implies that the grating and the surface must be separated by less than the coherence length of the laser beam. The frequencies of interest are much less than the distance between cavity modes of the laser; it is not necessary to use the laser in a monomode configuration. This means that the coherence length is $c/\Delta\omega$, where $\Delta\omega$ is the total laser spectral width; with $\Delta\omega \sim 3.10^{10}\ s^{-1}$, $c/\Delta\omega \sim 1$ cm. The best solution is to image the grating onto the surface. In this way, the wavefronts associated with surface scattering and with the local oscillator are also parallel. The image of the grating can be conveniently done with an afocal system made of two lenses; if the grating is placed at the focus of the first lens, the system gives an image of the same size as the object. In the common focal plane, the separation of the beams scattered by the grating is large enough and the attenuator can be conveniently mounted in this plane; the attenuator should be transparent to the transmitted beam and attenuate only the scattered beams (Fig. 4).

An important role of the grating is to reduce the influence of mechanical vibrations. Indeed, with such a grating, the local oscillator intensity is concen-

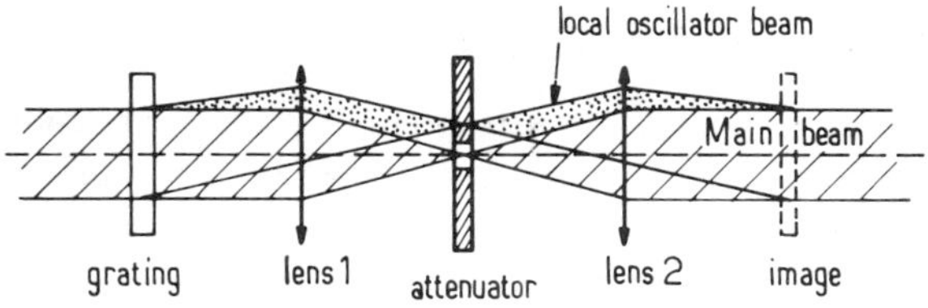

FIG. 4 Schematic representation of a local oscillator arrangement, involving a diffraction grating.

trated in well-defined scattering directions θ_{l_0}. When the detection pinhole is centered around one of these directions and the surface moves, the whole set of reflected and scattered beams is shifted by an angle $\delta\theta$. A_s does not vary very much, but if $\delta\theta$ is larger than the angular beam diameter $\delta\theta_0$, A_{l_0} drops to a very low value. When the surface returns to its initial position, the signal detected by the photomultiplier rises again. In summary, the measured average signal only contains appreciable contributions from surface scattering in the direction θ_{l_0} within $\pm\delta\theta_0$. Of course, if there are too many vibrations, the fraction of time during which the beam is in its equilibrium position is small, and the measured signal is noisy and increases only slowly with time. It is therefore important to reduce mechanical vibrations, even when a grating is used.

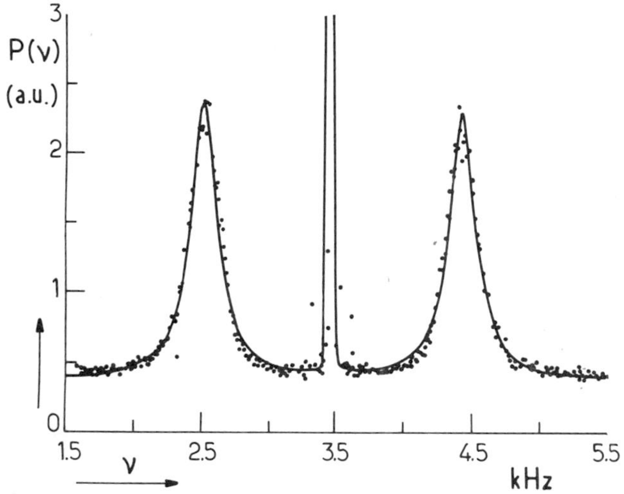

FIG. 5 Spectrum of the light scattered by the free surface of water obtained by using a local oscillator that introduces a frequency shift. (After Ref. 2.)

VI. OTHER TYPES OF LOCAL OSCILLATOR

Classical oscillators in optical experiments are made by splitting the laser beam in two. The reference beam serves as local oscillator and can be frequency shifted by acoustooptic crystals. This has been used by Mazur et al. to observe the frequency doublet ($\omega_0 \pm \omega_q$) for water [2] (Fig. 5).

VII. CHOICE OF THE ANGLE OF INCIDENCE

In reported studies, the light scattered at small angles around the reflected beam is analyzed. The scattered intensity per unit solid angle $dI/d\Omega$ is proportional to $R(\theta_i) \cos^3\theta_i$ where $R(\theta_i)$ is the reflection coefficient (Chap. 3, Eqs. 12 and 13).

When the incident beam travels in the medium of smaller refractive index, $n = n_i/n_t < 1$, and $dI/d\Omega$ varies smoothly with θ_i (Chap. 3, Fig. 3). When $n > 0.8$, it is maximal for normal incidence. But when the incident beam travels in the medium of larger refractive index, $n > 1$ and $dI/d\Omega$ exhibits a pronounced maximum close to the limit angle θ_l. The ratio of the scattered intensities for incidences $\theta_i = \theta_l$ and $\theta_i = 0$ (normal incidence) is $\cos^3\theta_l\,(n + 1)^2/(n - 1)^2$ and is shown in Fig. 6. The advantage of working near the limit angle is clear,

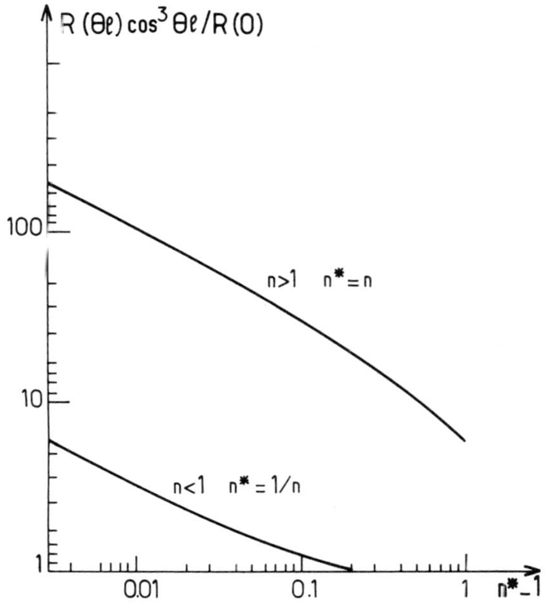

FIG. 6 Variation of the ratio of the scattered intensities for incidences $\theta_i = \theta_l$ and $\theta_i = 0$, versus the relative index of refraction n.

especially when the two media have similar refractive indices ($n \sim 1$). In the case of the free surface of water ($n = 1.33$, $\theta_l = 48°75$) the ratio of the intensities scattered in the two geometries is about 14. Fig. 7 illustrates the appreciable gain in the signal-to-noise ratio obtained by working near the limit angle.

When the medium of larger refractive index absorbs light (liquid crystals, crude oil), normal incidence is to be preferred. In other to avoid beam separation problems, a slightly tilted geometry ($\theta_i \sim 3°$) can be used. Exactly normal incidence is used however, by Dorshow and Swofford; the beam separation is achieved with a polarizing beam splitter and a quarter-wave plate [3].

Oblique geometries also are currently used, although the scattered intensity is slightly smaller than for $\theta_i = 0$. But the beam angular radius (at half-height) in the plane of incidence is smaller:

$$\delta\theta_0 = \frac{2}{\pi} \frac{\lambda\sqrt{\ln 2}}{l_0} \cos \theta_i \tag{1}$$

where l_0 is the beam diameter close to the surface. The angular resolution of the setup is thus better. As we shall see in the next section, it is particularly important to reduce instrumental broadening. A compromise used for water is $\theta_i \sim 60°$, for which $\delta\theta_0$ is about half its value at normal incidence [4]. But $R_\perp \cos^3 \theta_i (n + 1)^2/(n - 1)^2 \sim 0.7$, i.e., the scattered intensity is smaller than in the normal incidence geometry. Special care has to be taken to use laser polarization normal to the plane of incidence. Otherwise the intensity lost is much larger: $R_\parallel \cos^3 \theta_i (n + 1)^2/(n - 1)^2 \sim 0.03$.

When working close to normal incidence, the scattering wavevector q is independent of the azimuthal angle ϕ. For small scattering angles,

$$q = k\,\Delta\theta$$

where $\Delta\theta$ is the angular distance between reflected and scattered beams. This is no longer true when the incidence angle is large. From $\mathbf{q} = \mathbf{k}_{d\Sigma} - \mathbf{k}_{r\Sigma}$, one finds, again in the case of small $\Delta\theta$,

$$\begin{aligned} q &= k\,\Delta\theta && \text{for scattering in the plane perpendicular to the plane of incidence} \\ q &= k \cos \theta_i\,\Delta\theta && \text{for scattering in the plane of incidence} \end{aligned} \tag{2}$$

When θ_i is close to $\pi/2$, very small wavevectors can be studied by analyzing the scattering in the plane of incidence. Angular resolution is less limiting than for scattering in the plane perpendicular. The only inconvenience is that θ_i must be measured precisely to determine q. If the desired accuracy on q is better than 1%, Eqs. 2 are not sufficiently accurate. The exact q value can be calculated by

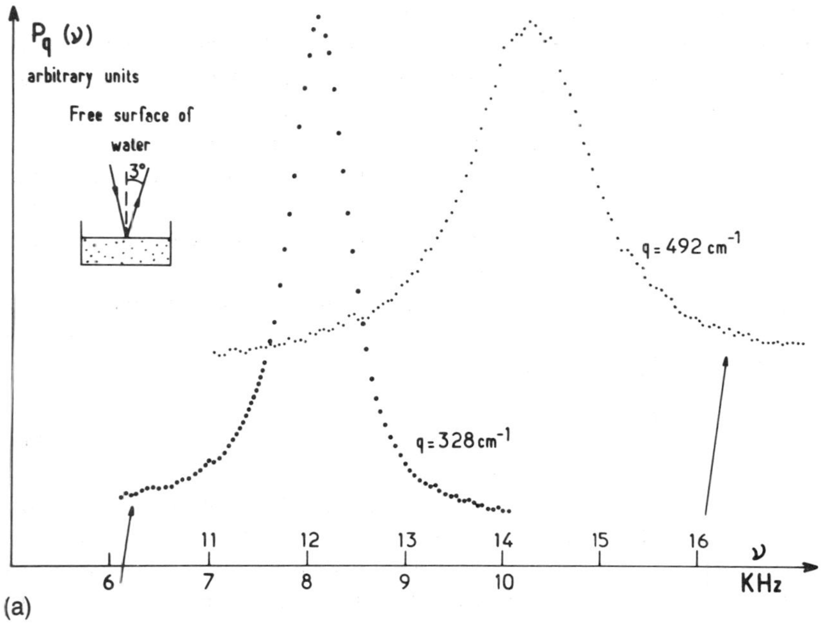

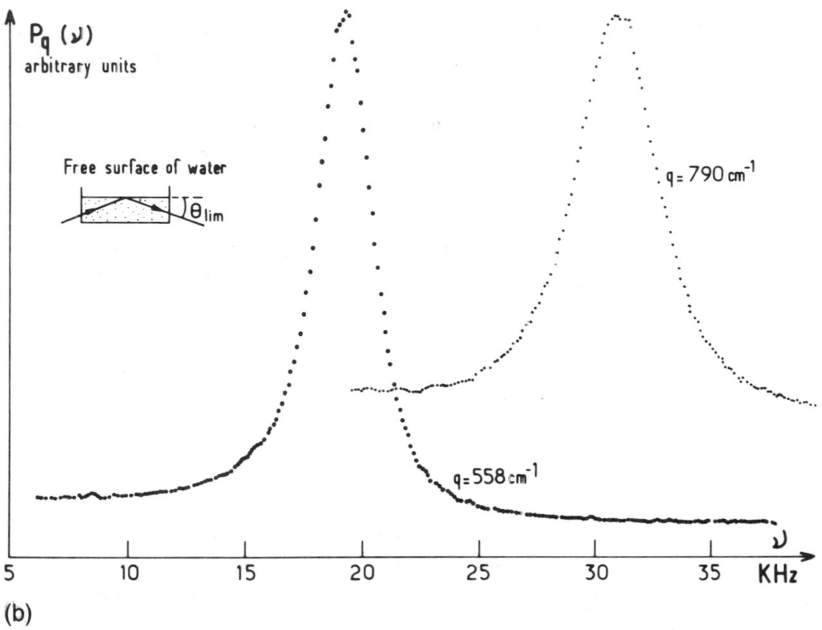

FIG. 7 Spectra of the light scattered by the free surface of water obtained in different geometries. (a) normal incidence; (b) limit incidence. (After Ref. 11.)

using Eq. 10 of Chap. 3. For scattering in the plane of incidence, the approximation $q = k \cos \theta_i \Delta\theta (1\text{-}\tan \theta_i \Delta\theta/2)$ has been used [5].

When $\theta_i \neq 0$, a pinhole detection must be used. Several authors have used annular diaphragms to study the free surface of water, selecting in this way a wide range of wavevectors, between k $\Delta\theta \cos \theta_i$ and k $\Delta\theta$ [6]. The peak frequencies of the corresponding spectra, which are proportional to $q^{3/2}$ (Chap. 2), spanned a proportionally still larger range, resulting in a very broad measured spectrum. The authors inferred from this large width that the water viscosity close to the free surface was several times larger than bulk water viscosity, whereas the effect was a pure artifact due to instrumental broadening.

VIII. OPTICAL ELEMENTS

Eq. 1 shows that better angular resolution is achieved when the illuminated surface area is larger. For this purpose, a spatial filter made of two lenses and a pinhole is used to remove laser stray light as well as to expand the beam (Fig. 8). The local oscillator arrangements have been described above. A set of two mirrors is generally used to adjust the incidence angle and to insure horizontally traveling beams. The length traveled by the reflected beam must be large to achieve good angular resolution: $\Delta\theta = d/D$, where d is the distance between the detection pinhole and the reflected spot on the focusing plane and D the distance between the surface and the focusing plane. Typically D is between 1 and 2 m. The focusing is achieved with the help of the spatial filter. The detection pinhole can be mounted on an automatic translation device to facilitate the scattering angle changes and the position adjustments of the grating scattering spots. A final lens is used to image the surface onto the photomultiplier cathode, in association with pinholes to remove stray light.

In order to minimize external local oscillator sources that reduce the grating efficiency, the optical elements should not be set closer than a few centimeters (laser coherence length) from the liquid surface. The cell windows can also be sources for local oscillators because they are close to the surface. These windows must be kept clean and free from gas condensation. A lens is then used to image the surface onto the photomultiplier, together with diaphragms to remove stray light. A typical arrangement is shown in Fig. 9.

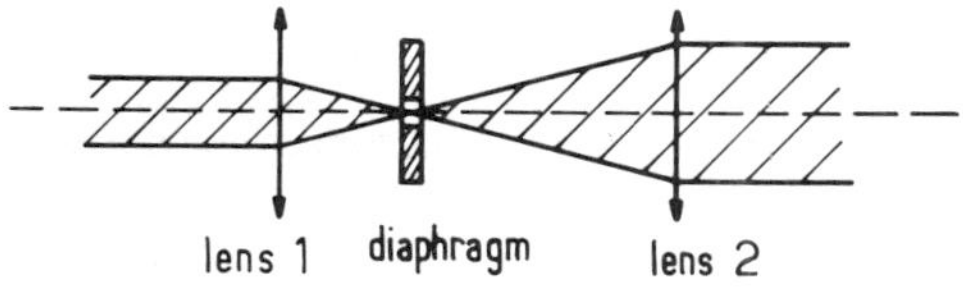

FIG. 8 Spatial filtering and beam expansion (the focal length of lens 2 is twice that of lens 1).

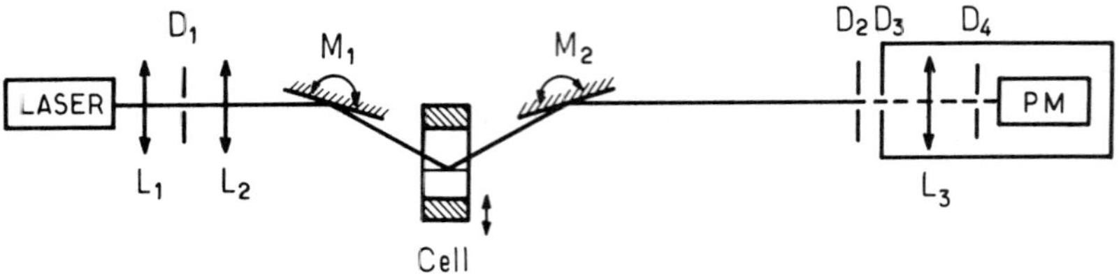

FIG. 9 Schematic representation of a device for the study of the light scattered by a liquid surface in the limit angle geometry. The local oscillator has not been represented.

IX. PHOTOMULTIPLIER DETECTION AND ANALYSIS

In heterodyne spectroscopy, one measures the second-order or intensity correlation function [7]:

$$G^{(2)}(\tau) = \langle E^*(t)\, E(t)\, E^*(t + \tau)\, E(t + \tau)\rangle \sim \langle I(t)\, I(t + \tau)\rangle \tag{3}$$

where $\langle\ \rangle$ denotes time average. The electric field contains two contributions, the scattered electric field

$$E_s(t) = f(t)\, E_i(t)$$

where $f(t)$ is a random function, proportional to $\zeta_q(t)$ as discussed in Chap. 3, and the local oscillator field

$$E_{l_0} = f_{l_0}\, E_i(t)$$

The laser incident field is generally of the form

$$E_i(t) = A_i(\mathrm{t})\, e^{i[\omega_0 t + \Psi(t)]}$$

We shall suppose that A_i is constant. $\Psi(t)$ is a random, slowly varying phase, whose characteristic time is the order of the inverse laser spectral width.

It can be seen from Eq. 3 that the intensity correlation function does not depend upon $\Psi(t)$; the spectral resolution is therefore not limited by the spectral width of the incident laser light.

The correlation function contains many terms. In the experiment, only the low frequency terms are analyzed, i.e.,

$$\langle E_{l_0}(\mathrm{t})\, E^*_{l_0}(t + \tau)\, E^*_s(t)\, E_s(t + \tau)\rangle + \mathrm{c.c.} \sim \langle f^*(t)\, f(t + \tau)\rangle$$

This part of $G^{(2)}$ is therefore proportional to the correlation function of the surface fluctuations

$$G(\mathbf{q},t) = \langle \zeta^*_{\mathbf{q}}(t)\, \zeta_{\mathbf{q}}(t + \tau)\rangle$$

whose Fourier transform is $P(\mathbf{q},\omega)$ (Chap. 4).

The experimental analysis is done by amplifying the photomultiplier signal and using either a correlator that measures directly $G^{(2)}$ or a wave analyzer that measures its Fourier transform

$$P^{(2)}(\omega) = \int_{-\infty}^{+\infty} \langle I(t)\, I(t + \tau)\rangle\, e^{i\omega\tau}\, d\tau \propto P(\mathbf{q}, \omega)$$

In the latter case, the photocurrent must be squared after the frequency analysis. A high-pass filter is also convenient to suppress the d.c. component of the photocurrent that could saturate the instruments. When spectrum analysers performing time compression procedures (after digitalization of the signal) are used, a low-pass filter is also necessary, because these analyzers fold the spectrum and superimpose the different frequency ranges 0-ω, ω-2ω, etc. [8]. Microcomputers can also be used after digitizing the photomultiplier signal; the signal is then sampled and stored in memories. A fast Fourier transform of the data leads to $P(\mathbf{q},\omega)$. The intensity scattered by the surface at small scattering angles can be too large to allow the use of photon counting devices without attenuation of the laser power. As a result, the signal-to-noise ratio can be lower than with analog techniques.

The data processing involves a comparison with the theoretical spectrum (Chap. 4) and eventually instrumental broadening corrections. This is more conveniently done with an experimental spectrum than with a correlation function. When a correlator is used, the fitting time is shorter when the correlograms are first Fourier transformed [5].

REFERENCES

1. R.V. Edwards, R.S. Sirohi, J.A. Mann, L.B. Shih, and L. Lading, *Appl. Optics 21*: 3555 (1982).
2. E. Mazur and D.S. Chung, *Physica 147A*: 387 (1987).
3. R.B. Dorshow and R.L. Swofford, *J. Appl. Phys. 65*: 3756 (1989).
4. S. Hard, Y. Hamnerius, and O. Nilsson, *J. Appl. Phys. 47*: 2433 (1976).
5. S. Hard and R.D. Neuman, *J. Colloid Interface Sci. 115*: 73 (1987).
6. D. McQueen and I. Lundstrom, *J. Chem. Soc. Faraday I, 69*: 694 (1973).
7. H.Z. Cummins and H.L. Swinney, in *Progress in Optics 8* (E. Wolf, ed.), Academic Press, New York, 1970, p. 133.
8. J. Meunier, thesis, Paris 1971.
9. J.-C. Herpin, J. Meunier, *J. Phys. Fr. 35:* 847 (1974).
10. A. Pouchelon, J. Meunier, D. Langevin, A.M. Cazabat, *J. Phys. Lett. 41:* L-239 (1980).
11. D. Langevin, *Colloid Surfaces 21:* 313 (1986).

6
Instrumental Broadening Problems

DOMINIQUE LANGEVIN Laboratoire de Physique Statistique, Ecole Normale Supérieure, Paris, France

I. ORIGIN OF INSTRUMENTAL BROADENING

Because of finite instrumental resolution, it is not possible to analyze the light scattered by a surface thermal fluctuation ζ_q of well-defined wavevector $\mathbf{q}$. One rather collects the light scattered by surface modes whose wavevectors are distributed in a range $q_0 \pm \Delta q$ around a particular $q = q_0$ value.

In the regime where the surface modes are damped ($y \ll 1$), the corresponding spectra are Lorentzian, centered at zero frequency. If $\Delta q/q$ is not too large (below 10%, as is usually the case), the measured spectrum is approximately Lorentzian, and its width is the average width

$$\Delta\omega = \Delta\omega_{q0} = \frac{\gamma\, q_0}{2(\eta_1 + \eta_2)}$$

In the regime where the surface modes propagate ($y \gg 1$), the different spectra are Lorentzian but they are *centered at different frequencies*. The measured spectrum is not in general Lorentzian, even when $\Delta q/q$ is small. Its width can be much larger than the width relative to a single q mode

$$\Delta\omega_q = \frac{2(\eta_1 + \eta_2)}{(\rho_1 + \rho_2)}\, q_0^2 \tag{1}$$

Recall that

$$\omega_q = \left(\frac{\gamma}{\rho_1 + \rho_2}\right)^{1/2} q^{3/2} \tag{2}$$

The frequency span for small $\Delta q/q$ is therefore

$$\Delta\omega_I = \frac{3}{2}\left(\frac{\gamma}{\rho_1 + \rho_2}\right)^{1/2} q_0^{1/2}\,\Delta q \tag{3}$$

Let us take the example of the free surface of water, and $\Delta q = 10\ \text{cm}^{-1}$ as commonly reported. One has for typical q values

$$q = 300\ \text{cm}^{-1} \begin{cases} \omega_{q0} = 4.4\ 10^4\ \text{s}^{-1} \\ \Delta\omega_q = 1.8\ 10^3\ \text{s}^{-1} \\ \Delta\omega_I = 2.2\ 10^3\ \text{s}^{-1} \end{cases}$$

$$q = 1000\ \text{cm}^{-1} \begin{cases} \omega_{q0} = 2.7\ 10^5\ \text{s}^{-1} \\ \Delta\omega_q = 2\ 10^4\ \text{s}^{-1} \\ \Delta\omega_I = 4\ 10^3\ \text{s}^{-1} \end{cases}$$

These numbers illustrate how crucial the instrumental broadening problem is.

In the general case, intermediate between $y \gg 1$ and $y \ll 1$, there are still instrumental broadening corrections when $\omega_q \neq 0$. In this case,

$$\Delta\omega_I = \frac{\partial\omega_q}{\partial q}\,\Delta q \approx \omega_q\,\frac{\Delta q}{q}$$

When $\omega_q/\Delta\omega_q$ approaches unity, $\Delta\omega_I$ becomes negligible compared to $\Delta\omega_q$ if $\Delta q/q$ is small. The major problems are thus found in the small damping regime, i.e., when $\Delta\omega_q \ll \omega_q$. But in this regime, the spectrum can be reasonably approximated by the Lorentzian function

$$P_q(\omega) = \frac{k_B T}{\pi\gamma}\,\frac{2(\eta_1 + \eta_2)}{(\rho_1 + \rho_2)}\left\{\frac{1}{(\omega - \omega_q)^2 + \Delta\omega_q{}^2}\right\} \tag{4}$$

This facilitates the deconvolution of the experimental spectrum from the instrumental function.

II. DECONVOLUTION PROCEDURE: LORENTZIAN APPROXIMATION

Several calculated spectra are compared with Lorentzian functions in Fig. 1 for y values of interest: $y = 180$ and $y = 600$ ($y = \gamma(\rho_1 + \rho_2)/4(\eta_1 + \eta_2)^2 q$, free surface of water, $q = 1000\ \text{cm}^{-1}$ and $300\ \text{cm}^{-1}$, respectively). The real spectrum is asymmetric, and the deviations from the Lorentzian approximation (Eq. 4) are slightly larger for $\omega < \omega_q$. For this reason, it can sometimes be convenient

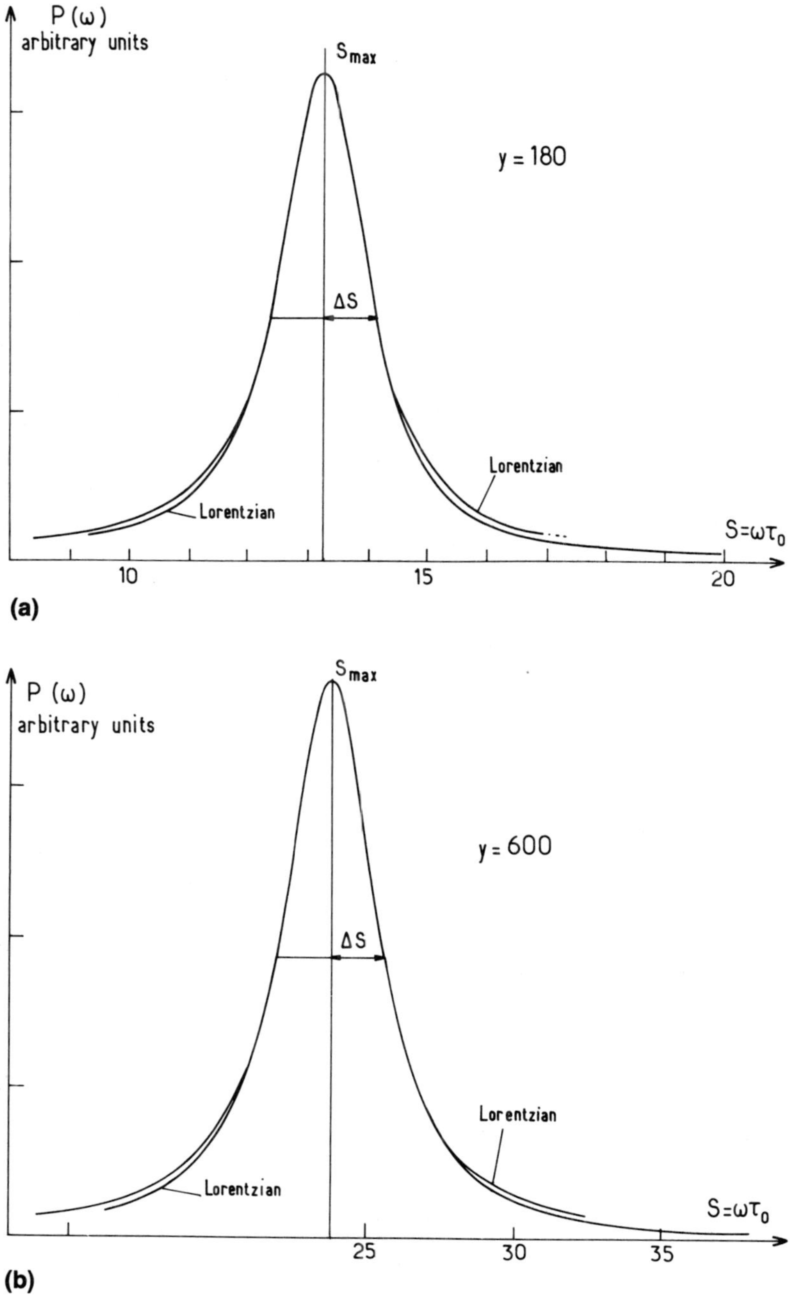

FIG. 1 Theoretical spectrum and its comparison with a Lorentzian spectrum having the same peak frequency and width. (a) $y = 180$; (b) $y = 600$.

to analyze only the region of frequencies larger than the peak frequency, if the Lorentzian approximation is used.

When the instrumental width $\Delta\omega_I$ is small compared to $\Delta\omega_q$, the instrumental function can be approximated by the Gaussian function

$$P_I(\omega) = \exp -\left[\ln 2 \left(\frac{\Delta\omega}{\Delta\omega_I}\right)^2\right] \tag{5}$$

$\Delta\omega_I$ is the half-width at half-maximum of the instrumental spectral shape. The experimental spectrum given by

$$P(\omega) = \int P_q(\omega - \omega')\, P_I(\omega')\, d\omega' \tag{6}$$

belongs to the family of Voigt functions, a Voigt function being the convolution of a Lorentzian and a Gaussian function [1]. The shape of a Voigt function is characterized by a parameter a: $a = \infty$ for a Lorentzian, $a = 0$ for a Gaussian (Fig. 2). If P_I is not strictly Gaussian but can be approximated by a Voigt function of parameter a_I, the convolution product of Eq. 6 is still a Voigt function

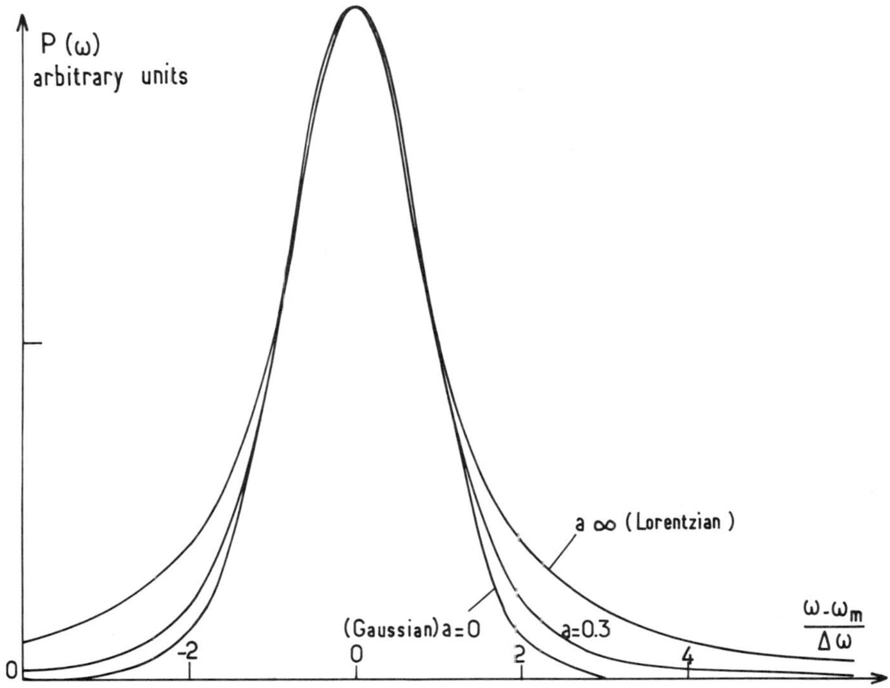

FIG. 2 Voigt functions corresponding to the parameters $a = 0$ (Gaussian), $a = 0.3$, and $a = \infty$ (Lorentzian).

of parameter a. The relations between a, a_I, the width $\Delta\omega$ of the experimental spectrum, the intrinsic width $\Delta\omega_{q0}$, and the instrumental width $\Delta\omega_I$ can be written as

$$\frac{\Delta\omega}{\Delta\omega_I} = \frac{a - a_I}{\xi(a_I)} + F\left(\frac{\Delta\omega}{\Delta\omega_I}, a_I\right)$$
$$\frac{\Delta\omega_q}{\Delta\omega_I} = \frac{a - a_I}{\xi(a_I)} \tag{7}$$

Tables for functions $\xi(a)$ and $F(\Delta\omega/\Delta\omega_I, a)$ can be found in Ref. 1. The variation of the ratio $\Delta\omega/\Delta\omega_I$ and $\Delta\omega_q/\Delta\omega_I$ with a are represented in Fig. 3 for $a_I = 0$ (Gaussian instrumental function) and $a_I = 0.1$. When $a_I = 0$ and $\Delta\omega_I \ll \Delta\omega_q$, the approximation

$$\Delta\omega_q = \Delta\omega - \frac{\Delta\omega_I^2}{\Delta\omega} \tag{8}$$

gives the $\Delta\omega_q$ value with good accuracy [2].

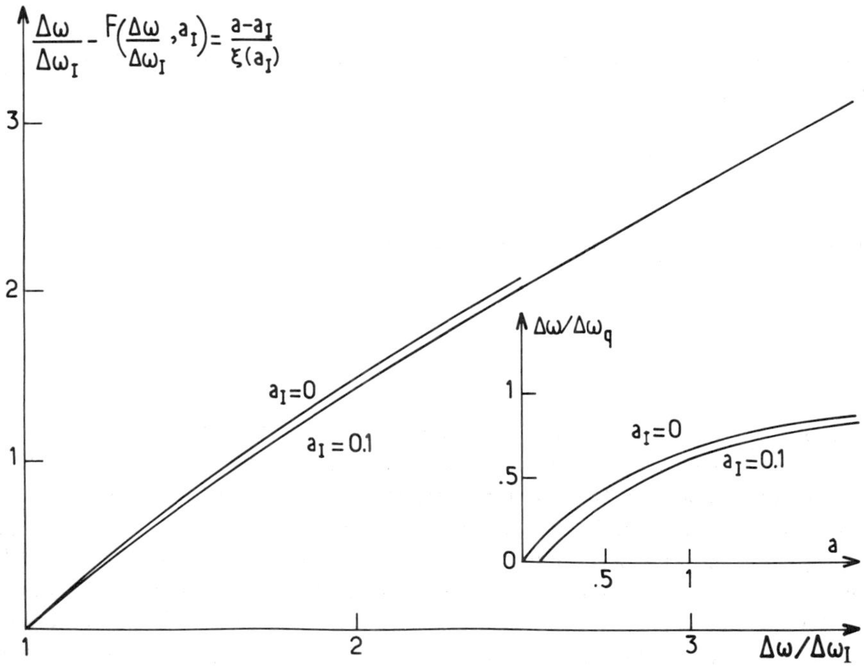

FIG. 3 Variation of the ratio of the measured width $\Delta\omega$ to the instrumental width $\Delta\omega_I$ (abcissa) versus $(a - a_I)/\xi(a_I)$. Insert: variation of the ratio of the measured width $\Delta\omega$ to the intrinsic width $\Delta\omega_q$ versus a for different a_I.

If instead of a spectrum analyzer a correlator is used, the Fourier transform of the convolution product (6) with P_q and P_I given by Eqs. 4 and 5 is

$$G(\tau) = e^{-\beta^2\tau^2} e^{-\tau/\tau_q} \cos \omega_q \tau$$

with $\beta = \Delta\omega_I/(2\sqrt{\ln 2})$. It has been shown [3] that if $G(\tau)$ is fitted with a damped cosine function $e^{-\tau/\tau'} \cos \omega_q\tau$, one has a relation between τ', τ_q, and β similar to Eq. 8:

$$\frac{1}{\tau_q} \sim \frac{1}{\tau'} - \frac{3}{2}\beta^2\tau' \qquad (9)$$

III. MEASUREMENT OF Δq AND OF THE INSTRUMENTAL SHAPE

As discussed in Chap. 5, when a grating is used as local oscillator, the instrumental width is due mainly to diffraction of the light beams. The diffraction contribution can be calculated explicitly by

$$\Delta q = k\, \delta\theta_0 \qquad (10)$$

where, for consistency, $\delta\theta_0$ will be taken as the half-angular-width at half-maximum intensity:

$$\delta\theta_0 = \frac{2\lambda}{\pi l} \sqrt{\ln 2} \qquad (11)$$

leading to $\Delta q = \dfrac{3.3}{l}$. Typically, $l \sim 2$ to 3 mm, and $\Delta q \sim 10\ \text{cm}^{-1}$. Using now Eqs. 2 and 3, one finds

$$\Delta\omega_I = \frac{5}{l}\frac{\omega_{q0}}{q}$$

If l_0 is the beam diameter in a plane close to the surface but normal to beam propagation, then $l = l_0$ for scattering in the plane perpendicular to the plane of incidence, and $l = l_0/\cos\theta_i$ for scattering in the plane of incidence. $\Delta\omega_I$ is therefore smaller in the last geometry as quoted in Chap. 5.

Δq can be measured directly by replacing the liquid surface by a mirror and by recording the variation of the intensity scattered by the grating in the spots in the focusing plane, moving for instance the pinhole in this plane. As noted by ourselves and others [4], the measured values are always slightly smaller than the ones deduced from the fits of the experimental spectra. For instance, in the setup used in Ref. 4, $\theta_i = 3°$, the half-width at half-maximum of the spots was $37 \pm 1\ \mu\text{m}$ with a mirror, and $48 \pm 2\ \mu\text{m}$ with the liquid surface. For the free surface of water and the first spot, $q = 167.8\ \text{cm}^{-1}$; this leads respectively to $\Delta\omega_I/2\pi = 81$ and 105 Hz, whereas for the value obtained by using the deconvolution procedure described thereafter one finds $\Delta\omega_I/2\pi = 113$ Hz. Probably residual vibrations affect the value of Δq and make it larger than given by Eqs. 10 and 11. It is thus safer to measure the size of the spot in the presence of the liquid surface, or to treat $\Delta\omega_I$ as an adjustable parameter, at least when $\Delta\omega_I$ becomes comparable with $\Delta\omega_q$. Δq can also be measured directly, with liquids of very small viscosity, for which $\Delta\omega \approx \Delta\omega_I$ (case of ethyl ether).

When $\Delta\omega_I$ becomes comparable to $\Delta\omega_q$, other problems arise. The instrumental function can no longer be approximated by a Gaussian. The different types of reported instrumental functions [2, 5–7] are shown in Fig. 4. Even when gratings are used, these functions are not exactly Gaussian. They can, however, be approximated by Voigt functions. The convolution of a Voigt function with a Lorentzian being still a Voigt function, the deconvolution procedure can be easily done using Eqs. 7.

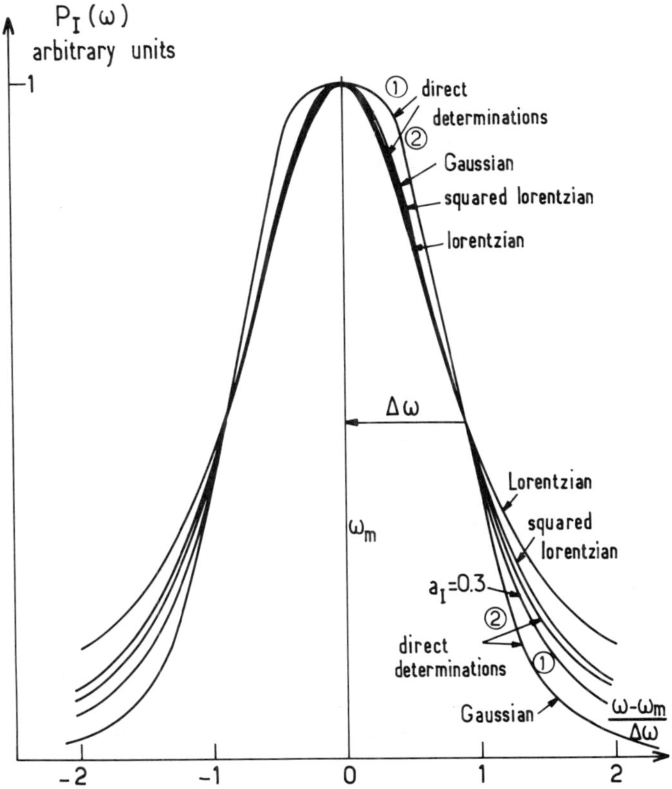

FIG. 4 Instrumental functions; direct determinations: (1) Ref. 6; (2) Ref. 2; Gaussian function, Ref. 2; Voigt function, $a = 0.3$, Ref. 5; squared Lorentzian, Ref. 7.

Some results obtained in this way for the free surface of water are summarized in Table 1, where the improvement brought about by a grating is demonstrated. a_I was found equal to 0.2 ± 0.1 (slightly larger without the grating $a_I \sim 0.3$ [5]. In these experiments, $\theta \sim 3°$, $l \sim 4$ mm, and $\Delta q \sim 8$ cm^{-1}, a value similar to that in other accurate experiments in which $\theta \sim 50°$ and $l \sim 2$ mm [8]. Examples of errors found when using a Gaussian instrumental function are also shown.

In order to determine the instrumental function, one can study the spectrum of the light scattered by a liquid of very small viscosity like ether where $\Delta\omega_q \ll \Delta\omega_I$. The shape of this spectrum reproduces exactly the instrumental function. One can also study a liquid of well-known properties like ethanol and determine a_I and $\Delta\omega_I$ from the experimental spectra. In these test experiments, the liquid surface tension being known, the position of the peak frequency gives also the value of q_0. This allows one to ensure that the distance D between the pinhole and the surface has been accurately measured, and that it does not change with time (because of liquid evaporation or imperfect table level control). When working in the plane of incidence, it allows the determination of q_0 and Δq from only the additional measurements of diffraction spot size and separation in the focusing plane (avoiding the long calibration process described in Ref. 8).

IV. DECONVOLUTION PROCEDURE: GENERAL CASE

The convolution product (6) and the fit with the experimental spectrum can also be done numerically using known algorithms. This method is especially helpful to account for small deviations of the spectrum $P_q(\omega)$ from a Lorentzian, or for differences between the approximated value of ω_q and $\Delta\omega_q$ and the real peak

TABLE 1 Parameters Deduced from the Spectra of the Light Scattered by the Free Surface of Water at 20° for Different Wavevectors[a]

q (cm^{-1})	167.8	335.6	503.4	671.2
ν_q (Hz)	2943	8318	15250	23430
ν_m	2960 ± 30	8335 ± 50	15290 ± 40	23400 ± 150
$\Delta\nu_q$ (Hz)	81.9	323	714	1257
$\Delta\nu$ (with grating)	185 ± 5	470 ± 20	925 ± 40	1620 ± 60
a	0.85 ± 0.05	1.85 ± 0.1	2 ± 0.2	2.5 ± 1
$\Delta\nu_q$ ($a_I = 0$)	115	400	796	1460
$\Delta\nu$ (without grating)	390	896	1279	2055

[a] ν_m and $\Delta\nu$ are the measured peak frequencies and half-widths; a is the shape parameter of the spectrum; $\Delta\nu_q(a_I = 0)$ is the width corrected for instrumental broadening using Eq. 8; ν_q and $\Delta\nu_q$ are calculated using Eqs. 4 of Chap. 7, which are better approximations than Eqs. 1 and 2.

frequency and width of $P_q(\omega)$ [9]. As we shall see later, these differences can be appreciable in monolayer covered surfaces.

When correlators are used, the deviations of the theoretical correlation function from a damped sinusoid can be described simply by using a phase shift Φ [10]:

$$G(\tau) = e^{-\tau/\tau_q} \cos(\omega_q \tau + \Phi) \tag{12}$$

If the instrumental function is assumed to be Gaussian, the functional form

$$G(\tau) = A\, e^{-\beta^2\tau^2}\, e^{-\tau/\tau_q} \cos(\omega_q \tau + \Phi) + B \tag{13}$$

fits the data very well. When ω_q and τ_q are used to solve the dispersion equation, the values found for γ and η agree well with the currently accepted values for water [11].

V. FREQUENCY SHIFTS

The spread in q values can also produce a frequency shift of the experimental spectra $\omega_m \neq \omega_{q0}$. This is because Lorentzian spectra of different integrated intensities are superimposed; when q increases, the intensity decreases as $1/q^2$. In general, this should result in peak frequencies ω_m smaller than ω_{q0}. This effect has been treated theoretically in detail in Refs. 8, 12–14. However, the shifts calculated are systematically larger than the ones measured, which are comparable to the standard deviations of the measured frequencies [4]. In fact, when q decreases, the integrated intensity increases, but the spectral width decreases, so that during the superposition process fewer wavevectors give significant contributions to the total spectrum at a given frequency. This can be demonstrated when the instrumental function is simply one in the range $\omega_{q0} \pm \Delta\omega_I$ and zero elsewhere. Then

$$P(\omega) \sim \int_{\omega_{q0}+\Delta\omega I}^{\omega_{q0}-\Delta\omega_I} \frac{d\omega_q}{(\omega - \omega_q)^2 + \Delta\omega_q^2}$$

If the variation of $\Delta\omega_q$ with q is neglected, one has

$$P(\omega) \sim \text{Arctg}\left(\frac{\omega - \omega_{q0} + \Delta\omega_I}{\Delta\omega_q}\right) - \text{Arctg}\left(\frac{\omega - \omega_{q0} - \Delta\omega_I}{\Delta\omega_q}\right)$$

The half-width of $P(\omega)$ is

$$\Delta\omega = \sqrt{\Delta\omega_q^2 + \Delta\omega_I^2}$$

and $P(\omega)$ is centered around $\omega = \omega_{q0}$. The frequency shifts arise therefore from the variation of $\Delta\omega_q$, i.e., they are small effects.

A similar frequency shift has also been predicted, in the opposite direction for scattering in the plane of incidence [8, 12, 14]. Indeed, the grating diffraction

TABLE 2 Interfacial Tensions Deduced from the Spectra for Different q Values in the Total Reflection Plane of Incidence Geometry

q (cm^{-1})	42	126	160	320	Literature value
γ water-butanol (mN/M)	2.2 ± 0.2	1.8	1.8	1.8	1.8
γ water-octane-SHBS (mN/m)	1.6	1.2 ± 0.1	1.2	1.2	—

Source: Ref. 15.

spots in the focusing plane are centered on a line contained in the plane of incidence, but they extend in directions normal to the plane of incidence where q is larger (Chap. 5, Sec. 7). Therefore the wavevectors contributing to the spectrum are larger than q_0 on the average. The result is to increase the center frequency with respect to ω_{q0}. Table 2 presents some data that are in agreement with the predictions of Ref. [8]. It must be noted that a further complication comes from the fact that the experimental spectrum is not the convolution of the instrumental function with $P_q(\omega)$ as given in Eq. 6. This is because $P_q(\omega)$ cannot be written as a function of $\omega - \omega'$. The corresponding differences between calculated $P(\omega)$ can be observed for large θ and $\phi = 0$ [12]. However, it is seen from Table 2 that these effects, together with the frequency shifts, are extremely small and can only be observed at very small q values.

REFERENCES

1. H.W. Leidecker, Jr. and J.T. Lamacchia, *J. Acoust. Soc. Amer. 43:* 143 (1968).
2. S. Hard, Y. Hamnerius, O. Nilsson, *J. Appl. Phys. 47:* 2433 (1976).
3. S. Hard and R.D. Neuman, *J. Colloid Interface Sci. 83:* 315 (1981).
4. D. Langevin, *Colloid Surfaces 21:* 313 (1986).
5. D. Langevin, *J. Chem. Soc. Faraday Trans. I, 70:* 95 (1974).
6. J.-C. Herpin and J. Meunier, *J. Phys. Fr. 35:* 847 (1974).
7. D. Jon, H. Rosano, and H.Z. Cummins, *J. Colloid Interface Sci., 114:* 330 (1986).
8. S. Hard and R.D. Neuman, *J. Colloid Interface Sci. 115:* 73 (1987).
9. J.C. Earnshaw, R.C. McGivern, A.C. McLaughlin, and P.J. Winch, *Langmuir 6:* 649 (1990).
10. H. Birecki and N.M. Amer, *J. Phys. Paris 40:* C3-433 (1979).
11. J.C. Earnshaw and R.C. McGivern, *J. Colloid Interface Sci., 123:* 36 (1988).
12. R.V. Edwards, R.S. Sirohi, J.A. Mann, L.B. Shih, and L. Lading, *Appl. Optics 21:* 3555 (1982).
13. J.F. Crilly and J.C. Earnshaw, *J. Phys. D 18:* 609 (1985).
14. J.C. Earnshaw, *Appl. Opt. 24:* 887 (1985).
15. D. Guest, Thesis, Paris, 1985.

II
Applications

7

Simple Liquids

DOMINIQUE LANGEVIN Laboratoire de Physique Statistique, Ecole Normale Supérieure, Paris, France

The study of the spectrum of the light scattered by the interface between two fluids allows the determination of the interfacial tension γ. It also allows the determination of the sum of the bulk viscosities if the dimensionless ratio $y = \gamma(\rho_1+\rho_2)/4(\eta_1+\eta_2)q^2$ is not too small; in this case, the determination of γ or $\eta_1 + \eta_2$ presupposes that the bulk densities are measured independently. When y is smaller than about 0.5, only the ratio $\gamma/(\eta_1+\eta_2)$ can be measured, and the determination of γ necessitates that the bulk viscosities be measured independently.

I. ACCURACY ON THE MEASURED PARAMETERS

For liquids of viscosities comparable to water, $y \gg 1$, the surface tension can be calculated from the peak frequency of the spectrum as

$$\omega_q^2 = \gamma q^3/(\rho_1 + \rho_2) \tag{1}$$

The relative accuracy on γ is twice the relative accuracy on ω_q plus three times the relative accuracy on q. In accurate measurements, ω_q is known within 0.25%, q within better than 0.1%, and the overall accuracy on γ is about 1%. For the free surface of water for instance, $\gamma = 72$ mN/m, and the corresponding accuracy is about 1 mN/m. Provided that a very careful calibration procedure is used, the accuracy can be made a little better: 0.5 mN/m [1]. There are many other more precise devices allowing the measurement of γ in this surface tension range.

The light scattering method only becomes competitive when the surface tension is very small. In this case, $y \ll 1$, $\omega_q = 0$, and the half-width (at half-height) of the spectrum is

$$\Delta\omega_q = \frac{\gamma q}{2(\eta_1 + \eta_2)} \tag{2}$$

The accuracy on γ is here better than 1%. When γ is below 0.1 mN/m, such an accuracy is difficult to achieve with other techniques such as the drop methods.

The accuracy on q plays a central role. From Chap. 5, Eqs. 2, we know that $q = k\,\Delta\theta \cos\theta_i$ or $q = k\,\Delta\theta$ according to whether the detection is made in the plane of incidence or perpendicular to the plane of incidence. $\Delta\theta$ and $\cos\theta_i$ can be measured directly or inferred from the measured ω_q on liquids of known properties. Caution has to be taken because these quantities may change with time. For instance, if the distance D between the liquid surface and the diaphragm is about 1 m and changes by 1 mm because of liquid evaporation or poor stabilization control of the table position, $\Delta\theta$ and q change by 0.1%, since $\Delta\theta \sim 1/D$. The corresponding error on γ is 0.2 mN/m in the case of water.

The viscosity can be independently deduced from the spectrum for $y \gtrsim 0.5$. When $y \gg 1$, ω_q is given by Eq. 1 and

$$\Delta\omega_q = \frac{2(\eta_1 + \eta_2)q^2}{(\rho_1 + \rho_2)} \tag{3}$$

For such large y values, the accuracy is poor because instrumental broadening corrections introduce large errors. In the very accurate measurements of Hard and Neuman ($\Delta q \sim 10$ cm^{-1}, $q > 800$ cm^{-1}), the reported accuracy on viscosity is 3% [1]. The accuracy is better when y is smaller because the instrumental broadening is less important. The optimal region is around $y \sim 1$.

Again, there are many other methods to measure viscosity more accurately. The study of light scattering by the surfaces of simple liquids serves rather as a check of the somewhat complex calibration procedures involved with this technique.

II. FREE SURFACE OF WATER

The light scattering from the free surface of water has been widely investigated because of its importance in relation to surfactant monolayer studies. It must be noted, however, that the free surface of water is one of the most difficult to study for the following reasons (1) the scattered intensity is small because the surface tension is large, (2) the instrumental broadening corrections are important because ω_q is large (large surface tension) and $\Delta\omega_q$ small (small viscosity): (3) the surface is very sensitive to contaminants that lower the tension. Very pure water has to be used, and the surface has to be swept (with hydrophobic teflon barriers, for instance) several times prior to measurements, to remove the residual surface active contaminants; (4) the liquid evaporates if the sample cell humidity is not controlled. In an ordinary room (relative humidity ~50%) the surface temperature could be about 2°C below room temperature. If this is so, the effective viscosity in the surface wave damping is about 4.5% larger than the room temperature value, because of the temperature dependence of water viscosity. Furthermore, the wavevector varies with time since D is varying due to the lowering of the water level: this can lead to serious errors in the surface tension determination (Sec. I).

Because of the difficulties involved in instrumental broadening corrections, widths larger than predicted were found in some cases [2, 3]. They were attributed to a water viscosity increase close to the surface. Although ordering of water molecules certainly occurs within a distance to the surface equal to a few molecular diameters, it must be recalled that the viscosity involved in the wave damping studied here is the average viscosity over a distance comparable to the damping depth of the waves. This depth is of the order of q^{-1}, i.e., typically 10 μm ($q = 1000\ \text{cm}^{-1}$). This is much larger than a molecular diameter. In order to observe anomalous damping, the viscosity close to the surface η_s should be such that $\eta_s a \sim \eta q^{-1}$, i.e., $\eta_s \sim \eta(qa)^{-1}$ where a is a molecular diameter. With $a \sim 1$ Å, this leads to $\eta_s \sim 10^5\eta$. Such large surface viscosities are observed when a condensed surfactant monolayer is present at the surface of water. As we shall see in Chap. 11, such a monolayer is able to produce changes both in the frequency and in the damping of capillary waves. But the ordering of water molecules close to the surface is unable to produce large local viscosities. It has been verified that when the suitable instrumental corrections are made, ω_q and $\Delta\omega_q$ follow the expectations of the standard hydrodynamic theory [1, 4, 5].

When $y \gg 1$, ω_q and $\Delta\omega_q$ are close to the roots of the dispersion equation for the free surface of a liquid and are given by Eqs. 1 and 3, where $\eta_1 \approx 0$, $\rho_1 \approx 0$, and $\eta = \eta_2$, $\rho = \rho_2$. Approximations better than Eqs. 1 and 3 for these roots are

$$\begin{aligned} \omega_q &= \sqrt{\frac{\gamma q^3}{\rho}}\left(1 - \frac{1}{2y^{3/4}}\right) \\ \Delta\omega_q &= \frac{2\eta q^2}{\rho}\left(1 - \frac{1}{2y^{1/4}}\right) \end{aligned} \tag{4}$$

The difference between limit Eqs. 1 and 3, approximated Eqs. 4, and spectral

TABLE 1 Reduced Frequencies and Widths

y	150	400	900
S_{i0}	12.25	20	30
S_i	12.10	19.89	29.90
S_m	12.07	19.86	29.89
S_{r0}	1	1	1
S_r	0.857	0.888	0.909
ΔS	0.858	0.888	0.909

S_{i0} and S_{r0} from Eqs. 1 and 3; S_i and S_r from Eqs. 4; S_m and ΔS from the spectrum, Eq. 7.

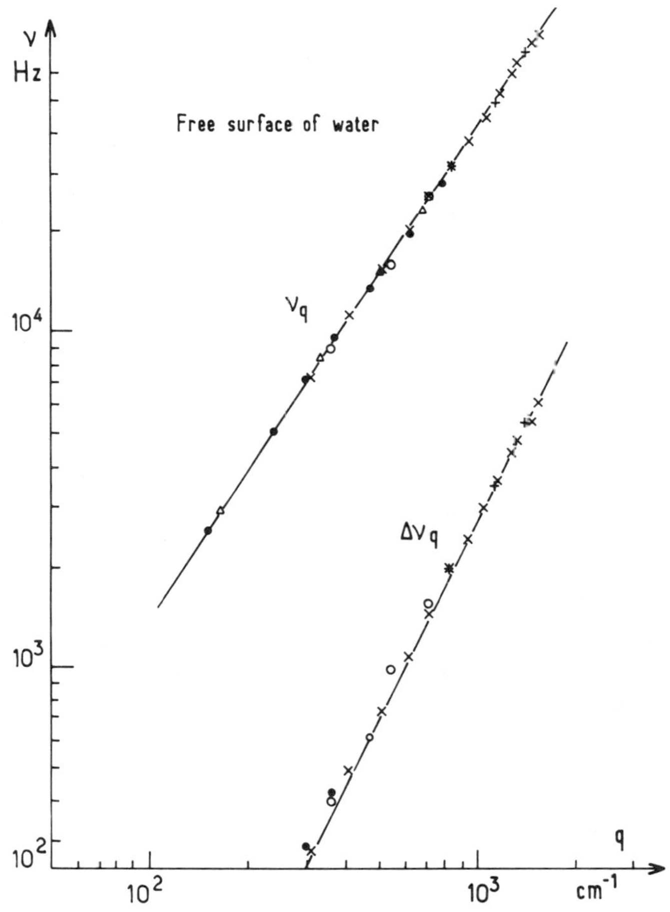

FIG. 1 Peak frequency and half-width of the spectrum of the light scattered by the free surface of water versus scattering wavevector. Points of: +, Ref. 1; ●, Ref. 4; Δ, Ref. 5; o, Ref. 6; x, Ref. 7. The slopes of the lines are $3/2(\nu_q)$ and $2(\Delta\nu_q)$.

peak frequency and width are illustrated in Table 1. The frequencies are given in reduced units, $S_i = \omega_q \tau_0$, $S_r = \Delta\omega_q \tau_0$, where $\tau_0 = \rho/2\eta q^2$. The spanned y range is $150 < y < 900$, corresponding to scattering wavevectors $200 < q < 1200\ \mathrm{cm}^{-1}$ for water.

It is seen in the table that Eqs. 4 are very good approximations for the peak frequency ω_m and half-width $\Delta\omega$ of the spectrum.

Existing experimental data are displayed in Fig. 1. The data are from references [1, 4–7]. When a very good accuracy is required, it may be worthwhile to account for nonzero air density and viscosity: 1.2 kg/m^3 and 1.8 10^{-5} kg/m · s respectively at 20°C. According to Eqs. 1 and 3, this leads to approximately a correction of 0.1% on the calculated peak frequency and 2% on the width.

III. FREE SURFACES OF OTHER LIQUIDS

The viscosity of ethyl-ether is about ten times smaller than the viscosity of water. y is large and the measured width of the spectra is mainly due to instrumental broadening. Typical spectrum and correlation functions are shown in Fig. 2.

Several liquids of viscosity similar to that of water but of smaller surface tension have been studied: ethanol, dodecane, carbon tetrachloride. The y values are smaller. The differences between the approximations used for ω_q and $\Delta\omega_q$ are shown in Table 2 for the range $5 < y < 50$. When $\gamma = 20$ mN/m, $\rho = 10^3$ kg/m^3, and $\eta = 10^{-3}$ kg/m · s, $y = 50$ corresponds to $q = 1000\ \mathrm{cm}^{-1}$.

In this y range, there are now appreciable differences even between the exact roots of the dispersion equation and the peak frequencies and widths of the spectrum.

Instrumental broadening corrections are less important than for water, and these liquids are less sensitive to surface contamination. They can therefore be used with profit to check the calibration procedures of the instrument.

When y is smaller, the spectra differ markedly from Lorentzians (Chap. 4, Fig. 3). They are closer to the spectrum of a thermally excited harmonic oscillator (Chap. 4, Eq. 20):

$$P_q^{\mathrm{HO}}(\omega) \sim \left\{ \frac{1}{(\omega - \omega_q)^2 + \Delta\omega_q^2} + \frac{1}{(\omega + \omega_q)^2 + \Delta\omega_q^2} + \frac{1 - \omega/\omega_q}{(\omega - \omega_q)^2 + \Delta\omega_q^2} + \frac{1 + \omega/\omega_q}{(\omega + \omega_q)^2 + \Delta\omega_q^2} \right\} \tag{5}$$

whose Fourier transform is simply (Chap. 4, Eq. 19):

$$G_q^{\mathrm{HO}}(\tau) \sim \exp(-\tau/\tau_q) \left(\cos \omega_q \tau - \frac{\Delta\omega_q}{\omega_q} \sin \omega_q \tau \right) \tag{6}$$

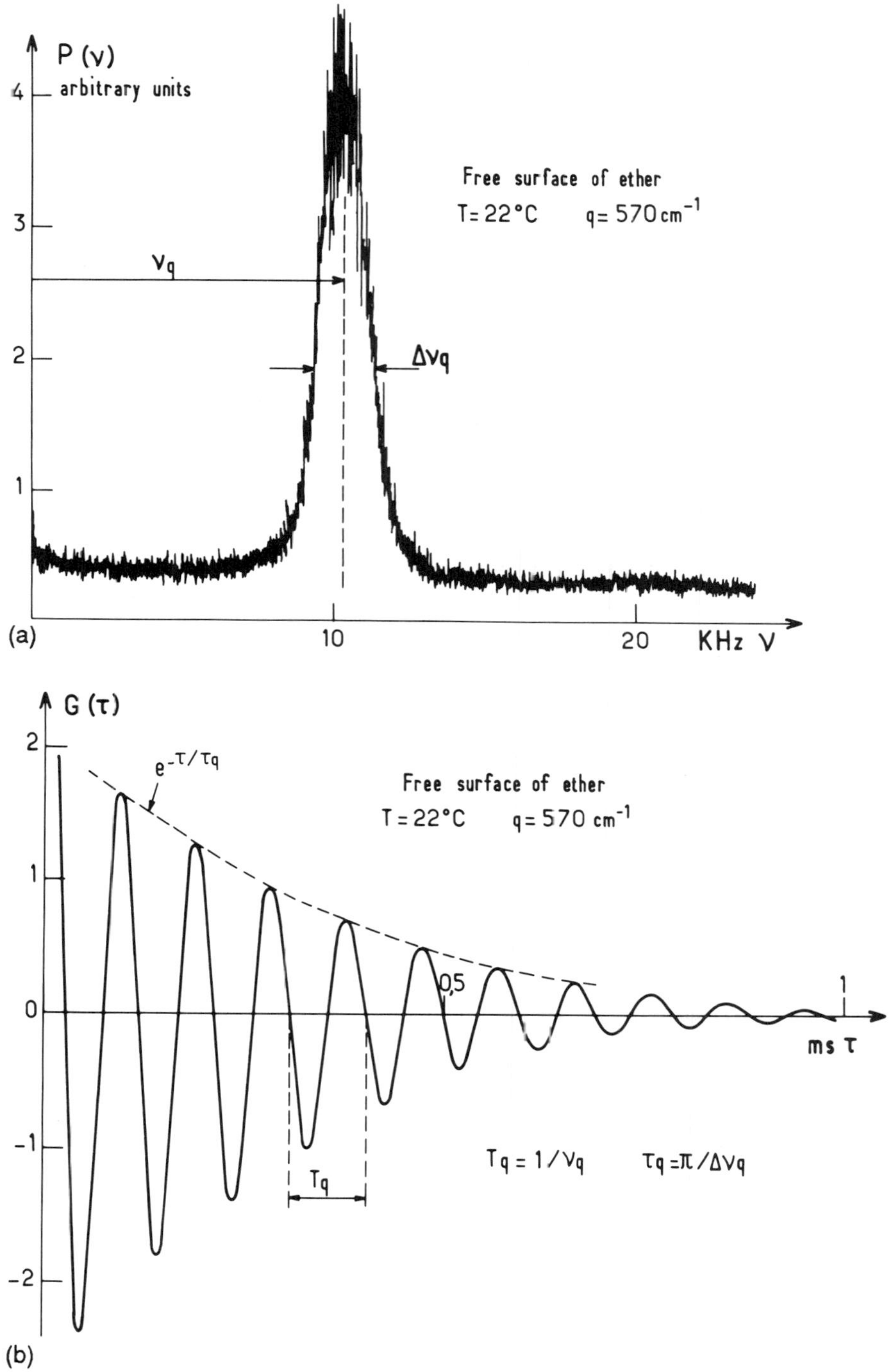

FIG. 2 Light scattering by the free surface of ethyl-ether: (a) spectrum; (b) correlation function.

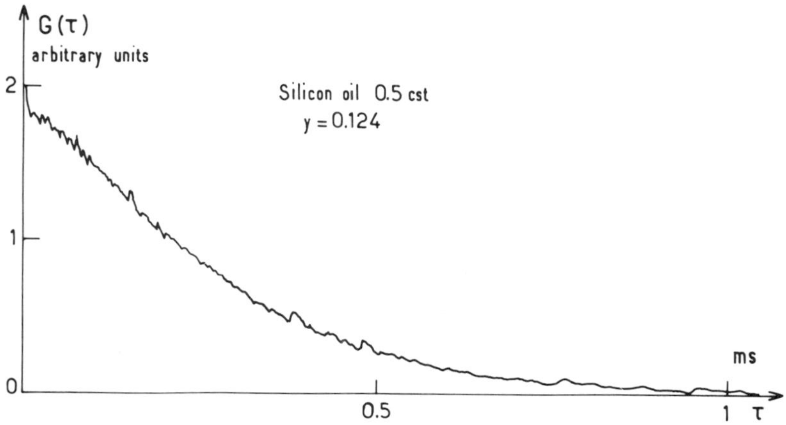

FIG. 3 Correlation function for the light scattered by the free surface of a silicon oil.

A typical correlation function is shown in Fig. 3: the initial slope of $G(\tau)$ is clearly much smaller than that of the first term of Eq. 6, $\exp(-\tau/\tau_q)\cos\omega_q\tau$, Fourier transform of the first two terms of Eq. 5. Indeed, for small τ, the slope of $\exp(-\tau/\tau_q)\cos\omega_q\tau$ is the same as the slope of the exponential envelope $\exp(-\tau/\tau_q)$. Equation 5 there better fits the experimental spectra than Lorentzian curves. However, it leads to large errors when used to extract γ and η from these experimental spectra. For instance, the surface tension deduced from the fit of the spectrum of Fig. 3 in Chap. 4, with $P_q(\omega)$ is 34.1 mN/m and from the fit with $P_q^{HO}(\omega)$ 36.4 mN/m. Similarly the viscosities are respectively 1.78 10^{-2} kg/m·s and 1.65 10^{-2} kg/m · s [8]. These errors are less important if y is larger, but still appreciable when $y < 10$ (Table 2).

TABLE 2 Reduced Frequencies and Widths

y	5	10	20	50
S_{i0}	2.23	3.16	4.47	7.07
S_{i1}	1.90	2.88	4.23	6.88
S_i	1.94	2.90	4.24	6.88
S_m	1.87	2.84	4.19	6.84
S'_m	1.84	2.82	4.18	6.84
S_{r1}	0.665	0.719	0.764	0.812
S_r	0.629	0.696	0.750	0.805
ΔS	0.660	0.720	0.765	0.810
$\Delta S'$	0.720	0.745	0.775	0.820

S_{i0} from Eq. 1; S_{i1} and S_{r1} from Eqs. 4; $\pm iS_i - S_r$ are exact roots of the dispersion equation; S_m and ΔS from the spectrum, Eq. 7; S'_m and $\Delta S'$ from the approximated spectrum, Eq. 5.

The approximate expression 5 gives an idea as to the difference between ω_q and the peak frequency ω_m:

$$\omega_m = \sqrt{\omega_q^2 - \Delta\omega_q^2} \qquad \text{for} \qquad y > 0.38$$

$$\omega_m = 0 \qquad \text{for} \qquad y < 0.38$$

These relations show that the peak frequencies become zero for y values where the capillary waves still propagate; ω_q becomes indeed zero for smaller y values: $y < 0.145$.

The exact values of ω_m and $\Delta\omega$ such as $P(\omega_m + \Delta\omega) = P(\omega_m)/2$ are shown in reduced units in Fig. 4 for the range $0.01 < y < 10$. They have been calculated from the exact spectrum $P_q(\omega)$:

$$P_q(\omega) = \frac{k_B T}{\pi\omega} \frac{q\tau_0^2}{\rho} \operatorname{Im}\left\{ \frac{1}{y + (1 + S)^2 - \overset{+}{\sqrt{1 + 2S}}} \right\}_{S = -i\omega\tau_0} \tag{7}$$

The roots of the dispersion equation are also shown in Fig. 4. An experimental spectrum obtained with the free surface of a silicon oil [9] is shown in Fig. 5, together with the fit with Eq. 7. The corresponding correlation function is that of Fig. 3. Instrumental broadening corrections are negligible here, and the agree-

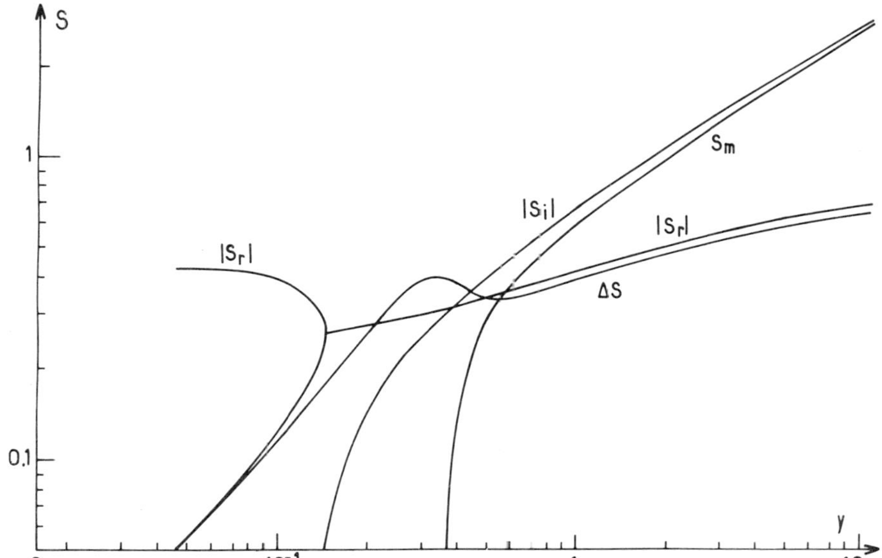

FIG. 4 Reduced peak frequencies S_m and widths ΔS versus the reduced parameter y. The real and imaginary parts of the roots of the dispersion equation have also been represented. ΔS is such as $P(S_m + \Delta S) = P(S_m)/2$.

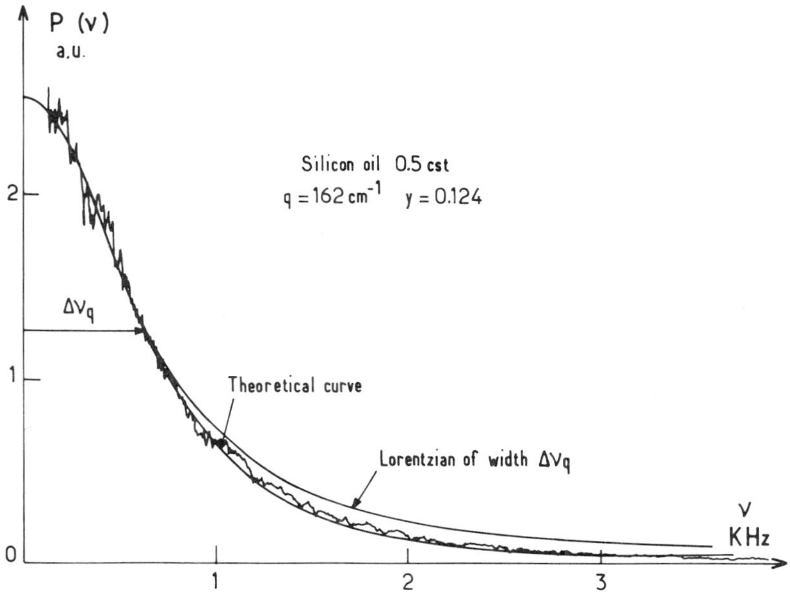

FIG. 5 Spectrum of the light scattered by the free surface of a silicon oil (same sample and same scattering angle as in Fig. 3).

ment found between theory and experiment when using independent determinations of η and γ is very good.

The shape of the spectra is now sensitive to γ and η. As a result, the accuracy in the determination of these parameters is optimal in this region.

IV. FREE SURFACES OF LIQUIDS OF LARGE VISCOSITIES

When y is smaller than 0.145, $\omega_q = 0$ and the capillary waves are completely overdamped. When y is smaller than 0.05, the difference between $\Delta\omega$ and $\Delta\omega_q$ (Eq. 2) is negligible (1% for $y = 0.05$, 0.3% for $y = 0.03$).

In order to check the validity of Eq. 2, the surface of glycerol was studied at room temperature and at 0°C. Typical spectrum and correlation function are shown in Fig. 6. The agreement found between theory and experiment was satisfactory (10%) in view of the uncertainties on γ (glycerol is rapidly contaminated by water, and the effective surface tension is larger than for pure glycerol) and on η (strongly temperature dependent).

In some cases, the light scattering technique is used to measure ultralow interfacial tensions. From Eq. 2, one sees that the measured $\Delta\omega_q$ will be very

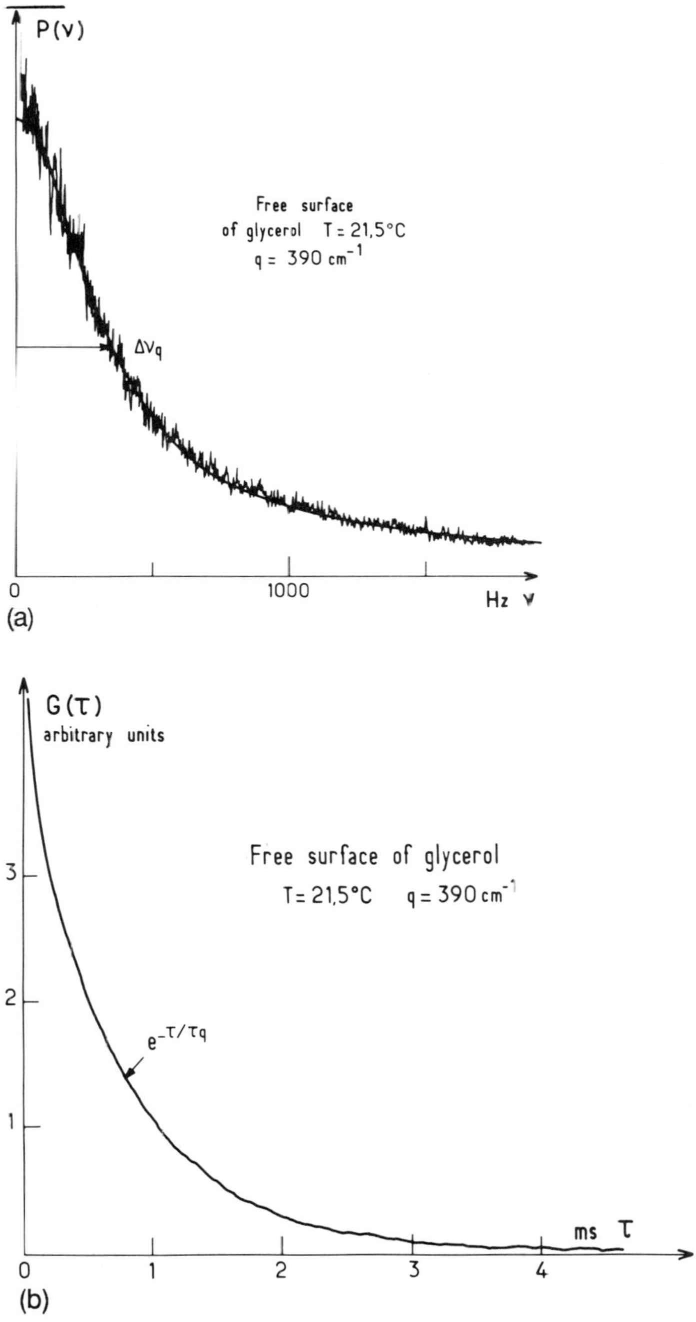

FIG. 6 Light scattering by the free surface of glycerol: (a) spectrum; (b) correlation function.

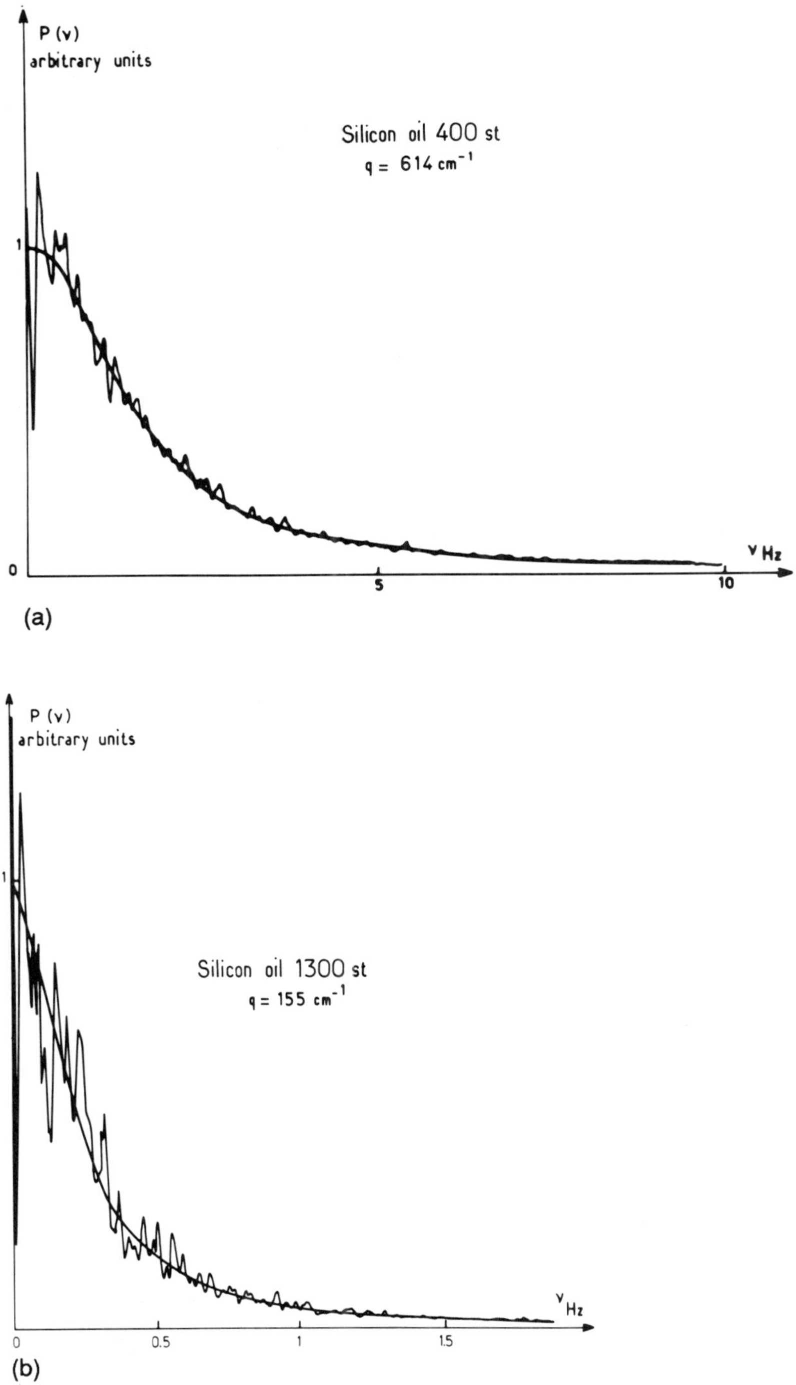

FIG. 7 Spectrum of the light scattered by the free surface of very viscous silicon oils.

small. In order to check that the fluctuations of the amplitude of the laser electric field were not a limiting factor, the surface of highly viscous silicon oils (η = 38.2 and 128 kg/m · s) was studied [10]. We first found that the motion of dust particles in the laser beam give a strong oscillatory contribution to the photocurrent in the hertz range. This signal was suppressed by enclosing the setup in a box and blowing filtered air into this box to immobilize the dust particles on the box walls. Typical spectra obtained in this way are shown in Fig. 7. The smallest measured widths were about 0.2 Hz. But there were no apparent limitations to going beyond this limit other than the very long time required for the data sampling procedure (about 30 min per single spectrum in the range 0–2 Hz with the wave analyser used; the second spectrum of Fig. 7 is an average of 2 single spectra). The measured and calculated widths (Eq. 2) with known η and γ values were in very good agreement.

Figure 8 illustrates the variation of measured peak frequencies and widths for several liquids of different viscosities allowing the spanning of a large range of y values.

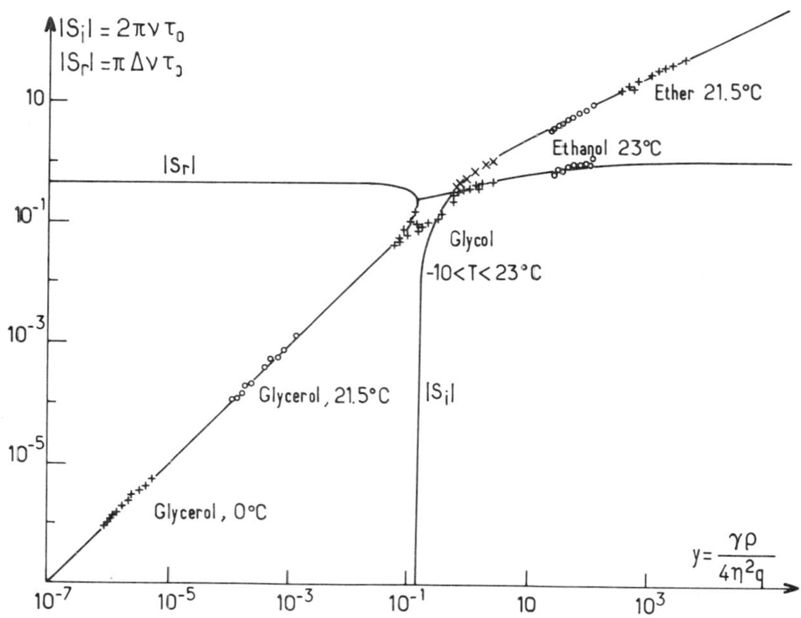

FIG. 8 Reduced measured peak frequencies S_m and widths ΔS versus the reduced parameter y for the free surface of different liquids. The lines are the real and imaginary parts of the roots of the dispersion equation.

V. LIQUID-LIQUID INTERFACES—THEORETICAL SPECTRUM

A reduced parameter $y = \gamma(\rho_1 + \rho_2)/4(\eta_1 + \eta_2)^2 q$ can be defined for liquid-liquid interfaces. When $y \gg 1$ or $y \ll 1$, Eqs. 1–3 can be used for the peak frequencies and the widths of the spectra. But when y takes intermediate values, the spectrum $P_q(\omega)$ is not the same as for a free liquid surface.

The calculation of $P_q(\omega)$ can be easily made by generalizing the calculation of Chap. 4. We will take for the velocity **v** and pressure p solutions of the form

$$z > 0 \qquad v_{z1} = A_1\, e^{-qz} + B_1\, e^{-m_1 z} \qquad m_1 = q\sqrt{1 - i\omega\rho_1/\eta_1 q^2}$$

$$v_{y1} = A_1'\, e^{-qz} + B_1'\, e^{-m_1 z}$$

$$v_{x1} = -i\, A_1\, e^{-qz} - i\frac{m_1}{q} B_1\, e^{-m_1 z}$$

$$p_1 = -i\frac{\omega}{q}\rho_1\, A_1\, e^{-qz}$$

$$z > 0 \qquad v_{z2} = A_2\, e^{qz} + B_2\, e^{m_2 z} \qquad m_2 = q\sqrt{1 - i\omega\rho_2/\eta_2 q^2}$$

$$v_{y2} = A_2'\, e^{qz} + B_2'\, e^{m_2 z}$$

$$v_{x2} = -i\, A_2\, e^{qz} - i\frac{m_2}{q} B_2\, e^{m_2 z}$$

$$p_2 = i\frac{\omega}{q}\rho_2\, A_2\, e^{qz}$$

The limiting conditions at the surface $z=0$ express the continuity of the velocity components and the balance between the strain tensor components $\sigma_{ij} = \eta(\partial v_i/\partial x_j) + \partial v_j/\partial x_i - p\delta_{ij}$.

$$\mathbf{v}_1(z=0) = \mathbf{v}_2(z=0)$$

$$\sigma_{zi1}(z=0) - \sigma_{zi2}(z=0) = F_i$$

Here, the force **F** per unit area is vertical and is equal to

$$F_z = \gamma\frac{\partial^2\zeta}{\partial x^2} = \frac{\gamma}{\omega} q^2\, v_z\,(z = 0)$$

As in the case of a free surface, it appears that the motion in the plane XOZ is completely decoupled from the motion in the direction OY.

The limiting conditions become in the presence of an external pressure $p_{ext}e^{-i\omega t}$

$$A_1 + B_1 = A_2 + B_2$$

$$qA_1 + m_1 B_1 = -qA_2 - m_2 B_2$$

$$\eta_1\{2q^2A_1 + (m_1^2+q^2)B_1\} - \eta_2\{2q^2A_2 + (m_2^2+q^2)B_2\} = 0$$

$$2\eta_1(qA_1 + m_1B_1) - i\frac{\omega}{q}\rho_1 A_1 + 2\eta_2(qA_2 + m_2B_2) - i\frac{\omega}{q}\rho_2 A_2$$

$$- \frac{\gamma}{i\omega} q^2 (A_1 + B_1) = p_{\text{ext}}$$

By setting $p_{\text{ext}} = 0$ one obtains the dispersion equation of surface phonons:

$$\Delta = q(m_1-q)(m_2 - q)\left\{(-i\omega(\eta_1 m_1 + \eta_1 q + \eta_2 m_2 + \eta_2 q)) \left(\gamma q^2 - i\omega(\eta_1 m_1 + \eta_1 q + \eta_2 m_2 + \eta_2 q) - \frac{\omega^2}{q}(\rho_1 + \rho_2)\right) - (-i\omega(\eta_2 m_2 - \eta_2 q - \eta_1 m_1 + \eta_1 q))^2\right\} = 0 \tag{8}$$

The response function $\chi(\mathbf{q}, \omega) = \zeta_q(\omega)/p_{\text{ext}}$ is related to the spectrum via the fluctuation-dissipation theorem

$$P(\mathbf{q}, \omega) = \frac{k_B T}{\pi\omega} \operatorname{Im} \chi(\mathbf{q}, \omega)$$

Finally one obtains

$$P(\mathbf{q}, \omega) = \frac{k_B T}{\pi\omega} \operatorname{Im} \frac{-i\omega(\eta_1(m_1 + q) + \eta_2(m_2 + q))}{\Delta'} \tag{9}$$

where $\Delta = \Delta'/q(m_1 - q)(m_2 - q)$

The spectrum given by Eq. 7 is reobtained by setting $\eta_1 = \rho_1 = 0$. Let us illustrate the differences between the cases of a free surface and of a liquid-liquid interface by choosing the simple case $\eta_1 = \eta_2 = \eta$, $\rho_1 = \rho_2 = \rho$ of two liquids having identical viscosities and densities. Then

$$P(\mathbf{q}, \omega) = \frac{k_B T}{\pi\omega} \operatorname{Im} \frac{1}{-2i\omega\eta(m + q) - 2\omega^2\rho/q + \gamma q^2}$$
$$= \frac{k_B T}{\pi\omega}\frac{q\tau_0^2}{2\rho} \operatorname{Im}\left\{\frac{1}{y + S\sqrt[+]{1 + 2S}\,(1 + \sqrt[+]{1 + 2S})}\right\}_{S=-i\omega\tau_0} \tag{10}$$

with $y = \gamma\rho/8\eta^2 q$ and $\tau_0 = \rho/2\eta q^2$. The difference between Eq. 7 and Eq. 10 appears clearly.

When the two fluid phases are different, but when the viscosity and density ratio are about the same, $\eta_1/\rho_1 = \eta_2/\rho_2$, the spectrum can also be simplified

[11] (Chap. 3, Eqs. 64–66):

$$P_q(\omega) = \frac{k_B T}{\pi\omega} \frac{q\tau_0^2}{(\rho_1 + \rho_2)} \operatorname{Im} \frac{1}{\Delta(S = -i\omega\tau_0)} \tag{11}$$

with

$$\Delta(S) = y + \frac{2\rho_1\rho_2}{(\rho_1 + \rho_2)^2} S \sqrt[+]{1 + 2S}\,(1 + \sqrt[+]{1 + 2S}) + \left(\frac{\rho_1 - \rho_2}{\rho_1 + \rho_2}\right)^2 ((1 + S)^2 - \sqrt[+]{1 + 2S})$$

Fig. 3 of Chap. 8 illustrates the variation of the spectra with the parameter $r = (\eta_1/\rho_1 - \eta_2/\rho_2)/[(\eta_1 + \eta_2)/(\rho_1 + \rho_2)]$. For a typical oil-water interface $r \sim 0.1$, and the differences between the exact and approximated spectra, Eqs. 9 and 11, are negligible. Figure 4 of Chap. 8 illustrates the variations of the spectra with the parameter $r' = (\rho_1 - \rho_2)^2/2\rho_1\rho_2$, which is typically 2% for an oil-water interface. Again the differences between the approximated spectra 10 and 11 are very small.

VI. EXPERIMENTAL RESULTS ON LIQUID-LIQUID INTERFACES

There have been many fewer test experiments done on liquid-liquid surfaces than on free surfaces, although they are easier to perform because the interface is less sensitive to vibrations. The existing data fully confirm the predictions of the theory presented in the former paragraph. These experiments were made on interfaces between water and ethyl acetate, benzene, carbon tetrachloride, and decane [12], on heptane-water interfaces [13], and on water-propanol-toluene mixtures, $0.3 < \gamma < 3$ mN/m [14]. In the last case, independent interfacial tension measurements done on the same systems with the pendent drop and the spinning drop method were in good agreement with the light scattering determinations. It was found, however, that the light scattering method has a standard deviation higher than the other methods. The method becomes valuable in the ultralow interfacial tension range. The scattered intensities are large, and since the capillary waves are overdamped, there are no instrumental broadening corrections. In the ultralow tension range, the spinning drop method is the only other available technique. It has been verified that the two methods give the same tension values [15]. The comparison has been made with interfaces between a microemulsion and oil or water phases. Although the microemulsions are not simple liquids, they behave as such, as far as the propagation of capillary waves at their surface is concerned. More details will be given in Chap. 13.

REFERENCES

1. S. Hard and R. D. Neuman, *J. Colloid Interface Sci. 115*: 73 (1987).
2. D. McQueen and I. Lundstrom, *J. Chem. Soc. Faraday I, 69*: 694 (1973).
3. J. C. Earnshaw, *Nature 292*: 138 (1981).
4. D. Langevin, *J. Chem. Soc. Faraday Trans. I., 70*: 95 (1974).
5. D. Langevin, *Colloid Surfaces 21*: 313 (1986).
6. S. Hard, Y. Hamnerius, and O. Nilsson, *J. Appl. Phys. 47*: 2433 (1976).
7. J. C. Earnshaw and R. C. McGivern, *J. Colloid Interface Sci., 123*: 36 (1988).
8. M.-A Bouchiat and D. Langevin, *Comptes Rendus Acad. Sci. 272B*: 1357 (1971).
9. J. Meunier, D. Cruchon, and M. A. Bouchiat, *Comptes Rendus Acad. Sci. 268B*: 92 (1969).
10. D. Langevin, J. Meunier, and M.-A. Bouchiat, *Optics Comm. 6*: 427 (1972).
11. J.-C. Herpin and J. Meunier, *J. Phys. Fr. 35*: 847 (1974).
12. H. Löfgren, R. D. Neuman, L. E. Scriven, and H. T. Davis, *J. Colloid Interface Sci. 98*: 175 (1984).
13. B. B. Sauer, Y. L. Chen, G. Zografi, and H. Yu, *Langmuir 2*: 683 (1986).
14. D. I. Jon, H. L. Rosano, and H. Z. Cummins, *J. Colloid Interface Sci. 114*: 330 (1986).
15. D. Chatenay, D. Langevin, J. Meunier, A.-M. Bellocq, D. Bourbon, and P. Lalanne, *J. Dispersion Sci. Tech. 3*: 245 (1982).

8
Surface Critical Opalescence

JACQUES MEUNIER Laboratoire de Physique Statistique*, Ecole Normale Supérieure, Paris, France

The study of thermal fluctuations of a flat liquid interface by dynamic surface light scattering gives information about the restoring forces at a microscopic scale (mainly the capillary force) and the viscous forces. It is a valuable method for the measurement of the surface tension, and in some cases the viscosities, whenever the classical methods fail. This is the case for interfaces close to a critical point, where the surface tension vanishes. The capillary rise method [1, 2] has been used, but it loses its accuracy close to the critical point because the capillary rise is too small to be measured accurately. Warren and Webb [3] studied the capillary rise along the walls of the cell containing a critical mixture and succeeded in making measurements up to 0.1°C from the critical point. The spinning drop technique [4] is able to measure very small surface tension but

*URA 1306 du CNRS, associée aux Universités Paris VI et VIII.

fails also for critical interfaces because it is not possible to control accurately the temperature in the spinning capillary tube and because, for liquid-gas interfaces, the large compressibility of the fluids introduces unmanageable corrections. Conversely, measurements with the light scattering technique are very easy close to critical points because the amplitude of the thermal fluctuations becomes large. Consequently, the scattered intensity is large: this is the surface critical opalescence first observed by Mandelstam [5]. The intensity of the light beam can be reduced to a very low value, especially if the optimal conditions for scattering are used (Sec. 2), minimizing the perturbation of the thermodynamic equilibrium by the light beam.

The thermal fluctuations at a microscopic scale are thermally excited capillary waves, which propagate if the ratio of the restoring forces to the viscous forces is large, and which are overdamped if this ratio is weak. In the first case, it is possible to measure the frequency of the capillary waves and their damping, giving two pieces of information: the interfacial tension and the sum of the bulk viscosities. In the second case, only one parameter, the damping of the surface fluctuations, can be measured, i.e., the ratio of the surface tension to the sum of the bulk viscosities.

Several critical interfaces [6–12] have been studied, with this technique, in order to test theories of the critical point: the critical behavior of the surface tension and, in some cases, the bulk viscosity have been measured. In the case of liquid-gas interfaces, the viscosities are small and it is possible to obtain the two parameters, interfacial tension and viscosity, as close to the critical point as $T_c - T = 0.01°C$, while for critical consolute points, it is possible to obtain only their ratio.

The spectrum of the thermal fluctuations (Sec. 3) is a function of five parameters: the densities ρ_1, ρ_2 and the viscosities η_1, η_2 of the two phases and the surface tension γ. The two densities are well known from other techniques, such as the determination of the rectilinear diameter $(\rho_1 + \rho_2)$ [13] and of the differences $(\rho_1 - \rho_2)$ by index of refraction measurements [14–16]. Viscosities close to a liquid-gas critical point are difficult to measure with classical methods, because of the high compressibility of the phases: large differences appear in the results of different authors [17, 18] The determination of the three parameters, γ, η_1, and η_2 by surface light scattering requires some approximation. Huang and Webb [11] have supposed that the two viscosities are equal $(\eta_1 = \eta_2)$ to reduce the number of parameters: this approximation is valuable for their system, which is a binary mixture close to a consolute critical point. The viscosities of the two components are close. In the case of liquid-gas interfaces of CO_2, an empirical relation [6–7],

$$\frac{\eta_L}{\rho_L} = \frac{\eta_G}{\rho_G} \tag{1}$$

has been deduced from the classical measurements of Ref. 18 and confirmed by surface light scattering results [7, 8]. The same relation has been used to reduce the number of unknown parameters in the studies of the liquid-gas interfaces of SF_6 [9] and Xe [10]. With this relation, the surface tension γ and the sum of the viscosities $\eta_1 + \eta_2$ are deduced from surface light scattering experiments. It was proved that deviations from this relation introduce insignificant corrections on the measurements [8] (Sec. 3).

The experimental results concerning the liquid-gas and consolute critical interfaces are given in Sec. 4. More complex systems (micellar systems) are described in Sec. 5. One of them is a lower critical consolute point [12]. The results are analysed in Sec. 6.

I. WHAT IS A CRITICAL POINT?

In a phase diagram of a mixture or of a pure component there are surfaces where two or more than two phases coexist. The dimension ν of one of these surfaces is given by the Gibbs phase rule: $\nu = C + 2 - \phi$ where C is the number of components and ϕ the number of phases. For instance, for a pure component ($C = 1$), the liquid-gas coexistence ($\phi = 2$) appears on a line ($\nu = 1$). A coexistence surface of dimension ν can be limited by a surface of critical points of dimension equal or less than $\nu - 1$: two phases coexisting on one side of a surface of critical points become identical on it. Each time the system moves in the phase diagram and crosses a surface of critical points, two of its phases become identical, or one of its phases separates into two phases of identical composition on the surface of critical points. For instance, a liquid-gas coexistence line in the phase diagram of a pure component ($\nu = 1$) is limited by a critical point ($\nu = 0$). It is given by the critical temperature and the critical pressure: T_c, p_c.

In the case of a two-component mixture, a surface of coexistence of two phases has the dimension $\nu = 2$. It is limited by a line of critical consolute points ($\nu = 1$), given by a function $f(p_c, T_c) = 0$. In practice, for a surface of critical consolute points, T_c is only slightly pressure dependent, and it is difficult to change the pressure on a large scale; the critical temperature T_c is fixed. This is not the case for more complex systems, such as the one studied by Abillon [12], which contains four components (water, salt, butanol, and SDS or sodium dodecyl sulfate). The surface of critical consolute points has the dimension $\nu = 3$. In practice, the pressure is fixed ($p_c = 1$ atm): $\nu = 2$. When the salinity of the water is fixed, $\nu = 1$. The critical temperature can be adjusted close to the room temperature by changing the ratio between brine, butanol, and SDS. In practice, it is difficult to obtain a mixture with an exact critical composition. If it is not an exact critical composition, the volume of one of the two phases that become identical on the surface of critical points vanishes before the system reaches the critical point, introducing an error in the determination of the position of the critical point. Since close to the critical points it takes a long time to reach thermal equilibrium, it takes a long time to obtain mixtures with a critical composition.

According to the theories of critical points, each thermodynamic property of a phase varies with a power law of the distance to the critical point, at least close to the critical point:

$$X = X_0\, \varepsilon^{\lambda} + X_c$$

where ε can be the reduced temperature $(T_c - T)/T_c$, the reduced pressure $(p_c - p)/p_c$, the reduced thermodynamical potential, etc. depending on how the critical point is approached. λ is the critical exponent, X_0 the scale factor of the thermodynamic property X.

According to theories of critical points, the exponent λ depends on how the critical point is approached and on the universal class of the studied system (which is characterized by the space dimension d and the dimension n of the symmetry group of the system): they are independent of the particular chemical composition of the studied system. There are universal relations between critical exponents and between scale factors. Critical systems have been studied in detail to test these universal predictions. The first theory allowing calculation of the exponents was a mean field theory given by van der Waals [19]. It qualitatively describes the experimental results, but the calculated exponents are not the experimental ones. The theory of the renormalization group [20], which takes into account the coupling between modes, gives good results, but the calculation is difficult and the results are not exact. In the case $n = 1$, these results agree with the exactly soluble Ising model.

Close to a critical point, there are universal equations connecting the thermodynamic properties. One which involves the surface tension γ is

$$\frac{\gamma(\xi)^{d-1}}{k_B T_c} = r \tag{2}$$

where ξ is the correlation length in the homogeneous phase and r is a number depending on the universal class of the critical point.

This equation yields a scaling law involving the exponent μ for the interfacial tension:

$$\mu = (d - 1)\nu \tag{3}$$

where $-\nu$ is the exponent for the correlation length in bulk.

The corresponding universal equations connecting the scale factors is

$$\frac{\gamma_0(\xi_0)^{d-1}}{k_B T_c} = r \tag{4}$$

where γ_0 is the scale factor for the surface tension and ξ_0 is the scale factor for the correlation length in the homogeneous phase.

Liquid-gas critical points and most of the consolute critical points are upper critical points: the two phases which become identical at T_c coexist below T_c. This is easy to understand: at low temperature, the two components of a mixture separate because the van der Waals attraction between two identical molecules is higher than between two different molecules. At high temperature, the entropy mixes the system. The case of lower consolute critical points is more complex. It has been explained by the hydrogen bonds [21–24], which are strongly directional forces: the temperature destroys the directional order. In the case of micellar systems such as the one studied by Abillon, the lower critical point is explained by the hydration forces [25]. The hydration of the micelles at low temperature maintains a large distance between micelles of SDS, decreasing the attractive forces.

As said before, surface light scattering allows the measurement of one thermodynamic parameter, the surface tension, and in some cases the viscosity (critical points of pure components). In most experiments, the distance to the critical point is the reduced temperature $\varepsilon = (T_c - T)/T_c$. Abillon [12] has made a study in function of the temperature and in function of the salinity of the brine. In microemulsion systems, which are complex systems (for instance a mixture of brine, toluene, SDS, and butanol), studies were made by varying the brine salinity (see Chap. 13).

II. OPTIMAL EXPERIMENTAL CONDITIONS

A. Scattered Intensity

The intensity I_d scattered to first order in h/λ by a sinusoidal interfacial deformation of wave vector **q** and amplitude h was first calculated by Gans [26] and by Andronov and Leontovich [27]. The equation simplifies for small q ($q\lambda \ll 1$):

$$\frac{I_d}{i_0} = k_1^2 R(\theta_i) \cos^2(\theta_i)\, h^2 \tag{5}$$

where I_0 is the intensity of the incident beam, $\mathbf{k}_1$ its wave vector (in the phase 1) and $R(\theta_i)$ the reflectivity of the interface at the angle of incidence θ_i. This equation can be used to calculate the light scattered by thermal modes. Counting the surface thermal modes ζ_q that scatter the light in the solid angle $d\Omega$, one obtain the intensity dI scattered in this solid angle $d\Omega$ by a liquid surface:

$$\frac{dI}{d\Omega} = I_0\, k_1^4\, R(\theta_i) \cos^3(\theta_i) \overline{|\zeta_q|^2} \tag{6}$$

with

$$\overline{|\zeta_q|^2} = \frac{k_B T}{\Delta\rho\, g + \gamma q^2}$$

The function $R(\theta_i)\cos^3(\theta_i)$ has a weak maximum if the incident beam propagates in the medium of lower refractive index. This maximum is obtained for $\theta_i = 0$ if $n_2/n_1 > 1.3$ and $\theta_i \neq 0$ if $n_2/n_1 < 1.3$ (the medium in which the incident beam propagates is noted 1). If the incident beam propagates in the medium of larger refractive index, $R(\theta_i)\cos^3(\theta_i)$ has a large peak at the limit angle for total reflection θ_L (Fig. 1). This peak is more pronounced when the difference between the refractive index is low. This can be seen in the ratio between the scattered intensity in the same solid angle $d\Omega$ but for the two incident angles $\theta_i = \theta_L$ and $\theta_i = 0$, for the case in which the incident medium has the larger refractive index:

$$\frac{dI(\theta_L)}{dI(0)} = \frac{(n_1 + n_2)^{7/2}}{(n_1 - n_2)^{1/2}\, n_1^3} \tag{7}$$

Close to the critical point $(n_1 - n_2) \ll 1$

$$\frac{dI(\theta_L)}{dI(0)} \simeq \frac{2^{3.5}}{(1 - n_2/n_1)^{1/2}} \tag{8}$$

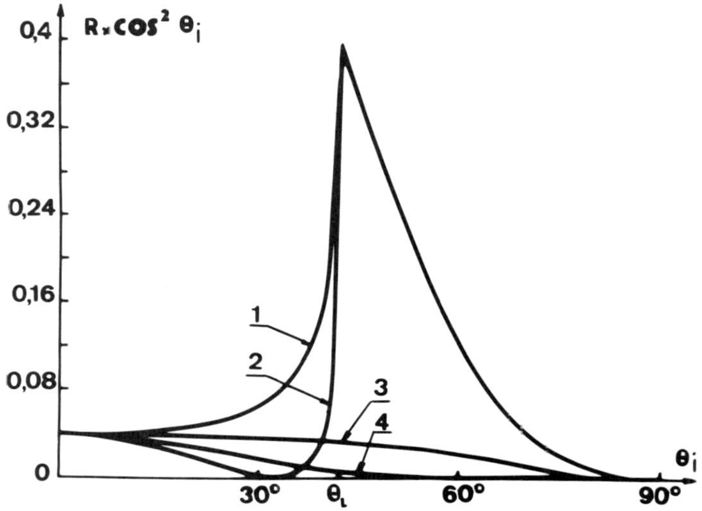

FIG. 1 $R\cos^3(\theta_i)$ versus θ_i. (1) $n_1/n_2 = 1.5$, s polarization; (2) $n_1/n_2 = 1.5$, p polarization; (3) $n_1/n_2 = 1/1.5$, s polarization; (4) $n_1/n_2 = 1/1.5$, p polarization. A large peak appears at the limit angle for total reflection θ_L when the incident beam propagates in the medium of largest refractive index (curves 1 and 2).

increasing as $(1 - n_2/n_1)^{-1/2} \sim (T_c - T)^{-\beta/2}$ (with $\beta = 0.33$ for a liquid-gas critical point). The gain in intensity is so large that it is very useful to use this angle in order to be able to reduce the beam intensity I_0. For instance, the gain in intensity is 100 for the liquid-gas interface of CO_2 at 0.01°C below the critical point.

The term $R(\theta_L)\cos^3(\theta_L) = (1 - n_1/n_2)^{3/2}$ decreases as $(T_c - T)^{3\beta/2}$ close to the critical point, but the mean square amplitude $\overline{|\zeta_q|^2}$ of a thermally excited capillary wave increases as $1/(\gamma q^2) \sim (T_c - T)^{-\mu}$, with $\mu \sim 1.3$. Consequently, the intensity scattered by a mode q at the limit angle for total reflection increases as $\sim (T_c - T)^{1.5\beta-\mu} \sim (T_c - T)^{-0.7}$ close to the critical point. It is the surface critical opalescence.

B. Scattering Direction

The spectrum of the light scattered in the direction $(\theta_r + \Delta\theta, \phi)$ around the direction $(\theta_r, 0)$ of the reflected light is analyzed by the heterodyne technique (Chap. 5). The surface scattering vector q is given by the projection of the scattering vector $k_d - k_r$ on the interface (Fig. 2):

$$q^2 = (k_1)^2\{[\sin(\theta_r + \Delta\theta)\cos(\varphi) - \sin(\theta_r)]^2 + [\sin(\theta_r + \Delta\theta)\sin(\varphi)]^2$$

To lowest order in $\Delta\theta$ and φ, one obtains

$$q^2 = (k_1)^2\{(\cos\theta_r\,\Delta\theta)^2 + (\sin\theta_r\,\varphi)^2\} \tag{9}$$

The most interesting case is obtained for $\Delta\theta = 0$:

$$q^2 = (k_1)^2(\sin\theta_r\,\varphi)^2 = (nk_0)^2(\Delta\Theta)^2 \tag{10}$$

where $\Delta\Theta$ is the angle between the reflected beam and the scattered beam. This quantity is conserved as the beams propagate through the windows of the sample cell, provided they are plane and perpendicular to the plane of incidence [8]. q

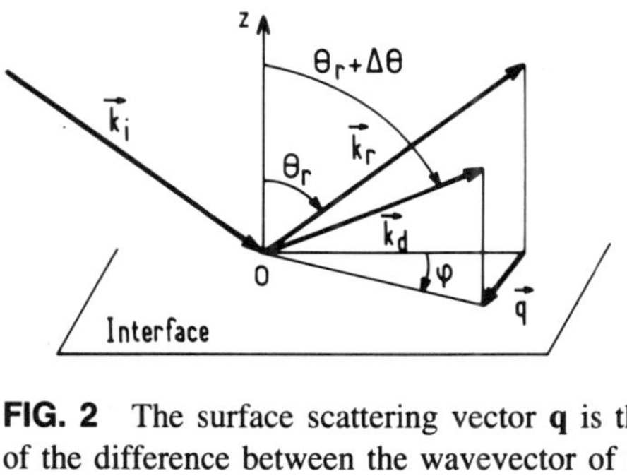

FIG. 2 The surface scattering vector **q** is the projection on the plane of the interface of the difference between the wavevector of the scattered light and that of the reflected light: $\mathbf{k}_d - \mathbf{k}_r$.

is known from the measurement of the angle $\Delta\Theta_0$ between the reflected beam and the scattered beam in air, outside the cell containing the two phases and by applying Eq. 10:

$$q^2 = (k_0)^2\,(\Delta\Theta_0)^2 \tag{11}$$

where k_0 is the modulus of the wavevector of the light in the air.

The case $\varphi = 0$ is useful for studying the scattering at very small q at grazing incidence ($\cos\theta_r$ is small close to the critical point), but the determination of q needs that of θ_r and $\Delta\theta$ inside the cell. The quantity $[\cos(\theta_r)\,(\Delta\theta)]$ is not conserved when the beams propagate through the plane windows of the cell.

III. THE SPECTRUM OF THE FLUCTUATIONS OF THE INTERFACE

The calculation of the spectral shape of the thermal fluctuations of wavevector q, at the free surface of a liquid [28] (Chap. 3), can be easily generalized to the calculation of the thermal fluctuations at the interface between two fluids [8]. This is the spectrum of surface fluctuations obtained when the interface is excited with white noise:

$$P_q(\omega) = \frac{Y}{\gamma_q}\frac{\tau_0}{q^2}\frac{k_B T}{\pi\omega\tau_0}\,\mathrm{Im}[D(i\omega\tau_0)]^{-1} \tag{12}$$

where

$$\gamma_q = \gamma + \frac{|\rho_1 - \rho_2|g}{q^2} \qquad \tau_0 = \frac{\rho_1 + \rho_2}{2(\eta_1 + \eta_2)q^2} \qquad Y = \frac{\gamma_q(\rho_1 + \rho_2)}{4(\eta_1 + \eta_2)^2 q}$$

D is the dispersion equation for the surface waves at the interface [29]:

$$\begin{aligned} D(S) = Y &+ \frac{\rho_1^2 - \rho_2^2 + 2R\,\rho_1\rho_2}{(\rho_1 + \rho_2)^2}\,\frac{\eta_1(m_1 - 1) - \eta_2(m_2 - 1)}{\eta_1(m_1 + 1) + \eta_2(m_2 + 1)}\,S \\ &+ \left(\frac{\rho_1\dfrac{\eta_1(m_1^2 + 1) + \eta_2[m_1 - 1 + m_2(m_1 + 1)]}{m_1 - 1}}{(\rho_1 + \rho_2)[\eta_1(m_1 + 1) + \eta_2(m_2 + 1)]} \right. \\ &+ \left. \frac{\rho_2\dfrac{\eta_2(m_2^2 + 1) + \eta_1[m_2 - 1 + m_1(m_2 + 1)]}{m_2 - 1}}{(\rho_1 + \rho_2)[\eta_1(m_1 + 1) + \eta_2(m_2 + 1)]} \right) S^2 \end{aligned} \tag{13}$$

where

$$m_1 = \sqrt[+]{1 + 2\frac{\rho_1 + \rho_2}{(\rho_1 + \rho_2) + R\,\rho_2}\,S}$$

$$m_2 = \sqrt[+]{1 + 2\frac{\rho_1 + \rho_2}{(\rho_1 + \rho_2) - R\,\rho_1}\,S}$$

$$R = \frac{\eta_1/\rho_1 - \eta_2/\rho_2}{(\eta_1 + \eta_2)/(\rho_1 + \rho_2)}$$

and g is the acceleration of gravity; γ is the surface tension. m_1 and m_2 are square roots with a positive real part (symbolized by a + before the square root). R is a measure of the discrepancy of the system to the empirical law $\eta_1/\rho_1 = \eta_2/\rho_2$.

The spectrum $P_q(\omega)$ is insensitive to the parameter R [8]. Figure 3 shows three spectra obtained for $R = 0$, and two large values of R to get a significant difference ($R = 0.8$ and $R = 1$), in the case $Y = 100$ and $\delta = (\rho_1 - \rho_2)/(\rho_1 + \rho_2) = 0.6$. They are very similar. $R = 0$ is a very good approximation as long as R is not too large. Figure 4 shows, in the case of propagating surface fluctuations ($Y \gtrsim 1$), the weak difference between the position of the maximum of the spectrum for $R = 0$ and $R = 0.1$, versus the parameter Y and for different values of the ratio $\delta = (\rho_1 - \rho_2)/(\rho_1 + \rho_2)$ (the corresponding ΔT is indicated on the figure for CO_2). The difference is maximum for $Y \sim 100$. Classical measurements for η_1 and η_2 indicate that $R < 0.1$ in the temperature gap $0.01 < \Delta T < 13°C$ [6–8] for the critical liquid-gas phase transition of CO_2. The approximation $R = 0$ is accurate for this system, and probably for any liquid-gas phase transition. For $R = 0$, the spectrum has a more simple expression and is only a function of ρ_1, ρ_2, $\eta_1 + \eta_2$, and γ.

$$D(S) = Y + \frac{2\rho_1\rho_2}{(\rho_1 + \rho_2)^2} S\sqrt{1 + 2S}\,(1 + \sqrt[+]{1 + 2S}) + \left(\frac{\rho_1 - \rho_2}{\rho_1 + \rho_2}\right)^2 [(1 + S)^2 - \sqrt[+]{1 + 2S}] \quad (14)$$

This equation allows the experimental determination of Y and τ_0 from which one deduces γ_q and $\eta_1 + \eta_2$. The values of $\eta_1 + \eta_2$ deduced by this method are accurate even if Eq. 1 ($\eta_1/\rho_1 = \eta_2/\rho_2$) is approximative. The interfacial tension of the critical interface of CO_2 [8], that of SF_6 [9] and Xe [10] and the sum of the bulk viscosities of the two phases, were deduced from the experimental spectrum of thermal fluctuations using the spectrum for surface thermal fluctuations Eq. 12 in the approximation Eq. 1. Before this theoretical exact spectrum was calculated, two experimental studies were published: one on the critical interface of CO_2 [6–7] and another one on the critical interface of the cyclohexane-methanol mixture [11]. The theoretical spectrum used for these studies is that of a harmonic oscillator with two modes S_1 and S_2, solutions of the dispersion equation:

$$D(S) = 0$$

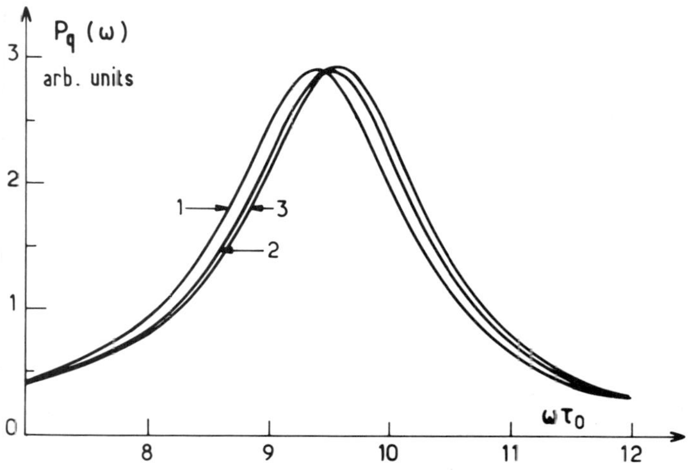

FIG. 3 The spectra $P_q(\omega)$ for $Y = 100$, $\delta = (\rho_1 - \rho_2)/(\rho_1 + \rho_2) = 0.6$ and three values of R: (1) $R = 0$; (2) $R = 0.8$; (3) $R = 1$. Large values of R were chosen to get a significant difference from the case $R = 0$.

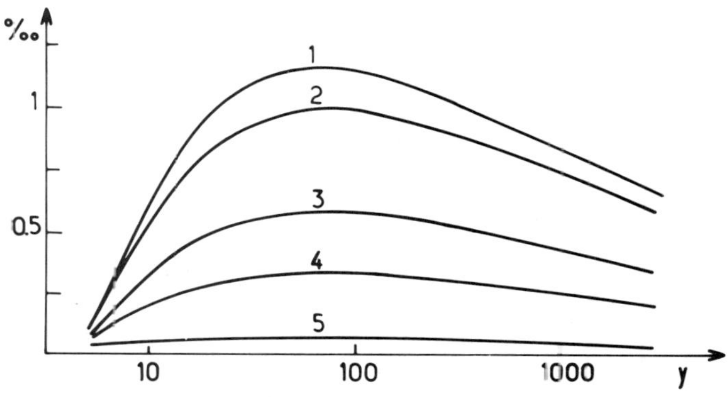

FIG. 4 Difference between the position of the maximum of the spectrum for $R = 0.1$ and that of the spectrum for $R = 0$ in the case of propagating thermal fluctuations ($Y \gtrsim 1$). (1) $\delta = 0.6$, corresponding to $T_c - T = 13°C$ for CO_2; (2) $\delta = 0.4$, corresponding to $T_c - T = 4°C$ for CO_2; (3) $\delta = 0.2$, corresponding to $T_c - T = 0.5°C$ for CO_2; (4) $\delta = 0.1$, corresponding to $T_c - T = 0.07°C$ for CO_2; (5) $\delta = 0$, corresponding to $T_c - T = 0°C$.

i.e., the spectrum $P'_q(\omega)$, which is the Fourier transform of the correlation function:

$$G'_q(t) = \frac{k_B T}{\gamma_q q^2 A} (S_1 - S_2)^{-1} (S_2 e^{-S_1 t/\tau_0} - S_1 e^{-S_2 t/\tau_0}) \quad (15)$$

This spectrum neglects the time of propagation and dissipation of the rotational flow in bulk on the two sides of the interface [28]. This effect can be neglected, except in the domain $Y \sim 1$ [28, 29] where the damping is close to the critical damping. This is the case for the critical interface of a fluid like CO_2, SF_6, or Xe close to the critical point ($\Delta T = T_c - T \sim 0.01°C$) and far from the critical point for the cyclohexane-methanol mixture ($\Delta T \sim 1°C$) (the viscosities of the phases of this mixture are larger than those of pure fluids close to the critical point). Wu and Webb [9] have made the comparison between the surface tension and the viscosity deduced from the exact theory and from the harmonic oscillator theory in the case of the critical interface of SF_6. The values of γ and $\eta_L + \eta_G$ they obtained with the two theories are the same far from the critical point ($\Delta T = 30°C$) but significantly different close to the critical point (a factor of 4 on γ for $\Delta T = 0.03°C$). This difference is probably much smaller (a few %) in

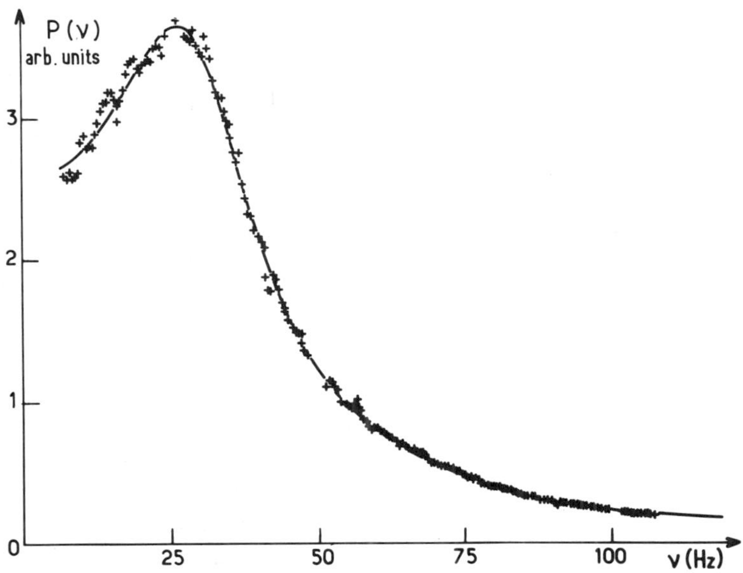

FIG. 5 A typical experimental spectrum (+) and the best fit obtained with the theoretical spectrum $P_q(\nu)$ (solid line). (Liquid-gas interface of CO_2, $T_c - T = 0.0250°C$, $q = 432$ cm^{-1}, $Y = 1.06$.)

the case where only γ is deduced from the experiment, $\eta_1 + \eta_2$ being fixed by another measurement as it was in the case of CO_2 [6–7].

For critical points of mixtures, Eq. 1 is not valid. Gilmer et al. have used the approximation $\rho_1 = \rho_2$ and $\eta_1 = \eta_2$ because the densities and viscosities of the two components of their mixture are close. In this case, the dispersion equation reduces to

$$D(S) = Y + \frac{1}{2} S\sqrt{1 + 2S}\,(1 + \sqrt{1 + 2S}) \tag{16}$$

Close to the critical point, the damping is large: $Y \ll 1$ and the spectrum reduces to a Lorentzian shape with a half-width of

$$\Delta\omega = \frac{\gamma q}{2(\eta_1 + \eta_2)} \tag{17}$$

An example of an experimental spectrum obtained from the critical interface of CO_2 at $T_c - T = 0.025°C$, in the case $Y = 1.06$, is given in Fig. 5.

IV. EXPERIMENTAL RESULTS: PURE FLUIDS AND BINARY MIXTURES

The experimental results are summed up in Table 1. Two cases must be distinguished. First, the case of liquid-gas critical points where the surface fluctuations are propagating waves because the viscosities are low. The frequency and the damping of the surface fluctuations can be measured. From those two parameters, one can deduce the surface tension and the sum of the bulk viscosities. This has been done for the critical interfaces of CO_2 [8], SF_6 [9], and Xe [10]. In another case (CO_2 [6, 7]), only the interfacial tension has been deduced; the sum of the viscosities has been taken as equal to the sum of the viscosities measured by another method. As theoretically predicted, the experimental results for the surface tension are in agreement with a scaling law:

$$\gamma = \gamma_0 \left(\frac{T_c - T}{T_c}\right)^{\mu}$$

The experimental points are on a straight line in a log-log plot (see Fig. 6), the slope of which is the exponent μ.

The scaling laws obtained on CO_2 by the two experiments, in Refs. 6 and 7 and in Ref. 8 are slightly different (Fig. 6). This can result from the difference between the analysis of the experimental spectrum, the analysis with the exact spectrum [8], or with the approximate one $P'_q(\omega)$ [6, 7] and from a slight difference between the estimation of T_c. $\eta_1 + \eta_2$ measured by classical techniques and that deduced from surface light scattering measurements are close; their differences (~8%) cannot explain the differences observed on γ.

TABLE 1 Experimental Results Obtained by Surface Light Scattering on Different Critical Interfaces in CGS

Liquid-gas critical points

	Approximations	$\eta_1 + \eta_2$	γ_o	μ	ΔT(°C)
CO_2 [7]	$P'_q(\omega)$ $\eta_1/\rho_1 = \eta_2/\rho_2$	No	77 ± 5	1.253 ± .010	0.034 – 34
CO_2 [8]	$\eta_1/\rho_1 = \eta_2/\rho_2$	Yes	95.1 ± 10.4	1.291 ± .023	0.012 – 12
SF_6 [9]	$P'_q(\omega)$ $\eta_1/\rho_1 = \eta_2/\rho_2$	Yes	55.1 ± 2.6	1.285 ± .016	0.039 – 22
Xe [10]	$\eta_1/\rho_1 = \eta_2/\rho_2$	Yes	62.9 ± 1.8	1.302 ± .006	0.070 – 5

Consolute critical point

	Approximations	Measurement	$[\gamma/(\eta_1 + \eta_2)]_o$	Ω^*	ΔT (°C)
Cyclohexane methanol [11]	Eq. 15 $\rho_1 = \rho_2$ $\eta_1 = \eta_2$	$\left(\dfrac{\gamma}{\eta_1 + \eta_2}\right)$	$\dfrac{10^4}{(9 \pm 0.1)}$	1.27 ± .03	0.09 – 0.7

Micellar system

	Approximations	$\eta_1 + \eta_2$	γ_o	μ	ΔT(°C)
Brine, water butanol [12]	Eq. 17	No	0.48 ± 0.09	1.24 + .05	0.6 – 21

Column 3 indicates whether $\eta_1 + \eta_2$ has been deduced from surface light scattering (except in the case of the consolute critical point, where it indicates the measured quantity).
Source: Refs. 7–12.

No divergence in the sum of the viscosities of the liquid and the vapor was deduced from these measurements. The results concerning CO_2 are given Fig. 7. They are in agreement with the values deduced by classical methods [30]. For SF_6, Wu and Webb give

$$\frac{\eta_1 + \eta_2}{2} = (425 + 14.5\ \Delta T) \pm 15 \text{ micropoises}$$

while for Xe, Zollweg et al. have found

$$\frac{\eta_1 + \eta_2}{\rho_1 + \rho_2} = (448 + 12\ \Delta T) \pm 15 \text{ microstokes}$$

where $\Delta T = T_c - T$.

The case of an interface close to a consolute critical point is different because the surface thermal fluctuations are overdamped. Only one parameter, the time

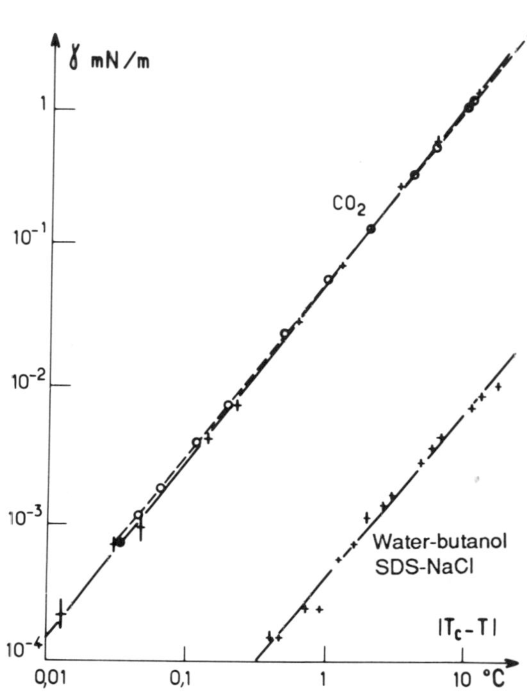

FIG. 6 The surface tension γ versus $T_c - T$ (using a logarithmic scale) for CO_2 (dashed line, Ref. 6 and 7; solid line, Ref. 8) and for the system water-butanol-SDS-NaCl of Ref. 12.

constant of the damping, can be deduced from the experimental spectrum, at least close to the critical point:

$$\tau_2 \simeq \frac{2(\eta_1 + \eta_2)}{\gamma_q q} \tag{18}$$

For large q, $\Delta\rho \cdot g$ can be neglected and $\gamma_q = \gamma$. The parameter that can be deduced from these systems by surface light scattering is $\gamma/(\eta_1 + \eta_2)$. Huang and Webb have studied the cyclohexane-methanol interface. Their results can be explained by a scaling law:

$$\frac{\gamma}{(\eta_1 + \eta_2)} = \left(\frac{\gamma}{\eta_1 + \eta_2}\right)_0 \left(\frac{T_c - T}{T_c}\right)^{\Omega^*}$$

$(\gamma/\eta_1 + \eta_2)_0$ and Ω^* are given Table 1. This result is deduced, in the temperature

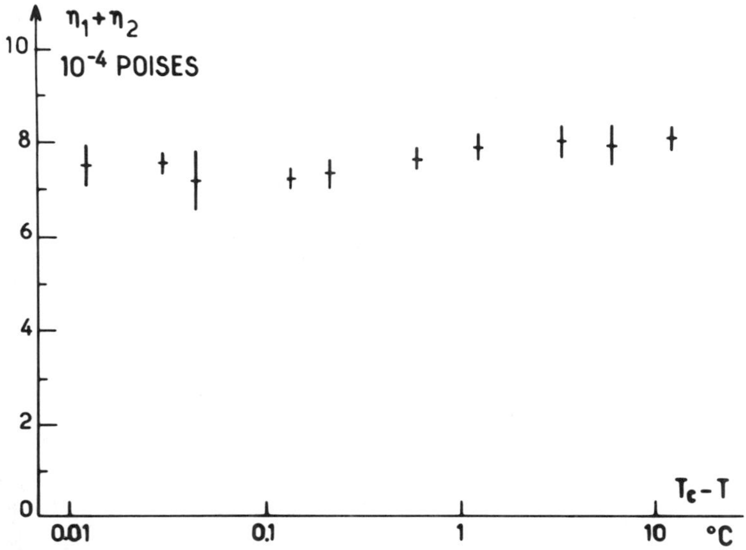

FIG. 7 The viscosity $(\eta_1 + \eta_2)/2$ versus $T_c - T$ for the critical interface of CO_2 (Ref. 8).

range $0.003 < T_c - T < 0.7°C$, from correlation functions obtained at only one wavevector $q = 220\ cm^{-1}$. Close to the critical point they used approximation 17; far from the critical point ($\Delta T = 0.7°C$, $Y \sim 1$) approximation 15. This approximation is crude in this case and introduces an unknown error on the scaling law.

Warren and Webb [3] have measured the capillary length on the same system by studying the capillary rise along the wall of the cell containing the two phases in the temperature range $0.09 < T_c - T < 9°C$. These data were combined with the density difference between the two phases, calculated from the coexisting compositions, to get the surface tension. They obtained

$$\gamma = \gamma_0 \left(\frac{T_c - T}{T_c}\right)^{\mu} \quad \text{with } \mu = 1.23 \pm 0.05 \quad \text{and} \quad \gamma_0 = 100 \text{ dyne/cm}$$

Huang and Webb, using this result and their measurements of the ratio $\gamma/(\eta_L + \eta_G)$, have deduced the critical behavior of the sum of the viscosities of the two phases of the consolute critical mixture:

$$\frac{\eta_L + \eta_G}{2} = \eta_0 \left(\frac{T_c - T}{T_c}\right)^{\mu - \Omega^*}$$

$$\mu - \Omega^* = 0.04 \pm 0.05 \quad \text{and} \quad \eta_0 = 0.034 \pm 0.02 \text{ poises}$$

The average shear viscosity of the critical binary mixture might exhibit an extremely weak divergence, but the accuracy is inadequate to prove its existence. The conclusion of this study is that the divergence, if any, is much less than that measured by classical methods [31] (15%) on a similar system.

V. MICELLAR CRITICAL POINTS

Amphiphilic molecules (or surfactants) are molecules made of two parts: one hydrophilic head that is soluble in water and a hydrophobic tail. An example of such a molecule is sodium dodecyl sulfate (SDS). It is soluble in water at very low concentration. Above the critical micellar concentration or CMC (2 g/L for SDS in pure water), the molecules associate in micelles, which are groups of amphiphilic molecules, more or less spherical, having their polar heads in contact with water, at the exterior of the micelle. This configuration avoids the contact of the hydrophobic tail with water and minimizes the energy. The micelles can be made of a mixture of surfactants, for instance SDS + butanol. The butanol is called a cosurfactant.

The heads of SDS molecules are polar groups. The addition of salt to the water screens the interactions between polar heads of two molecules of surfactant in one micelle or between two micelles, modifying the phase diagram. The phase diagram of this system (brine, SDS, butanol) contains four components and presents a demixion surface and a surface of lower critical points. It was studied by Abillon [12].

The critical composition was chosen to have a critical temperature close to the room temperature ($T_c = 23.66°C$), allowing good uniformity of temperature in the cell. The viscosities being large and the interfacial tension extremely low, approximation 17 is accurate for spectra obtained from surface light scattering. The measured quantity is $\gamma/(\eta_1 + \eta_2)$. The viscosities are deduced from the measurement of the velocity, induced by gravity, of the flow of the two phases in capillary tubes. The accuracy of the experimental scaling law is affected by the low accuracy on the sum of the viscosities and, more importantly, by the indetermination on the value of T_c. The result is given in Table 1.

Micelles of SDS and butanol (approximately one molecule of butanol for one molecule of SDS) can be swollen when an oil, such as toluene, is added. The oil is incorporated in the heart of the micelle, in contact with the tails, and the surfactant molecules form a monomolecular film between oil and water. The mixture obtained is an oil-in-water microemulsion. When more salt is added, the repulsive electrostatic forces between polar heads in micelles are more screened, the volume occupied by the polar head decreases, and one obtains reverse micelles, droplets of water surrounded by a monolayer of surfactant (water-in-oil microemulsion). At intermediate salinity, one obtains a bicontinuous structure (the mean curvature of the monolayer vanishes). The phase diagram of these systems exhibits surfaces of critical points, the surface tension of which has been studied by surface light scattering (Chap. 13).

VI. ANALYSIS OF THE RESULTS

The critical interfaces reported in Table 1 and studied by surface light scattering belong to the universal class $d = 3, n = 1$. The theoretical value for the exponent of the surface tension is $\mu = 1.26$ [32], in good agreement with the experiments, except in the case of Xe. In this last case, the experimental incertitude seems greatly underestimated.

The experimental values of the exponent μ are compared to a selection of experimental values of ν in Table 2 (scaling law 3). The agreement is not perfect, but a large dispersion appears in the different measurements of an exponent, for the same system (see for instance the exponent ν for Xe in Table 2).

The constant r that appears in Eq. 4 is calculated in Table 3 for the same systems (second column). As the experimental values of the critical exponents do not verify exactly scaling law 3, it is better to calculate $r(\varepsilon)$ by Eq. 2 rather than r by Eq. 4:

$$\frac{\gamma_0\,(\xi_0)^2\,\varepsilon^{\mu-2\nu}}{k_B T_c} = r(\varepsilon)$$

This has been calculated for two values of ε (10^{-3} and 10^{-4}) included in or close to the experimental domain of study of the critical systems. In the case where the exponents μ and 2ν are very different, the difference between r and $r(\varepsilon)$ is large and the values of columns 3 and 4 are certainly closer to the true value of r than that of column 2.

TABLE 2 Comparison of the Exponents μ and 2ν for Different Critical Systems of University Class $d = 3$, $n = 1$

System	μ	2ν
Theory [32] $d = 3$ $n = 1$	1.26	1.26
CO_2	1.253 ± .010 [6–7]	1.266 ± .020 [33]
	1.291 ± .023 [8]	
SF_6	1.285 ± .016 [9]	1.243 [34]
Xe	1.302 ± .006 [10]	1.280 ± .040 [35]
		1.140 ± .10 [36]
Cyclohexane methanol	1.23 ± .05[a]	
Brine, butanol SDS	1.24 ± .05 [12]	1.260 ± .020 [12]

[a]This value is deduced from capillary rise.
Source: Refs. 6–10, 12, 32–36.

TABLE 3 The Constant r Deduced form Experimental Results and $r(\varepsilon)$ Calculated for $\varepsilon = 10^{-3}$ and $\varepsilon = 10^{-4}$

System	r	$r(\varepsilon = 10^{-3})$	$r(\varepsilon = 10^{-4})$
Theory [32, 37, 38] $d = 3$ $n = 1$	0.186–0.221	0.186–0.221	0.186–0.221
CO_2 [7, 33]	0.412	0.451	0.465
[8, 33]	0.509	0.425	0.401
SF_6 [9, 34]	0.51	0.381	0.346
Xe [10, 35]	0.265	0.227	0.216
[10, 36]	1.35	0.44	0.30
Cyclohexane methanol			
Brine, butanol [12] SDS	0.20 ± 0.1		

Source: Refs. 7–10, 12, 32–38.

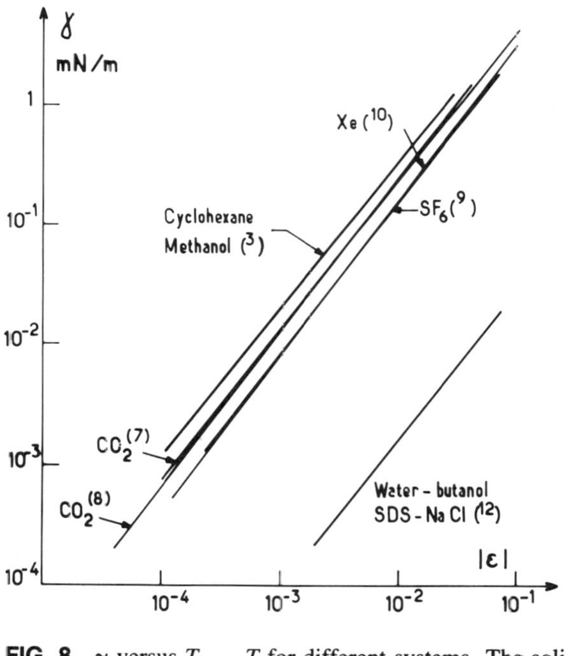

FIG. 8 γ versus $T_c - T$ for different systems. The solid lines indicate the best power laws deduced from experimental points. The length of a solid line indicates the domain of temperature that was studied.

The experimental value for r is systematically about two times larger than the theoretical one. This is not explained. There is one exception, the micellar system, but the accuracy of the results is small in this case.

It must be remarked that the scale factors γ_0 for the interfacial tension, given in Table 1, are in the domain 50–100 dyn/cm, except for the micellar system (~0.50 dyn/cm, about 100 times smaller; Fig. 8). This is easy to understand. This scale factor is given by Eq. 4 with $d = 3$: $\gamma_0 = rk_BT_c(\xi_0)^{-2}$, where ξ_0 is of the order of the range of the intermolecular forces, a few Å for classical systems. For the micellar system, the range of the forces between micelles or between micelles and solvent is 5 or 10 times larger than in classical systems.

VII. CONCLUSION

Dynamic surface light scattering is particularly suitable in the study of critical opalescence. Close to a critical point, the classical methods for the measurement of the surface tension fail because the capillary forces vanish. Conversely, the measurements with dynamic surface light scattering become more and more easy at the approach of a critical point. The thermal fluctuations and consequently the scattered light intensity increase. The intensity of the incident beam can be reduced to a fraction of a mW, reducing the perturbation to an ultra-low value. Only dynamic surface light scattering is able to measure the shear viscosity of a liquid-gas system close to the critical point. The classical methods introduce large perturbations in so compressible a system, and it is difficult to make the corrections on the measurements.

In the case of propagating surface fluctuations with large damping (the case of liquid-gas interfaces: $Y \sim 1$), the surface tension and the viscosity can be simultaneously measured. It is necessary to use the exact spectrum for reasonable accuracy. Its complexity is not a problem with the use of microcomputers.

REFERENCES

1. D. Atack and O.K. Rice, *Discussion Faraday Soc. 15:* 210 (1953).
2. A.M. Wims, J.V. Sengers, D. McIntyre, and J. Shereshefsky, *J. Chem. Phys. 52:* 3042 (1970).
3. C. Warren and W.W. Webb, *J. Chem. Phys. 50:* 3694 (1969).
4. B. Vonnegut, *Rev. Sci. Inst. 13:* 6 (1942).
5. L. Mandelstam, *Ann. der Physik 41:* 609 (1913).
6. M.A. Bouchiat and J. Meunier, *Phys. Rev. Lett. 23:* 752 (1969).
7. J. Meunier, *J. Phys.* (France) *30:* 933 (1969).
8. J.C. Herpin and J. Meunier, *J. Phys.* (France) *35:* 847 (1974). N.B.: First, two signs in Eq. 1 must be corrected to obtain the above Eq. 10; secondly, $\%_0$ in Fig. 1 must take the place of %. The accuracy of curves of this picture 1 in Ref. is low

because the calculation was made with too few digits. A better accuracy is given by Fig. 4 of this paper, but this does not change the conclusion of the study.
9. E.S. Wu and W.W. Webb, *Phys. Rev. 8A:* 2077 (1973).
10. J. Zollweb , G. Hawkins, and G.B. Benedek, *Phys. Rev. Lett. 27:* 1182 (1971).
11. J.S. Huang and W.W. Webb, *Phys. Rev. Lett. 23:* 160 (1969).
12. O. Abillon, thesis, Paris, 1984. O. Abillon, D. Chatenay, D. Langevin, and J. Meunier, *J. Phys. Lett.* (France) *45:* L-223 (1984).
13. A. Michels, B. Blaisse, and C. Michels, *Proc. Roy. Soc. 160A:* 358 (1937).
14. H.L. Lorentzen, *Acta Chem. Scand. 7:* 1336 (1953). H.L. Lorentzen, in *Statistical Mechanics of Equilibrium and Non-Equilibrium* (J. Meixner, ed.), North-Holland, Amsterdam, 1965.
15. D.A. Balzarini, *Can. J. Phys. 50:* 2194 (1972).
16. D. Balzarini and K. Ohrn, *Phys. Rev. Lett. 29:* 840 (1972).
17. A. Michels, A. Botzen, and W. Schuurmann, *Physica 23:* 95 (1957).
18. J. Kestin, J. H. Whiterlaw, and T. F. Zien, *Physica 30:* 161 (1964).
19. J. Van der Waals, *Z. Phys. Chem. 13:* 657 (1894).
20. M.N. Barber, *Phys. Reports 29 C* (1977).
21. J. Hirschfelder, D. Stevenson, and H. Eyring, *J. Chem. Phys. 5:* 896 (1937).
22. J.A. Barker and W. Fock, *Disc. Far. Soc. 15:* 188 (1953).
23. J.C. Wheeler, *J. Chem. Phys. 62:* 433 (1975).
24. G.R. Anderson and J.C. Wheeler, *J. Chem. Phys. 69:* 3403 (1978).
25. R. Kjellander, *J. Chem. Soc. Far. Trans. 2, 78:* 2025 (1982).
26. R. Gans, *Ann. der Phys.* (Leipzig) *79:* 204 (1926).
27. A. Andronov and M. Leontowiez, *Z. Phys. 38:* 485 (1926). A. Andronov, *Collected Works,* Izd. Akad. Nauk SSSR (1956).
28. M.A. Bouchiat and J. Meunier, *J. Phys.* (France) *32:* 561 (1971).
29. M.A. Bouchiat and J. Meunier, *J. Phys.* (France) *32:* C5a 181 (1971).
30. J.V. Sengers, "Critical Phenomena," Proceedings of a conference, Washington, D.C. (M. S. Green and J. V. Sengers, eds.), National Bureau of Standards Miscellaneous Publications N° 273, US Government Printing Office, Washington D.C., 1965.
31. T. Reed and T.E. Taylor, *J. Phys. Chem. 63:* 58 (1959).
32. A. Kumar, H.R. Krishnamurthy, and E.S.R. Gopal, *Phys. Rep.* (Netherlands) *98:* 57 (1983).
33. J.H. Lunacek and D.S. Cannell, *Phys. Rev. Lett. 27:* 841 (1971).
34. D.S. Cannell, *Phys. Rev. 12A:* 225 (1975).
35. H.L. Swinney, D.L. Henry, and H.Z. Cummins, *J. Phys.* (France) *33:* C1-81 (1972).
36. I.W. Smith, M. Giglio, and G.B. Benedek, *Phys. Rev. Lett. 27:* 1556 (1971). J. Zollweg, G. Hawkins, I.W. Smith, M. Giglio, and G.B. Benedek, *J. Phys.* (France) *33:* C1-135 (1972).
37. E. Brezin and S. Feng, *Phys. Rev. 29B:* 472 (1984).
38. A. Aharony and P.C. Hohenberg, *Phys. Rev. 13B:* 3081 (1976).

9

Anisotropic Fluids: Liquid Crystals

DOMINIQUE LANGEVIN Laboratoire de Physique Statistique, Ecole Normale Supérieure, Paris, France

The surface light scattering method has been applied to nematic liquid crystals by Langevin and Bouchiat in 1972 [1, 2], McQueen in 1973 [3, 4], and Shih, Mann, and Brown in 1983 [5]. The main purpose of these studies was to determine the surface tension without perturbing the surface alignment of the molecules, as well as the anisotropy of the bulk viscosity of the nematic liquid. An important question was the behavior of the surface tension at the nematic-isotropic transition: the change in surface tension at this transition is related to molecular ordering changes at the surface.

The technique also allowed the determination of the interfacial tension between nematic and isotropic phases (that coexist at the transition temperature) [6]. The divergence of some of the bulk viscosities close to a second-order nematic-to-smectic phase transition can be accurately studied with the technique [7]. Finally, partial measurements have been made at the free surface of smectic liquid crystals [4, 7]. This is particularly difficult to achieve, because the free surface of smectics is rarely flat: it is corrugated because of the presence of orientation defects.

In the following, we shall present the theoretical background necessary to analyze the data and then describe the existing experiments. In this chapter, cgs units will be used.

I. MOLECULAR ORIENTATION CLOSE TO THE SURFACE

Nematic liquid crystals are made of elongated molecules that spontaneously align parallel to each other [8]. In the absence of external constraints, the spontaneous orientation is established over typical distances of 0.1 μm. Orientation of large samples can be conveniently achieved by using external magnetic fields **H**, because of the unusually large anisotropy of the magnetic susceptibility of nematic liquids: $\chi_a = \chi_\parallel - \chi_\perp \sim 10^{-7}$ cgs units. χ_a is usually positive, so that the molecules orient parallel to the field **H**.

Free surfaces, as well as solid surfaces, impose particular orientations to the molecules. Frequently, the molecules are parallel or perpendicular to the free surface. However, the molecules can also lie at an angle θ from the normal to the surface, and θ can vary with temperature in the range of existence of the nematic phase. The molecular orientation at the surface can be conveniently determined by simple reflectivity studies [9, 10]. The reflection coefficient R of polarized light depends upon the directions of the electric field and of the molecules close to the surface. Optically, the liquid is uniaxial, with refractive indices n_o and n_e, o and e standing for ordinary and extraordinary respectively. Depending upon the sign of the dielectric susceptibility anisotropy ε_a, the optical axis is parallel ($\varepsilon_a > 0$) or perpendicular ($\varepsilon_a < 0$) to the molecules. If an external horizontal magnetic field **H** is applied, the horizontal projections of the molecules will be parallel to **H** at the surface. The molecules will become progressively oriented parallel to **H** in the bulk. The thickness of the zone where the molecular orientation is changing is of the order of the magnetic correlation length $\xi_H = \sqrt{K/\chi_a}\ 1/H$, K being an average elastic coefficient [8]. Typically $K \sim 10^{-6}$, $\chi_a \sim 10^{-7}$ and $H \sim 3000$ in cgs units. This gives $\xi_H \sim 10\ \mu m$, much larger than the optical wavelength λ. This means that the refractive indices to be considered in the reflection coefficient correspond to the molecular orientation at the surface. For a normal incidence, $R = (1 - n)^2/(1 + n)^2$ with $n = n_o\{(1 + \tan^2 \theta)/(\tan^2\theta + n_o^2/n_e^2)\}^{1/2}$ if the polarization is parallel to $\mathbf{H}$ and $n = n_o$ if the polarization is perpendicular to $\mathbf{H}$ [9].

It was found in this way that for para-azoxyanisole (PAA), a compound that is nematic between 119 and 125°C, the molecules are parallel to the free surface; for methoxybenzilidene butyl aniline (MBBA), nematic between room temperature and 46°C, the molecules are tilted at an angle $\theta \approx 70°$ from the surface at room temperature and become almost normal to the surface close to the nematic-isotropic transition [9] (Fig. 1).

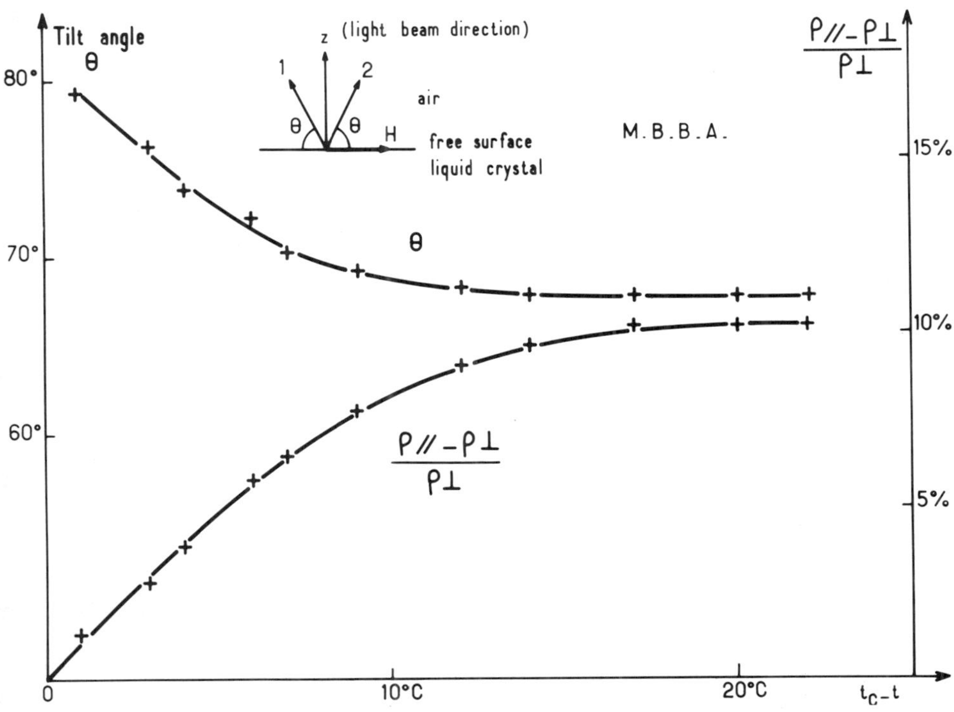

FIG. 1 Tilt angle θ of the MBBA molecules at the free surface versus temperature. The lower curve represents the relative difference between the reflection coefficients for polarizations of the incident electric field respectively parallel and perpendicular to the applied magnetic field.

II. THEORETICAL SPECTRUM FOR THE FREE SURFACE OF A NEMATIC LIQUID IN THE CASE OF A NEARLY UNIFORM MOLECULAR ORIENTATION

When the molecules are parallel to the free surface, and when a horizontal magnetic field is applied, the orientation is uniform and parallel to the field in the region where the probed capillary waves propagate (far from the cell boundaries).

The spectrum can be calculated following the method detailed in Chap. 4. The response function to a surface perturbation is calculated by using the hydrodynamic equations and the appropriate boundary conditions. The spectrum is then obtained by applying the fluctuation-dissipation theorem. In nematic liquid crystals, the hydrodynamic equations involve a new hydrodynamic variable, the molecular orientation, represented by a unit vector **n** [8]. The characteristic frequencies of the variations of **n** are of the order of $\omega \sim (Kq^2 + \chi_a H^2)/\eta$ where K and η are mean elasticity and viscosity coefficients. Typically $K \sim 10^{-6}$ dyn, $\chi_a \sim 10^{-7}$, $H \sim 3000$ G, $\eta \sim 0.1$ poise; for the wavevectors of interest in the surface light scattering experiments $q \sim 300\ \mathrm{cm}^{-1}$ and $\omega \sim 1 - 10\ \mathrm{s}^{-1}$. These frequencies are much smaller than the frequencies currently investigated. For this reason, the contributions corresponding to **n** variations will be neglected in the hydrodynamic equations. These equations are then the same as for isotropic fluids except for the viscous stress tensor, which is

$$\sigma_{ij} = \eta_3(\mathbf{n}\,\hat{\mathbf{A}}\,\mathbf{n})n_i n_j + 2\eta_2 A_{ij} + 2(\eta_1 - \eta_2)\,(n_i n_k A_{kj} + n_j n_k A_{ki}) \tag{1}$$

where $\hat{\mathbf{A}}$ is the tensor of components

$$A_{ij} = \frac{1}{2}\left(\frac{\partial v_j}{\partial x_i} + \frac{\partial v_i}{\partial x_j}\right) \tag{2}$$

and η_1, η_2, and η_3 are viscosity coefficients related to the five viscosity coefficients α_i introduced by Leslie [11]:

$$\eta_1 = \frac{\alpha_4 + \alpha_5}{2} - \frac{\alpha_2}{2}\frac{\gamma_2}{\gamma_1} \qquad \eta_2 = \frac{\alpha_4}{2} \qquad \eta_3 = \alpha_1 + \frac{\gamma_2^2}{\gamma_1} \tag{3}$$

where $\gamma_1 = \alpha_3 - \alpha_2$ and $\gamma_2 = \alpha_3 + \alpha_2$. η_1 and η_2 are bulk viscosities corresponding to a velocity gradient normal to **n** and to a hydrodynamic flow respectively parallel and normal to **n.** The ratio γ_1/γ_2 determines the angle ψ between the molecules and the flow in the absence of external alignment: $\cos 2\psi = -\gamma_1/\gamma_2$. The coefficient α_1 diverges when a smectic phase, where very different flow processes (permeation) take place, is approached.

Let us now consider a thermal fluctuation $\zeta_\mathbf{q}$ for which **q** is parallel to Ox. As in Chap. 4, we will use the Fourier transform of the hydrodynamic equations with respect to time t and coordinate x; labelling the Fourier transform of the hydrodynamic variables with $\sim$, we get

$$iq\,\tilde{v}_x + \frac{\partial}{\partial z}\tilde{v}_z = 0 \tag{4}$$

$$-i\omega\rho\tilde{v}_x = iq\,\tilde{\sigma}_{xx} + \frac{\partial}{\partial z}\tilde{\sigma}_{xz} - iq\tilde{p} \tag{5}$$

$$-i\omega\rho\tilde{v}_y = iq\,\tilde{\sigma}_{yx} + \frac{\partial}{\partial z}\tilde{\sigma}_{yz} \tag{6}$$

$$-i\omega\tilde{v}_z = iq\,\tilde{\sigma}_{zx} + \frac{\partial}{\partial z}\tilde{\sigma}_{zz} - \frac{\partial}{\partial z}\tilde{p} \tag{7}$$

$\tilde{v}$ and $\tilde{p}$ being respectively the Fourier transforms of the fluid velocity and the pressure. The limit conditions are:

$$\mathbf{v} \to 0 \qquad \text{when} \qquad z \to -\infty \tag{8}$$

and

$$\tilde{\sigma}_{xz}(z = 0) = 0 \tag{9}$$

$$\tilde{\sigma}_{yz}(z = 0) = 0 \tag{10}$$

$$\tilde{\sigma}_{zz}(z = 0) - \tilde{p}(z = 0) = -\gamma q^2\tilde{\zeta} + \tilde{p_{ext}} \tag{11}$$

$\tilde{p_{ext}}$ being the external pressure.

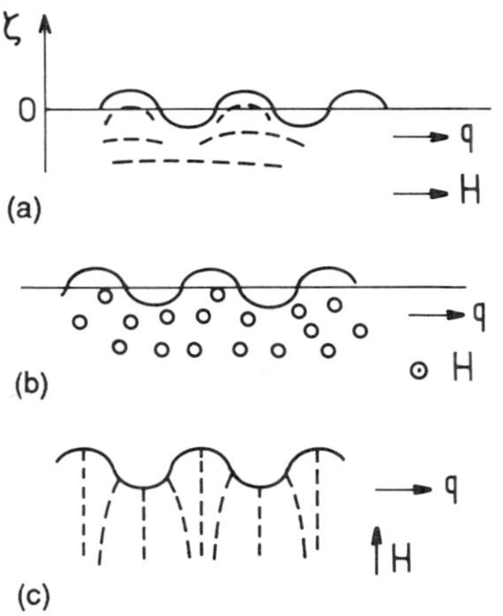

FIG. 2 Schematic representation of the molecular alignement close to the free surface for the three investigated geometries.

We now must specify the orientation of **n** with respect to **q** in order to write the components of $\hat{\sigma}$. We will consider two simple cases: **q** parallel and perpendicular to **H** (Fig. 2a and b).

A. Magnetic Field Parallel to Ox

The molecules must remain parallel to the free surface; they will not be rigorously parallel to **q** because of the surface distortion. But the angle between the surface and the horizontal plane is of the order of $q\zeta_q$, i.e., extremely small. We will then assume that **n** is parallel to Ox in Eq. 1. This gives

$$\begin{aligned}
\tilde{\sigma}_{xx} &= (\eta_3 + 4\eta_1 - 2\eta_2)\, iq\tilde{v}_x \\
\tilde{\sigma}_{yy} &= 0 \\
\tilde{\sigma}_{zz} &= 2\eta_2 \frac{\partial \tilde{v}_z}{\partial z} \\
\tilde{\sigma}_{xz} &= \eta_1\left(iq\tilde{v}_z + \frac{\partial \tilde{v}_x}{\partial z}\right) \\
\tilde{\sigma}_{xy} &= \eta_1\, iq\tilde{v}_y \\
\tilde{\sigma}_{yz} &= \eta_1 \frac{\partial \tilde{v}_y}{\partial z}
\end{aligned} \tag{12}$$

The solution of Eqs. 6, 8, and 10 is then $\tilde{v}_y = 0$ everywhere. Solving for $\tilde{v}_x$ and $\tilde{v}_z$, one finds exponential solutions of the type e^{mz}. Taking into account Eq. 8, m can take only two values, m_1 and m_2, such that

$$\begin{aligned}
m_1 m_2 &= \varepsilon q^2 \sqrt[+]{1 - i\omega\rho/\eta_1 q^2} \\
m_1 + m_2 &= q \sqrt[+]{2 + \alpha - i\omega\rho/\eta_1 q^2 + 2\varepsilon \sqrt[+]{1 - i\omega\rho/\eta_1 q^2}}
\end{aligned} \tag{13}$$

where $\sqrt[+]{\ }$ means the root determination having a positive real part,

$$\varepsilon = \text{sign Re}\left(\sqrt[+]{2 + \alpha + 2S + 2\sqrt[+]{1 + 2S}} - \sqrt[+]{2 + \alpha + 2S - 2\sqrt[+]{1 + 2S}}\right)_{S = \frac{-i\omega\rho}{2\eta_1 q^2}}$$

and

$$\alpha = \eta_3/\eta_1 \tag{14}$$

The general solution for $\tilde{v}_z$ is of the form $A_1 e^{m_1 z} + A_2 e^{m_2 z}$. The constants A_1 and A_2 can be calculated from the limit conditions Eqs. 9 and 11. The response function $\tilde{\chi}_1(\omega)$ is finally

$$\tilde{\chi}_1(\omega) = \frac{\tilde{\zeta}}{\tilde{p_{ext}}} = -\frac{1}{i\omega}\frac{A_1 + A_2}{\tilde{p_{ext}}} = \frac{\tau_o^2 q}{\rho\, \Delta_1(\omega)} \tag{15}$$

$\Delta_1(\omega) = 0$ being the dispersion equation. As in Chap. 4, we shall introduce the reduced parameters

$$y_1 = \frac{\gamma\rho}{4\eta_1^2 q} \qquad \tau_1 = \frac{\rho}{2\eta_1 q^2} \qquad S = -i\omega\tau_1 \tag{16}$$

Then

$$\Delta_1(S) = y_1 + \frac{S^2 + \varepsilon\{S + (2 + \alpha/2)\}S \overset{+}{\sqrt{1 + 2S}}}{\overset{+}{\sqrt{2 + \alpha + 2S + 2\varepsilon \overset{+}{\sqrt{1 + 2S}}}}} \tag{17}$$

The spectrum is finally

$$P_1(\omega) = \frac{k_B T}{\pi\omega}\frac{\tau_1^2 q}{\rho}\, \mathrm{Im}\left(\frac{1}{\Delta_1(-i\omega\tau_1)}\right) \tag{18}$$

It can be shown that $\varepsilon = +1$ for all ω if $\alpha > -2$. The viscosity coefficients η_1, η_2, and η_3 must be such that the production of entropy $T\frac{dS}{dt}$ is always positive. Since $T\frac{dS}{dt} = \sigma_{ij}A_{ij}$, in a system of coordinates where **n** is parallel to Ox,

$$T\frac{dS}{dt} = (\eta_3 + 4\eta_1 - 2\eta_2)A_{xx}^2 + 4\eta_2(A_{yy}^2 + A_{zz}^2 + A_{yz}^2) + 4\eta_1(A_{xy}^2 + A_{xz}^2)$$

This is the most general form describing the irreversible loss of kinetic energy in a uniaxial incompressible medium. The loss has to be positive whatever $\hat{A}$ is, so that

$$\eta_1 > 0 \qquad \eta_2 > 0 \qquad \eta_3 + 4\eta_1 - 2\eta_2 > 0 \tag{19}$$

The condition $\alpha > -2$ is therefore satisfied if $\eta_2 > \eta_1$. This is usually the case for nematic liquids. ε can then be replaced by $+1$ in Eq. 17.

Remark: We have neglected the role of the orientation variations in hydrodynamic motion. Let us also show that these variations produce only very little change in the surface energy. In the approximation where the elastic constants for splay, bending, and twist deformations are all the same, the free energy of distorsion *per unit volume is $\frac{1}{2}K(\Delta n)^2$. The corresponding surface contribution is*

$$\Delta F = \frac{1}{2}\int_{-\infty}^{0} K(\Delta n)^2\, dz \sim \frac{Kq}{2}\, |\mathbf{\delta n}(z = 0)|^2$$

$|\mathbf{\delta n}(z = 0)| \sim q\zeta_q$. With $F = \frac{1}{2}\gamma q^2 \zeta_q^2$, $\Delta F/F \sim Kq/\gamma$. Taking $K \sim 10^{-6}$ dyn, $\gamma \sim 30$ dyn/cm and $q = 300\ \mathrm{cm}^{-1}$; $\Delta F/F \sim 10^{-5}$. The magnetic field also contributes to the surface energy with a term $\Delta F_m/F \sim \chi_a H^2/\gamma q \sim 10^{-4}$.

B. Magnetic Field Parallel to Oy

In this case **n** remains parallel to **H** even at the surface (Fig. 2b). The nonzero components of $\hat{\sigma}$ contain only the viscosity η_2:

$$\tilde{\sigma}_{xx} = 2\eta_2 \tilde{A}_{xx} \qquad \tilde{\sigma}_{zz} = 2\eta_2 \tilde{A}_{zz} \qquad \tilde{\sigma}_{zx} = 2\eta_2 \tilde{A}_{zx}$$

n is normal to both the velocity **v** and to the velocity gradient, and there is no coupling between the orientation and the flow. The nematic liquid behaves like

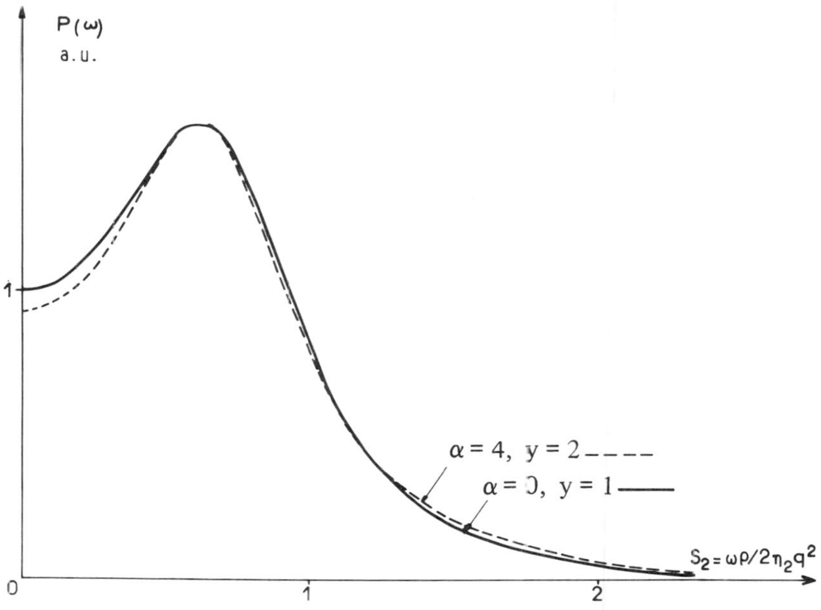

FIG. 3 Comparison between the shape of the theoretical spectra for $\mathbf{q} \parallel \mathbf{H}$ ($\alpha = 4$) and $\mathbf{q} \perp \mathbf{H}$. The y values are different and were adjusted to have the same peak frequency-width ratio.

a simple liquid of viscosity η_2. The spectrum is therefore as in Chap. 4:

$$P_2(\omega) = \frac{k_B T}{\pi\omega}\frac{\tau_2^2 q}{\rho}\,\mathrm{Im}\left(\frac{1}{\Delta_2(-i\omega\tau_2)}\right) \tag{20}$$

with

$$\Delta_2(S) = y_2 + (1 + S)^2 - \sqrt[+]{1 + 2S} \tag{21}$$

using

$$y_2 = \frac{\gamma\rho}{4\eta_2^2 q} \qquad \tau_2 = \frac{\rho}{2\eta_2 q^2} \qquad S = -i\omega\tau_2 \tag{22}$$

P_2 is also equal to P_1 (Eq. 18) if one sets $\eta_3 = 0$ and $\eta_1 = \eta_2$.

Figure 3 illustrates the differences in shape of P_2 and P_1 in the y domain where they are the largest. When $y \gg 1$ or $y \ll 1$, both P_1 and P_2 are Lorentzian within a good approximation. One sees from Fig. 3 that even for $y \sim 1$, the shape differences are very small. It is therefore difficult to determine γ, η_1, and η_3 simultaneously. A good compromise is to compare the two spectra measured for $\mathbf{q} \perp \mathbf{H}$ and $\mathbf{q} \parallel \mathbf{H}$: the first allows the determination of γ and η_2, and the second η_1 and η_3, γ being known.

C. Vertical Magnetic Field

In this case, when the molecules are normal to the surface, the orientation distortions are also negligibly small (Fig. 2c). The stress tensor components are

$$\tilde{\sigma}_{xx} = \eta_2 iq\tilde{v}_x$$

$$\tilde{\sigma}_{zz} = (\eta_3 + 4\eta_1 - 2\eta_2)\frac{\partial\tilde{v}_z}{\partial z}$$

$$\tilde{\sigma}_{xz} = \eta_2\left(\frac{\partial\tilde{v}_z}{\partial z} + iq\tilde{v}_x\right)$$

When solving for $\tilde{v}_x$ and $\tilde{v}_z$, the equations become identical to case A. and the final spectrum identical to P_1, Eq. 18.

III. ROLE OF A DISTORTED CONFIGURATION CLOSE TO THE FREE SURFACE

There are many types of configurations where the bulk and surface orientations can be different. However, the role of the distortion being always similar, we will give here the result for one of the simplest cases: molecules vertical at the free surface and horizontal magnetic field perpendicular to **q**. We will assume that the **n** variations with z are of the form

$$n_z = e^{2z/\xi} \qquad n_y = \pm\sqrt{1 - e^{z/\xi}} \qquad n_x = 0$$

The two n_y values are equally probable, so that the products $n_y n_i$ with $i \neq y$ in the stress tensor σ are 0 in average. The motion along Oy is as before entirely decoupled from the motion in the plane zOx, and $\tilde{v}_y$ is zero everywhere. The relevant components of σ are then

$$\tilde{\sigma}_{xx} = 2\eta_2\, iq\tilde{v}_x$$

$$\tilde{\sigma}_{zz} = [\eta_3 n_z^4 + 2\eta_2 - 4(\eta_2 - \eta_1)n_z^2]\, \frac{\partial \tilde{v}_z}{\partial z}$$

$$\tilde{\sigma}_{zx} = [\eta_2 - 2(\eta_2 - \eta_1)n_z^2] \left(iq\tilde{v}_z + \frac{\partial \tilde{v}_z}{\partial z} \right)$$

In typical experiments, $q \sim 300\ \text{cm}^{-1}$ and $q\xi \sim 0.1$. When $q\xi$ is small, the problem can be solved to first order in $q\xi$. The detailed calculation is given in the appendix. It is shown in particular that $\tilde{v}_z$ is the sum of the zeroth order solution ($\xi = 0$) and of $\delta\tilde{v}_z$ such that

$$\delta\tilde{v}_z = e^{qz} \sum_{n=1}^{\infty} A_n\, e^{nz/\xi} + e^{mz} \sum_{n=1}^{\infty} B_n\, e^{nz/\xi}$$

with $\quad m = q \sqrt[+]{1 - i\omega\rho/\eta_2 q^2}$

The existence of a zone of thickness $\xi \ll q^{-1}$, close to the surface where the molecular orientation rotates, creates a supplementary z variation of the velocity **v** and of the viscous stresses. The ratio $\delta\tilde{v}_z/\tilde{v}_z$ is of order $(q\xi)^3$; $\delta\tilde{v}_x$ is then of order $(q\xi)^2$ (because div v $= 0$) and $\delta\sigma_{xx}$ and $\delta\sigma_{zz}$ are also of order $(q\xi)^2$, whereas $\delta\tilde{\sigma}_{xz}$ is of order $q\xi$. As a result, the derivative of $\tilde{\sigma}_{xz}$ with respect to z is discontinuous at the surface. This is equivalent to a net force per unit area parallel to Ox. The distortion plays a role similar to that of a surfactant monolayer deposited on a liquid surface (Chap. 11). More precisely, taking these orders of magnitude, it can be easily checked that the corrections to the different terms of Eq. 11:

$$\int_{-\infty}^{0} \tilde{v}_z\, dz \qquad \int_{-\infty}^{0} \tilde{\sigma}'_{xz}\, dz \qquad \tilde{\xi}$$

are respectively of the order of $(q\xi)^4$, $(q\xi)^3$, and $(q\xi)^2$ and can be safely neglected. Modifications of order $q\xi$ are only found in the limit condition Eq. 9. It is shown in the appendix to this chapter that the correction is equivalent to the deposition of a layer of surface viscosity

$$\eta_s = \left(\frac{\eta_3}{2} - 4\eta_2 + 4\eta_1\right)\xi$$

$\eta_s = 0$ in particular if $\eta_3 = 0$ and $\eta_2 = \eta_1$, a case where the nematic liquid behaves as if it were isotropic.

Let us note that η_s can be negative here, as it cannot be for monolayers. This means that the molecular orientation at the surface is such that the viscous dissipation in the distorted zone is smaller than the dissipation corresponding to molecular orientation parallel to the field in this zone.

It can be checked numerically that this surface viscosity only leads to significant modifications of the spectrum when $\eta_s q/\eta_2 \geqslant 2$, i.e.,

$$\frac{\eta_3 - 8\eta_2 + 8\eta_1}{2\eta_2} q\xi \geqslant 1$$

For many nematic liquids, the viscosity anisotropy is not large enough to fulfill this condition when $q\xi \leqslant 0.1$. The role of the distorted zone can then be neglected. Everything happens as if the molecules were parallel to the magnetic field everywhere.

IV. THEORETICAL SPECTRUM FOR THE NEMATIC-ISOTROPIC INTERFACE

The nematic-isotropic interface also imposes a preferred orientation to the molecules. For MBBA for instance the molecules are approximately parallel to the interface [6]. The calculation of the spectrum in the presence of a horizontal magnetic field can be done following the same procedure as in Sec. II, but the hydrodynamic equations also have to be solved in the isotropic phase. The vertical component of the fluid velocity is

$$\tilde{v}_z = A_1 e^{m_1 z} + A_2 e^{m_2 z}$$

where m_1 and m_2 are given by Eq. 13 in the geometry $\mathbf{q} \parallel \mathbf{H}$. In the geometry $\mathbf{q} \perp \mathbf{H}$, $m_1 = q$, $m_2 = q \sqrt[+]{1 - i\omega\rho/\eta_2 q^2}$. For the upper isotropic phase, $m_1 = -q$ and $m_2 = -q \sqrt[+]{1 - i\omega\rho/\eta q^2}$ whatever the angle between $\mathbf{q}$ and $\mathbf{H}$ is, because the liquid is not oriented by the field; η is the corresponding bulk viscosity. The real parts of m_1 and m_2 are positive in the nematic phase, negative in the isotropic phase to achieve the condition $\mathbf{v} \to 0$ when $|z| \to \infty$; they are all of the order of q. This leads to a considerable simplification of the problem. In the experiments, typically $q \sim 1000\ \text{cm}^{-1}$, and the waves are damped at a distance less than 10 μm from both sides of the interface. The interface can be stabilized in vertical temperature gradients of the order of 10°/mm. The viscosities and densities can thus considered to be constant and equal to their values at the transition temperature $t = t_{NI}$. Moreover, although the nematic-isotropic transition is (weakly) first order, the anisotropy of the viscosity disappears at t_{NI}: $\eta_1 = \eta_2 = \eta$ and $\eta_3 = 0$. The spectrum is then the same as for the interface of two liquids of identical densities and viscosities. According to Chap. 7,

$$P(\omega) = \frac{k_B T}{\pi\omega} \operatorname{Im}\left(\frac{1}{\gamma q^2 - 2\omega^2 \rho m / q(m - q)}\right) \tag{23}$$

γ being small, $P(\omega)$ is approximately Lorentzian, of half-width

$$\Delta\omega_q = \frac{\gamma q}{4\eta} \tag{24}$$

V. THEORETICAL SPECTRUM FOR THE FREE SURFACE OF A SMECTIC LIQUID CRYSTAL

Smectics frequently orient with the molecules perpendicular to the free surface, i.e., the smectic layers parallel to the surface. This leads to the formation of stepped drops, when the liquid is spread on a solid surface. The light scattering experiment can be performed only on samples having large and flat free surfaces. This is never the case: there are always many step defects at the surface, which are very stable and do not disappear even after long times. Some smectic liquids have a nematic phase at a higher temperature, a feature that is known to facilitate the smectic orientation via thermal cycles between the nematic and the smectic phase. The free surface of the smectic is indeed flat close to the nematic-smectic transition temperature, after cooling the liquid from the nematic phase. However, if a horizontal magnetic field is applied to investigate the viscosity anisotropy, the conflict between surface and bulk orientation results in a surface distortion (Fig. 4). The vertical amplitude of the distortion is small close to the transition temperature t_{SN} but becomes of the order of 100 Å a few degrees below t_{SN}: this produces a strong static scattering and does not allow the analysis of the inelastic scattering produced by the thermal surface waves [12]. For the studied compounds, similar distortions were observed in vertical magnetic fields: this indicates that the molecules are neither parallel nor perpendicular to the free surface.

For the above reasons, there have been only very few attempts to study this type of surface, and the information obtained is limited.

In the following we shall not present the detailed calculations of the spectrum [13, 14] but just recall the results useful to describe the existing experimental data.

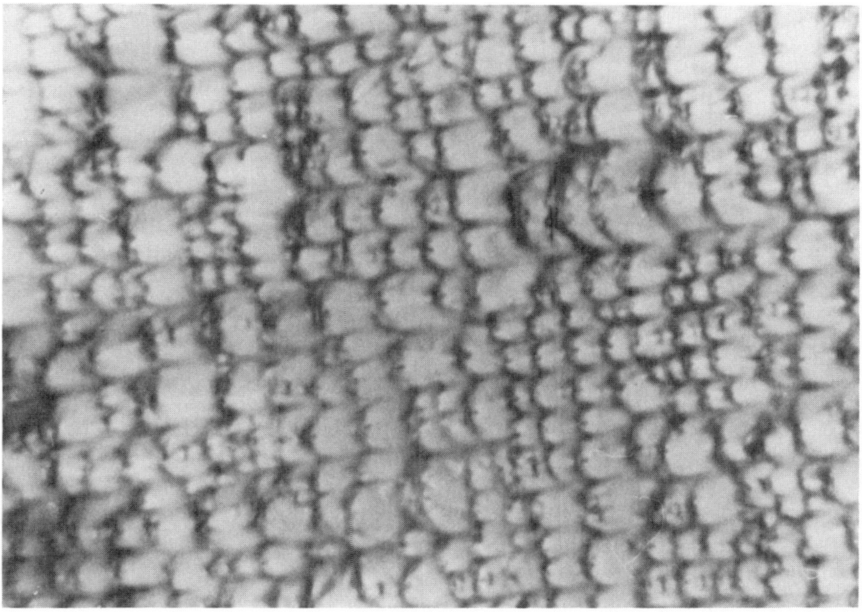

FIG. 4 Aspect of the free surface of M24 in the smectic phase under a polarizing microscope.

A. Smectic Layers Parallel to the Surface [13]

Surface deformations of the type $\zeta_{\mathbf{q}}\, e^{i\mathbf{q}\cdot\mathbf{r}}$ are bending modes of the layers. These modes are overdamped; their characteristic time is $\tau_0 = \eta_1/Kq^2$, where K is the bending elastic constant and η_1 the viscosity introduced in Sec. II: the viscous stress tensor has the form of Eq. 1 [8, 14]. τ_0 is the viscous relaxation time for the orientation, and is the same as in nematic phases. Typically, for $q \sim 300$ cm^{-1}, $K \sim 10^{-6}$, $\eta_1 \sim 0.1$, $\tau_0 \sim 1$ sec. The expected spectrum is then very narrow.

B. Smectic Layers Perpendicular to the Surface [15]

If a horizontal magnetic field is used, a very simple result is obtained for $\mathbf{q} \perp \mathbf{H}$. Indeed, as in the nematic phase, there is a complete decoupling between the orientation and the flow. The spectrum is then given by Eq. 20. When $\mathbf{q} \parallel \mathbf{H}$, the situation is very different: the flow produces a motion of the molecules across the smectic layers, a phenomenon called permeation. It produces a strong damping of the waves characterized by the permeation length κ^{-1}: typically $\kappa^{-1} \sim$

100 Å. The spectrum is centered around $\omega = 0$, it is non-Lorentzian, and its width is

$$\Delta\omega_q \sim (\tau_1\tau_2)^{-1/2}$$

where $\tau_1 = \dfrac{\eta_1\kappa^2}{Bq^2}$ and $\tau_1 = \dfrac{\eta_1\kappa}{\gamma q^2}$, B being the compressibility modulus of the smectic layers. Typically, $B \sim 10^6$ and $\Delta\omega_q \sim 10\ \mathrm{s}^{-1}$. The vertical damping length of the dominant mode is of the order of $\kappa/q^2 \sim 1$ cm: if the molecules at the surface are not horizontal, the thickness of the distorted zone $\xi_H = \sqrt{\dfrac{K}{\chi_a}\dfrac{1}{H}}$ is however much smaller than κ/q^2. The role of this distorted zone is then negligible.

VI. EXPERIMENTAL RESULTS—FREE SURFACE OF A NEMATIC LIQUID

Table 1 describes some of the compounds studied and the temperature ranges of their liquid crystalline phases.

A. Para-Azoxyanisole

The viscosity of para-azoxyanisole in its nematic phase is only about four times larger than the viscosity of water. The y parameter is larger than 10 if $q > 200$

TABLE 1

Compound	Formula	Temperature range
PAA para-Azoxyanisole	CH_3O ⟨φ⟩ N═N ⟨φ⟩ OCH_3	C $\xrightarrow{118°}$ N $\xrightarrow{135°}$ I
MBBA Methoxy benzilidene butyl aniline	CH_3O ⟨φ⟩ CH═N ⟨φ⟩ C_4H_9	C $\xrightarrow{20°}$ N $\xrightarrow{46°}$ I
CBOOA Cyanobenzylidene octyloxyaniline	$C_8H_{17}O$ ⟨φ⟩ N═CH ⟨φ⟩ C≡N	C $\xrightarrow{73°}$ S $\xrightarrow{83°}$ N $\xrightarrow{108°}$ I
M24 Octyloxycyanobiphenyl	$C_8H_{17}O$ ⟨φ⟩ — ⟨φ⟩ C≡N	C $\xrightarrow{54°5}$ S $\xrightarrow{67°}$ N $\xrightarrow{80°}$ I
K24 Octylcyanobiphenyl	C_8H_{17} ⟨φ⟩ — ⟨φ⟩ C≡N	C $\xrightarrow{21°}$ S $\xrightarrow{32°5}$ N $\xrightarrow{40°}$ I

cm^{-1}. When $y \gg 1$, the spectra given by Eqs. 18 and 20 are Lorentzian, centered at a frequency $\omega_q^2 = \gamma q^3/\rho$ and of half-width $\Delta\omega_q = 2\eta q^2/\rho$ with $\eta = \eta_2$ for $\mathbf{q} \perp \mathbf{H}$ and $\eta = \eta_1 + \eta_3/8$ for $\mathbf{q} \parallel \mathbf{H}$. An experimental spectrum taken at 122°C is shown in Fig. 5 together with a fit with a Lorentzian and with the exact spectrum. The low q scattering was identified as being due to bulk scattering around the beam reflected by the bottom of the liquid cell [2]. The spectra for $\mathbf{q} \parallel \mathbf{H}$ and $\mathbf{q} \perp \mathbf{H}$ are indistinguishable: there is no observable anisotropy. The peak frequency and half-width variations with wavevector q are shown in Fig. 6. The slope of the frequency curve in a log-log representation is 3/2 as predicted for large y. Here $9 < y < 52$. The slope of the width curve is smaller than 2, because the asymptotic law is reached at larger y values: one needs to correct for the exact dispersion equations and corresponding theoretical spectra, and also for instrumental broadening. This leads to, for $t = 122°C$ [16],

$$\gamma = (37 \pm 2) \text{ dyn/cm}$$

$$\eta_2 = \eta_1 + \frac{\eta_3}{8} = (3.7 \pm 0.3) \text{ centipoise}$$

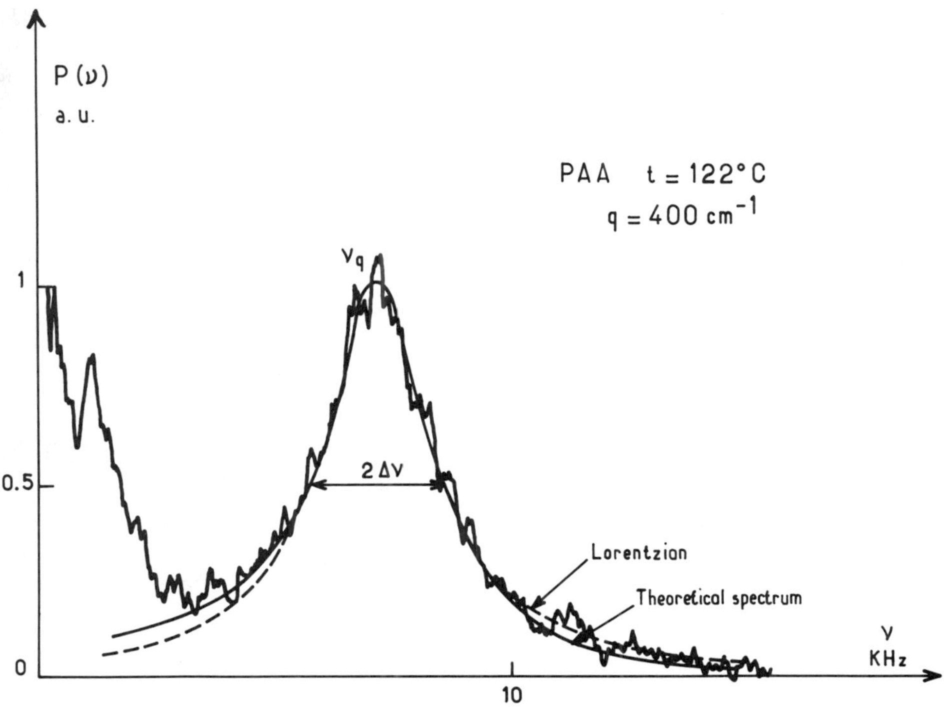

FIG. 5 Spectrum of the light scattered by the free surface of PAA.

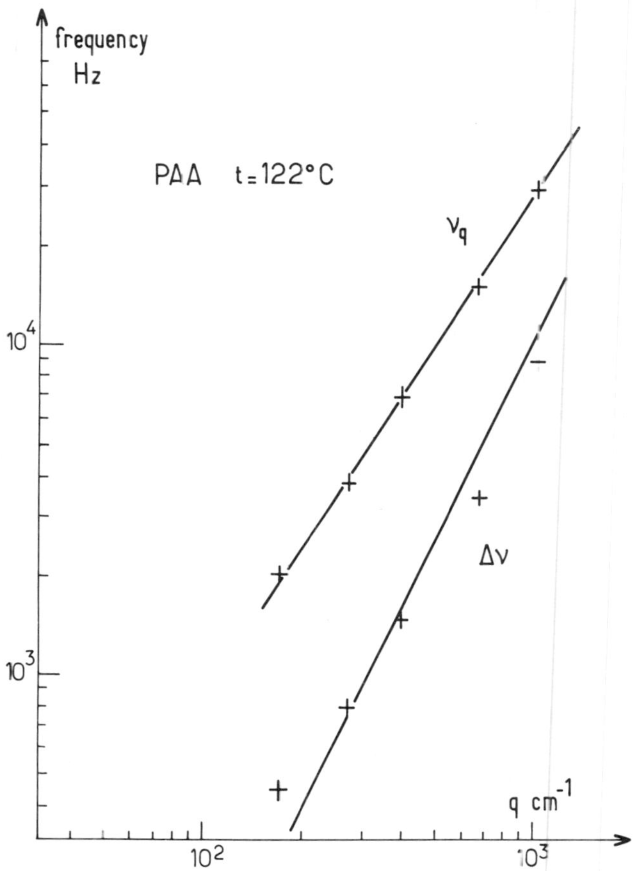

FIG. 6 Peak frequency and half-width of the spectrum of the light scattered by the free surface of PAA versus q.

The temperature variations of these quantities are shown in Fig. 7 [3, 5, 16]. The γ values are in agreement with determinations with classical methods [17, 18]. The data of Ref. 16 do not show any anomaly at the nematic-isotropic transition. Those of Ref. 5, where the transition was approached closer, do suggest a tension increase close to t_{NI}, but the shape of the anomalies seems to depend on the method used. Further work is needed to clarify the issue.

The η_2 viscosity values are in agreement with those of Miesowicz using a capillary viscosimeter in a magnetic field [18]. Miesowicz also measured η_1. From $\eta_2 \sim \eta_1 + \eta_3/8$ one gets $\alpha = \eta_3/\eta_1 \sim 4$ for $t = 122°C$. This value is similar to the MBBA value as will be shown later.

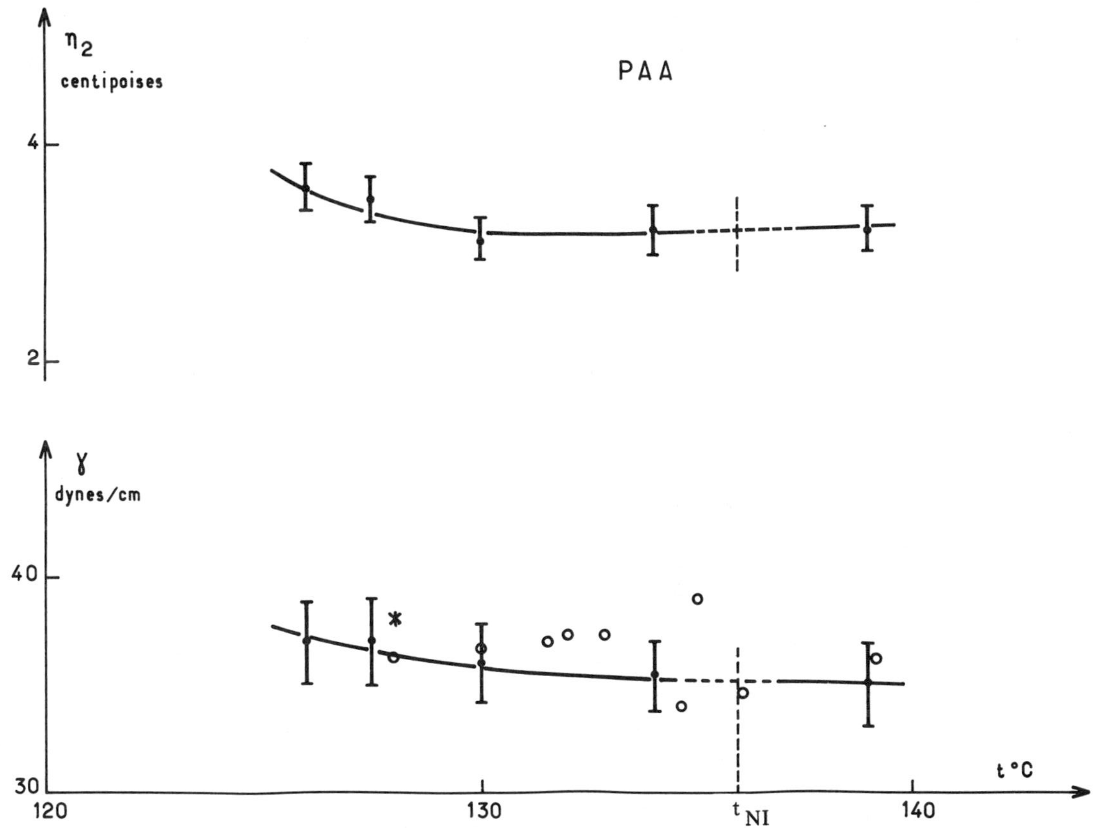

FIG. 7 Surface tension and bulk viscosity η_2 of PAA deduced from the spectra as a function of temperature. Points of: ●, Ref. 16; ○, Ref. 5; *, Ref. 3.

B. Methoxybenzelidene Butylaniline

Methoxybenzelidene butylaniline is about ten times more viscous than PAA in its nematic range. It becomes possible to determine the four parameters γ, η_1, η_2, and η_3. Typical spectra for a horizontal magnetic field parallel and perpendicular to **q** and for a vertical magnetic field are shown in Figs. 8–10. The lines are fit with the theoretical spectra. An anisotropy is clearly observable in a horizontal magnetic field, even for $\alpha \sim 4$. The spectra for a horizontal magnetic field parallel to **H** and a vertical magnetic field are the same. This confirms the absence of influence of the distorted regions below the surface, which are different in the two cases. It was also verified that this was the case by varying the amplitude of the horizontal magnetic field between 1000 and 3500 G ($15 > \xi_H > 4$ μm): no spectral modifications were observed.

The measured surface tensions and viscosities are reported in Fig. 11 [2]. They correspond to average values for wavevectors in the range $138 < q < 509$ cm^{-1}. The γ and η_2 values were deduced from the spectra for $\mathbf{q} \perp \mathbf{H}$ and the η_3 and η_1 values from the spectra $\mathbf{q} \parallel \mathbf{H}$: the signal-to-noise ratio is not good enough to allow the determinations of γ, η_3, and η_1 values from these spectra and we have used the average γ value determined from the $\mathbf{q} \perp \mathbf{H}$ spectra. The results for $t - t_{NI} < -10°$ are not shown, because the viscosities are large and

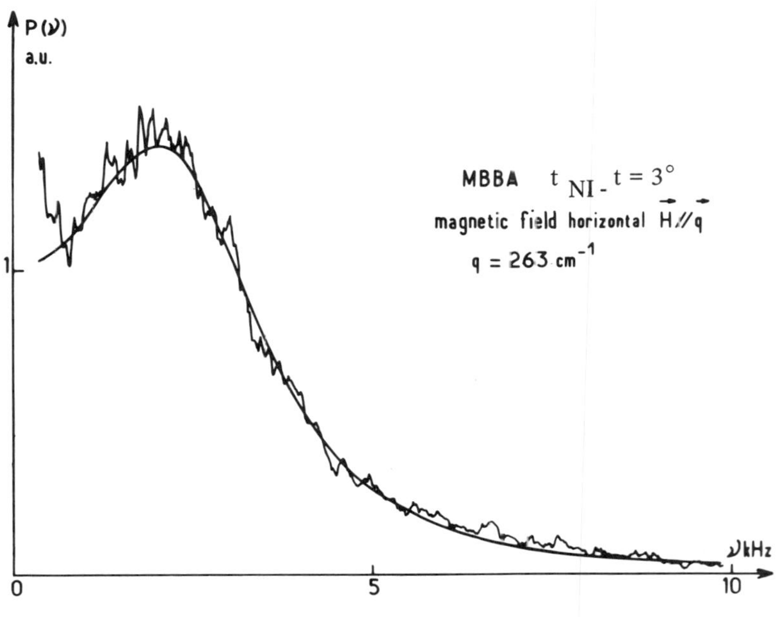

FIG. 8 Spectrum of the light scattered by the free surface of MBBA for $\mathbf{q} \parallel \mathbf{H}$.

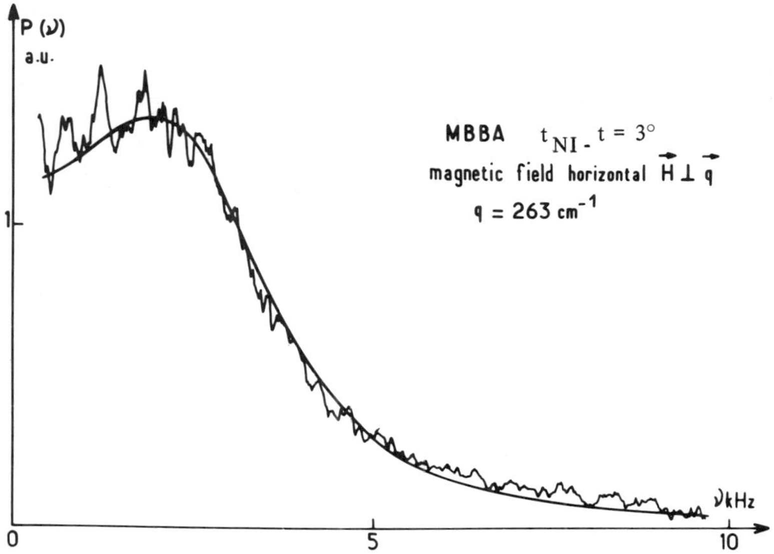

FIG. 9 Spectrum of the light scattered by the free surface of MBBA for **q** ⊥ **H**, and **H** horizontal.

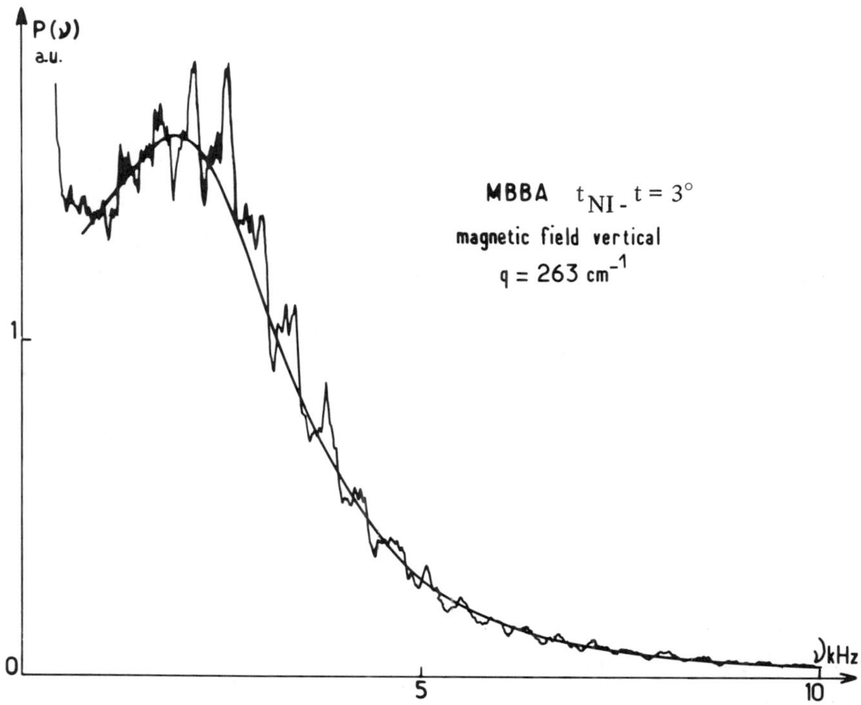

FIG. 10 Spectrum of the light scattered by the free surface of MBBA for a vertical magnetic field.

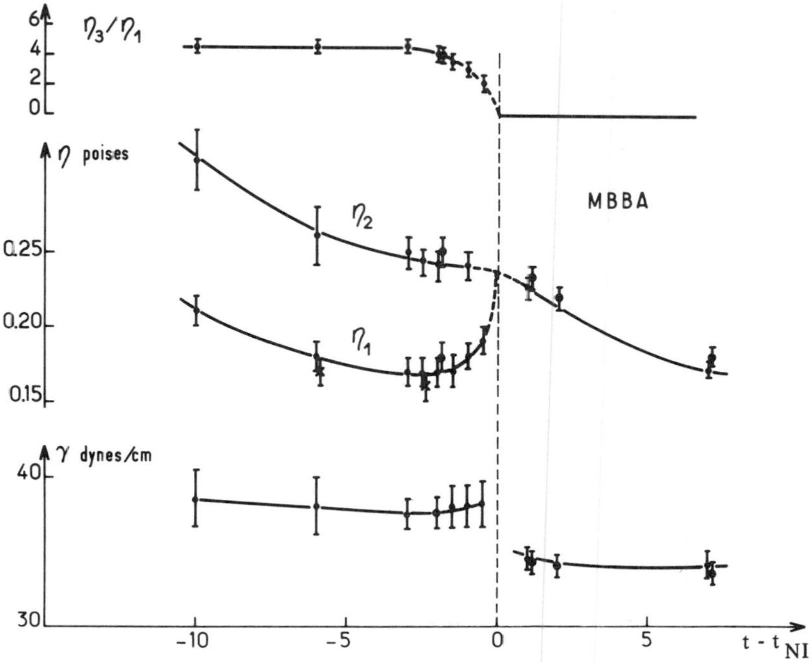

FIG. 11 Surface tension and bulk viscosities of MBBA versus temperature.

the frequency peak of the spectra is gone to $\nu = 0$: the surface waves are overdamped. It becomes difficult to deduce four parameters from these spectra.

The η_1 values are in agreement with measurements using capillary viscometers [19, 20]: in this case one measures $(\alpha_4 + \alpha_6 + \alpha_3)/2$, whereas $\eta_1 = (\alpha_4 + \alpha_6 - \alpha_3\gamma_2/\gamma_1)/2$; but $\gamma_2/\gamma_1 \sim -1$ for MBBA [21]. η_1 and η_2 are slightly different from the values obtained using ultrasonic attenuation [19]: this can be due in fact to relaxation processes between ultrasonic and capillary wave frequencies. η_1,η_2, and η_3 are finally in agreement with combined flow and birefringence measurements in a magnetic field [21]. For $t - t_{NI} < -3°$, the ratios η_2/η_1 and η_3/η_1 are respectively 1.47 and 4.5, close to the corresponding values of PAA in its nematic phase.

Measurements without magnetic fields were also performed [16]. They lead to effective surface tensions and viscosities that noticeably depend on q: for instance at $t = t_{NI} - 2°$, for which $\gamma = 38$ dyn/cm, γ_{eff} increases between 31 and 35.5 dyn/cm, and η_{eff} between 0.23 and 0.26 poise. This is probably due to an orientation distortion below the free surface, whose depth is much larger than when a magnetic field is applied. This points out the importance of the magnetic field in the determination of correct surface tension and viscosity values.

There should also be light scattered by orientation fluctuations at the surface. The corresponding characteristic frequencies are small: $\omega \sim (Kq^2 + \chi_a H^2)/\eta$, and the corresponding intensities are also very small [16, 22]. They have never been observed experimentally.

C. Vicinity of a Second-Order Nematic-Smectic A Transition

When a second-order nematic-smectic A transition is approached, several static (elastic constants) [8] and dynamic (viscosities) [23, 24] properties diverse. Close to the transition, two correlation lengths can be defined in the nematic phase: $\xi_{\|}$ and $\xi_{\perp}$ that can be viewed as dimensions of transient smectic domains respectively parallel and perpendicular to the molecular axis. These correlation lengths diverge at the critical temperature with the exponent ν: $\xi = \xi_0((t - t_{NS})/t_{NS})^{-\nu}$; $\nu = 1/2$ in mean field theory, 2/3 in the helium analogy [8]. A new viscous force appears when the liquid moves across the smectic layers; the force increases when the transition is approached because the size of the smectic domains increases, as well as their lifetime. The corresponding (critical) contribution to the viscosities diverges with the exponent $\nu = 0.5$ being the mean field approximation, $\nu/2 = 1/3$ in the helium analogy.

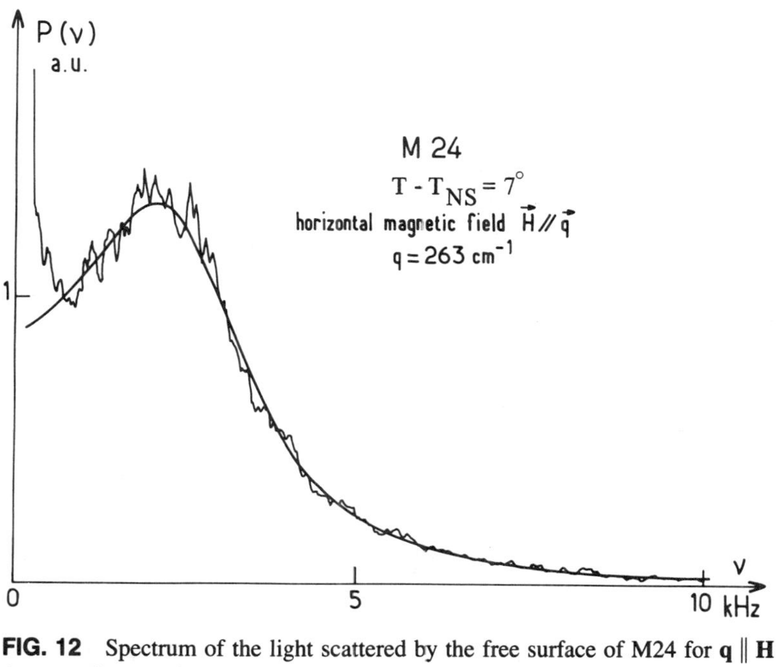

FIG. 12 Spectrum of the light scattered by the free surface of M24 for **q** ∥ **H** far from the nematic-smectic phase transition.

Among the quantities measured in surface light scattering experiments, only η_3 has a critical behavior: $\eta_3 = \eta_{3B} + \tilde{\eta}_3$ where B stands for background and $\sim$ for the critical part. From $\eta_3 = \alpha_1 + \gamma_2^2/\gamma_1$, we find

$$\tilde{\eta}_3 = \tilde{\alpha}_1 + \tilde{\gamma}_1 = (\xi_{\parallel}/\xi_{\perp})^2 \, \tilde{\gamma}_1$$

$\xi_{\parallel}/\xi_{\perp}$ is typically about 5, so the amplitude of the critical part of η_3 is large, contrary to the one of most other viscosity coefficients: this facilitates the critical exponent determination.

The experiments have been done on CBOOA and M24 (Table 1) [7]. Figures 12 and 13 show two typical experimental spectra for $\mathbf{q} \parallel \mathbf{H}$ far and close to the transition. Close to t_{NS}, the spectrum for $\mathbf{q} \perp \mathbf{H}$ is similar to the one of Fig. 12. Since η_3 is very large, the anisotropy here is extremely pronounced. The

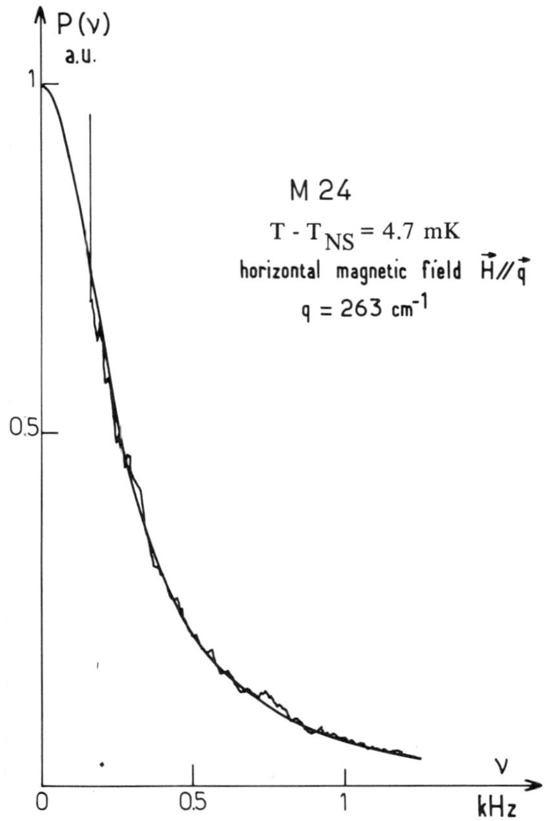

FIG. 13 Spectrum of the light scattered by the free surface of M24 for $\mathbf{q} \parallel \mathbf{H}$ very close to the nematic-smectic phase transition.

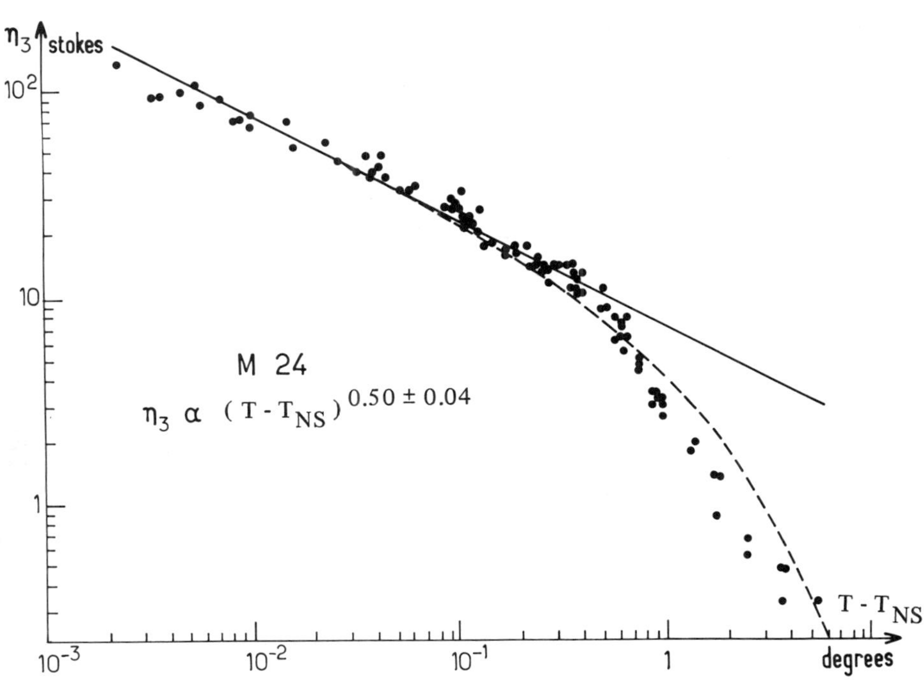

FIG. 14 Bulk viscosity η_3 versus temperature for M24.

data analysis confirms that η_1, η_2, and γ do not have critical behavior. The variation of η_3 with temperature is shown in Fig. 14 for M24. The divergence of the critical part is compatible with mean field behavior: exponent 0.5. At the time when these experiments were performed there was a controversy on this particular point: working on the same compound, several authors found mean field exponents, whereas others found helium exponents. Further information on the subject can be found in Ref. 25.

VII. NEMATIC ISOTROPIC INTERFACE OF MBBA

The molecules are approximately parallel to the nematic-isotropic interface: their small angle [6] with the interface may vary between different batches (due to impurities) [26]. The interfacial tension is relatively small, because the nematic-isotropic transition is weakly first order; it is close to a second-order transition (within about 1°). It can be shown that [8]

$$\gamma = \sqrt{3A}\,\xi \int_0^{Q_0} \sqrt{\Phi(Q)}\, dQ \tag{25}$$

where Φ is the Landau free energy

$$\Phi(Q) = \frac{1}{2}AQ^2 + \frac{1}{3}BQ^2 + \frac{1}{4}CQ^2 \tag{26}$$

ξ is the correlation length and Q is the order parameter; $Q = \frac{3}{2} < \cos^2 \phi - \frac{1}{3} > \phi$, being the angle between the molecules and the mean orientation $\mathbf{n}$. At the transition, $Q = Q_0$, and $\Phi(Q_0) = d\Phi/dQ(Q_0) = 0$. Finally, from Eqs. 25 and 26,

$$\gamma = \sqrt{\frac{1}{12}}\, Q_0^2\, A\, \xi \tag{27}$$

For MBBA, $Q_0 \sim 0.4$ [27], $A \sim 6.2\ 10^{-2}$ J/cm, and $\xi \sim 150$ Å [28], thus $\gamma \sim 3\ 10^{-2}$ syn/cm.

The y parameter is of the order of 10^{-4} and the spectrum given by Eq. 23 is Lorentzian to very good approximation; its width is given by Eq. 24. Because

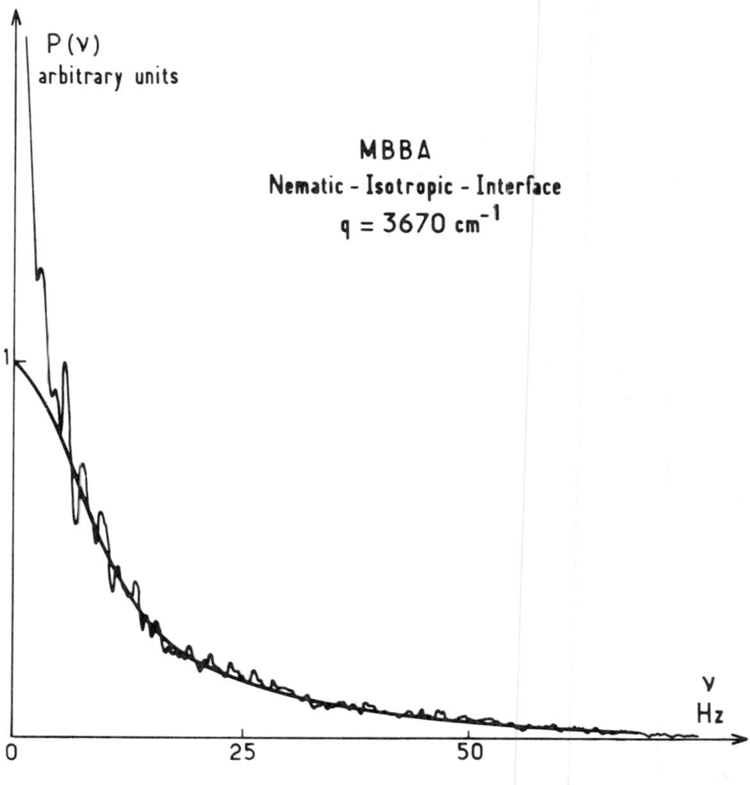

FIG. 15 Spectrum of the light scattered by the nematic-isotropic interface of MBBA.

γ is so small, one may wonder about the role of surface forces resulting from possible orientation distortions and magnetic field orientation (Sec. II.A). Although they are still smaller than γ, they should appear as a surface tension anisotropy in the experiments. It was checked that this was not the case: the spectra for $\mathbf{q} \parallel \mathbf{H}$ and $\mathbf{q} \perp \mathbf{H}$ were identical within experimental accuracy.

A typical spectrum is shown in Fig. 15. The signal-to-noise ratio is not very good because of the large amount of light scattered elastically in the forward direction. This is due to small planarity defects of the interface, because the plane $t = t_{NI}$ is not exactly horizontal: in order to have interfacial planarity defects smaller than 10 Å, the temperature gradient should be vertical within 10^{-4} rad in the range of studied scattering angles. Fortunately, γ is small, and scattering angles larger than in the case of free surfaces could be studied; the stray light intensity still produces a strong signal at $\omega \sim 0$ in the spectrum but does not wash out the higher frequency part of this spectrum (Fig. 15).

The q domain studied was $1000 < q < 8000$ cm^{-1}, and the data are shown in Fig. 16 for two different samples (for which the angles θ of the molecules at the interface are slightly different) [6]. The interfacial tensions are the same within instrumental accuracy. From Eq. 24 and $\eta = 0.24$ poise, one obtains

$$\gamma = (2.2 \pm 0.4)\ 10^{-2}\ \text{dyn/cm}$$

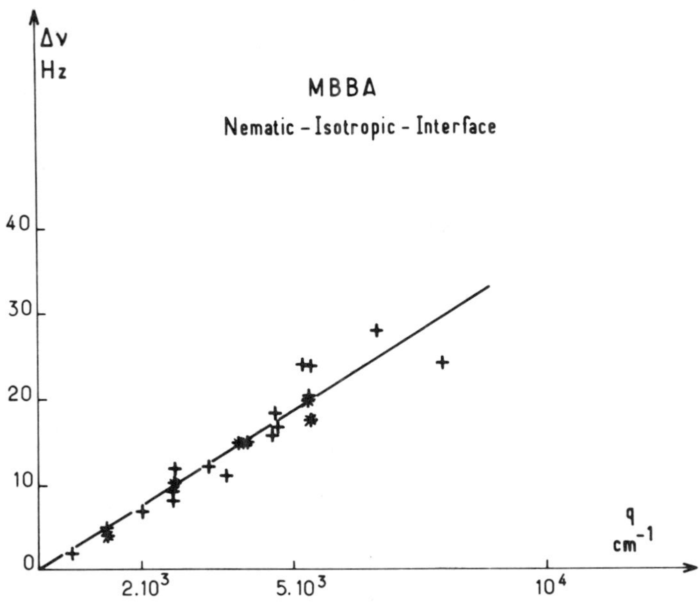

FIG. 16 Half-width of the spectrum of the light scattered by the nematic-isotropic interface of MBBA versus q. The different symbols are for different samples.

in good agreement with the predictions of Eq. 27 and the values of Q (from refractive index data), and of ξ and A (from bulk light scattering data).

VIII. FREE SURFACE OF SMECTIC LIQUID CRYSTALS

There have been very few measurements done on the free surface of smectics, because the free surface is not flat due to the high energy associated with orientation defects. The only reported measurements deal with compounds that have a nematic phase above the smectic phase. The first reported experiment was done with ethyl methoxybenzelidene aminocinnamate [4] without using a magnetic field. The molecules were assumed to be normal to the surface, and the spectra were analyzed using the expression valid for nematics in a vertical magnetic field (Eq. 18). Although no experimental spectra are shown in the paper, according to the values found for γ, η_1, and η_3 these spectra are not much narrower than in the nematic phase, although they were taken far from the nematic-smectic transition (8° away). Theory predicts that the spectra are in fact extremely narrow because capillary waves are coupled to high-energy orientation distorsions [13]. It is possible that the measurements have been made in the nematic phase because of transition temperature shifts due to impurities.

Another reported study was made on CBOOA (Table 1). The molecules of this compound are neither parallel nor perpendicular to the magnetic field: except close to t_{NS}, the free surface is strongly distorted when horizontal or vertical magnetic fields are applied [12]. Light scattering measurements were done in the temperature range $0.1 < t_{NS} - t < 1°$ in a horizontal magnetic field. The spectra for $\mathbf{q} \perp \mathbf{H}$ were found very similar to those measured in the nematic phase close to t_{NS} as expected (Sec. V.B): γ and η_2 are not discontinuous at the transition. The spectra for $\mathbf{q} \parallel \mathbf{H}$ were too narrow to be analyzed: this is also as expected (Sec. V.B). The anisotropy of the spectrum in a horizontal magnetic field is therefore extremely large.

IX. CONCLUSION

The study of the spectrum of the light scattered by the free surface of nematic liquid crystals in suitable magnetic fields allows the determination of the surface tension and three bulk viscosities without perturbing the molecular alignment. The surface tension is isotropic and can be discontinuous at the nematic-isotropic transition (case of MBBA). Two of the bulk viscosities correspond to those measured in capillary viscometers when the molecules are respectively parallel (η_1) and perpendicular (η_2) to the flow and perpendicular to the velocity gradient. The third viscosity (η_3) is associated to permeation in presmectic layering and strongly increases close to a weakly first-order nematic-to-smectic A transition.

The surface tension and the viscosity η_2 were found continuous at the nematic-to-smectic transition of CBOOA. But in most geometries, the surface waves are too strongly damped in the smectic phase to be measured. Finally, the nematic-isotropic interfacial tension has been measured in the case of MBBA using a vertical temperature gradient: γ is about three orders of magnitude below the surface tension values. This is because the nematic-isotropic transition is close to a second-order transition (correlation length of about 100 Å).

It must be stressed that the role of the magnetic field is extremely important to ensure that the data analysis is correct: q dependent γ and η values can be obtained when no magnetic fields are applied because of the orientation distortions close to the surface. It is also of use to measure the molecular orientation at the surface to learn when these distortions will play a role.

REFERENCES

1. D. Langevin and M.-A. Bouchiat, *J. Physique 33:* 101 (1972); *33:* C1–77 (1972).
2. D. Langevin, *J. Physique 33:* 249 (1972).
3. D.McQueen, *J. Phys. D: Appl. Phys. 6:* 2273 (1973).
4. D. McQueen and V. K. Singhal, *J. Phys. D: Appl. Phys. 7:* 1983 (1974).
5. L. B. Shih, J. A. Mann, and S. H. Brown, *Mol. Cryst. Liq. Cryst. 98:* 47 (1983).
6. D. Langevin and M.-A. Bouchiat, *Mol. Cryst. Liq. Cryst. 22:* 317 (1973).
7. D. Langevin, *J. Physique 36:* 745 (1975); *37:* 901 (1976).
8. P. G. de Gennes, *The physics of Liquid Crystals,* Oxford Univ. Press, 1974.
9. M. A. Bouchiat and D. Langevin-Cruchon, *Phys. Lett. 34A:* 331 (1971).
10. P. Chiarelli, S. Faetti, and L. Fronzoni, *J. Physique 44:* 1061 (1983).
11. F. M. Leslie, *Quart. J. Mech. Appl. Math. 19:* 357 (1966).
12. D. Langevin, *Phys. Lett. 56A:* 61 (1976); *J. Physique 37:* 755 (1976).
13. A. Rapini, *Can. J. Phys. 53:* 968 (1975).
14. P. C. Martin, O. Parodi, P. S. Perhan, *Phys. Rev. A 6:* 2401 (1972).
15. D. Langevin, *J. Physique 37:* 737 (1976).
16. D. Langevin, thesis, Paris, 1974.
17. I. G. Chystiakov, *Sov. Phys. 9:* 551 (1967).
18. M. Miesowicz, *Nature 158:* 343 (1946).
19. P. Martinoty and S. Candau, *Mol. Cryst. Liq. Cryst. 14:* 243 (1971).
20. D. Berchet, A. Hopchafel, and R. Viovy, *Comptes Rendus Acad. Sci. 270C:* 70 (1965).
21. C. Gahwiller, *Mol. Cryst. Liq. Cryst.20:* 301 (1973).
22. M.-A. Bouchiat and D. Langevin, *J. Coll. Int. Sci. 63:* 193 (1978).
23. F. Jahnig and F. Brochard, *J. Physique 35:* 301 (1974).
24. W. L. McMillan, *Phys. Rev. A9:* 1720 (1974).
25. P. S. Pershan, *The Structure of Liquid Crystals,* World Scientific, Singapore, 1988.
26. D. Langevin, and M.-A. Bouchiat, *Comptes Rendus Acad. Sci. 277* B: 731 (1973).
27. W. Maier, A. Saupe, *Z. Natursforsch A15:* 287 (1960).
28. I. W. Stinson and J. D. Litster, *Phys. Rev. Lett. 25:* 503 (1970).

APPENDIX

We calculate here the effect on the spectrum of an orientation distortion close to the free surface of a nematic liquid. We consider the particular case where the magnetic field is horizontal and perpendicular to **q**, and where the molecules are vertical at the free surface (this is almost the case for MBBA close to the nematic-isotropic transition). By using Eqs. 23 for the stress tensor, the equation for $\tilde{v}_z$ can be rewritten in the form

$$-i\omega\rho\left(\frac{\partial^2}{\partial z^2} - q^2\right)\tilde{v}_z = -q^2\frac{\partial}{\partial z}[\eta_3 n_{oz}^4 + 4\eta_2 + 4(\eta_1 - \eta_2)n_{oz}^2]\frac{\partial \tilde{v}_z}{\partial z} + \left(\frac{\partial^2}{\partial z^2} + q^2\right)[\eta_2 + (\eta_1 - \eta_2)n_{oz}^2]\left(\frac{\partial^2}{\partial z^2} + q^2\right)\tilde{v}_z \quad \text{(A.1)}$$

using $n_{oz}^2 = e^{z/\xi}$, and setting $\eta_2 - \eta_1 = \eta_4$:

$$\left[-i\omega\rho\left(\frac{\partial^2}{\partial z^2} - q^2\right) - \eta_2\left(\frac{\partial^2}{\partial z^2} - q^2\right)^2\right]\tilde{v}_z = e^{z/\xi}\left\{4\eta_4 q^2\frac{\partial}{\partial z}\left(\frac{\partial}{\partial z} + \frac{1}{\xi}\right) - \eta_4\left(\frac{\partial^2}{\partial z^2} + q^2\right)\left[\left(\frac{\partial}{\partial z} + \frac{1}{\xi}\right)^2 + q^2\right]\right\}\tilde{v}_z - e^{2z/\xi}\eta_3 q^2\frac{\partial}{\partial z}\left(\frac{\partial}{\partial z} + \frac{2}{\xi}\right)\tilde{v}_z$$

The general solution of this differential equation is

$$\tilde{v}_z = \sum_{\nu}\sum_{n=0}^{\infty} A_n^{\nu}\, e^{m_\nu z + nz/\xi}$$

Σ_ν meaning the sum over the roots m_ν of the equation $f(m) = 0$, where

$$f(m) = -i\omega\rho(m^2 - q^2) - \eta_2(m^2 - q^2)^2$$

In order to satisfy the limit condition (8), only two roots are physically acceptable, those whose real part is positive:

$$m_1 = q \qquad m_2 = q\sqrt[+]{1 - i\omega\rho/\eta_2 q^2}$$

The coefficients A_ν obey the recurrence relation

$$f\left(m_\nu + \frac{n+2}{\xi}\right)A_{n+2}^{\nu} = g\left(m_\nu + \frac{n+1}{\xi}\right)A_{n+1}^{\nu} + h\left(m_\nu + \frac{n}{\xi}\right)A_n^{\nu} \quad \text{(A.2)}$$

where

$$g(m) = 4\eta_4 q^2 m\left(m + \frac{1}{\xi}\right) - \eta_4(q^2 + m^2)\left[\left(m + \frac{1}{\xi}\right)^2 + q^2\right]$$

$$h(m) = -\eta_3 q^2 m\left(m + \frac{2}{\xi}\right)$$

A_1^ν and A_0^ν are related by

$$f\left(m_\nu + \frac{1}{\xi}\right)A_1^\nu = g(m_\nu)A_0^\nu \qquad \text{(A.3)}$$

The solution obtained in this way depends on two parameters, A_0^1 and A_0^2, that are determined from the limit conditions for $z = 0$ (Eqs. 9 and 11):

$$-i\omega\rho \int_{-\infty}^{0} \tilde{v}_z\, dz - iq \int_{-\infty}^{0} \tilde{\sigma}'_{zx}\, dz + \frac{\gamma q^2}{-i\omega}\tilde{v}_z\,(z = 0) = \widetilde{p_{ext}}$$

$$\tilde{\sigma}'_{zx}\,(z = 0) = 0$$

that become here

$$\sum_\nu \sum_{n=0}^{\infty} \left\{ \frac{-i\omega\rho}{m_\nu + \frac{n}{\xi}} + \eta_2\left[\left(m_\nu + \frac{n}{\xi}\right) + \frac{q^2}{m_\nu + \frac{n}{\xi}}\right] - \eta_4 \frac{\left(m_\nu + \frac{n}{\xi}\right)^2 + q^2}{m_\nu + \frac{n+1}{\xi}} + \frac{\gamma q^2}{-i\omega} \right\} A_n^\nu = \widetilde{p_{ext}} \qquad \text{(A.4)}$$

$$\sum_\nu \sum_{n=0}^{\infty} \left[\left(m_\nu + \frac{n}{\xi}\right)^2 + q^2\right] A_n^\nu = 0 \qquad \text{(A.5)}$$

We are going to assume that $q\xi \ll 1$ and make an expansion to first order in $q\xi$. Within this approximation, the recurrence relation A.2 becomes, for $n \geqslant 1$,

$$-\eta_2\left(m_\nu + \frac{n+2}{\xi}\right)^4 A_{n+2}^\nu = -\eta_4\left(m_\nu + \frac{n+2}{\xi}\right)^2 \times \left(m_\nu + \frac{n+1}{\xi}\right)^2 A_{n+1}^\nu$$

and, using the notation $B_n^\nu = \left(m_\nu + \frac{n}{\xi}\right)^2 A_n^\nu$

$$\eta_2 B_{n+1}^\nu = \eta_4 B_n^\nu \qquad \text{for} \qquad n \geqslant 2$$

The new coefficients B_ν form a geometrical series: $B_n^\nu = (\eta_4/\eta_1)^{n-2} B_2^\nu$; B_1^ν is determined according to Eq. A.3:

$$-\eta_2\left(m_\nu + \frac{n}{\xi}\right)^2 B_n^\nu = \left\{4\eta_4\, q^2\, m_\nu\left(m_\nu + \frac{n}{\xi}\right) - \eta_4(m_\nu^2 + q^2)\left(m_\nu + \frac{n}{\xi}\right)^2\right\} A_0^\nu$$

giving

$$B_1^\nu = \frac{\eta_4}{\eta_2}(m_\nu^2 + q^2)\, A_0^\nu - \frac{4\eta_4}{\eta_2} q^2\, m_\nu\, \xi\, A_0^\nu$$

B_2^ν is given by Eq. A.2:

$$B_2^\nu = \frac{\eta_4}{\eta_2} B_1^\nu + \frac{\eta_3}{2\eta_2} q^2\, m_\nu\, \xi\, A_0^\nu$$

The limit equation A.5 becomes

$$\sum_\nu \left\{(m_\nu^2 + q^2)A_0^\nu + \sum_{n=1}^{\infty} B_n^\nu\right\} = 0$$

or

$$\sum_{n=2}^{\infty} B_n^\nu = \frac{B_2^\nu}{1 - \eta_4/\eta_2}$$

(if the ratio of the series is such as $|\eta_4/\eta_2| < 1$). One finds finally

$$\sum_\nu (m_\nu^2 + q^2)\, A_0^\nu + \left(\frac{\eta_3 - 8\eta_4}{2\eta_2}\right) \xi q^2\, m_\nu\, A_0^\nu = 0 \tag{A.6}$$

The condition A.4 is written

$$\sum_\nu \left\{ -i\omega\rho\frac{A_0^\nu}{m_\nu} + \eta_2\frac{m_\nu^2 + q^2}{m_\nu} A_0^\nu + \gamma q^2 \frac{A_0^\nu}{-i\omega} - \eta_4 \frac{(m_\nu^2 + q^2)}{m_\nu + \dfrac{1}{\xi}} A_0^\nu + \sum_{n=1}^{\infty}\left[\eta_2\left(m_\nu + \frac{n}{\xi}\right) - \eta_4 \frac{\left(m_\nu + \dfrac{n}{\xi}\right)^2}{\left(m_\nu + \dfrac{n+1}{\xi}\right)}\right] A_n^\nu \right\} = \tilde{p_{\text{ext}}}$$

We have

$$A_n^\nu = \frac{B_n^\nu}{\left(m_\nu + \dfrac{n}{\xi}\right)^2} = \left(\frac{\eta_4}{\eta_2}\right)^{n-1} \frac{B_1^\nu}{\left(m_\nu + \dfrac{n}{\xi}\right)^2}$$

To first order:

$$\sum_{n=1}^{\infty} \left[\eta_2\left(m_\nu + \frac{\eta}{\xi}\right) - \eta_4 \frac{\left(m_\nu + \dfrac{\eta}{\xi}\right)^2}{\left(m_\nu + \dfrac{n+1}{\xi}\right)} \right] \times \left(\frac{\eta_4}{\eta_2}\right)^{n-1} \frac{B_1^\nu}{\left(m_\nu + \dfrac{n}{\xi}\right)^2} = \eta_2 \, B_1^\nu \frac{1}{m_\nu + \dfrac{1}{\xi}}$$

To this order, one can replace B_1^ν by $\dfrac{\eta_4}{\eta_2}(m_\nu^2 + q^2) A_0^\nu$. One obtains finally

$$\sum_\nu \left\{ -i\omega\rho \frac{A_0^\nu}{m_\nu} + \eta_2 \frac{m_\nu^2 + q^2}{m_\nu} A_0^\nu + \gamma q^2 \frac{A_0^\nu}{-i\omega} \right\} = \tilde{p}_{\text{ext}} \tag{A.7}$$

The limit conditions A.6 and A.7 allow the calculation of the response function of the system to an external perturbation p_{ext} and finally of the spectrum. This problem is identical to the calculation of the spectrum for a liquid of viscosity η_2 covered by a monolayer (Chap. 11) on surface viscosity:

$$\eta_s = \left[\frac{\eta_3}{2} - 4(\eta_2 - \eta_1)\right]\xi$$

10
Polymer Solutions

DOMINIQUE LANGEVIN Laboratoire de Physique Statistique, Ecole Normale Supérieure, Paris, France

At present, there have been very few experiments in which the light scattered by the surface of polymer solutions has been studied. However, these experiments have demonstrated the interesting potential of the technique in the area. This chapter will then be very short and will briefly describe the existing experiments and their interpretation.

I. CELLULOSE SOLUTIONS

The first experiments were performed by McQueen and coworkers on aqueous solutions of carboxymethylcellulose in water [1, 2]. It was observed that when the polymer concentration increases, the spectrum of the light scattered by the solution broadens progressively; for the scattering wavevectors studied, $q = 353$ and $530\ \mathrm{cm}^{-1}$, above a concentration of about 2 wt%, the peak frequency becomes 0. This happens in principle when the reduced parameter $y = \gamma\rho/4\eta^2 q$ becomes of the order of 0.4. The surface tension of the solution does not depend appreciably on the polymer concentration c; it decreases by 1 mN/m for $c = 1\%$. The capillary waves damped regime should therefore be reached when the bulk viscosity attains a critical value η_c of about 30 times that of water. But when the bulk viscosity was measured (with a rotational viscosimeter), it was found that $\eta = \eta_c$ for $c \sim 0.5\%$, much smaller than the concentrations at which the peak disappears.

In the first paper [1], the spectral width measured for pure water was found to be too large, but the authors did not realize that this was due to the instrumental broadening (see Chap. 6). In order to explain this feature, they invoked the presence of a surface region of thickness $h \approx 10$ μm, which viscosity was four times the water bulk viscosity. Similarly, they explained the fact that the damping observed in their polymer solutions was smaller than expected by invoking the presence of a surface zone of viscosity different from the bulk polymer solution viscosity. The thickness h was of the order of 25–15 μm and depends upon the polymer concentration. In the second paper [2], they realized that the measured width in the case of water was larger than predicted simply because of instrumental broadening and that the existence of a very thick surface zone was unlikely. They also realized that the viscosity of polymer solution is frequency dependent, and that the inconsistencies observed when comparing the data to the theories were due to the fact that the viscosity probed during the propagation of capillary waves is smaller than the viscosity measured with mechanical devices; indeed, the capillary waves frequency (8–14 kHz for $c \sim 0.5\%$) is much larger than the mechanical frequency of the rotational viscometer (0.3–30 Hz).

In order to reinterpret their data, they used the Kirkwood-Auer model describing the frequency dependence of the viscosity of a solution of rigid rods. Cellulose molecules were thus regarded as stiff rods with a length L given by the degree of polymerization: $L \sim 200$ nm for the molecular weight of the studied polymer $M \sim 100{,}000$. The Kirkwood model predicts that the solution is viscoelastic, with a complex viscosity given by [3]

$$\eta^* = \eta_s \left\{ 1 + c\,[\eta]\,\frac{1 + i\omega\tau/4}{1 + i\omega\tau} \right\} \tag{1}$$

with $\tau = \pi\eta_s L^3/18\, k_B T \ln(L/b)$. In these expressions, η_s is the solvent viscosity, $[\eta]$ the zero frequency intrinsic viscosity, and b the monomer length. According to them, τ is predicted to be of the order of 10^{-4} s, i.e., such as $\omega\tau \sim 1$ in the surface light scattering experiments. This makes the technique particularly well suited to probe the viscoelastic behavior of these solutions. By using the theoretical spectrum of the light scattered by the free surface of a simple liquid (Chap. 4), but allowing the viscosity to depend on frequency as in Eq. 1, the authors found a good qualitative agreement between their data and the theory. The best fits occured for $\tau \sim 1\text{–}3 \times 10^{-4}$ s, and the effective viscosities were reduced by a factor of 4 with respect to the low frequency values. This explains why the probed capillary waves propagate until $c \sim 2\%$.

II. POLYSTYRENE SOLUTIONS

A more complete study was performed later by Meunier on ethylmethylcetone solutions of polystyrene ($M = 2 \times 10^{+6}$, $M_w/M_n < 1.2$) [4]. For polymer concentrations c of 1%, the viscosity showed no appreciable dependence on frequency, but for c larger than 2%, the dependence was large enough to be seen in the light scattering experiments. This is illustrated in Fig. 1, where the zero frequency value was measured with a capillary viscosimeter. Polystyrene is a flexible polymer, and the expression for the frequency dependence of the viscosity is not the same as for polyelectrolytes (Eq. 1). Meunier found that the introduction of one or two relaxation times was not sufficient to fit the data. He used a distribution of relaxation times such as

$$\eta = \eta_0 + A \sum_n \frac{1}{n^2\,(1 + i\omega\tau_n)} \qquad \text{with } \tau_n = \frac{\tau_1}{n^2} \tag{2}$$

In this model, the relaxation times decrease as $1/n^2$; this is the free draining limit in the Zimm theory [5]. However, the fit was not very sensitive to the power of n in τ_n (which is different in other flexible polymer theories). Toluene solutions of polystyrene were also studied, both at the free surface of the solution and at the interface with a second liquid [6].

A general treatment for surface modes on concentrated polymer solutions and gels has been published recently [7]. Several interesting regimes are predicted, including a propagating regime controlled by the elasticity of the solution $\omega \sim c_s\, q$ with $c_s = (E/\rho)^{1/2}$, E being the elastic modulus ($\eta^* = \eta + iE/\omega$). It will certainly be very interesting to probe this behavior in more detail. It must be noted that the experiments on gels are likely to be difficult to perform, because the surface needs to be extremely flat.

In conclusion, surface light scattering is a particularly well suited technique for the measurement of the frequency dependence of the bulk viscoelasticity of polymer solutions. It explores a frequency range (1–100 kHz) intermediate between the mechanical frequencies and the ultrasonic frequencies where other

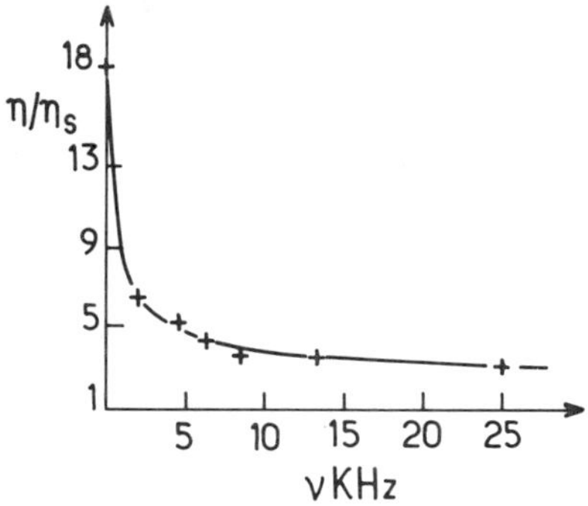

FIG. 1 Ratio between solution and solvent viscosities versus frequency for a 2.2 wt% polystyrene solution in ethylmethylcetone. The line is the fit with the theory.

devices are operating. The corresponding time scale (10^{-4}–10^{-6} s) frequently falls in the range of the relaxation times found in these systems.

REFERENCES

1. N. Hammarlund, L. Ilver, I. Lundström, and D. McQueen, *J. Chem. Soc, Faraday I 69*: 1023 (1973).
2. I. Lundström and D. McQueen, *J. Chem. Soc, Faraday I, 70*: 2351 (1974).
3. J. D. Ferry, *Viscoelastic Properties of Polymers*, Wiley, New York, 1970, Chap. 9.
4. D. Langevin and J. Meunier in *Photon Correlation Spectroscopy and Velocimetry* (H. Z. Cummins and E. R. Pike, eds.), Plenum, New York, 1977.
5. B. H. Zimm, *J. Chem. Phys. 24*: 256 (1956).
6. L. Léger, unpublished results.
7. J. L. Harden, H. Pleiner, P. A. Pincus, *Langmuir 5*: 1436 (1989).

11

Adsorbed Monolayers

DOMINIQUE LANGEVIN Laboratoire de Physique Statistique, Ecole Normale Supérieure, Paris, France

I. INTRODUCTION

When monolayers are present at a liquid-gas or a liquid-liquid interface, surface light scattering allows the determination not only of the surface tension but, in addition, of some rheological parameters characteristic of the monolayer [1]. The rheology of monolayers is an important subject that is thought to control many practical applications: foam and emulsion stability, Langmuir-Blodgett deposition, liquid-liquid extraction, two-phase flow in porous media, etc. On the other hand, it is very difficult to investigate the rheological behavior of monolayers without disturbing the system. Surface light scattering is therefore a particularly well suited method, because of the total absence of external perturbation. This explains why most of the experiments performed with the technique are done on monolayers.

Monolayer rheology can be characterized by various elastic and viscous parameters. Condensed monolayers frequently exhibit viscoelastic behavior at low frequency: both the elastic and the viscous parameters are frequency dependent. The frequency range of the capillary waves studied by light scattering is relatively limited because of the small q range available: typically between 3 and 50 kHz for the free surface of water. It is therefore interesting to combine light scattering with other techniques (Chap. 16) to enlarge the frequency range of the measurements.

II. VISCOELASTICITY OF MONOLAYERS

A. Elasticity

Monolayers exert a resistance against various types of surface deformation. This has been described in many ways. In the Kramer treatment [2], the monolayer is assumed to be isotropic in the surface plane. The elastic free energy density is then given by [2, 3]

$$f = \frac{1}{2} k_1 (u_{xx} + u_{yy})^2 + \frac{1}{2} k_2 u_{zz}^2 + k_3 (u_{xx} + u_{yy}) u_{zz} \\ + \frac{1}{2} k_4 [(u_{xx} - u_{yy})^2 + 4u_{xy}^2] + 2k_5 (u_{xz}^2 + u_{yz}^2) \tag{1}$$

where $\mathbf{u}$ is the surface displacement and $u_{ij} = (1/2)(\partial u_i/\partial x_j + \partial u_j/\partial x_i)$; the k_i are the elastic coefficients. The characteristic distances over which u varies are of the order of q^{-1}, i.e., much larger than the monolayer thickness h; therefore $u_{zz} \sim 0$. By integration of f over z, one obtains the elastic free energy per unit area:

$$\mathcal{F} = \frac{1}{2} \varepsilon (u_{xx} + u_{yy})^2 + 2s(u_{xy}^2 - u_{xx} u_{yy}) - 2\Pi(u_{xz}^2 + u_{yz}^2) \tag{2}$$

with $\varepsilon = (k_1 + k_4)h$, $s = k_4h$, $\Pi = -k_5h$. As will be seen later ε, s, and Π are positive coefficients.

The following will treat the case where $\mathbf{u}$ does not vary along y and is only a function of x. Then

$$\mathcal{F} = \frac{1}{2}\,\varepsilon\left(\frac{\partial u_x}{\partial x}\right)^2 + \frac{1}{2}\,s\left(\frac{\partial u_y}{\partial x}\right)^2 - \frac{1}{2}\,\Pi\left(\frac{\partial u_z}{\partial x}\right)^2 \tag{3}$$

The stress tensor is $\sigma_{ij} = \partial\mathcal{F}/\partial u_{ij}$, and the force per unit area is $F_i = \partial\sigma_{ij}/\partial x_j$. Its components are

$$F_x = \varepsilon\,\frac{\partial^2 u_x}{\partial x^2} \qquad F_y = s\,\frac{\partial^2 u_y}{\partial x^2} \qquad F_z = -\Pi\,\frac{\partial^2 u_z}{\partial x^2} \tag{4}$$

B. Physical Significance of the Elastic Moduli

When there is no monolayer at the surface of the liquid, there is already a force per unit area exerted on the surface, $F_z = \gamma_0\,\dfrac{\partial^2 u_z}{\partial x^2}$, where γ_0 is the "bare" surface tension. The total vertical force is then

$$F_z = (\gamma_0 - \Pi)\,\frac{\partial^2 u_z}{\partial x^2} = \gamma\frac{\partial^2 u_z}{\partial x^2} \tag{5}$$

The surface tension in the presence of the monolayer is decreased by Π: Π is called the surface pressure. Indeed, the free energy variation when the surface area varies by dA is $dF = \gamma\, dA = (\gamma_0 - \Pi)\, dA$. The contribution of the monolayer is $-\Pi\, dA$, analog to the $-p\, dV$ term in a bulk fluid, p being the pressure and dV the volume variation. Π is positive, because the surface energy must decrease during the monolayer adsorption (otherwise, there will be a desorption layer, as in salt solutions, where the ions are repelled by the surface).

Pure shear deformations are represented by traceless u_{ij} tensors [3]. Any deformation can be written as the sum of pure shear and pure compression terms; for a motion in the surface plane ($u_z = 0$):

$$u_{ij} = \frac{1}{2}(u_{xx} + u_{yy})\delta_{ij} + \left[u_{ij} - \frac{1}{2}(u_{xx} + u_{yy})\delta_{ij}\right]$$

$\mathcal{F}$ can be rewritten as

$$\mathcal{F} = \frac{1}{2}(\varepsilon - s)(u_{xx} + u_{yy})^2 + s\sum_{ij}\left[u_{ij} - \frac{1}{2}(u_{xx} + u_{yy})\delta_{ij}\right]^2$$

where it is seen that the compressional modulus is $\varepsilon - s$, and the shear modulus is s. The modulus acting in the horizontal force parallel to Ox is then the sum of the pure compression and pure shear moduli: this is because uniaxial compression always involves shear. The shear modulus is nonzero only in solid monolayers [1]. This is not a frequent case, and in practice $s = 0$ most of the time. The above distinction between uniaxial and pure compression moduli is then not necessary: both moduli are the same.

By analogy with the three-dimensional case, the compressional modulus can be written as

$$\varepsilon = -A \frac{\partial \Pi}{\partial \mathrm{A}} \tag{6}$$

When the molecules cannot exchange between the monolayer and the bulk liquid, the modulus can be also calculated in the following way [4]; the horizontal force F_x is a Marangoni force $F_x = \partial\gamma/\partial x = -\partial\Pi/\partial x$; this force is similar to the bulk force in a medium where the pressure is not uniform, $\mathbf{F} = -\mathbf{grad}\, p$. The surface force F_x is directed towards the high surface tension regions: this shows again that the surface pressure has to be positive, otherwise the monolayer would not spread spontaneously. The surface concentration Γ depends here on x as all the other surface variables: $F_x = \partial\gamma/\partial x = (\partial\gamma/\partial\Gamma)\,(\partial\Gamma/\partial x)$. Since the monolayer is insoluble, the conservation of matter in the surface gives

$$\frac{\partial \Gamma}{\partial t} + \Gamma \frac{\partial \mathrm{v}_x}{\partial x} = 0 \tag{7}$$

with $\mathrm{v}_x = \partial u_x/\partial t$; differentiating with respect to x and integrating with respect to time, we obtain

$$\frac{\partial \Gamma}{\partial x} = -\Gamma \frac{\partial^2 u_x}{\partial x^2} \qquad \text{and} \qquad F_x = -\Gamma \frac{\partial \gamma}{\partial \Gamma} \frac{\partial^2 u_x}{\partial x^2}$$

The compressional modulus is therefore

$$\varepsilon = -\Gamma \frac{\partial \gamma}{\partial \Gamma} = \Gamma \frac{\partial \Pi}{\partial \Gamma} \tag{8}$$

For an insoluble monolayer, $\Gamma \sim A^{-1}$, and Eqs. 6 and 8 are equivalent.

When the monolayer is soluble, i.e., when it can exchange matter between surface and bulk, Eq. 8 can be used only when the rate of compression is faster than the rate of dissolution. When this is not the case, models have been proposed to relate ε and $\varepsilon_o = \Gamma\partial\Pi/\partial\Gamma$: ε_o is the limit value for infinitely slow dissolution rates. The model proposed by Lucassen [5] assumes that desorption and adsorption proceed freely, without energy barriers. Equation 7 can then be easily generalized to

$$\frac{\partial \Gamma}{\partial t} + \Gamma \frac{\partial \mathrm{v}_x}{\partial x} = -D\left(\frac{\partial c}{\partial z}\right)_{z=0} \tag{9}$$

where c is the bulk concentration and D the diffusion coefficient. The conservation equation for c is the usual diffusion equation

$$\frac{\partial c}{\partial t} + D\frac{\partial^2 c}{\partial z^2} = 0 \tag{10}$$

Integration of Eqs. 9 and 10, for solutions of the type $e^{i\omega t}$ and for linear variations $\delta c(z = 0) = (dc/d\Gamma)\delta\Gamma$, leads to $F_x = \varepsilon\, \partial^2 u_x/\partial x^2$ with

$$\varepsilon = \Gamma\frac{d\Pi}{d\Gamma}\frac{1}{1 + (1 - i)\Omega} = \varepsilon_o\frac{1}{1 + (1 - i)\Omega} \tag{11}$$

where Ω is the reduced frequency

$$\Omega = \sqrt{\frac{D}{2\omega}}\frac{dc}{d\Gamma} \tag{12}$$

Equation 11 can be split into real and imaginary parts:

$$\varepsilon_r = \varepsilon_o\frac{1 + \Omega}{1 + 2\Omega + 2\Omega^2} \tag{13}$$

$$\varepsilon_i = \varepsilon_o\frac{\Omega}{1 + 2\Omega + 2\Omega^2} \tag{14}$$

When $\omega \to \infty$, $\Omega \to 0$, $\varepsilon_r \to \varepsilon_o$, and $\varepsilon_i \to 0$: the monolayer behaves as if it were insoluble as expected. When $\omega \to 0$, $\Omega \to \infty$, both ε_r and ε_i go to zero: this is because if the rate of compression is slow compared to the rate of dissolution, there is no resistance to compression: the equilibrium between bulk and surface concentration has time to establish during the compression. When ω takes intermediate values, $\varepsilon_r < \varepsilon_o$ and $\varepsilon_i \neq 0$. The meaning of ε_i will be discussed hereafter, in relation to surface viscosity.

C. Surface Viscosity

In classical viscoelasticity theory, when the deformations are sinusoidal in time with a frequency ω, the response of the system, i.e., the stress-deformation relation, can be described by complex moduli: $\tilde{K} = K + i\omega\eta$, where K is the elastic modulus and η the viscosity [3]. In an anisotropic system, there are many elastic and viscous coefficients, and the formalism of the dissipation function can be used conveniently. The dissipation function Ψ has the same general form as the elastic energy. In the problem discussed here,

$$\Psi = \frac{1}{2}\kappa\left(\frac{\partial v_x}{\partial x}\right)^2 + \frac{1}{2}\eta_s\left(\frac{\partial v_y}{\partial x}\right)^2 + \frac{1}{2}\mu\left(\frac{\partial v_z}{\partial x}\right)^2 \tag{15}$$

κ, η_S, and μ are respectively compressional (uniaxial), shear, and transverse shear surface viscosities. As before, κ is the sum of the pure compressional and of the shear surface viscosities. Contrary to the shear modulus s that is zero except for solid monolayers, the surface shear viscosity is nonzero for fluid monolayers; the distinction between pure compressional and uniaxial compression viscosities is always relevant.

The force per unit area is now the sum of an elastic and a viscous term:

$$F_i = \frac{\partial}{\partial x_j}\left\{\frac{\partial \mathcal{F}}{\partial u_{ij}} + \frac{\partial \Psi}{\partial v_{ij}}\right\} \tag{16}$$

For deformations of the type $e^{i\omega t}$, the force components are still given by Eq. 4, provided ε, s, and Π are replaced by $\varepsilon + i\omega\kappa$, $s + i\omega\eta_S$, $\Pi + i\omega\mu$.

The imaginary part of ε found in Sec. B, Eq. 14 for soluble monolayers can now be interpreted as a contribution of the adsorption-dissolution process to the compressional viscosity: $\varepsilon_i = \omega\kappa$.

It might seem surprising to obtain a complex surface pressure. The surface tension is indeed an equilibrium property and cannot be treated as a response function. The physical meaning of μ, found here from pure symmetry considerations, should lie in the fact that an independent hydrodynamic variable was hidden in the above treatment; the equation of evolution of such a variable can perfectly produce terms of the type $i\omega\mu$ to be added to the surface tension term. Physically, μ is related to relaxation processes associated with the vertical displacement of the surface. One possible mechanism is the following (Fig. 1): just after the distortion, the angle of the molecules with the surface has changed slightly; when their equilibrium position is perpendicular to the surface as in Fig. 1, they become slightly tilted and will turn back to their equilibrium position after a time τ; τ is very fast in general, and the corresponding viscous loss is probably observable only in condensed monolayers ($\omega\tau \sim 1$). This is as observed experimentally (Sec. IV). The tilt angles are very small, and the viscosity μ is also expected to be very small. Let us show below a very simple model to estimate μ. The surface tension will be assumed to depend on the tilt angle θ

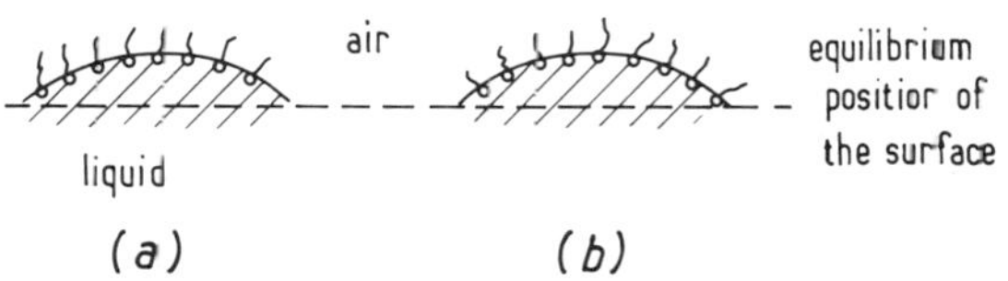

FIG. 1 Possible mechanism giving rise to a shear transverse viscosity; position of the surfactant chains (a) just after the surface displacement, (b) after chain reorientation.

through: $\gamma = \gamma_0 + \gamma_1 \theta^2$, in order to be minimum for $\theta = 0$. If the surface is distorted at $t = 0$, $\theta(t = 0) = \theta_0$ and $\theta^2 = \theta_0^2 e^{-t/\tau}$: the surface tension is then a function of time. After a Fourier transform, the corresponding viscosity at the frequency ω is

$$\mu = \frac{\tau}{1 + \omega^2 \tau^2} \gamma_1 \theta_0^2$$

Here $\theta_0^2 = (\partial\zeta/\partial x)^2 + (\partial\zeta/\partial y)^2$ and $\zeta = \sum_q \zeta_q e^{i\mathbf{q}\cdot\mathbf{r}}$. Keeping in mind that ζ is a gaussian statistical variable, the mean surface free energy can be written as

$$F = \frac{1}{2} \gamma_0 \sum_q q^2 \langle|\zeta_q|^2\rangle + \gamma_1 \left(\sum_q q^2 \langle|\zeta_q|^2\rangle \right)^2$$

$$= \frac{1}{2} (\gamma_0 + 2\gamma_1 \sum_q q^2 \langle|\zeta_q|^2\rangle) \sum_q q^2 \langle|\zeta_q|^2\rangle$$

θ_0^2 can then be identified with twice the mean squared interfacial slope $\sum_q q^2\langle|\zeta_q|^2\rangle$.

$$\theta_0^2 = \frac{2}{(2\pi)^2} \int_0^{q_m} \frac{q^2 k_B T}{\gamma q^2 + \rho g} 2\pi q \, dq$$

For monolayers, the upper cutoff q_m is related to the bending elastic modulus K of this monolayer (see Chap. 17): $q_m = \sqrt{\gamma/K}$. After integration, one finds $\theta_0^2 = k_B T/2\pi K$, and

$$\mu = \frac{\tau}{1 + \omega^2 \tau^2} \gamma_1 \frac{k_B T}{2\pi K} \qquad (17)$$

Let us take $\tau \sim 10^{-4}$ s, $\omega \sim 10^4$ s^{-1}, $\gamma_1 \sim 100$ mN/m, K $\sim 10 k_B T$. This leads to $\mu \sim 10^{-4}$ gs^{-1}, in agreement with values measured in condensed monolayers (see Sec. IV).

The expression found for the force per unit area in the Kramer treatment can be derived more rigorously by using the formalism or surface excess properties [6] or with molecular hydrodynamic treatments [7]. The results are identical to those obtained with the simple treatment. Let us note that isotropic bulk systems are characterized by only two types of viscoelastic parameters: compression and shear. Because of the uniaxial symmetry, there is an additional surface viscosity in monolayers, associated with transverse shear. The compression viscoelasticity of bulk fluids does not play a role at frequencies smaller than sound frequencies, i.e., in most practical cases. In monolayers, the sound (compression wave) frequencies are much smaller, i.e., these media are more compressible, and the compressional viscoelasticity plays an important role.

III. THEORETICAL SPECTRUM

A. Surface Modes

Let us consider a surface deformation mode of wavevector **q** parallel to Ox:

$$\mathbf{u}(x, y, z, t) = \mathbf{u}(t)\, e^{iqx} \tag{18}$$

The total surface energy is

$$\mathcal{F} = \frac{1}{2}\, \varepsilon\, q^2\, u_x^2 + \frac{1}{2}\, s\, q^2\, u_y^2 + \frac{1}{2}\, (\gamma q^2 + \rho g)\, u_z^2$$

The deformations u_x, u_y, u_z are statistically independent and their mean square values are

$$\langle u_x^2 \rangle = \frac{k_B T}{\varepsilon q^2} \qquad \langle u_y^2 \rangle = \frac{k_B T}{s q^2} \qquad \langle u_z^2 \rangle = \frac{k_B T}{\gamma q^2 + \rho g} \tag{19}$$

Among the three types of deformation, only u_x and u_z are associated with local density variations and scatter light due to the corresponding refractive index variations. The scattered intensities were calculated in Chap 3; it was shown in particular that only capillary waves (u_z) scatter an appreciable amount of light. The two-dimensional sound or compression waves (u_x) scatter an intensity lower by a factor of$(qh)^2$. The two-dimensional shear waves (u_y) do not produce local density variations and do not scatter light at all. They have to be detected by other than optical means [8].

The theoretical spectra for both capillary and compression waves have been derived by Bouchiat and Langevin [9] in the case of a monolayer at a liquid-gas interface. The spectrum in the more general case of a monolayer at the interface between two liquids was calculated independently by Loudon [10], Langevin et al. [11], and Hajiloo and Slattery [12]. In the calculation of Hajiloo and Slattery, the shear elastic modulus is assumed to be zero and the uniaxial compressional modulus is therefore identified with the pure compressional modulus. The case of a liquid crystalline monolayer was treated by C. Fan [13] and involves additional surface modes; but this special case will not be discussed here, in view of the lack of experimental data. It must be noted that in his paper, Kramer proposes also expressions for the theoretical spectra [2]. However, these expressions are not correct, because of an improper statistical treatment.

B. Dispersion Equation

Let us take the case of a monolayer at a liquid-liquid interface. The bulk hydrodynamic equations and their solutions will be the same as in Chap. 4 and Chap. 7, Sec. V. The limit conditions for the velocity will also be the same, but additional surface forces must be considered (Eq. 4). These limit equations, in the presence of an external force $\mathbf{F}_{\text{ext}}\, e^{-i\omega t}$ are

$$A_1 + B_1 = A_2 + B_2$$

$$qA_1 + m_1B_1 = -qA_2 - m_2B_2$$

$$\eta_1\{2q^2A_1 + (m_1^2 + q^2)B_1\} - \eta_2\{2q^2A_2 + (m_2^2 + q^2)B_2\} - \frac{\tilde{\varepsilon}}{i\omega} q^2 (qA_1 + mB_1) = F_x^{\text{ext}}$$

$$2\eta_1(qA_1 + m_1B_1) - i\frac{\omega}{q}\rho_1 A_1 + 2\eta_2(qA_2 + m_2B_2) - i\frac{\omega}{q}\rho_2 A_2 - \frac{\tilde{\gamma}}{i\omega} q^2 (A_1 + B_1) = F_z^{\text{ext}}$$

where $\tilde{\varepsilon} = \varepsilon + i\omega\kappa$, and $\tilde{\gamma} = \gamma + i\omega\mu$. The dispersion is obtained by setting the external force equal to zero:

$$\Delta = q(m_1 - q)(m_2 - q) \Big\{(\varepsilon\tilde{q}^2 - i\omega(\eta_1 m_1 + \eta_1 q + \eta_2 m_2 + \eta_2 q)) \left(\tilde{\gamma}q^2 - i\omega(\eta_1 m_1 + \eta_1 q + \eta_2 m_2 + \eta_2 q) - \frac{\omega^2}{q}(\rho_1 + \rho_2)\right) - (i\omega(\eta_2 m_2 - \eta_2 q - \eta_1 m_1 + \eta_1 q))^2 \Big\} = 0 \tag{20}$$

As is the case of simple liquids, the motion along y evolves independently from the motion along x and z. Shear waves are then completely decoupled from capillary and compression waves: both their mean square amplitude and their time evolution are independent. Although the displacements associated with capillary and compression waves are statistically independent (Eq. 19), their evolution is coupled: even if $u_x\ (t = 0) = 0$ when a capillary wave is excited: $u_z\ (t = 0) \neq 0$, u_x becomes nonzero at later times. The coupling also appears in the dispersion equation, which is common to the two types of waves. This coupling disappears when $\eta_1 = \eta_2 = \eta$ and $\rho_1 = \rho_2 = \rho$ ($m_1 = m_2 = m$), i.e., when the monolayers sit at the interface between identical fluids. In practice this is never the case, but when the two fluids are for instance water and an alkane, the decoupling is already significant [5]. The case of perfect decoupling arises for bilayers immersed in water, where the above theoretical description also applies (the only assumption done in relation to the layer thickness was

$qh \ll 1$). In this case, the dispersion equation splits into different pieces: $m_1 = m_2 = q$, which are nonphysical solutions, and

$$\tilde{\varepsilon} q^2 - 2i\omega\eta\,(m + q) = 0 \tag{21}$$

$$\tilde{\gamma} q^3 - 2i\omega\eta\,(m + q)\,q - 2\omega^2\rho = 0 \tag{22}$$

The first is the dispersion equation for pure compression waves, the second for pure capillary waves.

In the case of a liquid-gas-interface, $\eta_1 \ll \eta_2 = \eta$, $\rho_1 \ll \rho_2 = \rho$, and the dispersion equation becomes

$$\Delta = [\tilde{\varepsilon} q^2 - i\omega\,(m + q)]\,[\tilde{\gamma} q^2 - i\omega\eta\,(m + q) - \omega^2\rho/q] + \omega^2\eta^2\,(m - q)^2 = 0 \tag{23}$$

The coupling between capillary and compression waves is the largest. This points out that if capillary waves are used to measure compressional viscoelasticity, the accuracy will be better for monolayers at liquid-gas interfaces, becoming very poor for monolayers at oil-water interfaces. This is illustrated in Fig. 2, where capillary waves electrically induced have been used to probe compressional viscoelasticity [5]. The damping of waves is maximum at a surfactant concentration for which the frequency of the capillary and compression waves are about equal; this effect is not even visible at the alkane-water interface.

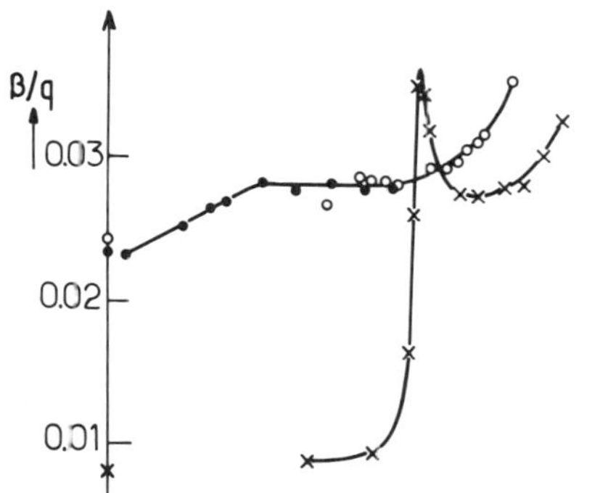

FIG. 2 Relative spatial damping of capillary waves for dodecylammonium chloride aqueous solutions versus surfactant concentration: crosses, water free surface; circles, heptane-water interface. (Data from Ref. 5.)

C. Spectrum

Two response functions can be calculated here: $\chi_z = u_z/F_z^{ext}$ and $\chi_x = u_x/F_x^{ext}$, related to the vertical and horizontal displacement respectively. The fluctuation-dissipation theorem allows the calculation of the corresponding spectra $P_i = k_BT/\pi\omega \text{ Im } \chi_i$. One obtains

$$P_z(q, \omega) = \frac{k_BT}{\pi\omega} \text{Im} \frac{-i\omega(\eta_1(m_1 + q) + \eta_2(m_2 + q)) + \tilde{\varepsilon}q^2}{\Delta'} \tag{24}$$

where $\Delta' = \Delta/q(m_1 - q)(m_2 - q)$ and

$$P_x(q, \omega) = \frac{k_BT}{\pi\omega} \text{Im} \frac{\tilde{\gamma}q^2 - i\omega(\eta_1 m_1 + \eta_1 q + \eta_2 m_2 + \eta_2 q) - \omega^2(\rho_1 + \rho_2)/q}{\Delta'} \tag{25}$$

In the case of a free surface, these expressions simplify. By introducing the reduced parameters

$$S = -i\omega\,\tau_0, \quad \tau_0 = \frac{\rho}{2\eta q^2}, \quad y = \frac{\gamma\rho}{4\eta^2 q}, \quad \alpha = \frac{\varepsilon}{\gamma}, \quad \beta = \frac{\kappa q}{2\eta}, \quad \delta = \frac{\mu q}{2\eta}$$

one obtains

$$P_z(q, \omega) = \frac{k_BT}{\pi\omega}\frac{\tau_0^2 q}{\rho} \text{Im}\left\{\frac{S^2 + (\alpha y + \beta S)[\sqrt[+]{1 + 2S} - 1]}{\Delta(S)}\right\} \tag{26}$$

and

$$P_x(q, \omega) = \frac{k_BT}{\pi\omega}\frac{\tau_0^2 q}{\rho} \text{Im}\left\{\frac{S\sqrt[+]{1 + 2S} + (y + S)\sqrt[+]{1 + 2S - 1}}{\Delta(S)}\right\} \tag{27}$$

The dispersion equation is $\Delta(S) = 0$ with

$$\Delta(S) = S^2[(1 + S)^2 + y - \sqrt[+]{1 + 2S}] + (\alpha y + \beta S) [S^2\sqrt[+]{1 + 2S} + (y + \gamma S)\sqrt[+]{1 + 2S} - 1] + \delta S^3 \tag{28}$$

This was the result derived in Ref. 9 with the Laplace transform method.

In the case of an interface between two liquids of the same density and

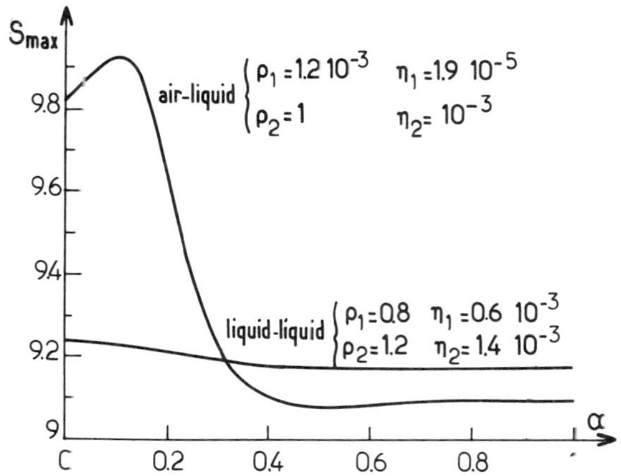

FIG. 3 Reduced peak frequency of the spectrum versus reduced compressional elasticity for the free surface of water and for an oil-water interface. Densities are relative to water. Viscosities are in kg m^{-1} s^{-1}.

viscosity $\rho_1 = \rho_2 = \rho$, $\eta_1 = \eta_2 = \eta$:

$$P_z(q,\omega) = \frac{k_B T}{\pi\omega} \operatorname{Im}\left\{\frac{1}{-2i\omega\eta(m + q) - 2\omega^2 \rho/q + \tilde{\gamma} q^2}\right\} \tag{29}$$

$$P_x(q, \omega) = \frac{k_B T}{\pi\omega} \operatorname{Im}\left\{\frac{1}{\tilde{\varepsilon} q^2 - 2i\omega\eta(m + q)}\right\} \tag{30}$$

P_z depends only on $\tilde{\gamma}$ and P_x on $\tilde{\varepsilon}$: capillary and compression waves are entirely decoupled.

In the intermediate case of an interface between two liquids, the decoupling is important. This is illustrated by Fig. 3 showing the variation of the reduced peak frequency S_{max} of the spectrum P_z with reduced elastic modulus in the cases of a monolayer at the free surface of water and at a liquid-liquid interface. S_{max} varies by about 10% when ε goes from zero to infinity in the free surface case. It varies by less than 1% at the liquid-liquid interface. The maximum of S_{max} reflects the resonance phenomenon: the frequencies of capillary and compression waves are about equal.

D. Special Case of Monolayers on Water

From the above discussion, it follows that if surface compressional viscoelasticity needs to be measured, the surface light scattering method is much better suited to the study of monolayers on water. Capillary waves on monolayers at oil-water interfaces or bilayers in water are barely sensitive to this particular viscoelastic behavior. Only the transverse shear viscosity μ can be measured. This was indeed achieved for lipid bilayers (Chap. 12).

The spectra for monolayers on water are given by Eqs. 26–28. The y parameter is large and P_z is approximately Lorentzian. Figure 4 shows the shape of P_z and P_x for $y = 400$ (free surface of pure water for $q \sim 450\ \text{cm}^{-1}$) and $\alpha = 1$ ($\varepsilon = \gamma$). P_x is very broad (the compression waves are strongly damped) and has a dispersion-shaped variation around the maximum value of P_z: although for $\alpha = 1$, one is far from resonance, the coupling between capillary and compression waves still affects P_x.

Very roughly, assuming $y \gg 1$, $\alpha \geqslant 1$, $\beta \sim \delta \sim 0$ and dropping smaller order terms, $\Delta \sim (S^2 + y)(S^2 + \alpha y \sqrt{2S})$; Δ is about zero for $S^2 \sim -y$ and for $S^{3/2} \sim \sqrt{2}\, \alpha y$, the roots corresponding respectively to capillary and longitudinal waves. The frequency of propagation and damping of compression waves are such that $S' \sim \sqrt[3]{2\alpha^2 y^2}\, e^{\pm 2i\pi/3}$: the real and imaginary parts are comparable, meaning that the waves are strongly damped, as evidenced in Fig. 4. The root $S = i\sqrt{y}$ is also approximately a root of N_x such as $P_x \sim \text{Im}(N_x/\Delta)$, and the root S' of N_z such as $P_z \sim \text{Im}(N_z/\Delta)$. These spectra are therefore approximately maximum for the frequency of propagation of the surface wave to which they correspond. For $y = 400$, $S_i' \sim 68$ and $S_r' \sim 34$, in good qualitative agreement

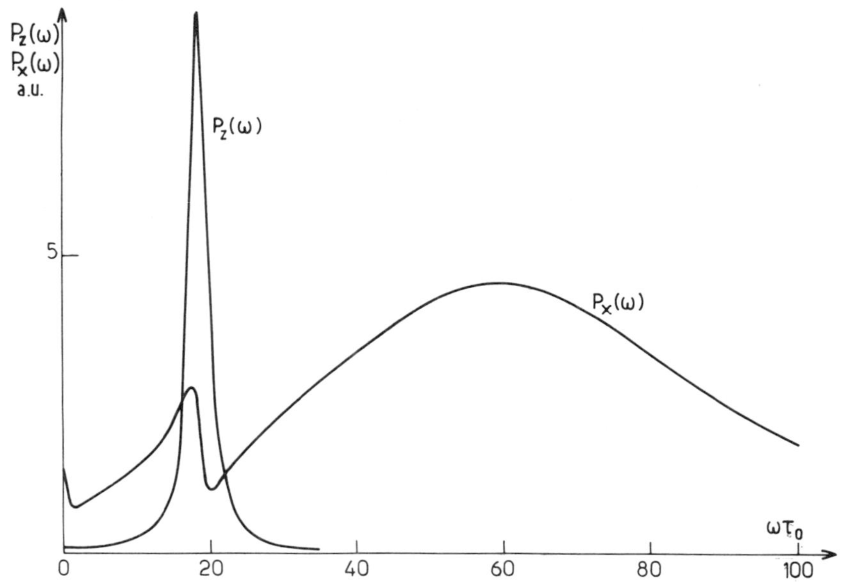

FIG. 4 Spectrum of the light scattered by thermally excited capillary and compression waves; $y = 400$; $\alpha = \varepsilon/\gamma = 1$.

with the aspect of P_x in Fig. 4. The anomaly observed around $|S| \sim \sqrt{y}$ comes probably from the fact that the roots of N_x and Δ close to $i\sqrt{y}$ are not exactly the same. Unfortunately P_x is hardly observable experimentally (Chap. 3).

Let us now focus on P_z. For pure water, better approximations to the solution of the dispersion equation $\Delta = y + (1 + S)^2 - \sqrt[+]{1 + 2S} = 0$ can be found easily:

$$S_i = \sqrt{y} - \frac{1}{2y^{1/4}} \tag{31}$$

$$S_r = 1 - \frac{1}{2y^{1/4}} \tag{32}$$

The two first terms correspond to the limit values for $y \gg 1$:

$$\omega_{q0} = \sqrt{\frac{\gamma q^3}{\rho}} \qquad \text{and} \qquad \Delta\omega_{q0} = \frac{2\eta q^2}{\rho}$$

as discussed earlier (Chaps. 2 and 4). For $y = 400$, one finds

$$\omega_q = 0.995\ \omega_{q0} \qquad \text{and} \qquad \Delta\omega_q = 0.89\ \Delta\omega_{q0}$$

The values of the peak frequency ω_m and of the half-width $\Delta\omega$ of the theoretical spectra are in excellent agreement with the above values. This agreement is less good for smaller y values (larger q). For $y = 100$ ($q = 1800\ \text{cm}^{-1}$ for pure water), for instance, the values calculated from Eqs. 31 and 32 and the values corresponding to the peak and width of the spectrum are respectively

$$\omega_q = 0.984\ \omega_{q0} \qquad \Delta\omega_q = 0.842\ \Delta\omega_{q0}$$

$$\omega_m = 0.982\ \omega_{q0} \qquad \Delta\omega = 0.840\ \Delta\omega_{q0}$$

Eqs. 31 and 32 are very good approximations of the solutions of the dispersion equation. The difference obtained between these values, and those deduced from spectral shapes, points out the importance of using the exact spectrum when analyzing the data. When the dispersion equation is used, errors of about 0.2% on frequencies and 0.2% on widths are made if $y \sim 100$. This lead to errors of 0.1% on surface tension and 0.2% on viscosities. These errors are small, but they become appreciable when monolayers are present: up to 1–2% for frequency and 10–20% for widths around $y \approx 100$. These deviations are similar to those produced by the viscoelastic parameters: large errors on these parameters can therefore be made if the dispersion equation is used instead of the complete spectrum.

The effect of the viscoelastic parameters is shown in Fig. 5. The compressional elasticity ε produces effects similar to the compressional viscosity κ: frequency shift and broadening. These modifications are plotted in reduced units versus

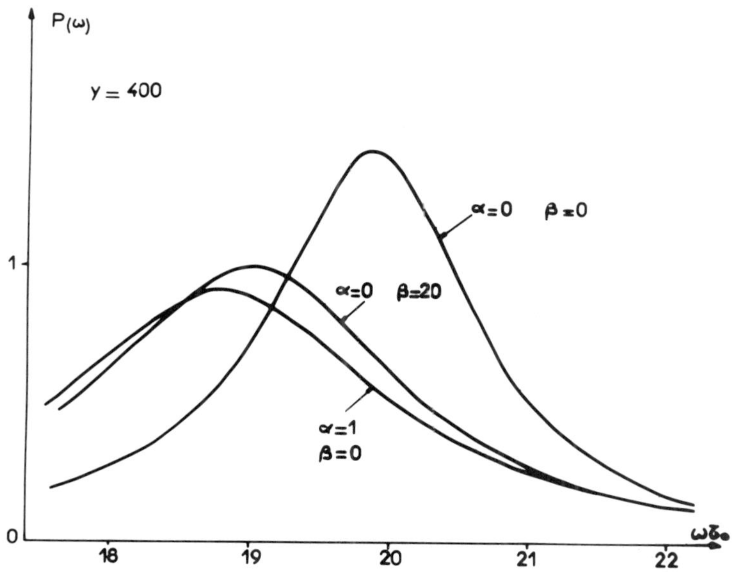

FIG. 5 Influence of the compressional elasticity and viscosity on the spectrum.

reduced elasticity for different reduced compressional viscosities in Fig. 6 and for transverse shear viscosities in Fig. 7. A maximum of the peak frequency and of the width are observed for $\alpha \sim 0.1$–0.2 value for which capillary and compression wave frequencies are about equal ($\sqrt{y} \sim \sqrt[3]{2\alpha^2 y^2} \sin 2\pi/3$ and $y \sim 400$ lead indeed to $\alpha \sim 0.2$). For large α ($\alpha > 0.5$) or large β ($\beta > 10$) values, both $S_{\max}$ and ΔS become independent of α and β. The physical origin of this effect is the following[4]: the limit condition relative to horizontal surface force is

$$\eta[2q^2 A + (m^2 + q^2) B] - \frac{\tilde{\varepsilon}}{i\omega} q^2 (qA + mB) = 0$$

When α or β is large, $\tilde{\varepsilon}$ is large and the limit condition becomes $qA + mB = 0$; this sum is proportional to v_x ($z = 0$) ($\operatorname{div} \mathbf{v} = 0$ leads to $v_x = \frac{i}{q}\frac{\partial v_z}{\partial z} = \frac{i}{q}(qAe^{qz} + mBe^{mz})e^{iqx}$. Then when $\tilde{\varepsilon}$ is large, the horizontal component of the fluid velocity vanishes. The monolayer behaves as if it were incompressible: when the surface is distorted, the monolayer is simply bent like a solid plate, and the frequency and damping of the capillary waves do not depend any more on the values of α and β. This is a common result for monolayers; the same phenomenon

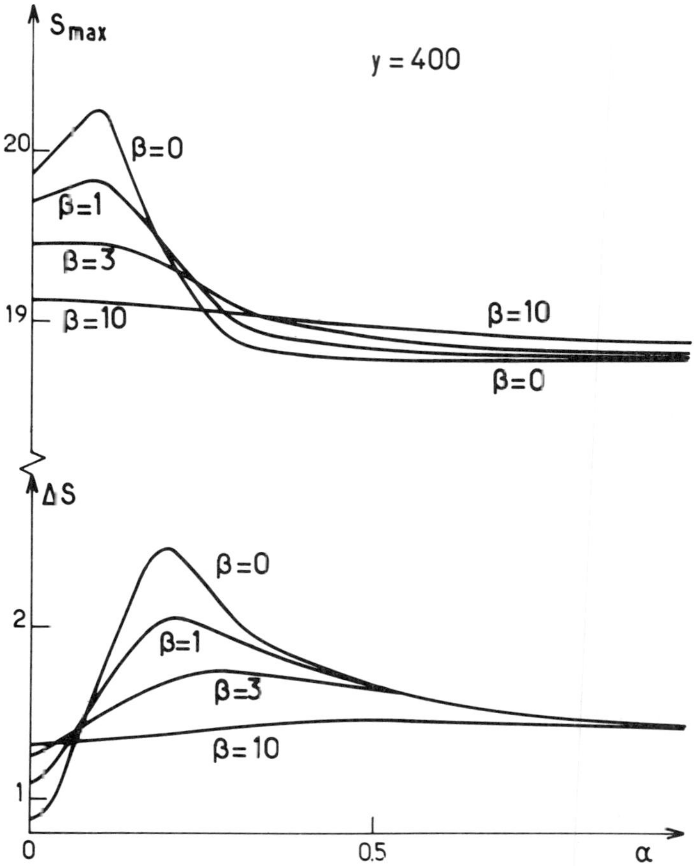

FIG. 6 Reduced peak frequency and half-width versus reduced compressional elasticity for different reduced compressional viscosities.

is encountered for instance in the flow of gas bubbles in liquids, that behave like solid spheres when dense monolayers are adsorbed at their surface.

This discussion shows that in order to determine ε and κ from the capillary wave propagation characteristics, one should not be too far from the maximum coupling conditions between capillary and compression waves: $\alpha < 0.5$ and $\beta < 10$. When the compressional elasticity and viscosity are too large, the capillary waves are no longer affected by these parameters.

Fig. 7 shows that this is not the case for the shear transverse viscosity that affects more directly the capillary wave propagation. There is no upper limit of the width increase and frequency decrease in this case. When δ is large, the

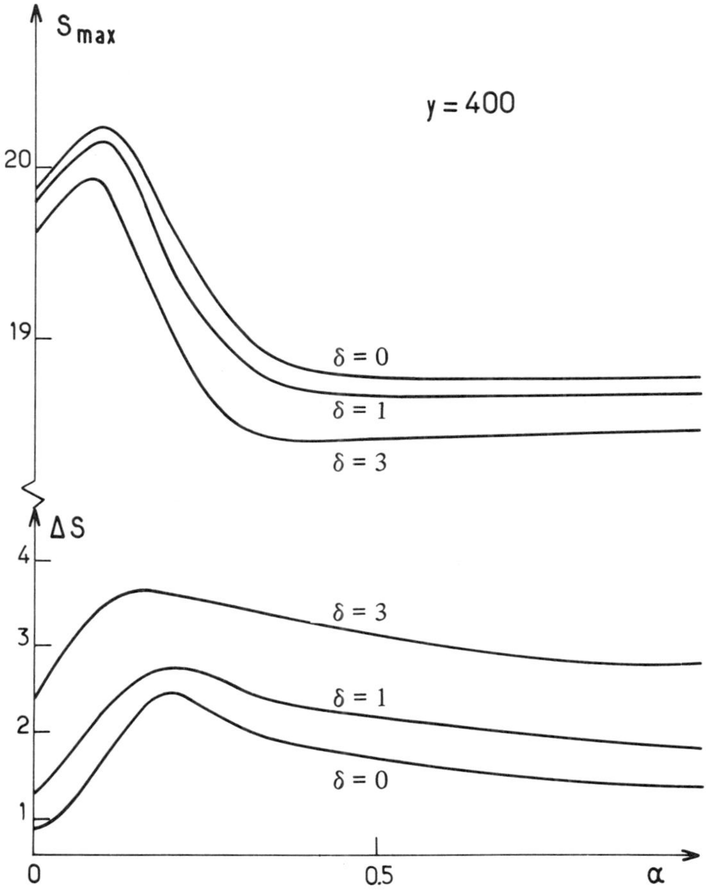

FIG. 7 Reduced peak frequency and half-width versus reduced compressional elasticity for different reduced transverse shear viscosities.

capillary waves can even become completely damped. This has never been observed experimentally, however; μ is never much larger than 10^{-4} gs^{-1} ($\delta \sim 1$).

IV. EXPERIMENTAL RESULTS

A. Monolayer Preparation

The free surface of water is rapidly contaminated by surface active impurities if stringent precautions are not taken. Pure water must be used, either at least doubly distilled or ultrapure water from special filtration systems (Elga, Millipore). The water surface must be cleaned in the liquid container prior to the experiments. Convenient devices are rectangular Teflon (or Teflon-coated) troughs, where the liquid surface can be swept with Teflon barriers: the impurities are compressed and sucked away with a clean glass pipette; when this procedure is repeated a few times, the surface tension does not change measurably upon further compressions. It is essential to use a device like a Wilhelmy plate to measure the surface tension in situ during the experiments. Possible contaminations or leakage can be rapidly detected. Furthermore, it is very useful to know the static value of the surface tension to analyze the scattering data. The container must be enclosed in a Plexiglas-type box to protect the liquid against dust and to control the temperature and the humidity. The apertures left for the laser beams on the box and on the trough must be filled by optical glass windows. The trough and the glassware should be cleaned thoroughly with chromic acid prior to the experiments.

Insoluble monolayers are generally spread from solvent solutions. The solvent should be purified prior to use: the purity could be checked by spreading the pure solvent on water and measuring the surface tension after evaporation. High purity surface active substances should also be used. The surface pressure is extremely sensitive to contamination by trace impurities. For instance, the surface pressure curves versus surface area at constant temperature called surface isotherms exhibit linear parts in the region where different two-dimensional phases coexist in the monolayer: gas-liquid expanded, liquid expanded-liquid condensed [1]. These linear parts are usually not horizontal as in three-dimensional systems, except in extremely pure compounds (>99.95% for pentadecanoic acid monolayers [14]).

B. Data Analysis

The theoretical spectrum depends on four parameters: γ, μ, ε, and κ. It is approximately Lorentzian, i.e., only two parameters are available from the experimental side, ω_m and $\Delta\omega$. Furthermore, the capillary wave regime is weakly damped and the instrumental broadening corrections are important. Along with the contamination problems, this explains why the data reported in the literature are not always in agreement. In practice, if the instrumental broadening is properly corrected, the small deviations between the theoretical spectrum and Lorentzian shapes lead to sufficient information to deduce the four parameters γ, μ, ε, and κ [15]. It is not always necessary to use the corresponding complex data fitting procedure. In a first approximation, γ must be assumed to be equal to the static value γ_{st} as measured with the Wilhelmy plate; as a consequence, the transversal shear viscosity μ should be zero. ε and κ can then be more easily determined. For this purpose, diagrams giving the variation of ω_m and $\Delta\omega$ with

ε and κ such as those of Fig. 6 can be used. This is somewhat tedious, because these diagrams have to be constructed for each wavevector studied, and in addition they change slightly with surface pressure. When $\varepsilon/\gamma > 0.5$ or $\kappa q/2\eta > 10$, S_m and ΔS do not depend appreciably on ε and κ any more. The assumption $\gamma = \gamma_{st}$ and $\mu = 0$ can be put aside, and these last two quantities can be determined. The four parameters are subject to constraints: if the monolayer is insoluble, $\varepsilon > \varepsilon_{st} = d\gamma/d \ln A$. Π, ε, κ, and μ should be continuously increasing functions of the surface concentration except in regions where the monolayer is inhomogeneous (2-phase regions) and close to relaxation frequencies.

In most of the early published works, the data analysis was not done properly, either because the instrumental broadening was not corrected at all [17] or insufficiently [18], or because ω_m and $\Delta\omega$ were compared with the solutions of the dispersion equation [19, 20] or assimilated to the oscillation frequency and damping rate of measured correlation functions [18, 20].

The case of Ref. 21 deserves some comments. The spectrum used in this reference is more complicated than in Sec. 2, but the hydrodynamic limit conditions at the surface are incorrect: the compressional viscosity is assimilated to the imaginary part of γ, and an unphysical slip condition is assumed (here $qh \gg 1$ and the velocities in the fluid and in the monolayers are equal); this leads to an additional horizontal surface force $\Gamma\partial^2\xi/\partial t^2$, a force which is in fact a factor qh smaller than the Marangoni force. A possible surface diffusion is included, but the frequency of the corresponding modes is expected to be much smaller than the capillary wave frequency and thus not to affect the spectrum in the studied frequency range. Although the data analysis of Ref. 21 is limited, it was done with a theoretical spectrum very similar in fact to Eq. 26, because during the analysis the surface viscosities were taken to be zero.

Because of easy contamination problems, the data obtained without a simultaneous measurement of the static surface tension might not be free of systematic errors, especially if independently measured γ_{st} values are used to analyze these data [17, 18, 21]. In the following, we shall show only examples of measurements where the data acquisition and analysis are correct from these points of view.

Because of the limited resolution of the light scattering technique, gaseous monolayers cannot be studied. In two-dimensional gaseous states, the viscoelastic parameters are too small to be measured. However, in the coexistence regions between two-dimensional gas and liquid states, fluctuations of these parameters between the values of pure gas and pure liquid can be observed when the size of the domains is large enough [16, 19, 21, 22]. When it is much smaller than the size of the illuminated area, effective values are measured that represent complicated averages. This problem is the one of rheology of composite media. The best domain for the technique is the liquid-expanded type of monolayers. In liquid-condensed or solid monolayers, ε and κ are large and can no longer be determined accurately. Only γ and μ are accessible.

C. Fatty Substance Monolayers

Well-known fatty substance monolayers have been studied by several authors. Figures 8 and 9 show surface isotherms of various compounds at room temperature: myristic (tetradecanoic), pentadecanoic, stearic (octadecanoic) acids and propyl stearate, on acid water (10^{-2} M HCl) to avoid monolayer dissolution. The kinks in the isotherms of myristic and pentadecanoic acids mark the onset of the liquid expanded-liquid condensed transition. In the two-phase region, the isotherms are horizontal for very pure pentadecanoic acid as discussed earlier. Stearic acid and, at low temperatures (below the triple point), pentadecanoic acid monolayers do not show liquid-expanded phases: the monolayer goes directly from gaseous to liquid-condensed phases. The liquid-expanded phase of propyl stearate has an especially large range. This behavior can be rationalized in terms of chain flexibility. The flexibility decreases when chain length increases [23]: long chain substances form rather condensed phases; short chain substances show expanded phases. Substances with two chains of different length are quite flexible and hardly form condensed phases. Typical experimental spectra are shown in Fig. 10: the frequency shift and broadening produced by the monolayer are clearly visible [24].

Examples of the variation of ω_m and $\Delta\omega$ versus surface pressure are given in Fig. 11 [25]; ω_m and $\Delta\omega$ are maximum around $\Pi \sim 1$–2 dyn/cm: this corresponds to the resonance phenomenon between capillary and compression waves predicted theoretically.

The values of the compression modulus ε and the viscosity κ for myristic acid at room temperature are shown in Fig. 12 together with the static values calculated from isotherm data $\varepsilon_{st} = -A\ d\Pi/dA$ [16, 19, 24]. The observed

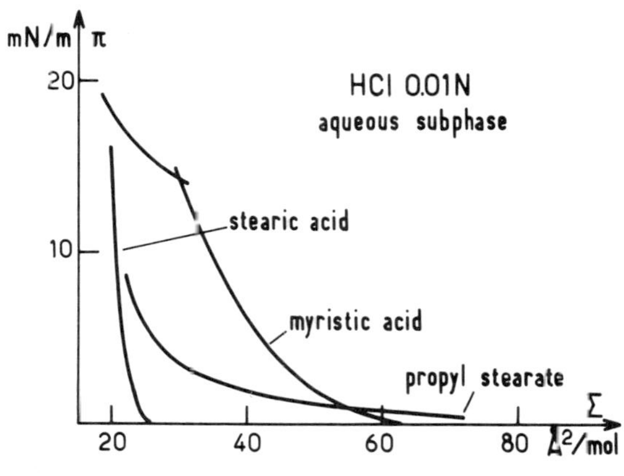

FIG. 8 Surface pressure versus area per molecule for different fatty substances at room temperature.

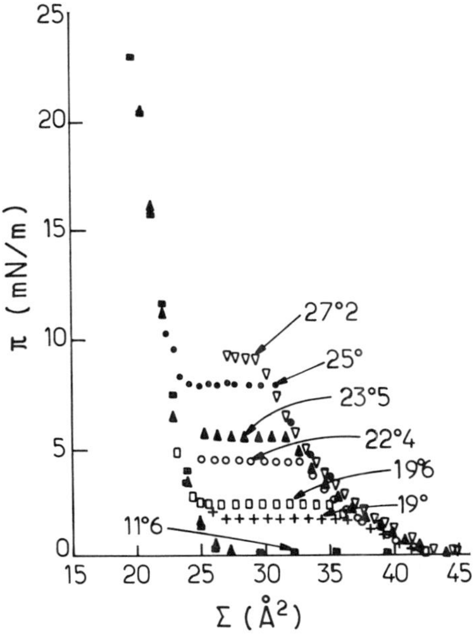

FIG. 9 Surface pressure versus area per molecule for pentadecanoic acid at different temperatures. (Data from Refs. 28 and 29.)

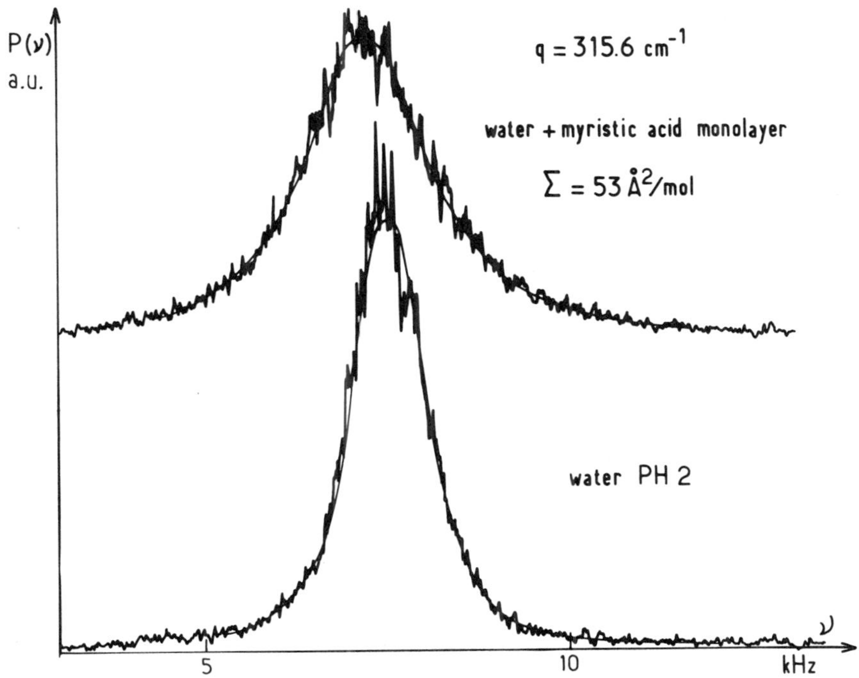

FIG. 10 Spectrum of the light scattered by the free surface of water and by water covered with a monolayer of myristic acid.

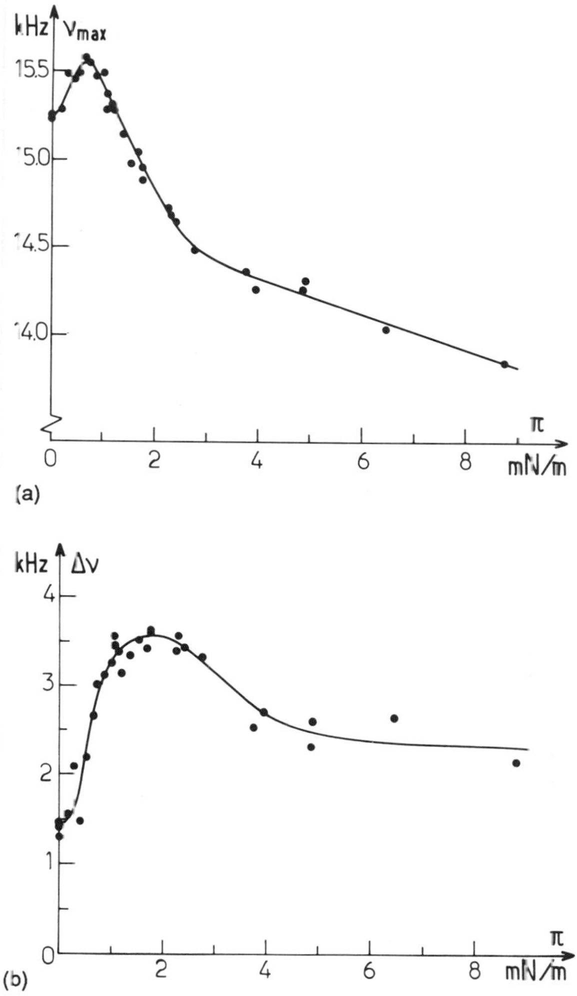

FIG. 11 (a) Peak frequency and (b) half-width of the spectrum of the light scattered by water covered by propyl stearate monolayers versus surface pressure. The lines are to guide the eye. (Data from Ref. 25.)

differences between different sets of data could be due to relaxation effects. Indeed they correspond to different wave vectors and therefore to different frequencies: from $4 \times 10^4\ s^{-1}$ to $4 \times 10^5\ s^{-1}$. ε is larger at high frequency and κ smaller, as expected. For $\Pi = 10$ mN/m and the smallest frequency $\varepsilon - \varepsilon_{st} \sim 10$ mN/m and $\kappa \sim 2 \times 10^{-4}$ g/s: $\omega\kappa \sim 8$ mN/m $\sim \varepsilon - \varepsilon_{st}$; viscous and elastic effect are comparable, indicating that the frequencies of the measurement

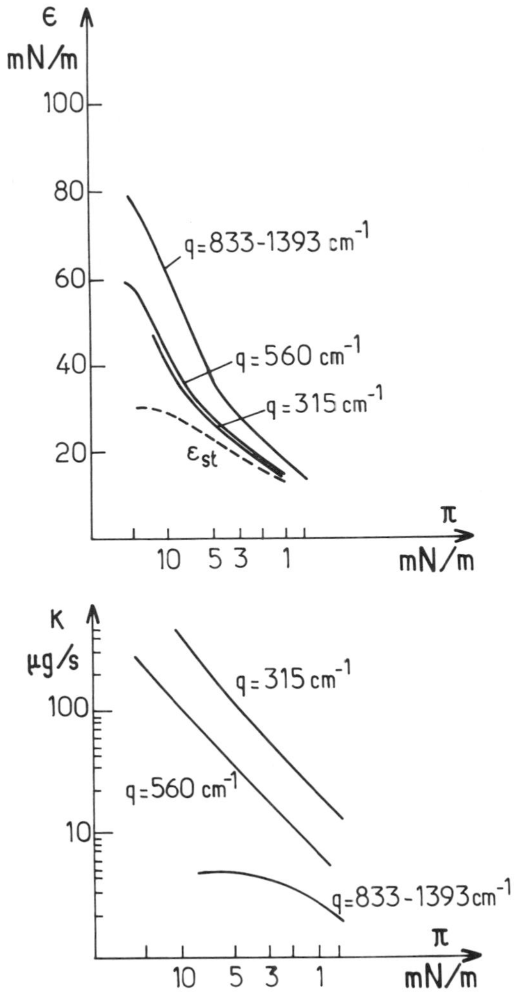

FIG. 12 Compressional elasticity and viscosity versus surface pressure for myristic acid monolayers; data taken at different wavevectors (and frequencies): 833–1393 cm^{-1} (Ref. 16); $q = 560\ cm^{-1}$ (Ref. 19); $q = 315\ cm^{-1}$ (Ref. 24).

fall in the range of relaxation frequencies. The largest κ values are of the same order of magnitude as the shear viscosities η_S measured at very low frequencies with mechanical devices [26] (let us recall that κ is at least equal to η_S).

In the low-frequency measurements ($4 \times 10^4\ s^{-1}$), nonzero μ values associated with differences between Π and Π_{st} were found above $\Pi = 2$ mN/m. These effects were very small, however, $\mu < 50$ μg/s and $\Pi - \Pi_{st} \sim 1$ mN/m, comparable to the instrumental accuracy. Again, in the frequency range studied, $\omega\mu \sim \Pi - \Pi_{st}$; μ was observed to decrease to zero at large surface pressure (>14 mN/m), as predicted by Eq. 17 if one assumes that τ decreases when Π increases. More accurate μ measurements were made on lipid mono- and bilayers and will be described later.

Data for ε and κ are shown in Fig. 13 for myristic and stearic acids, and for propyl stearate [27]. They illustrate the different types of viscoelastic behavior that can be observed according to the nature of molecular packing in the layer.

Pentadecanoic acid monolayers were thoroughly investigated recently both

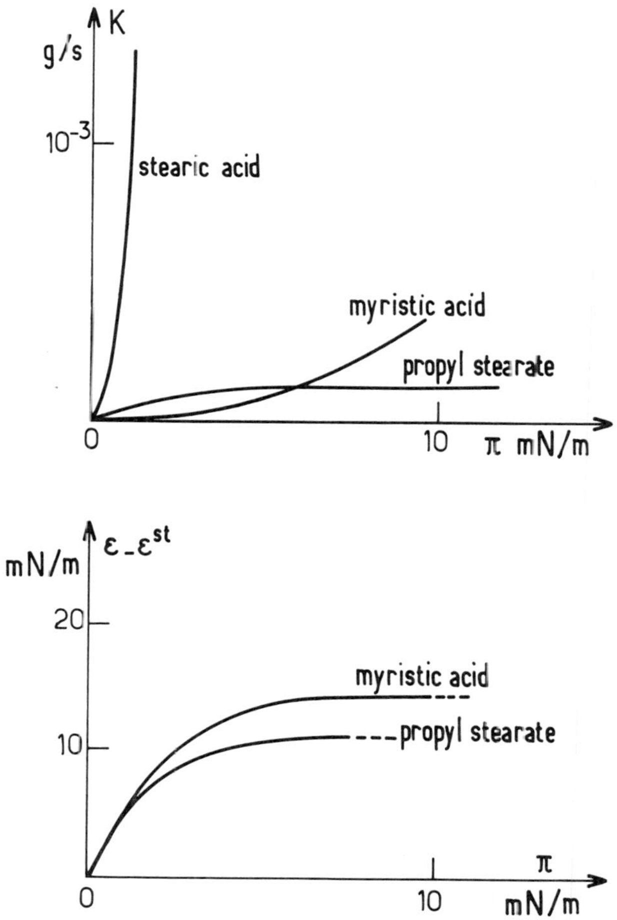

FIG. 13 Compressional viscosity and difference between static and dynamic compressional elasticities for the substances of Fig. 8.

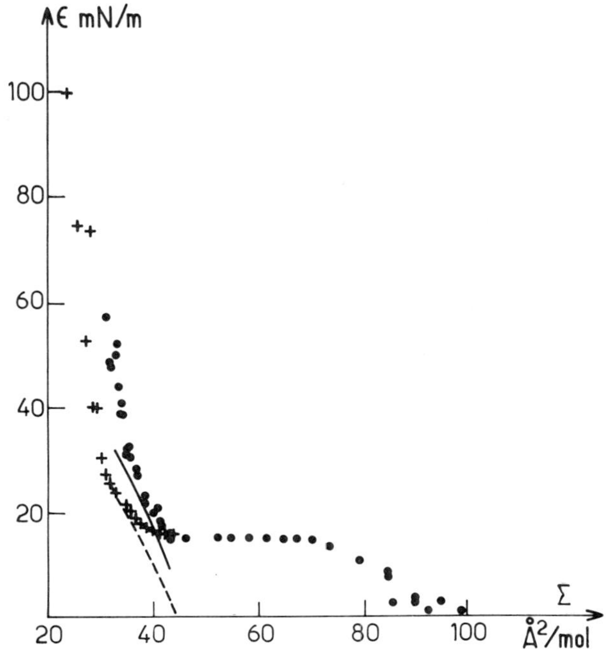

FIG. 14 Compressional elasticity of pentadecanoic acid versus area per molecule; circles (data from Ref. 28), $t = 25°C$; crosses (data from Ref. 29), $t = 22°4$ C. The lines are the static elasticities, as determined from the curves of Fig. 9.

above and below triple point [16, 28, 29]. Above the triple point, they behave much like myristic acid: $\varepsilon \gg \varepsilon_{st}$ below the gas-liquid expanded and above the liquid expanded-liquid condensed transitions (Fig. 14). κ is very small in the liquid expanded region (below 10 μg/s) and increases in the liquid expanded-liquid condensed coexistence region. μ is nonzero in the dilute regions (up to 50 Å^2/mol) and vanishes at the liquid condensed-liquid expanded transition. Below the triple point, pentadecanoic acid behaves rather like stearic acid.

D. Lipid Monolayers

Dipalmitoylphosphatidylcholine (DPPC) [20] and dipalmitoyl-L-α-lecithin (DPL) [19] monolayers were studied at room temperature. They behave much as myristic acid monolayers, although $\gamma = \gamma_{st}$ and $\mu = 0$ in the range of scattering angles investigated.

Glycerol monooleate (GMO) monolayers [30] behave more like propyl stearate: they exhibit a large liquid expanded region and collapse before forming a liquid condensed phase. The compression modulus is larger than the static value at small coverages Γ, but surprisingly it becomes smaller close to the collapse pressure. This was already quoted for propyl stearate monolayers [25]. Nonzero transversal shear viscosities were found in all the studied area range; μ is maximum around 100 Å^2/mol ($\mu \sim 4 \times 10^{-5}$ g/s) and decreases until the collapse is reached at 28.5 Å^2/mol. This is accompanied by an increasing surface tension deviation with respect to the static value. The γ and μ values of the fully compressed monolayers are shown as a function of frequency in Fig. 15. The frequency range corresponds to the studied wavevector range $220 < q < 1500$ cm^{-1}. The lines in the figure represent a fit with a Maxwell type of relaxation model:

$$\gamma = \gamma_0 + \gamma_2 \frac{\omega^2\tau^2}{1 + \omega^2\tau^2} \qquad \mu = \frac{\gamma_2\tau}{1 + \omega^2\tau^2} \tag{33}$$

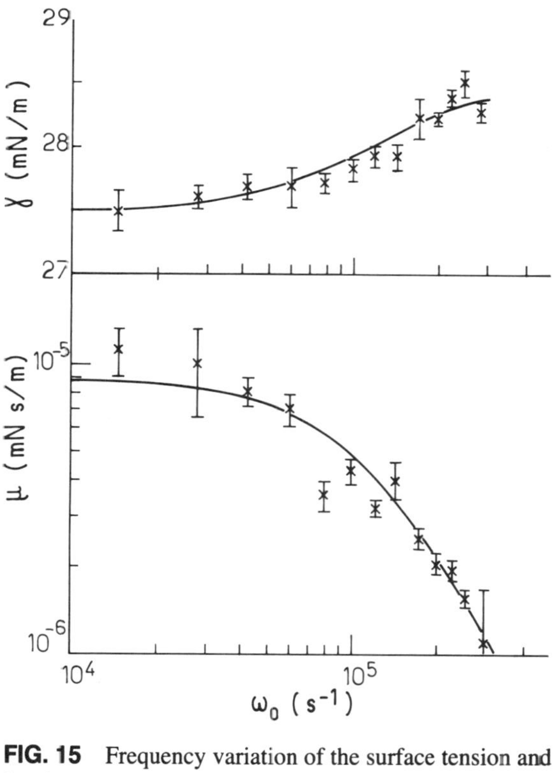

FIG. 15 Frequency variation of the surface tension and of the shear transverse viscosity for GMO monolayers; the lines are a fit with the Maxwell relaxation model. (Data from Ref. 30.)

The fit gives $\tau \sim 9\ \mu s$. Experiments and simulations on vesicles concur in assigning a time scale of the order of $5\ \mu s$ to local chain melting favoring free rotational isomerization, that could be at the origin of the observed relaxation process. Similar light scattering studies on GMO bilayers have led to $\tau \sim 37$ μs (Chap. 12). Let us note that Eqs. 33 and 17 are equivalent. Similar fits done at different concentrations were also in agreement with the data. The variation of the relaxation time versus surface pressure is shown in Fig. 16. As expected intuitively, τ decreases when Π decreases.

DPPC monolayers were also studied with the light scattering technique at a heptane-water interface [20]. As expected, no influence of the compressional viscoelasticity could be seen, because of the almost complete decoupling between capillary and compression waves. γ was found equal to γ_{st} and μ to zero. The net information was merely the γ value. In some cases, γ can become very small and difficult to measure. The surface light scattering is interesting to use in these cases. This type of measurement will be presented in Chap. 13 in connection with the microemulsion systems.

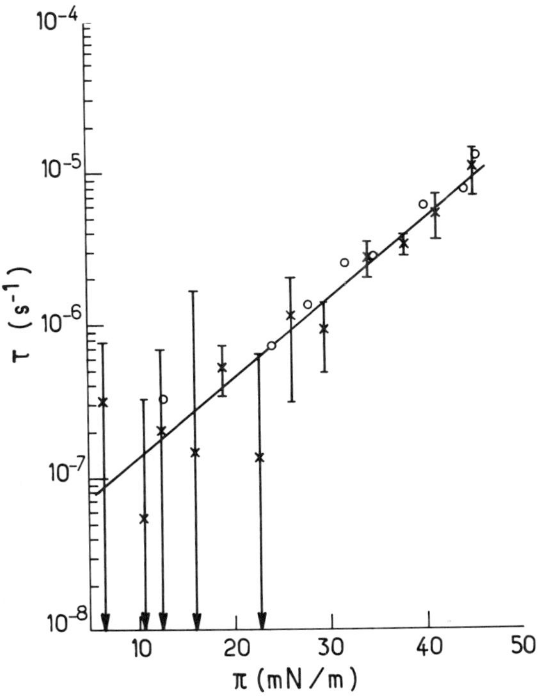

FIG. 16 Relaxation time for the transverse motion versus surface pressure for GMO monolayers; o, q = 968.5 cm^{-1}; × q = 1160 cm^{-1}. (Data from Ref. 30.)

E. Soluble Monolayers

There are only very few reported studies of soluble monolayers in comparison with the relatively large amount of data on insoluble monolayers.

Octanoic acid was studied because low-frequency viscoelasticity data were available. Fig. 17 shows the compression elasticity data obtained with the various techniques: (a) light scattering $\nu \sim 50$ Hz [16], (b) electrically induced capillary waves $\nu = 200$ Hz [31], (c) mechanically induced compression waves $\nu \sim 1$ Hz [32]. ε is a decreasing function of the frequency as usual. But the curves (b) and (c) exhibit a maximum as a function of the concentration and decrease afterwards. This is a specific effect of soluble monolayers: shortcircuiting of the compression by monolayer dissolution. It can be explained satisfactorily by the Lucassen model: Eqs. 13, 14.

Sodium dodecyl sulfate (SDS) and dodecyltrimethyl ammonium bromide (DTAB) aqueous solutions were also investigated [33]. Typical spectra are shown in Fig. 18. The compressional viscoelastic coefficients are shown in Fig. 19. At the critical micellar concentration (cmc), micelles begin to form in the bulk liquid; the cmc is evidenced as a break followed by a plateau in the surface tension curve (Fig. 20). Close to the cmc, ε and κ slightly vary with q. This variation transformed into a frequency variation is shown in Fig. 21, where data at lower frequencies measured with electrically excited waves are also shown [34]. In order to compare the data with the Lucassen theory, it is necessary to know the derivative $dc/d\Gamma$ of bulk to surface concentration. This was achieved

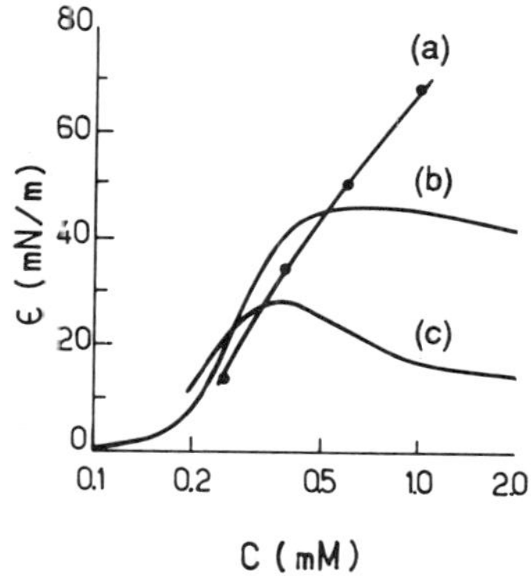

FIG. 17 Compressional elasticity of octanoic acid monolayers versus bulk acid concentration; (a), light scattering data (Ref. 16); (b), excited capillary wave data (Ref. 31); (c), compression wave data (Ref. 32).

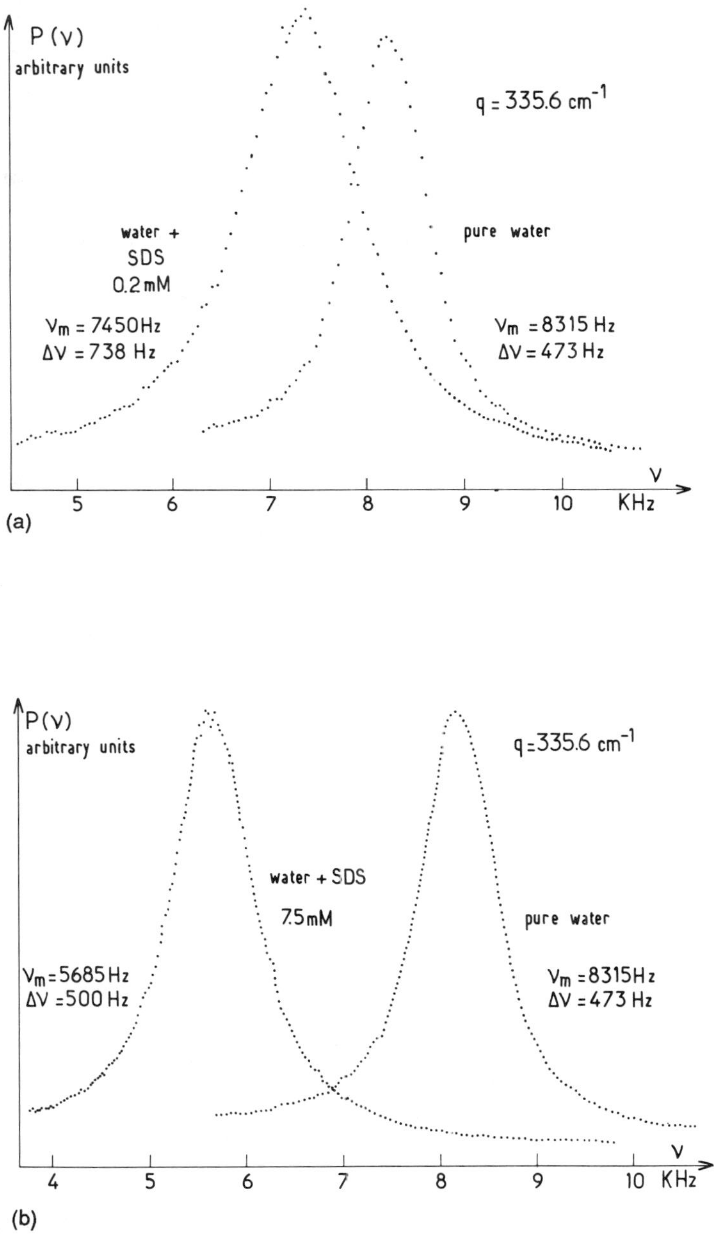

FIG. 18 Spectrum of the light scattered by the free surface of water and SDS aqueous solutions.

by using the Frumkin isotherm [35]

$$\pi = k_B T \, \Gamma_\infty \ln\left(1 - \frac{\Gamma}{\Gamma_\infty}\right) - \frac{H\Gamma^2}{k_B T} \tag{34}$$

$$1 - \frac{\Gamma}{\Gamma_\infty} = \left(1 - \frac{c}{a}\right) \exp - \left(\frac{H\Gamma^2}{k_B T}\right) \tag{35}$$

where a is the Szykowsky concentration, Γ_∞ the limit surface concentration, and

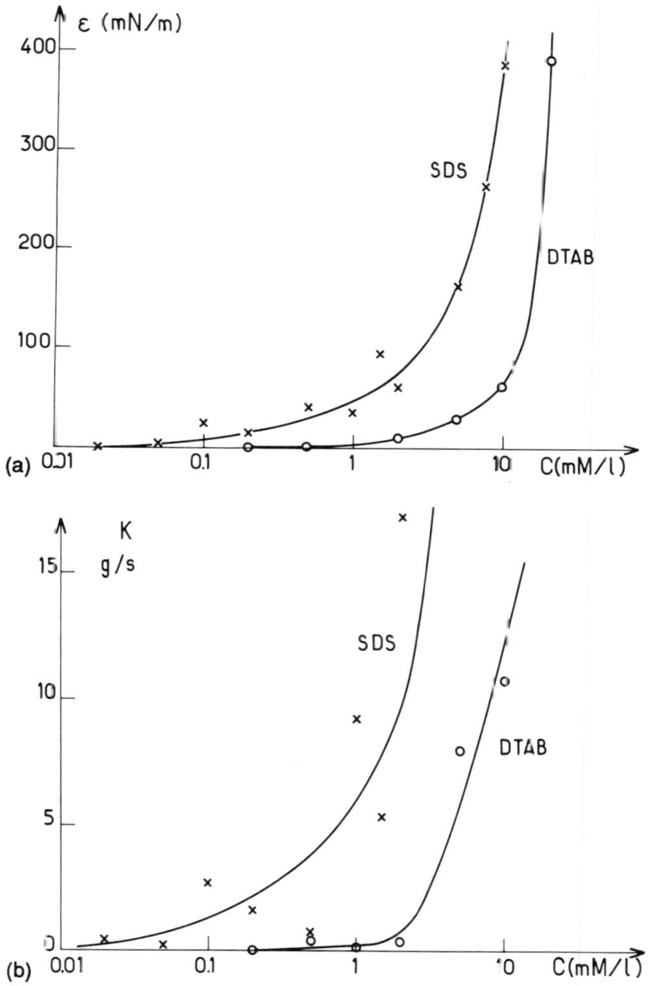

FIG. 19 (a) Compressional elasticity and (b) viscosity of SDS and DTAB aqueous solutions versus bulk surfactant concentration.

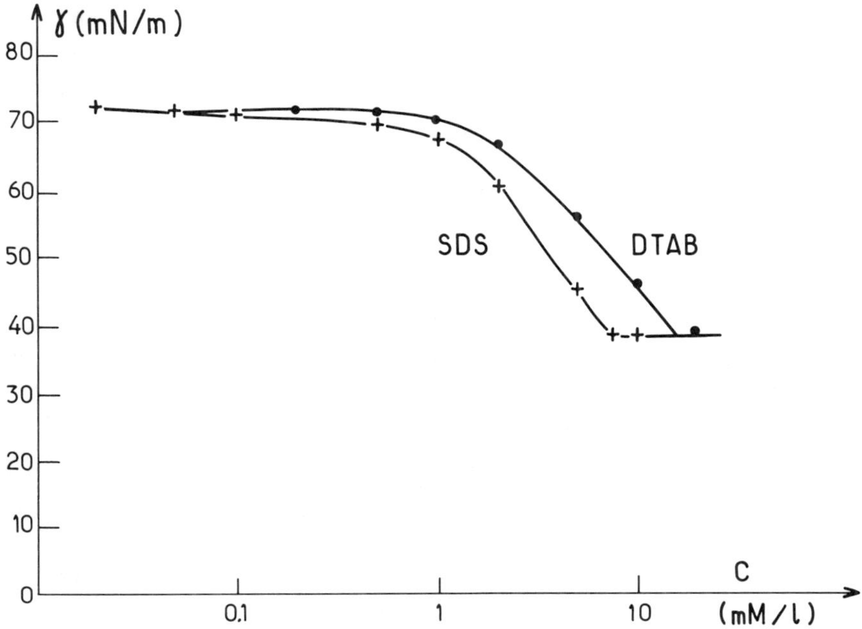

FIG. 20 Surface tension of SDS and DTAB aqueous solutions versus bulk surfactant concentration. The lines are to guide the eye.

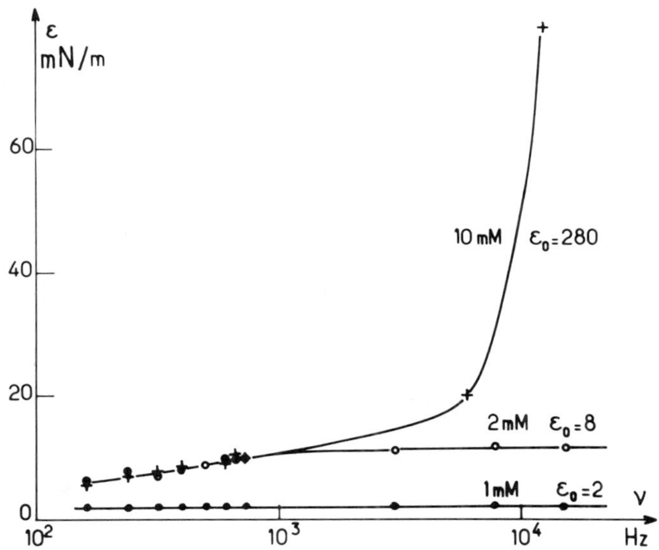

FIG. 21 Compressional elasticity of DTAB aqueous solutions versus frequency for different bulk concentrations. Data from light scattering experiments (high frequencies) and excited capillary wave experiments (lower frequencies).

H a nonideality parameter. For $H = 0$, Eqs. 34 and 35 reduce to the Langmuir isotherm. The parameters a, Γ_∞, and H can be deduced from the curve $\gamma(c)$ (Fig. 20). The values $\varepsilon_0 = -\partial\gamma/\partial \ln \Gamma$ obtained in this way are indicated in Fig. 21. The model leads to a satisfactory agreement with the data [34].

V. POLYMER MONOLAYERS

Light scattering allows the study of the dynamic behavior of polymer monolayers at a liquid surface. Such behavior is of importance to check the validity of theoretical models [36] and of course because of the large variety of applications like colloidal stabilization (protein-stabilized foams and emulsions in the food industry, depletion stabilization with flexible polymers), coating, lubrication, and adhesion of solid surfaces. The conformations of polymer chains at solid and liquid surfaces are similar, provided that the molecular interactions between the substrate and the polymer segments are similar. In the case of liquid surfaces, it is very easy to change these interactions by changing the liquid or adsorbing a surfactant at the surface: in the case of the free surface of water, for instance, the surface changes from hydrophilic to hydrophobic. Spreading is faster on liquid surfaces, and dynamical behavior more easily accessible.

There have been several studies making use of excited capillary waves (Chap. 16) on polymer monolayers [37–39]. We shall discuss here the experiments made with the light scattering technique. All these experiments deal with flexible polymers.

Like surfactants, polymers can be grouped according to the shape of their pressure-area curves. At very low concentrations they form gaseous-like layers [40], which we shall not describe here because their viscoelastic behavior is not accessible to the technique ($\varepsilon \sim \kappa \sim \mu \sim 0$). At larger concentrations, the less hydrophobic polymers form liquid-expanded monolayers, the more hydrophobic, liquid-condensed monolayers. This is as for the short chain amphiphiles: in the case of fatty acids, for instance, increasing the chain length from 15 (pentadecanoic acid) to 18 (stearic acid) carbon atoms transforms a liquid-like monolayer into a condensed one. The molecular interpretation for the formation of liquid-expanded polymer monolayers is the following: at low concentrations Γ, the polymer chains are isolated and folded as pancakes in the surface, the equivalent of coils in bulk; the monolayer is gaseous-like. When Γ increases, it reaches a value Γ^* where the polymer chains are touching each other. Typically $\Gamma^* \sim 1$ mg/m^2. Above Γ^*, the chains interpenetrate progressively. Like the osmotic pressure of a bulk solution [41], the surface pressure varies with Γ following a power law $\Pi \sim \Gamma^x$ with $x \sim 3$ [42] and is independent of molecular weight. These are called two-dimensional good solvent conditions. The above scaling law is in good agreement with experiments [43]. Examples of these systems are polyvinyl acetate and polyethylene oxide on water. The liquid-condensed type of monolayers are thought to form when above Γ^* the chains do not interpenetrate: the pancakes behave like rather incompressible hard discs, and the surface pressure rises very rapidly when Γ increases above Γ^*. The data can still be fitted with a power law $\Pi \sim \Gamma^x$, but x is very large (above 10). This situation is the two-dimensional analog of a polymer at or above the Θ point [44]. Examples of these systems are polymethylmetacrylate and polydimethylsiloxane on water.

At present, there is no evidence of systems showing liquid expanded-liquid condensed or liquid condensed-solid phase transitions. Above a certain pressure, the monolayer merely collapses. In certain cases (as with the fatty acids) evidence for multilayer formation is found [39]. The collapse pressures seems higher at oil-water interfaces than at the water free surface [45].

In the following, we will describe the viscoelastic behavior investigated with surface light scattering for the different types of monolayers. The experimental procedure and the data analysis is the same as for surfactant monolayers.

A. Liquid Expanded Monolayers

Polyvinyl acetate monolayers (PVA, molecular weight of about 10^5) have been studied in detail. The surface pressure curve at 20°C is shown in Fig. 22 together with that of polymethylmetacrylate (PMMA) to illustrate the difference between liquid expanded and liquid condensed monolayers.

The compressional elasticities and viscosities are shown in Fig. 23 versus surface pressure together with the data for PMMA and for fatty substances. It is seen in the figure that there is also a striking similarity between the viscoelastic behavior on the liquid-expanded type of layers, and of the liquid-condensed ones as well [27].

The behavior close to the collapse pressure was also investigated [45, 46].

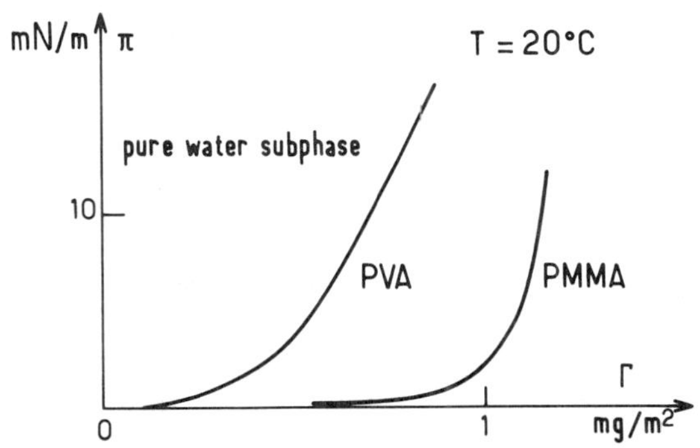

FIG. 22 Surface pressure of PVA and PMMA polymers spread on water versus surface concentration.

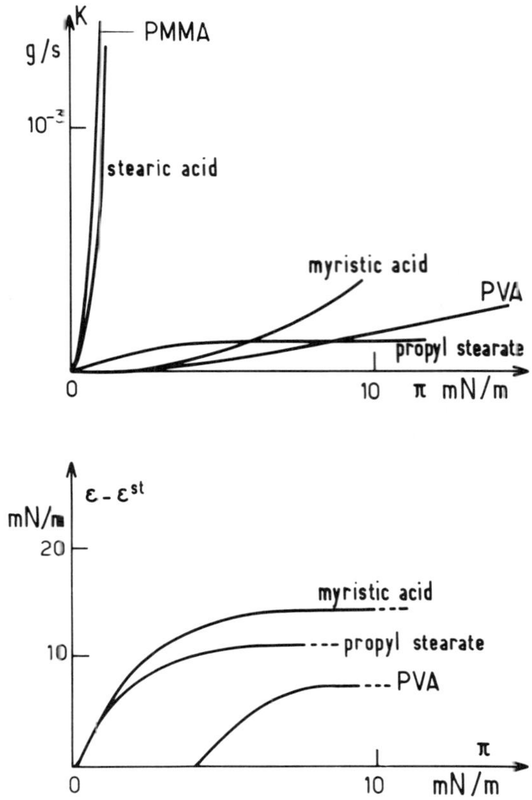

FIG. 23 Compressional viscosity and difference between static and dynamic compressional elasticities for the substances of Figs. 8 and 22, illustrating the similarities between fatty substances and polymer monolayers.

Since the surface pressure begins to round off at these large surface concentrations, the elasticity drops, and the peak frequencies and half-widths of the light scattering spectra exhibit a second maximum (Fig. 24). This is because the condition for maximum coupling of capillary and compression waves, $\varepsilon/\gamma \sim 0.15$, is reached twice as Γ is varied. In the case of multilayer formation, it can be reached more than two times. Several maxima were indeed observed (with the excited wave technique) with these particular monolayers [37].

In some of the papers [27, 46] it was reported that at large enough Γ, the shear transverse viscosity μ was nonzero and that γ, as determined from light scattering experiments, was different from the static value γ_{st}. This is as observed in several monolayers made with fatty substances. Yu and coworkers did series of experiments with the same polymer at the same temperature, and interpreted

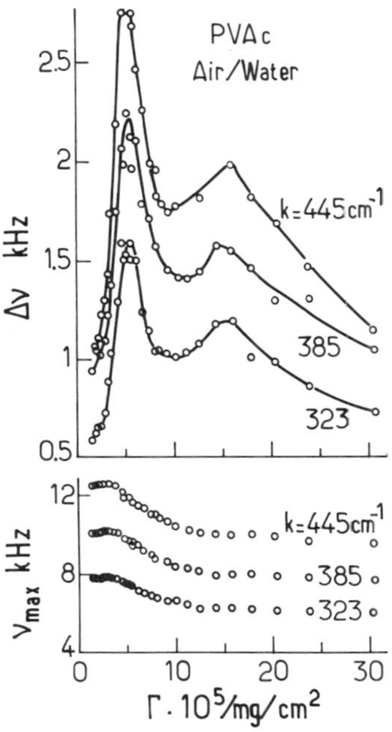

FIG. 24 Peak frequency and half-width of the spectrum of the light scattered by water covered by PVA monolayers versus surface coverage for different wavevectors. The lines are to guide the eye. (Data from Ref. 45.)

them in different ways. In a first paper, the peak frequencies ν_m and half-width $\Delta\nu$ were compared with those of the theoretical spectrum to extract the four parameters ε, κ, γ, and μ. They found in this way that for $\Gamma \geq 0.5$ mg/m^2, $\mu \neq 0$ and $\gamma \neq \gamma_{st}$ [46]. In a second paper, they compared ν_m and $\Delta\nu$ with the roots of the dispersion equation to extract the same four parameters. They found in this way $\mu = 0$ and $\gamma = \gamma_{st}$ for all Γ [45]. But this procedure, as discussed earlier, is not correct. In the range of parameters studied, values of ν_m and $\Delta\nu$ obtained from the theoretical spectrum and by solving the dispersion equation can differ by as much as 1%. Roughly speaking, an error on ν_m of 1% is equivalent to an error of 2% on γ, i.e., for $\gamma = 50$ mN/m an error of 1 mN/m. This is the order of magnitude of the reported differences $\gamma - \gamma_{st}$. This discussion illustrates how important it is to use proper data analysis in order to obtain reliable information on the transverse shear viscoelasticity.

PVA was also studied at the heptane-water interface [45]. The surface pressure curve is shown in Fig. 25. The collapse pressure is much higher than at the free surface of water. The variation of ν_m and $\Delta\nu$ with Γ is markedly different (Fig. 26). This is because compression and capillary waves are here almost completely

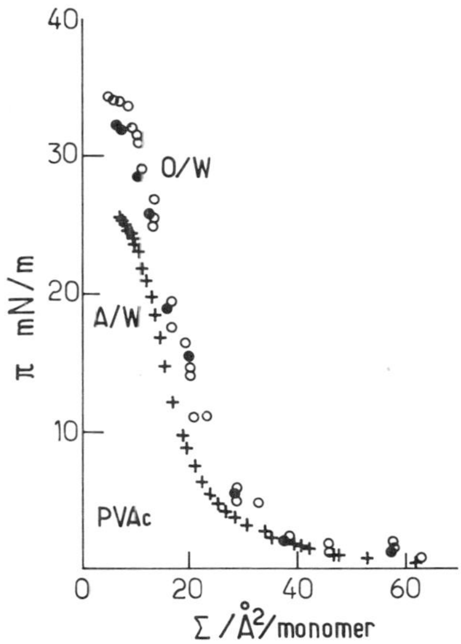

FIG. 25 Surface pressure of PVA monolayers at the free surface of water (+) and at the heptane-water interface(open circles, light scattering measurements; closed circles, static data) versus area per monomer. The lines are to guide the eye. (Data from Ref. 45.)

decoupled: ν_m and $\Delta\nu$ are practically unaffected by ε and κ. The data allow the determination of γ and μ only. It was found that $\gamma = \gamma_{st}$ and $\mu = 0$.

The results obtained for polyethylene oxide monolayers at the air-water and heptane-water interface are qualitatively similar [45]. The main difference lies in the low value of the compressional viscosity κ (air-water interface), about 10 times smaller than for PVA at the surface concentration Γ_m for which ε_{st} is maximum. The PVA κ values are themselves about 10 times smaller than the shear viscosities measured with mechanical devices [47]. This difference is probably due in part to relaxation processes. The values of ε and κ measured with light scattering were found to be independent of the molecular weight for these two polymers. It must be noted that PEO is a water-soluble polymer and that the monolayers can also be obtained by adsorption. It was checked that the results obtained with spread and adsorbed monolayers were the same [48].

The results obtained for other liquid-expanded monolayers are qualitatively similar [29]. These were, including the two previous ones, in the order of increasing hydrophobicity, PEO, PTHF (polytetrahydrofuran), PVA, and PMA (polymethyl metacrylate). $\varepsilon \sim \varepsilon_{st}$ for $\Gamma < \Gamma_m$ (before collapse) and $\varepsilon > \varepsilon_{st}$ for $\Gamma > \Gamma_m$. The maximum elasticity increases with increasing hydrophobicity: 12 mN/m for PEO, 18 for PTHF, 25 for PVA and PMA. κ increases also with

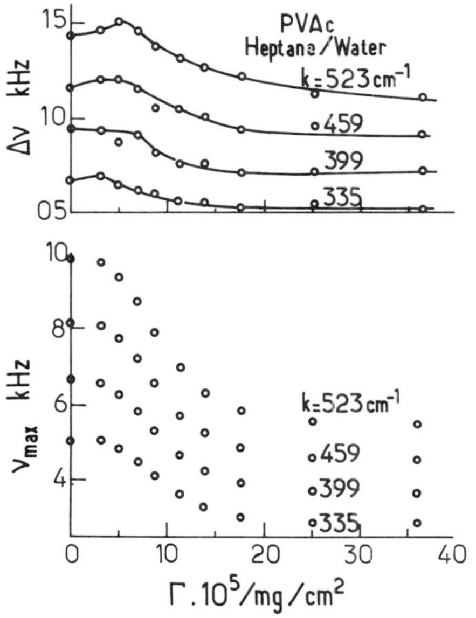

FIG. 26 Peak frequency and half-width of the spectrum of the light scattered by the heptane-water interface covered by PVA monolayers versus surface coverage for different wavevectors. The lines are to guide the eye. (Data from Ref. 45.)

increasing hydrophobicity: for $\Gamma \sim \Gamma_m$, $\kappa \sim 18$ μg/s for PEO, 60 μg/s for PTHF, 150 μg/s for PVA and PMA. For the two last polymers, κ continues to increase after the collapse.

B. Liquid Condensed Monolayers

Different types of polymers forming liquid condensed monolayers have been studied at the air-water interface: polymethylmetacrylate (PMMA) [46, 16, 49, 27], polydimethylsiloxane (PDMS) [16], polybutylmethacrylate (PBMA) [49]. The viscoelastic behavior is similar (Fig. 23). Because the surface pressure is a steep function of Γ above Γ*, ε also shows a steep increase above Γ* and eventually drops when collapse is approached. κ also exhibits a sharp increase above Γ*. In the PMMA case, small but nonzero values for $\gamma - \gamma_{st}$ and μ were observed above Γ* for $q \sim 300$ cm^{-1} [27]. For $q \sim 400$ cm^{-1}, $\gamma - \gamma_{st} \sim 1$ mN/m, but μ was found to be zero [49]. At larger wavevectors (>800 cm^{-1}), μ and $\gamma - \gamma_{st}$ were reported to be zero [16]. As in the case of fatty substance monolayers, more work is certainly needed to clear the problem of the transverse shear viscoelasticity in polymer monolayers. The compressional surface viscosities found for PMMA are the smallest for the largest q (largest frequencies).

The shear viscosities obtained with mechanical devices are orders of magnitude larger: 2g/s instead of $\sim$ 1 mg/s for $\Gamma = 0.75$ mg/m^2 [50]. Similarly, low-frequency shear elasticities are much smaller than the light scattering determinations of ε: 0.05 mN/m instead of $\sim$ 50 mN/m at the same value of Γ.

For these polymers, at the collapse Γ_m, $\varepsilon_{st} \sim 50$ mN/m. For PBMA, $\varepsilon > \varepsilon_{st}$ for $\Gamma > \Gamma_m$, but for PMMA and PDMS, $\varepsilon > \varepsilon_{st}$ in the whole concentration range. This is attributed to the fact that for $\Gamma < \Gamma_m$, the monolayer is inhomogeneous [49]. For PMBA and PMMA, $\kappa(\Gamma_m) \sim 1$ mg/s. In contrast, PDMS has a very small compressional surface viscosity, below 10 μg/s and too small to be measured even with light scattering techniques. This shows that the surface viscosity depends not only on the hydrophobicity of the polymer but also on the chain flexibility (which is the largest for PDMS in the above series).

C. Block Copolymers

Polyethylene oxide (PEO)–polystyrene (PS) block copolymers have been studied at both the air-water and the heptane-water interface [51]. The results are qualitatively similar to those of the PEO homopolymer. The surface compressional viscosity is, however, about ten times larger. Let us recall that for PEO, κ was very small.

VI. CONCLUSION

The surface light scattering technique is a nonperturbative technique ideal for the study of the viscoelastic behavior of monolayers. It gives information about (uniaxial) compressional viscoelasticity. The information is significant only for monolayers at the free surface of water. This type of viscoelasticity has a very small effect on the propagation of capillary waves at oil-water interfaces. A second type of viscoelasticity can be investigated, the shear transverse, but it is characterized by very small coefficients and is negligible in most cases.

The light scattering technique (like all capillary wave techniques) better applies to liquid expanded types of monolayers as far as the determination of compressional viscoelasticity is concerned: the modulus should not be larger than the surface tension and the viscosity not larger than 1 mg/s.

The viscoelastic behavior of monolayers is very strongly marked, for fatty substances and for polymers as well. The viscoelastic parameters are strongly frequency dependent, making the data analysis specially difficult. It is indeed dangerous to assume that ε and κ for instance are frequency independent and try to deduce γ and μ from the data taken at different scattering angles. Furthermore, $\gamma - \gamma_{st}$ and μ being small, they can be measurable for certain q values and fall below instrumental accuracy for others q. Except for the extensive work on GMO monolayers, the question of the existence of the shear transverse viscoelasticity remains open in most studied systems and still deserves more experimental work to be clarified.

The viscoelastic behavior of polymer monolayers is quite similar to that of the liquid-expanded and liquid-condensed monolayers of fatty substances. The viscoelastic parameters are of the same order of magnitude. The surface viscosities of some polymers may even be much smaller than that of fatty acids for an equivalent surface concentration. It must also be stressed that above the chain contact concentration Γ^*, the monolayer properties do not depend any more on the molecular weight; they are only functions of the surface concentration Γ.

The scattering wavevector range, and consequently the frequency range of the technique, is limited. It is therefore useful to combine the measurements with those of other techniques leading to the same parameters to investigate their frequency dependence in more detail. These techniques are, for instance, electrically induced capillary waves and mechanically induced compression waves. It should also be stressed that the viscoelastic coefficients measured with the technique are not shear coefficients as in better known devices: canal viscosimeters, oscillating disk devices, deep channel viscosimeters. In this sense, complementary information can also be gained by combining the different types of devices.

The relaxation processes giving rise to the frequency dependence of the viscoelastic parameters in insoluble monolayers are mostly unknown. In soluble monolayers the adsorption-dissolution process driven by monolayer compression is better understood. Much work remains to be done, however, to clarify this problem. It must be stressed that in a particular phenomenon, the relevant viscoelastic coefficients are those measured at the characteristic frequency of the phenomenon. If the phenomenon is fast, the coefficients measured at low frequencies (with mechanical devices) can be very different from those that affect the phenomenon. Light scattering is especially well suited to the measurements in the kilohertz range.

REFERENCES

1. G. Gaines, *Insoluble Monolayers at Liquid-Gas Interfaces,* Interscience, New York, 1966.
2. L. Kramer, *J. Chem. Phys. 55:* 2097 (1971).
3. L. Landau and E. Lifshift, *Theory of Elasticity,* Addison-Wesley, 1959.
4. V. G. Levich, *Physico-Chemical Hydrodynamics,* Prentice Hall, 1962.
5. E. H. Lucassen-Reynders and J. Lucassen, *Adv. Colloid Interface Sci. 2*: 347 (1969).
6. F. C. Goodrich, *Proc. R. Soc. London A 374*: 341 (1981).
7. M. Baus and C. F. Tejero, *J. Chem. Phys. 78*: 483 (1983).
8. J. A. de Feijter and J. Benjamins, *J. Colloid Interface Sci. 70*: 375 (1979).

9. D. Langevin and M. A. Bouchiat, *Comptes Rendus Acad. Sci. 272B*: 1422 (1971).
10. R. Loudon in *Surface Excitations* (V. M. A. Agranovich and R. Loudon, eds.), Elsevier, 1984.
11. D. Langevin, J. Meunier, and D. Chatenay in *Surfactants in Solution* (K. L. Mittal and B. Lindman, eds.), Vol. 3, 1984, p. 1991.
12. A. Hajiloo and J. Slattery, *J. Colloid Interface Sci. 112*: 325 (1986).
13. C. Fan, *J. Colloid Interface Sci. 44*: 369 (1973).
14. S. R. Middleton, M. Iwahashi, N. R. Pallas, and B. A. Pethica, *Proc. R. Soc. Lond. A 396*: 143 (1984).
15. J. E. Earnshaw, R. C. McGivern, A. C. McLaughlin, and P. J. Winch, *Langmuir 6: 649 (1990).*
16. S. Hard and R. D. Neuman, *J. Colloid Interface Sci. 120*:15 (1987).
17. D. McQueen and I. Lundstrom, *J. Chem. Soc. Faraday Transactions I 69*: 694 (1973).
18. D. Byrne and J. C. Earnshaw, *J. Phys. D: Appl. Phys. 12*: 1145 (1979).
19. S. Hard and R. D. Neuman, *J. Colloid Interface Sci. 83*: 315 (1981).
20. B. B. Bauer, Y. L. Chen, G. Zografi, and H. Yu, *Langmuir 4:* 111 (1988).
21. H. Birecki and N. M. Amer, *J. Physique 40*: C3-433 (1979).
22. P. J. Winch and J. C. Earnshaw, *J. Phys. Condens. Matter 1*: 7187 (1989).
23. I. Szleifer, D. Kramer, A. Ben-Shaul, D. Roux, and W. M. Gelbart, *Phys. Rev. Lett. 60*: 1966 (1988).
24. D.Langevin and C. Griesmar, *J. Phys. D: Appl. Phys. 13*: 1189 (1980).
25. S. Hard and H. Löfgren, *J. Colloid Interface Sci. 60*: 529 (1977).
26. A. Poskanzer and F. C. Goodrich, *J. Colloid Interface Sci. 52*: 213 (1975).
27. D. Langevin, *J. Colloid Interface Sci. 80*: 412 (1981).
28. Y. L. Chen, M. Sano, M. Kawaguchi, H. Yu, and G. Zografi, *Langmuir 2*: 349 (1986).
29. J. C. Earnshaw and P. J. Winch, *J. Phys. Condensed Matter, 2*: 8499 (1990).
30. J. C. Earnshaw, R. C. McGivern, and P. J. Winch, *J. Physique 49*: 1271 (1988).
31. J. Lucassen and R. S.Hansen, *J. Colloid Interface Sci. 22*: 32 (1966).
32. L. Ting, D. T. Wasan, K. Miyano, and S. Q. Wu, *J. Colloid Interface Sci. 102*: 248 (1984).
33. V. Thominet, C. Stenvot, and D. Langevin, *J. Colloid Interface Sci. 126*: 54 (1988).
34. C. Stenvot and D. Langevin, *Langmuir 4*: 1179 (1988).
35. E. H. Lucassen Reynders in *Anionic Surfactants* (E. H. Lucassen Reynders, ed.), Surfactant Science Series, Vol. 11, Marcel Dekker, New York, 1981. E. H. Lucassen Reynders, *Prog. Surface Membrane Sci. 10*: 253 (1976).
36. P. G. de Gennes, *Adv. Colloid Interface Sci. 27*: 189 (1987).
37. W. D. Garrett and W. A. Zisman, *J. Phys. Chem. 74*: 1796 (1970).
38. K. Ito, B. B. Sauer, R. J. Skarlupka, M. Sano, and H. Yu, *Langmuir 6*: 1379 (1990).
39. E. K. Mann and D. Langevin, *Langmuir, 7*: 1112 (1991).
40. R. Vilanove, D. Poupinet, and F. Rondelez, *Macromolecules 21*: 2880 (1988).
41. P. G. de Gennes, *Scaling Concepts in Polymer Physics,* Cornell Univ. Press, Ithaca, N. Y., 1979.

42. J. C. Le Guillou and J. Zinn-Justin, *Phys. Rev. Lett. 39*: 95 (1977).
43. R. Vilanove and F. Rondelez, *Phys. Rev. Lett. 45*: 1502 (1980).
44. B. Duplantier and H. Saleur, *Phys. Rev. Lett. 59*: 539 (1987).
45. B. B. Sauer, M. Kawaguchi, and H. Yu, *Macromolecules 20*: 2732 (1987).
46. M. Kawaguchi, M. Sano, Y. L. Shen, G. Zografi, and H. Yu, *Macromolecules 19*: 2606 (1986).
47. M. Joly in *Surface and Colloid Science* (Matijevic, ed.), Interscience, New York, 1972, Vol. 5, p. 91.
48. B. B. Sauer, W. G. Griffin, and Y. Yu, *Macromolecules 22*: 786 (1989).
49. M. Kawaguchi, B. B. Sauer, and Y. Yu, *Macromolecules 22*: 1735 (1989).
50. K. Motomura and R. Matuura, *J. Colloid Interface Sci. 18*: 295 (1963).
51. B. B. Sauer, H. Yu, C. F. Tien, and D. F. Hager, *Macromolecules 20*: 393 (1987).

12

Bilayer Lipid Membranes

J. C. EARNSHAW Department of Pure and Applied Physics, The Queen's University of Belfast, Belfast, Northern Ireland

I. INTRODUCTION

Biomembranes are fascinating supramolecular physicochemical systems whose biological activities arise directly from their molecular architecture and chemical composition. It is unnecessary to discuss this composition and structure in detail here. The basic lipid bilayer, apart from its inherent and vital functions, provides an appropriate environment for the membrane proteins. At the molecular level all is turmoil [1]: lipids and proteins diffuse in the membrane plane, proteins may rotate to present active sites to the inner and outer faces of the membrane, lipid acyl chains flex and straighten, "solid" and "fluid" lipid domains form and disperse, molecules bob up and down. This ceaseless activity is both inherent in the membrane and vital to its biological functioning.

The chemical and physical heterogeneity of real biomembranes poses real problems for the interpretation of experimental data from novel techniques. This complexity has encouraged the development of various chemically and physically more homogeneous model systems. The different membrane models reflect different aspects of the membrane behavior, in particular of the range of dynamic processes described. The physical uniformity of such controlled systems makes it easier to design experiments to detect particular modes of fluctuation. Similarly, control of the chemical composition permits investigation of the role of molecular structure upon the observed fluctuations. This chapter concentrates upon one specific model system, the bilayer lipid membrane (BLM) or black lipid film.

Quasi-elastic light scattering has been shown to be a useful probe of certain of the fluctuations present in BLM. To date only bilayers of the simplest molecular constitution have been studied, to avoid possible complication due to chemical composition (e.g., lipid-protein interactions). Such simple physicochemical models not only reflect many of the dynamic processes characteristic of biomembranes but also constitute interesting supramolecular systems in their own right.

II. THEORETICAL BACKGROUND

In general, lipid membranes sustain many modes of fluctuation [2, 3, 4], some of which scatter light. These range from motions of the membrane as a bulk system to intramembrane molecular modes (Fig. 1). Various theories describing the thermally excited fluctuations of thin molecular films have been developed. The molecular forces involved in the fluctuations must be highly complex, and so it is convenient to summarize their effects in terms of phenomenological viscoelastic properties [5]. These express the effects of microscopic forces as molecules are deformed from their equilibrium conformations and positions. There are few, if any, theoretical calculations of the viscoelastic properties of lipid bilayers. However, the experimentally observable fluctuations governed by the various membrane moduli have been discussed theoretically. Of the theoretical accounts to date, most [3, 4] follow the general lines laid out in Chapter 11, although an approach more reminiscent of the description of thin soap films has also been used [6, 7]. The former theories have sufficed to account for the observations to date. As in any physical system, each modulus corresponds to one mode of fluctuation. Some of these modes do not couple to the optical field and hence do not scatter light, but there remains a rich variety of membrane modes that are potentially observable. Only part of this potential has been exploited to date.

The multiplicity of membrane modes that may scatter light (Fig. 1) suggests

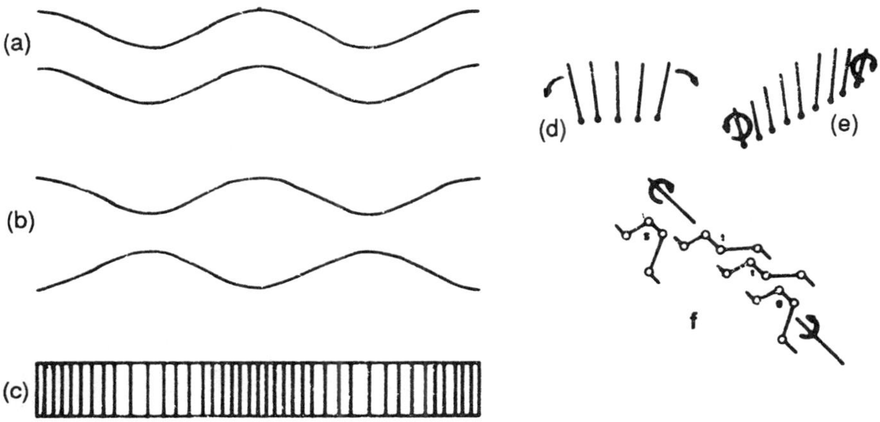

FIG. 1 Bilayer modes that can scatter light. Hydrodynamic modes that can scatter light include (a) capillary, (b) thickness, and (c) compression. Modes of molecular fluctuation include (d) splay of a row of molecules, (e) twist of such a row, and (f) fluctuations of "order" involving *trans* to *gauche* rotations of the C—C bonds of the acyl chains.

that the signal received will be complicated, but fortunately various symmetries simplify this apparent complexity by causing various modes to decouple. For example, the usual assumption of isotropy (or hexagonal symmetry) within the plane of the membrane causes the shear fluctuations in the membrane plane to decouple from the other modes [3]. As the shear modes do not couple to the optical field, no further discussion is necessary. For BLM, which are effectively interfaces between identical fluids, the transverse or capillary modes further decouple from the longitudinal or compression modes. There are, indeed, density waves in the two apposed monolayers comprising the BLM that are coupled to the transverse waves, but they are in antiphase. Those longitudinal modes in the two monolayers which *are* coherent couple to thickness fluctuations of the BLM. The latter will, in principle, scatter light, but for real systems are restricted to wavenumbers far above the experimentally accessible range [8]. Further, except perhaps near a critical point of the membrane, scattering by the compression modes is negligible [9].

The predominant scattering thus occurs from the transverse or capillary waves on the BLM. The scattered intensity is not large, depending as it does upon the difference of the dielectric constant (ε_b) of the BLM and that of the ambient fluid (ε_0) and upon the BLM thickness ($h \sim 4$ nm):

$$I_s \propto \frac{h^2(\varepsilon_b - \varepsilon_0)^2}{\gamma q^2} \tag{1}$$

where the dependence upon q indicates the thermal nature of the excitation of the capillary waves [9]. Despite the low scattered intensity, it has proved possible to detect and measure the spectrum of the scattering.

The temporal evolution of a capillary wave of wavenumber q is characterized by a frequency ω. The dispersion relation connects ω to q, involving the material properties of the system—the membrane tension γ, the viscosity η, and the density ρ of the ambient fluid:

$$i\omega - \frac{\gamma}{2\rho}\frac{q^2}{m}\frac{q-m}{i\omega} = 0 \tag{2}$$

where

$$m = \sqrt{q^2 - \frac{i\omega\rho}{\eta}} \qquad Re(m) > 0. \tag{3}$$

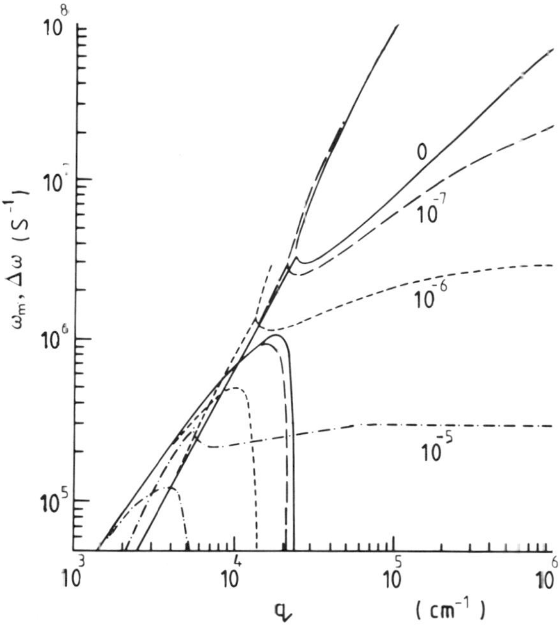

FIG. 2 The dispersion behavior of capillary modes upon a membrane separating identical aqueous fluids. The membrane tension was taken to be 3 mN/m. As q increases, the wave frequency and damping increase until, at a critical value of q, overdamping sets in, leaving two overdamped modes. The broken lines indicate the effects of transverse shear viscosity of various magnitudes in mN·s/m. As μ increases, the critical q falls, and the faster of the overdamped modes ceases to exist at progressively lower $\Delta\omega$. (Reproduced from the *Biophysical Journal*, 1987, *52*, 87–94 by copyright permission of the Biophysical Society.)

Our BLM were formed in a vertical plane, so that gravitational effects can be neglected. The dispersion behavior is sketched in Fig. 2.

At low q the roots of the dispersion equation are complex conjugates ($\omega = \omega_q + i\Delta\omega_q$): the waves evolve as damped oscillations. At larger q the dissipation increases relative to the driving forces, and above a critical value of q the fluctuations are overdamped, Eq. 2 having two imaginary roots. Approximate expressions for ω are useful: in the propagating regime,

$$\omega = \sqrt{\frac{\gamma q^3}{2p}} + \frac{i\eta q^2}{\rho} \tag{4}$$

whereas in the overdamped region,

$$\Delta\omega_1 = \frac{\gamma q}{4\eta} \tag{5}$$

$$\Delta\omega_2 = \frac{2\eta q^2}{\rho}. \tag{6}$$

$\Delta\omega_1$ governs the slow recovery of the membrane at a rate determined by capillary forces, whilst $\Delta\omega_2$ depends upon the inertia of the system.

A. Membrane Viscoelasticity

Membrane dissipation may be introduced into the above formalism by expanding the tension as a positive definite response function [3, 5]:

$$\tilde{\gamma} = \gamma + i\omega\mu \tag{7}$$

where γ is the membrane tension and μ is that membrane viscosity appropriate to shear stress applied normal to the membrane plane. This is *not* the conventional "membrane viscosity," which applies to shear *within* that plane. The exact nature of this viscosity, or indeed of other interfacial viscosities, is not well understood. While it may be regarded as an interfacial excess quantity [10, 11], for molecular films such as the present BLM it is probably best thought of as a macroscopic property of the system. The main effect of the extra dissipation represented by μ is to increase the wave damping, by $\mu q^3/4\rho$ to first order. The increased dissipation will also lower the q value at which critical damping occurs. These effects are sketched in Fig. 2 for various constant values of μ.

The assumptions of constant membrane tension and viscosity, corresponding to the Voigt or viscoelastic solid, may be unnecessarily restrictive—viscoelastic relaxation may occur. There is as yet no theory concerning viscoelastic relaxation of membranes comprising lipid molecules, and so we resort to the other simple linear viscoelastic model: the Maxwell fluid. In this case there is an exponential relaxation of stress with a single time constant τ, and we have (cf, Ref. 12)

$$\gamma = \gamma_e + G\frac{\omega^2\tau^2}{1 + \omega^2\tau^2} \tag{8}$$

$$\omega\mu = G\frac{\omega\tau}{1 + \omega^2\tau^2} \tag{9}$$

where G is the strength of the relaxation process and γ_e is the thermodynamic, equilibrium membrane tension. Here ω is the frequency of the applied perturbation, which in the present case, for propagating waves, is ω_q.

Such viscoelastic relaxation will, of course, modify the effects of $\tilde{\gamma}$ upon the propagation of the capillary waves. The general effects are sketched in Fig. 3. At $\omega_q \ll 1/\tau$ we have $\gamma = \gamma_e$ and $\mu = G\tau$, while for $\omega_q \gg 1/\tau$, $\gamma = \gamma_e + G$ and $\mu \sim 0$. Thus the dispersion behavior changes from that characteristic of constant tension and membrane viscosity to that for a higher tension and zero μ. Interesting effects may occur at $\omega_m \sim 1/\tau$.

B. Further Scattering Processes

Turning to intramembrane molecular modes, fluctuations of twist, splay, and ordering are predicted to occur and may scatter light [4]. Splay couples to capillary waves, while the ordering fluctuations couple to the compression waves. The main effects for BLM thus seem likely to arise from the coupling of the splay to the transverse modes. Assuming that the lipid molecules tend to orient themselves normal to the local membrane plane, this coupling affects the light scattering: the frequency and intensity of the scattered light may both be modified. The potential effect upon the frequency arises from an increase in the effective tension:

$$\gamma_{\text{eff}} = \gamma + Kq^2 \tag{10}$$

where γ is the "bare" BLM tension and K is the splay (or bending) modulus. However, typical values of K for lipid bilayers [13] are so low ($K \sim 10^{-19}$ J for phospholipid vesicles) that the Kq^2 term would be completely negligible except for quite inaccessible values of q (experimentally $q < 2000$ cm^{-1}). The splay modes themselves scatter light, and this may interfere with the light scattered by the capillary modes [4]. The intensity effects depend upon the combined dielectric anisotropy and order parameter of the BLM ($\varepsilon_a S_0$) relative to ε_b.

Various other effects may arise: for polar lipids, for example, or when electric fields are applied across the BLM [6, 7]. To date it has not proved necessary to go beyond the theory sketched above. Certain membrane modes that might scatter light are, in practice, not readily observed.

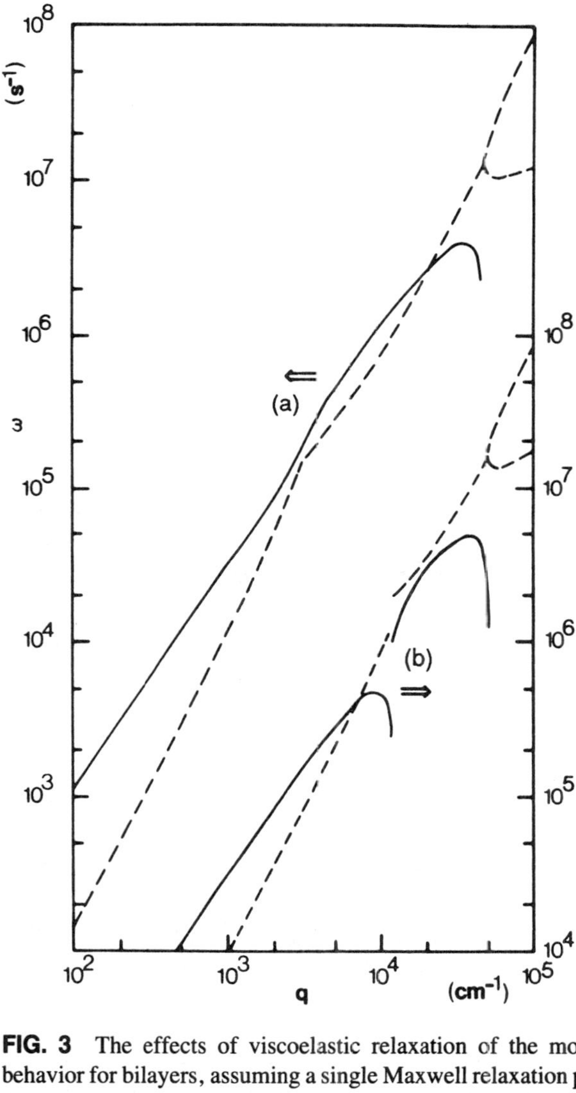

FIG. 3 The effects of viscoelastic relaxation of the modulus $\tilde{\gamma}$ upon the dispersion behavior for bilayers, assuming a single Maxwell relaxation process. The full lines indicate the wave frequency, while the broken lines indicate $\Delta\omega$. Parameters used were $\gamma_e = 3$ mN/m, $G = 2.7$ mN/m, and (a) $\tau = 3.7$ μs or (b) $\tau = 0.37$ μs. (Reproduced from the *Biophysical Journal*, 1987, *52*, 87–94 by copyright permission of the Biophysical Society.)

III. PRACTICAL CONSIDERATIONS

Bilayer lipid membranes comprise a bimolecular film of lipid molecules separating two aqueous phases. Unfortunately, they are not in thermodynamic equilibrium, and it has not proved easy to form BLM suitable for light scattering studies. Black lipid films tend to be both relatively small and nonplanar. The bimolecular films are surrounded by relatively thick Plateau-Gibbs borders that contribute to the lack of flatness; they are also "puckered and convoluted" [14]. Light scattering requires that a locally flat and stable portion of the BLM be available, at least as large as the region illuminated by the laser beam, and separated from the Plateau-Gibbs border. These requirements of flatness and stability have been almost insuperable. However it has proved possible [15] to form suitable membranes of one lipid, glycerol monooleate (GMO). This neutral lipid appears to be sterically suited to bilayer formation and conveniently has a transition temperature rather below room temperature [16]. As will be seen below, it has proved possible to vary the physical state and chemical composition of the BLM for this material. Apart from one early study, involving BLM formed from oxidized cholesterol [14], all published work has been on GMO membranes.

Even when a locally flat region could be found, it transpired that the local orientation of GMO membranes changed randomly and erratically [17]. While small, these changes sufficed to alter the q value associated with a fixed detector position. It appeared that temperature changes promoted these effects, and a servocontrolled optical system was devised to permit studies of thermotropic transitions in BLM [18]. It is possible that these effects might not be so severe for a different lipid.

In all our studies the temporal evolution of the thermally excited capillary waves upon BLM was measured by photon correlation. Details of the experimental refinements necessary for work with these thin and delicate structures are to be found in the original papers.

The spectrum of light scattered by thermally excited capillary waves of given q upon a bilayer is just the power spectrum of the waves. Within experimental precision the form

$$G(\tau) = B + A \cos(\omega_m|\tau| + \phi) \exp(-\Delta\omega\tau) \exp \frac{\Delta\omega_i^2\tau^2}{4} \tag{11}$$

has been found adequate to fit the observed correlation functions in the propagating regime. The phase term ϕ allows for the deviations of the spectrum from an exact Lorentzian form, and the final Gaussian accounts for instrumental linebroadening effects. Figure 4 shows a typical experimental correlation function. The agreement with the above form indicates that other scattering processes do not contribute significantly to the observed data. This has been significant in

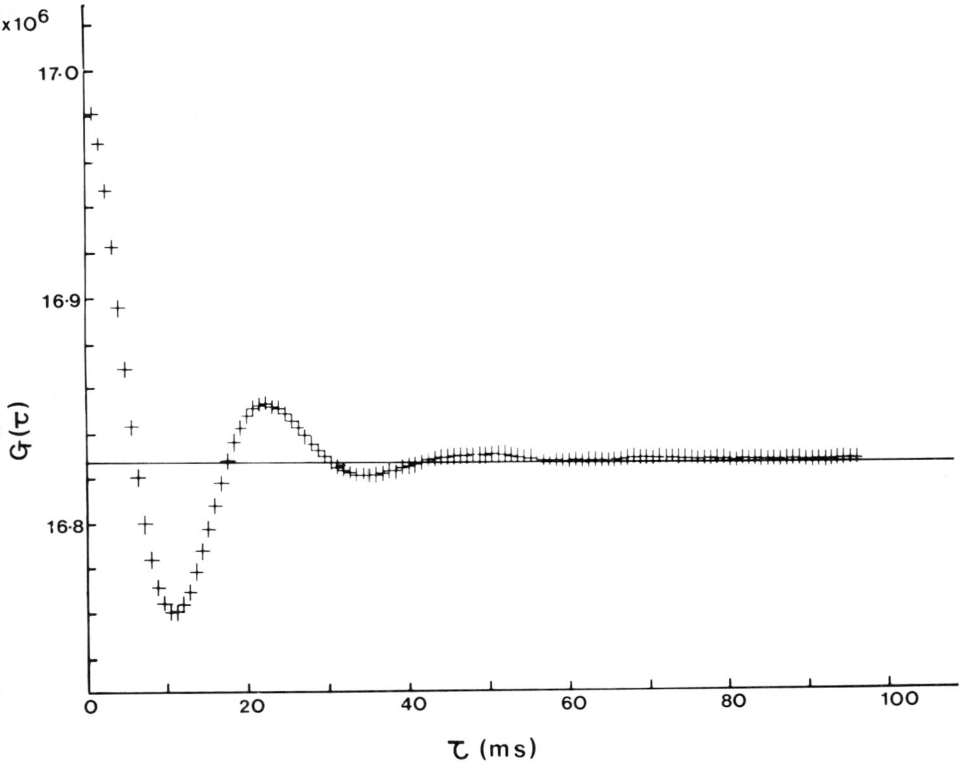

FIG. 4 A typical observed correlation function for $q = 922.3\ \text{cm}^{-1}$. The line is the calculated background level. (From Ref. 20.)

certain studies. The wave propagation frequency ω_m and the temporal damping constant $\Delta\omega$ thus estimated can be shown to be unbiased estimates of the corresponding theoretical quantities predicted from the dispersion equation for known system properties [19].

Under appropriate circumstances, overdamped correlation functions have been observed. Figure 5 shows a function displaying two exponential modes; the two decay constants agree with expectation (Eqs. 5 and 6).

It is further possible to estimate from the same data the relative intensities of both the scattered light I_s and the heterodyne reference beam I_r. Together these two quantities can provide a considerable amount of information. The I_r values depend upon the specular reflectivity of the membrane, which is proportional to h^2.

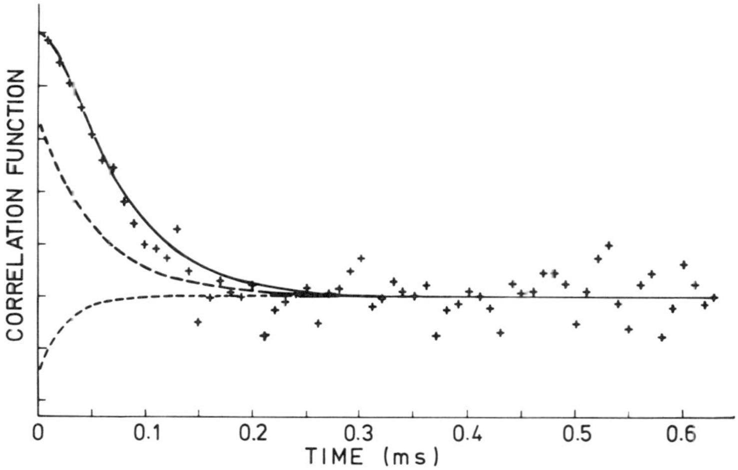

FIG. 5 An example of an overdamped correlation function observed for a bilayer immersed in a 40% aqueous glycerine solution. The function comprises two exponentially decaying modes, which are indicated. (Reproduced from the *Biophysical Journal,* 1983, *41,* 197–210 by copyright permission of the Biophysical Society.)

IV. EXPERIMENTAL RESULTS

For bilayers formed from GMO dissolved in decane, the variations with q of ω_m and $\Delta\omega$ deduced by photon correlation were in accord with the predictions of the hydrodynamic theory mentioned above (Fig. 6). The membrane tension and the viscosity of the ambient fluid were estimated by solving the dispersion equation using the observed values of ω_m and $\Delta\omega$ for a given q value. For both quantities the results were in good accord with established values. The curves of Fig. 6 represent the theoretical dispersion behavior, based on the average tension and the known viscosity of the aqueous phase: there is no fitting involved. The data agree excellently with the predicted variations.

Loading the ambient medium with glycerine at various concentrations demonstrated clearly that the capillary wave damping on the GMO/decane membranes was essentially determined entirely by viscous dissipation within this medium (Fig. 7). These results were significant because they demonstrated the correctness of the experimental determination of $\Delta\omega$ and of the treatment of the instrumental linebroadening in the data analysis.

These GMO/decane membranes are known to incorporate quite large quantities of hydrocarbon solvent in their bilayer structure. This has several effects upon light scattering experiments. Firstly, the intermolecular interactions appear to be reduced, particularly those aspects of energy dissipation within the BLM

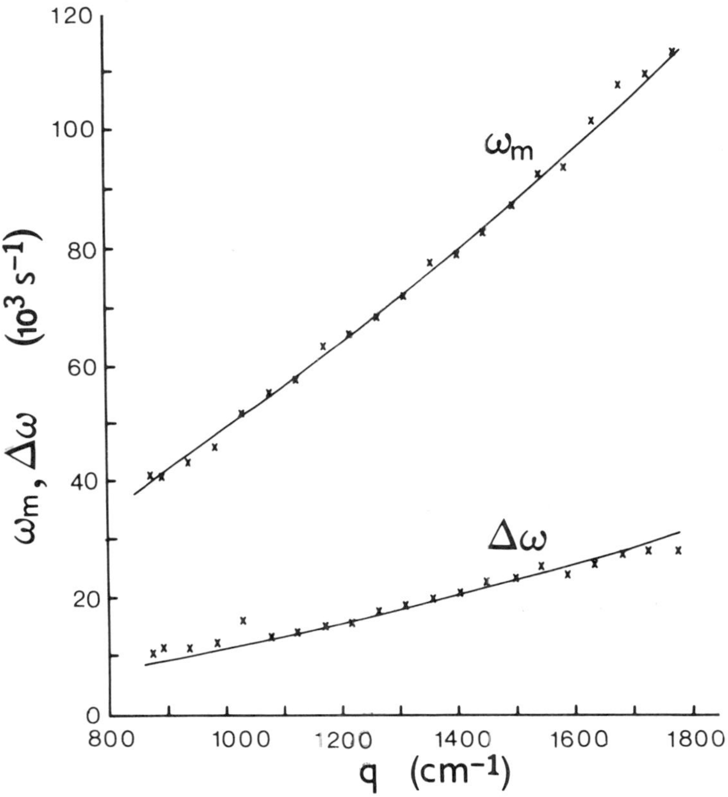

FIG. 6 Experimental ω_m and $\Delta\omega$ values for a GMO/decane bilayer, over a range of wavenumbers. The lines indicate theoretical variations based upon the average γ and the accepted η.

that manifest themselves via the membrane viscosity μ. Secondly, the scattered intensity from such membranes fluctuated considerably, apparently because of variations in local membrane thickness and tension at points where the excess solvent disproportionated into lenses. This made it difficult to determine the scattered intensity from the photon correlation records.

"Solvent-free" bilayers can be formed from GMO dispersed in the long-chain hydrocarbon squalane ($C_{30}H_{62}$). As the quotes suggest, these BLM are not exactly solvent free, but they do incorporate substantially less hydrocarbon than those formed from GMO dissolved in short chain solvents.

A study of the variation of the intensity of light scattered by such membranes (Fig. 8) nicely confirmed the thermal excitation of the capillary waves on the

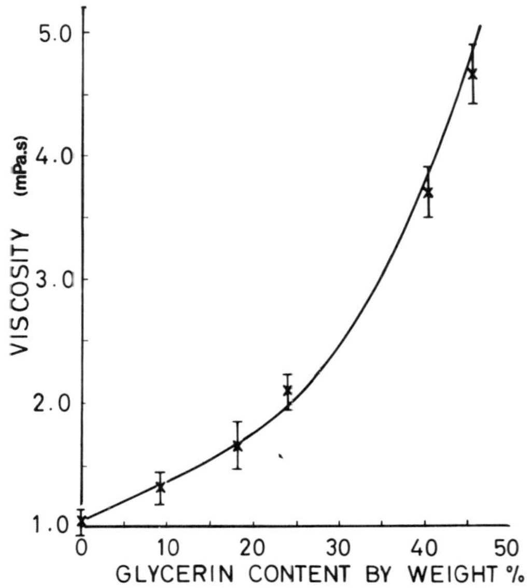

FIG. 7 The viscosity of the ambient medium inferred from bilayers immersed in aqueous solutions of glycerine, compared to the accepted variation. (Reproduced from the *Biophysical Journal*, 1983, *41*, 197–210 by copyright permission of the Biophysical Society.)

BLM [20]: the best-fit power law was of exponent -1.87 ± 0.20, close to the expected value of -2.

These initial exploratory experiments thus demonstrated that the process scattering the light was indeed the thermally excited capillary or transverse waves upon the BLM. The physical properties of the system inferred from the observations were in good accord with literature values. Comparison values of the bilayer tension can be taken as approximately twice the tension of an oil-water interface supporting a lipid monolayer, as BLM tensions *per se* are rarely measured. The viscosity of the ambient fluid deduced for the case of those BLM containing solvent within their structure (for which the wave damping is apparently entirely due to the fluid viscosity) agreed excellently with the accepted values. The scattered intensities clearly reflected the thermal origin of the transverse fluctuations.

Changes in the chemical composition of the system induced detectable changes in the observable properties. Increasing the lipid concentration in the film-forming solution led to an increased lipid packing density in the two apposing monolayers, causing a measurable decrease in BLM tension [15]. The results (Fig. 9) were in reasonable accord with the results for studies of similar monolayers.

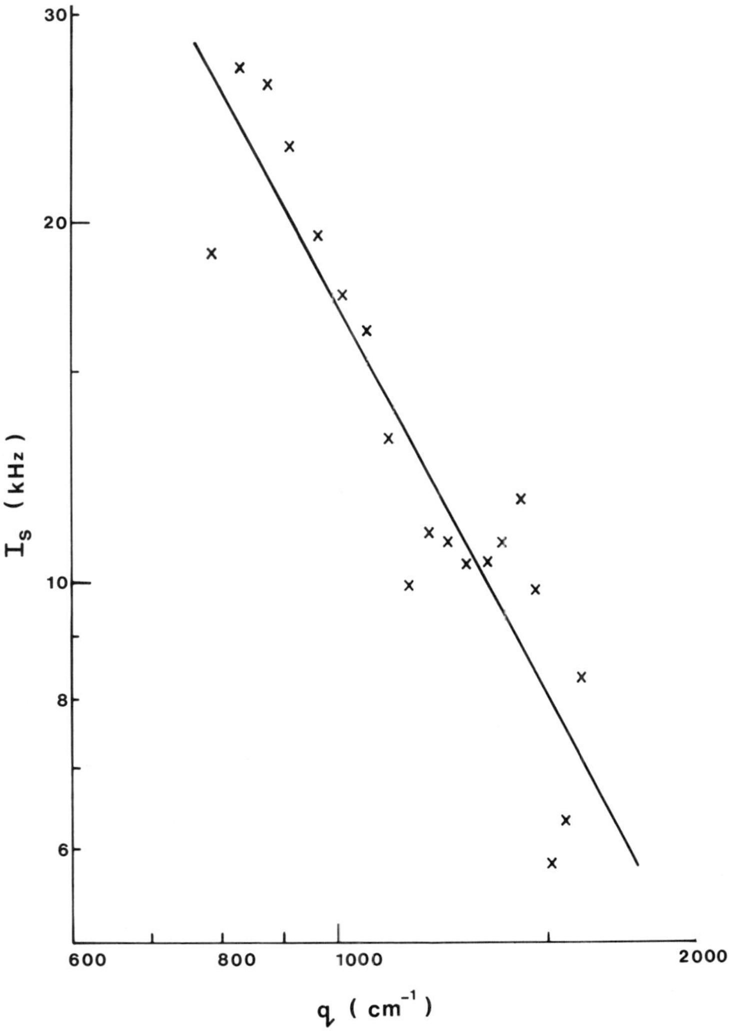

FIG. 8 The variation of I_s with q for a "solvent-free" GMO bilayer. The line is the best-fit power law, having exponent equal to -1.87 ± 0.20. (From Ref. 20).

For certain bilayer systems the measured $\Delta\omega$ values exceeded those predicted from the ambient viscosity. In such cases there must clearly be further dissipative processes: the one most directly affecting the capillary waves on a symmetric membrane is the transverse shear interfacial viscosity μ.

Examples of systems in which excess dissipation has been observed include

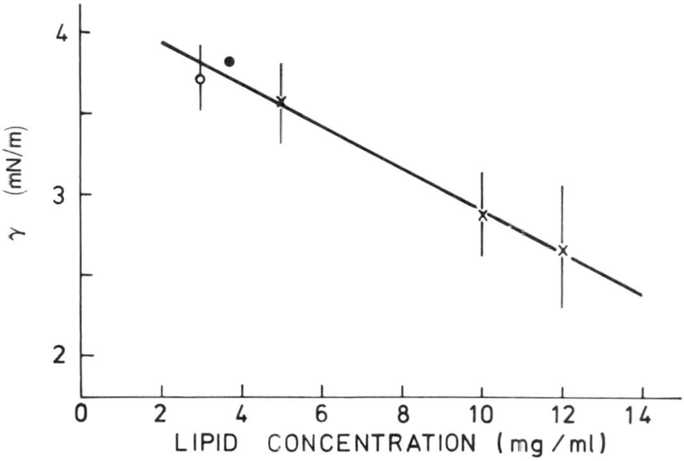

FIG. 9 The variation of γ with lipid concentration of the BLM-forming solution. The line is a fit to all the data, including literature values (○, Ref. 21; and ● Ref. 16). (Reproduced from the *Biophysical Journal*, 1983, *41*, 197–210 by copyright permission of the Biophysical Society.)

"solvent-free" bilayers [15] and those formed from solutions of GMO and cholesterol in decane [22]. In the latter case both the BLM tension and the transverse viscosity rose with the cholesterol content in the film-forming solutions (Fig. 10), apparently saturating at concentrations similar to those at which many other membrane properties also saturate. However, it was notable that the maximum value of μ for GMO/cholesterol membranes was similar to values found for "solvent-free" BLM ($\sim$34 nN·s/m). This presumably indicates that the free "interactive volume" available to the lipid acyl chains is similarly reduced from that in GMO/decane bilayers in both cases. Unfortunately the exact compositions of these bilayers were uncontrolled, and remain unknown, so quantitative deductions of the effects of cholesterol are somewhat difficult. The detailed implications of the above correspondence remain to be elucidated.

A. Thermotropic Transitions

All of the experiments described above were restricted to room temperature, well above the reported transition of GMO. Further insight into the membrane processes underlying novel properties such as the transverse shear viscosity μ might be expected from their temperature dependence. We have thus studied the transitional behavior of GMO bilayers [23]. While all of the BLM types discussed to date have been investigated, the fullest studies have been those upon "solvent-free" membranes.

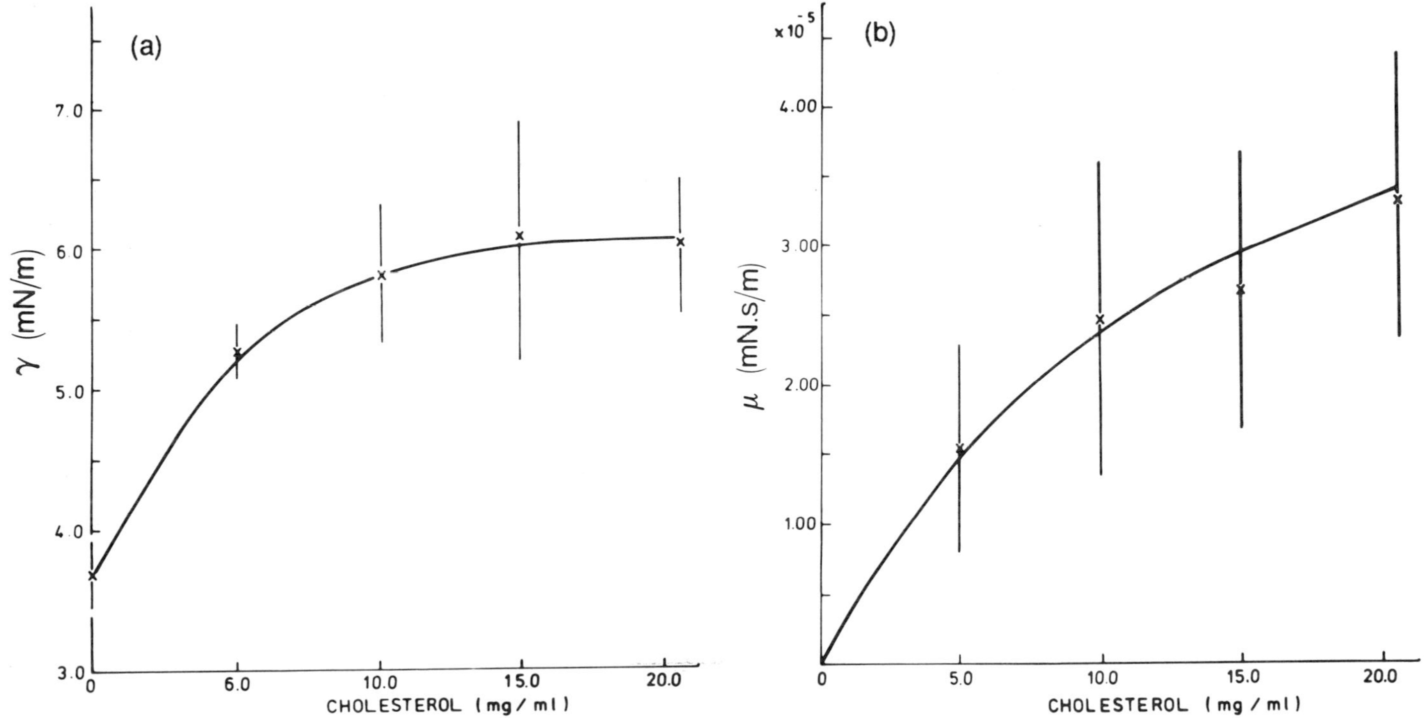

FIG. 10 The dependence of γ and μ upon concentration of cholesterol in the film-forming solutions. The lines are to guide the eye. (Reproduced from the *Biophysical Journal*, 1983, *41*, 211–216 by copyright permission of the Biophysical Society.)

For GMO/decane bilayers, the temperature variation of the tension changed slightly at about 12 and 18°C. Over the range 10–22°C no significant hysteresis was observed. The damping of the capillary waves observed for these BLM was entirely consistent with that expected for the ambient viscosity η over the entire range of temperature.

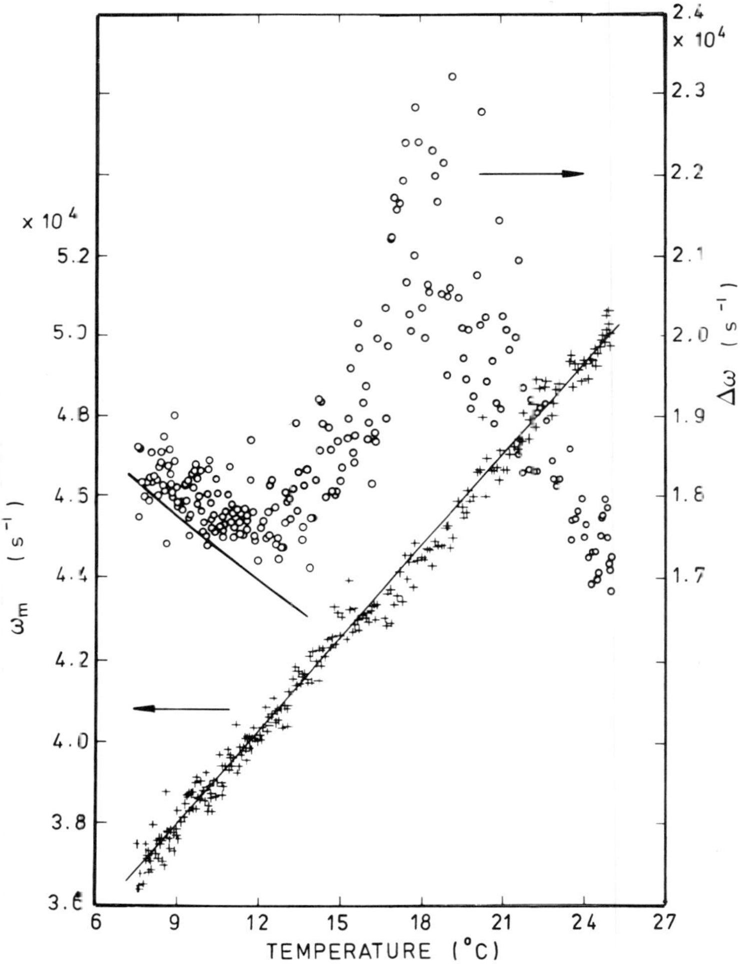

FIG. 11 The frequency and damping of capillary waves on a "solvent-free" GMO bilayer (q = 1275 cm^{-1}). The line through ω_m is a linear fit, while that with $\Delta\omega$ indicates the variation expected due to η (this line extrapolates smoothly to higher T). (Reproduced from the *Biophysical Journal*, 1986, *49*, 869–889 by copyright permission of the Biophysical Society.)

This was not the case for "solvent-free" bilayers, for which the $\Delta\omega$ values above about 12°C clearly exceeded those expected for this viscosity (Fig. 11). The variations of ω_m and of the difference between $\Delta\omega$ and that expected translate into the temperature variations of γ and μ shown in Fig. 12. The behaviors of both properties change at two temperatures, 12.5 and 17°C. At the lower temperature μ increases from near zero, apparently independent of temperature, then rises until 17°C, above which temperature it falls again. This behavior is suggestive of pretransitional fluctuations in the response function of a system undergoing a first-order phase transition at a temperature T_t close to a critical point T_c of the system [24]. The rather small accompanying change in γ is compatible with the slight change expected in the "order parameter" relevant to the modulus $\tilde{\gamma}$

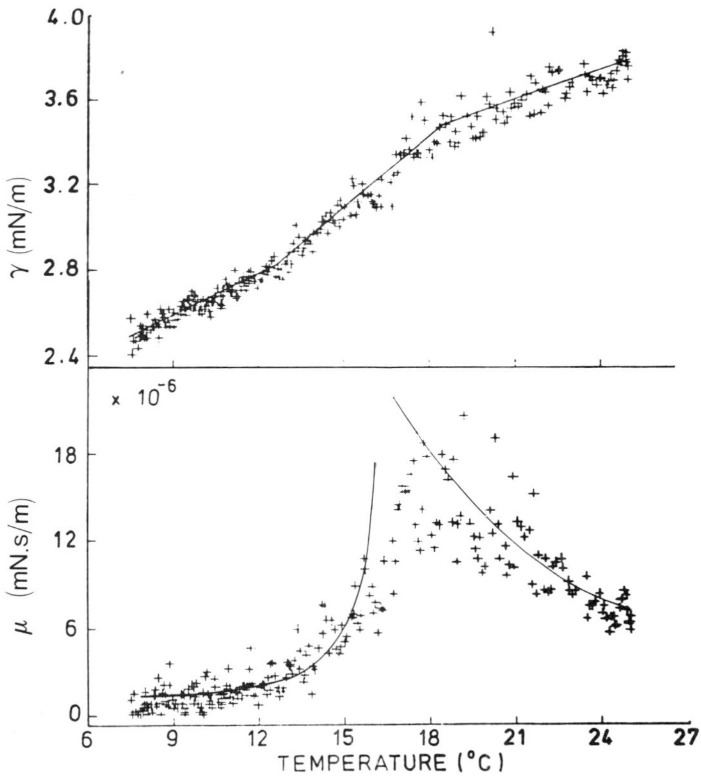

FIG. 12 Membrane tension and viscosity as functions of temperature, derived from the data of Fig. 11. The curves indicate the classical precritical fluctuations of μ. They are asymptotic to 12.7 and 16.8°C. (Reproduced from the *Biophysical Journal,* 1986, *49,* 869–889 by copyright permission of the Biophysical Society.)

under such circumstances. The transition is *relatively* far from T_c and so a Landau expansion should be valid: the response function should show classical exponents. The lines drawn in Fig. 12 show such variations fitted to the μ data. The model seems to fit the data self-consistently.

As already explained, the same correlation data that provide values of γ and μ yield estimates of the intensity I_s of the scattered light and that of the heterodyne reference beam I_r. The latter depends upon the membrane reflectivity and hence upon the mean square membrane thickness h^2, using the near constancy of the refractive index of the BLM [25]. With a literature value for h at one particular temperature [26] as a calibration datum, the temperature variation of h could be inferred (Fig. 13). The large peak in the apparent I_r between 12 and 14°C appears to derive from an extraneous scattering process and does not imply a change in h. Two main features are apparent: about 16.6°C the rms thickness changes by about 8 Å, and away from this temperature h tends to increase with T.

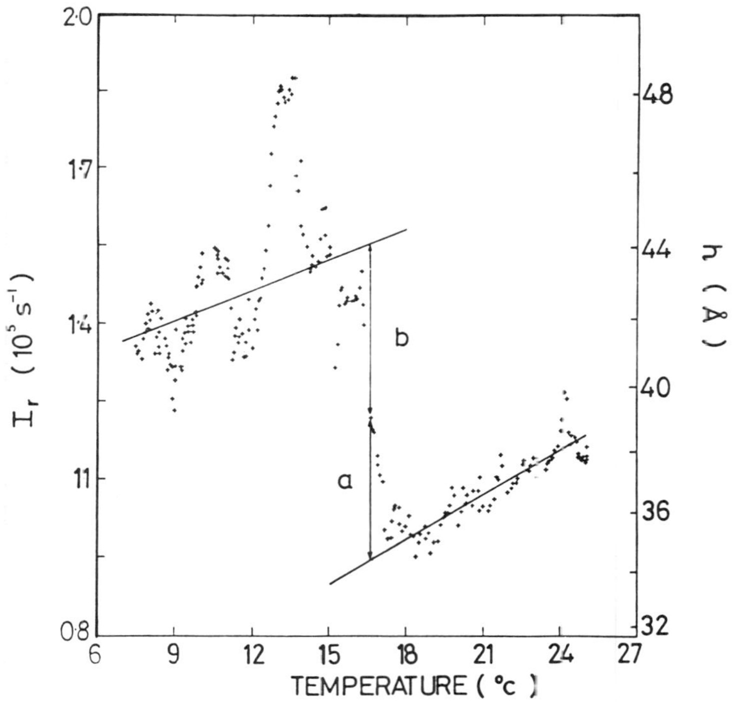

FIG. 13 The reference beam intensities for the bilayer of Fig. 11. The lines indicate linear fits to the data away from the transition, neglecting the peak between 12 and 14°C. The right-hand scale shows the corresponding rms membrane thickness. (Reproduced from the *Biophysical Journal*, 1986, *49*, 869–889 by copyright permission of the Biophysical Society.)

The first point is as expected: the appearance of some seven *gauche* bonds in the acyl chains of the lipid molecules in a bilayer at a chain-melting transition would lead to an 8 or 9 Å reduction in the membrane thickness. The midpoint of the variation of h^2 yields a precise value of 16.6°C for T_t. Just below the transition the rms thickness is 44 Å, close to twice the length of an all-*trans* GMO molecule. This variation of h^2 is closely paralleled by a rather small increase in γ that is superposed on the broader triphasic behavior of Fig. 12. This similarity is just a reflection of the effect of a decrease in the lipid packing density accompanying the chain-melting process. The second clear feature of Fig. 13 is the positive temperature gradient of h both above and below the T_t. Naive expectation suggests a constant thickness below the transition and a negative gradient above it. Above T_t our data are supported by observations [16] of a decrease with T of membrane capacitance ($\propto 1/h$). It seems possible that an

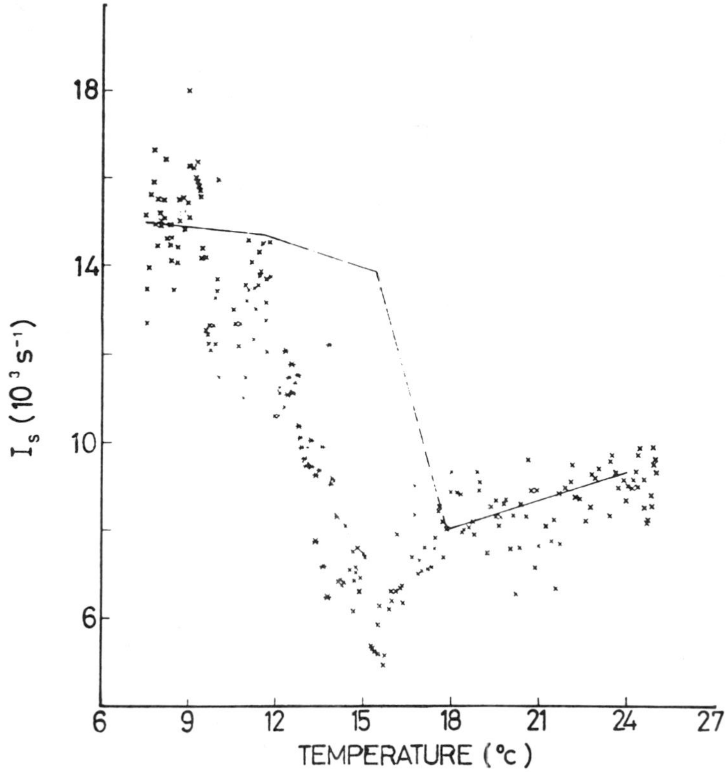

FIG. 14 I_s data for the same bilayer as Fig. 13. The lines show the variations expected from Eq. 1, using the data of Figs. 12 and 13. (Reproduced from the *Biophysical Journal*, 1986, *49*, 869–889 by copyright permission of the Biophysical Society.)

increase in h with T could arise from an averaging of thickness fluctuations having large q.

The membrane reflectivity, and thus h^2, shows quite large fluctuations below T_t, leaving aside the apparent peak between 12 and 14°C. These probably reflect the coexistence within the membrane of regions of all-*trans* lipid and areas comprising chain-melted lipid. Such a model certainly accords with observation of fluorophore diffusion in GMO bilayers [27], in which the diffusion constant changed only slightly across T_t (cf. much larger changes in lecithin membranes, in which such coexistence does not occur).

The scattered intensity I_s depends upon both h^2 and γ, and so, given the established temperature dependences of these quantities (Figs. 12 and 13), the expected form of variation of I_s can be predicted. This is shown in Fig. 14, normalized to the observed value at 24°C. Clearly, well above T_t the agreement is good, but below 18°C a major discrepancy is evident. To summarize briefly a rather complex argument, we believe that the scattered intensities were here reduced by interference with scattering from different fluctuations. However, the observed I_s values derived from the amplitudes of the correlation functions, which showed no evidence of any other processes in their time dependence. We conclude that the second process must be closely coupled to the capillary waves, so that it does not materially alter the spectrum of the scattered light. These criteria are met by fluctuations of molecular splay in the bilayer, which couples to the capillary waves. As the bilayer dielectric constant is basically independent

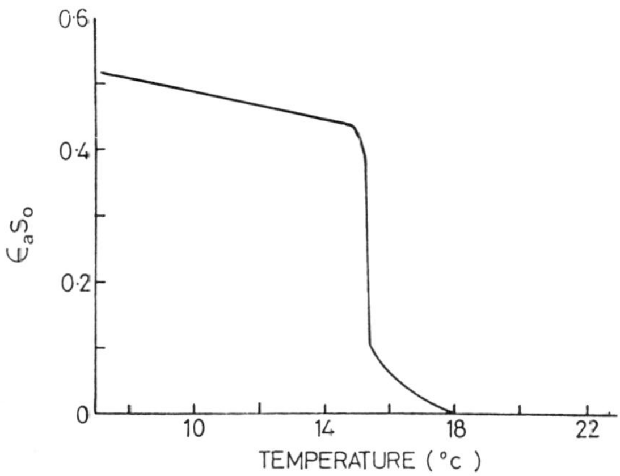

FIG. 15 The variation of $\varepsilon_a S_0$ derived from the discrepancies between the data and the line of Fig. 14. As the line is normalised to those data at 25°C, the present zero value above 18°C is purely instrumental. However, the form of the variation appears assured. (Reproduced from the *Biophysical Journal,* 1986, *49,* 869–889 by copyright permission of the Biophysical Society.)

of T, comparisons of the observed deviations of I_s from the expected behavior with the predictions of Fan's theoretical arguments [4] permitted the temperature variation of the product of the anisotropic part of the dielectric constant and the lipid order parameter ($\varepsilon_a S_0$) with T to be estimated (Fig. 15). The arguments are too lengthy to be rehearsed here: see Ref. 23. Certain instrumental uncertainties affected the absolute values of $\varepsilon_a S_0$. However, the variation of the relative values seems much better assured and indeed accords well with both intuition and the rather scanty available information on molecular order in GMO bilayers.

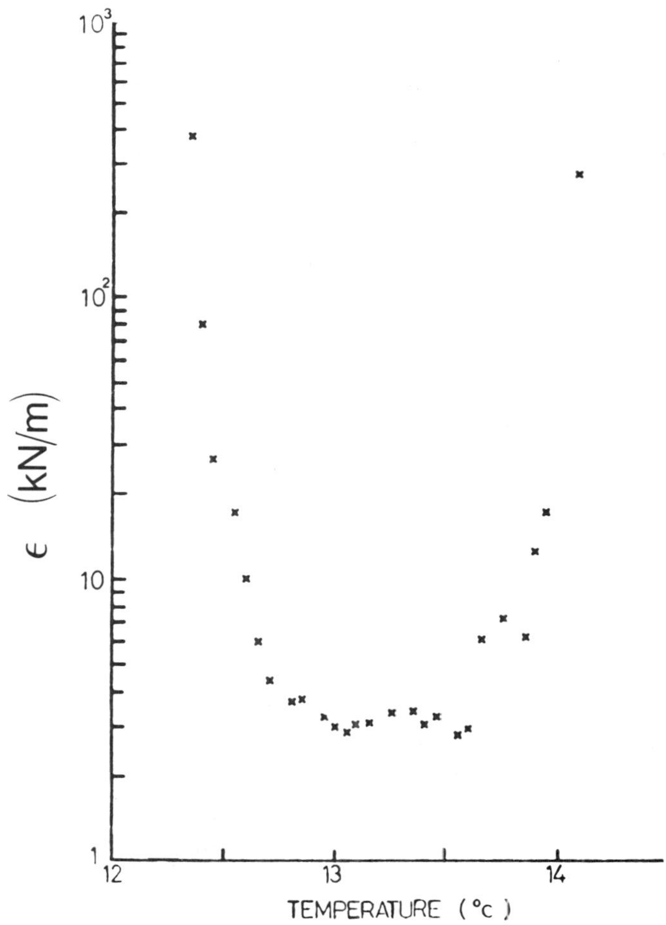

FIG. 16 The apparent variation of the bilayer compression modulus inferred from the peak in I_r at 12–14°C of Fig. 13. (Reproduced from the *Biophysical Journal*, 1986, *49*, 869–889 by copyright permission of the Biophysical Society.)

A final piece of information came from the large peak between 12 and 14°C in the apparent I_r (Fig. 13), which does not form part of the heterodyne reference beam intensity but rather seems to arise from another process that scatters light independently. The associated time dependence of the scattered light was apparently averaged out in the first delay time of the observed correlation functions (~ 5 μs), and so the process must be fast. The most likely process seems to be compression waves, which would be both fast and highly damped. The intensity of this peak relative to the I_s expected for capillary waves (line of Fig. 14) permitted the temperature variation of the membrane dilational modulus ε to be determined (Fig. 16). Again the absolute values are not certain, because of instrumental uncertainties, but the form of the variation shown seems established.

These experiments revealed a remarkable amount of information on the transitions of "solvent-free" GMO bilayers. The simultaneous determination of five different membrane properties permitted correlation of transitional effects such as the parallel changes in h^2 and γ at T_t. The molecular anisotropy ($\varepsilon_a S_0$) collapsed at 15.5°C, just as h started to fall. These aspects of the results strongly suggest that the transition involves chain-melting of the lipid acyl moieties. Indeed this model is consistent with all of the present data.

We do not fully understand why the transitional changes seem to cluster about two temperatures: 12.5 and 16.6°C. It may signal true two-stage melting, as for two-dimensional systems [28], or the two temperatures may be just the upper and lower limits of metastability in the transition. More work is required to clarify these questions, as well as to reduce the uncertainties of some properties that have been noted.

B. Effects of Cholesterol upon the Transitions

The addition of rather large quantities of cholesterol to GMO bilayers produced certain remarkable effects on the membrane transitions [18]. As in the earlier studies of the effects of cholesterol, these studies involved GMO/decane BLM which, in the absence of cholesterol, showed no membrane viscosity. Figure 17 shows the temperature variations of both ω_m and $\Delta\omega$ for a bilayer saturated with cholesterol. Above about 17°C both varied smoothly, the damping being substantially greater than expected from the ambient fluid viscosity. Below the transition considerable fluctuations were apparent in both frequency and damping; the lower bound upon $\Delta\omega$ in this region approximately coincided with the variation expected for dissipation due to η. Above the transition the capillary waves upon the "fluid" bilayer were evidently affected by excess dissipation, as described by the transverse shear viscosity μ.

At low temperatures the illuminated area of the bilayer evidently comprised phase-separated regions or domains of differing properties (γ and μ). Now close inspection of Fig. 17 (particularly of ω_m) shows that in several cases a sequence

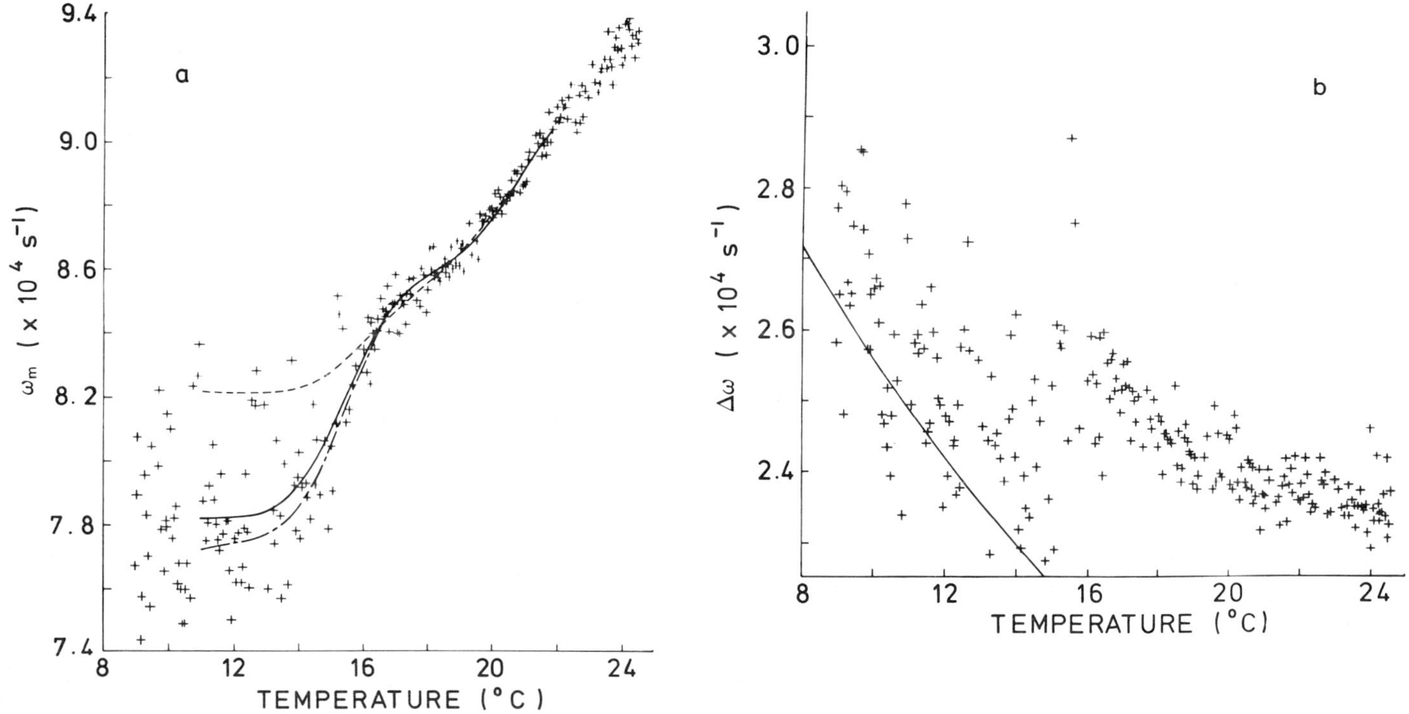

FIG. 17 Capillary wave frequency (a) and damping (b) for a cholesterol saturated GMO bilayer. The curves for frequency indicate different spline fits: — for all data, — - — omitting points above the full line at low *T*, --- omitting points below the full line at low *T*. For the damping, the line indicates the variation expected from the *T* behavior of η. (From Ref. 18.)

of successive observations adjacent in temperature lie together, close to a line extrapolated from the high-temperature "fluid" behavior. It is natural to assume that in these cases the illuminated area essentially comprised a single "fluid" domain, of diameter $\gtrsim 300$ μm. It is not clear how these domains evolved with time, whether by formation and dissolution or by gross movement within the bilayer. Similar but more marked observations on insoluble monolayers of pentadecanoic acid [29] suggest the latter hypothesis. Further work is required to elucidate the mechanism whereby cholesterol induces such macroscopic domains, whether by "fluidizing" the bilayer or by edge-stabilizing phase-separated regions [30].

C. Viscoelastic Relaxation

As we have seen, the transverse shear modulus of the membrane may exhibit viscoelasticity, the low-frequency viscosity transforming into a contribution to the elastic modulus at frequencies above the reciprocal of the relaxation time. This change will be apparent in the capillary wave propagation.

Just such effects have been observed, constituting the first observations of frequency dependence of surface viscoelasticity, apart from effects due to mass transfer between bulk and surface for surfactant solutions. In the early studies described above, the capillary wave dispersion seemed, within errors, to agree with the theoretical expectation for constant viscoelastic properties. However, later, more precise experiments [31] showed that while this seems to be true for GMO/decane bilayers, it is not appropriate for the "solvent-free" case. The q dependence of the wave frequencies for both types of BLM (well above T_t) are plotted in Fig. 18; the $\Delta\omega$ values behaved similarly, with greater errors. The lines drawn in the figure represent the variations predicted from the dispersion equation for constant system properties (γ and μ). For the "solvent-free" case, the measured data clearly deviate from the predicted line. In fact, when the measured wave frequencies and damping constants were analyzed, both γ and μ changed systematically with q. In particular, γ increased by about 0.5 mN/m, whilst μ fell by about an order of magnitude over the frequency range $2.5 \times 10^4 < \omega_0 < 7 \times 10^4$ s^{-1} (Fig. 19).

Various explanations were eliminated. One such involved an increase in the apparent tension due to the membrane bending modulus: this would not account for the observed fall in μ, and also the value of K inferred was many orders of magnitude greater than those found for lipid films by other methods [13].

The variations of both γ and μ were compatible with a single Maxwell relaxation process of time constant ~ 37 μs. The equilibrium membrane tension was compatible with accepted values. The latter point is significant, in that it demonstrates that there is no slower relaxation in the system affecting the modulus $\tilde{\gamma}$. The general conclusions were well verified in a later study of a fully com-

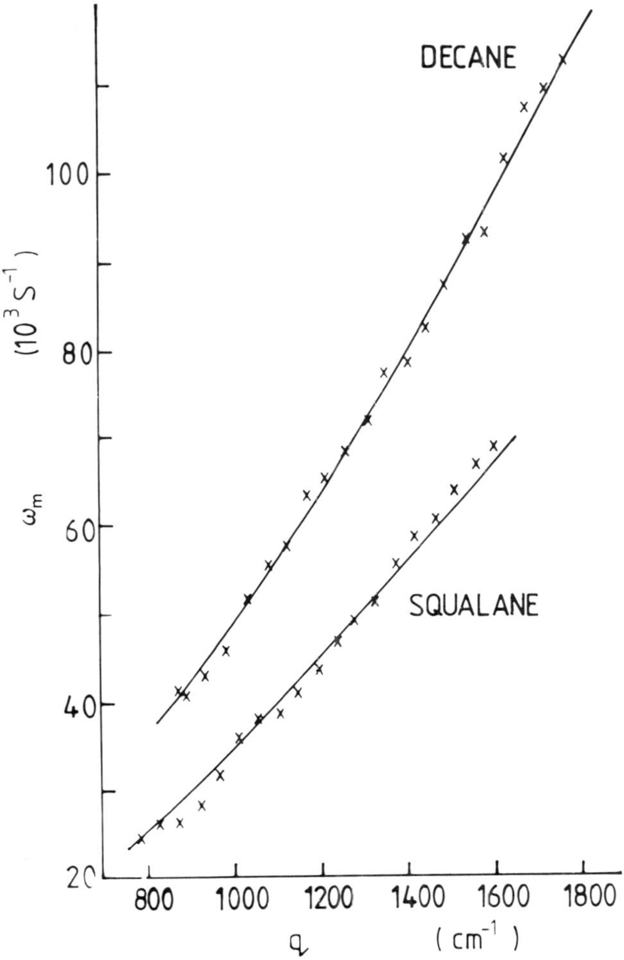

FIG. 18 Measured capillary wave frequencies for GMO/decane and "solvent-free" bilayers. The lines indicate the dispersion behavior expected for constant viscoelastic properties. (Reproduced from the *Biophysical Journal,* 1987, *52,* 87–94 by copyright permission of the Biophysical Society.)

pressed monolayer of GMO at the air/water surface [33]. In that case the relaxation time was rather smaller, about 10 μs. As the monolayer was expanded, the relaxation time fell, decreasing approximately exponentially with molecular area. The equilibrium tensions deduced from the light scattering data agreed with the values measured using a Wilhelmy plate.

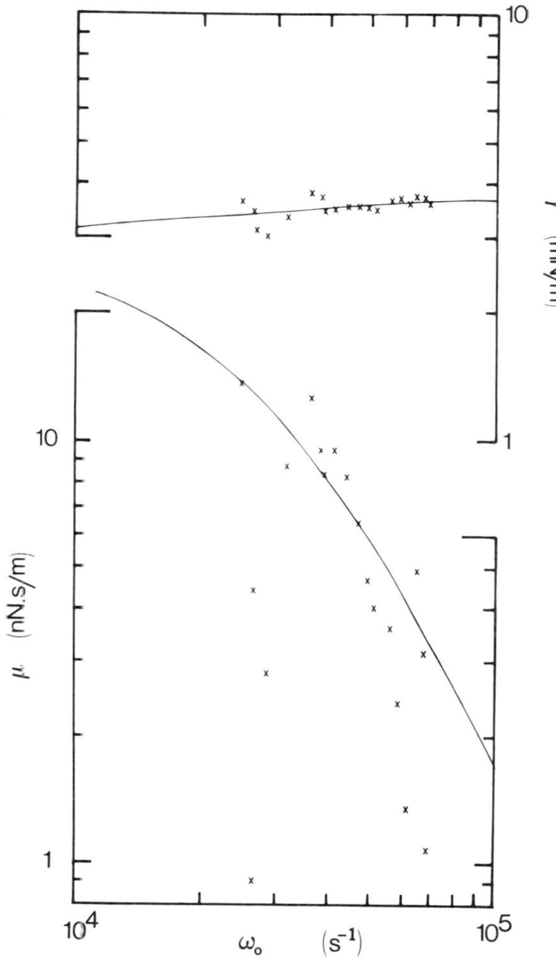

FIG. 19 BLM tension and viscosity as functions of ω_{π}. The lines are the best-fit Maxwell relaxation functions, as Eqs. 8 and 9. (Reproduced from the *Biophysical Journal,* 1987, *52,* 87–94 by copyright permission of the Biophysical Society.)

Because of a lack of any direct relation between the hydrodynamic modes of the membrane and microscopic motions, it is not possible to link this relaxation phenomenon directly to molecular processes. However, several arguments suggest an association with chain melting of the lipid molecules. Computer simulations have shown capillary wave-like motions of bilayers associated with transverse motions of lipid molecules, involving chain melting [34]. A time scale $\gtrsim$ μs has been suggested both experimentally [35] and from simulations [36] for chain melting, in reasonable accord with the present values for BLM or fully compressed monolayers. The speeding up of the relaxation as a monolayer is expanded is entirely plausible, because of the increase in conformational freedom of the hydrocarbon chains at larger molecular areas. The slower relaxation in BLM could be a natural consequence of the steric hindrance exerted by the apposed monolayers.

The temperature dependence of the viscoelastic relaxation would be interesting, particularly around the membrane transition. Unfortunately, a study of the dispersion behavior of the capillary waves at a series of temperatures through

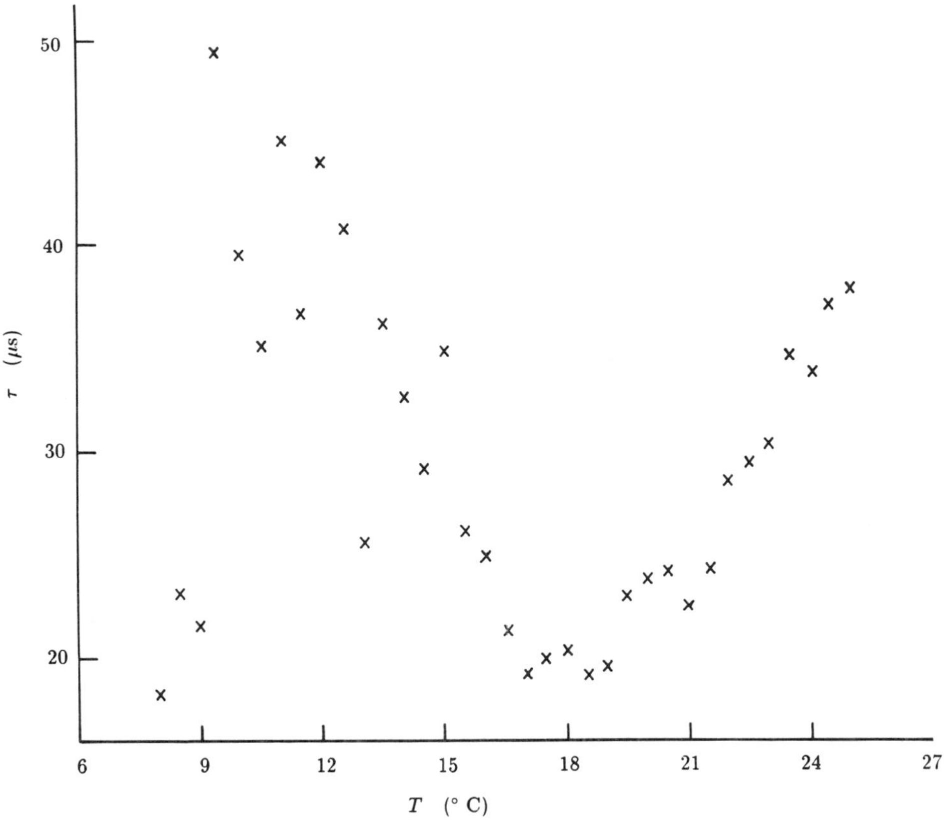

FIG. 20 The temperature variation of the relaxation time for the "solvent-free" GMO bilayer of Figs. 11 to 16. The relaxation is clearly fastest about T_t. The fluctuations at much lower T arise from the smallness of μ and $(\gamma—\gamma_e)$ in that region. (Reproduced from the *Biophysical Journal,* 1989, *55,* 1017–1021 by copyright permission of the Biophysical Society.)

the transition would be impossibly time-consuming: BLM tend not to last for many hours. However, the experiments discussed above on the transitions of "solvent-free" membranes do provide the temperature dependence of $\tilde{\gamma}$. Now, provided one assumes that a single Maxwell process is involved in the relaxation of the modulus concerned, γ and μ at a single q and the equilibrium tension γ_e, together with the wave frequencies ω_0, suffice to determine the relaxation time τ (cf. Eqs. 8 and 9). The assumption is known to be correct at 24.7°C (see Fig. 19) and seems plausible at other temperatures. Assuming that the temperature variation of γ_e follows that determined for a GMO monolayer at a water/oil interface, the data of Fig. 12 permit the T variation of τ to be determined (Fig. 20). Remarkably, the relaxation of the transverse shear stress appeared to be fastest in the neighborhood of the transition [32]. This contrasts with several previous studies of bilayer relaxation, but the light scattering approach addresses hydrodynamically well-defined modes of motion, unlike most other methods used, so that the conflict may be more apparent than real. The strength G of the relaxation was very roughly constant above T_t, at about 1 mN/m, but fell below the transition, approaching zero at about 9°C.

"Bobbing up and down" molecular motions, such as those in capillary waves, are associated with permeation through membranes [1, 34], which is known to exhibit a transitional maximum [37]. This has been explained by differences between "solid," "liquid," and "interfacial" (between solid and liquid domains) regions of the bilayer [38]. To date such differences are purely speculative and unexplained, but if they arise from different rates of transverse motion of the molecules, the observed minimum in τ at T_t might be explained quite easily.

V. CONCLUSIONS

These light scattering studies of bilayer lipid membranes have shown considerable potential for interesting science. The method is particularly suited to investigation of changes in membrane property, whether caused by changes in chemical composition, by thermotropic transitions, or as a function of wavenumber. A variety of completely novel effects have been observed, amongst which the molecular relaxation is perhaps particularly stimulating.

The experiments to date suggest several new areas which would advance insight into molecular fluctuations in membranes. Carefully designed studies of the depolarized scattering should lead to better understanding of the molecular splay and twist modes [4]. Various membrane moduli should be accessible, giving a wealth of detail on the molecular interactions involved. As stated in the introduction, molecular fluctuations induce or modify certain membrane functions. Studies involving the correlation of these fluctuations with membrane functions such as conductance or pore formation would be most illuminating.

The studies described have, in several ways, indicated the presence of phase separated domains of lipid molecules within the membranes below the phase transition. The existence of such domains within biomembranes would profoudly affect various membrane functions. Further investigation of the structure and dynamics of these molecular associations would be useful. The presence of cholesterol in the bilayers seems to enhance these domains, or rather stabilize them so that their effect is more marked [18]. The mechanism involved is as yet a matter for speculation.

As noted, molecular relaxation effects upon the viscoelastic properties of molecular films have been observed in other systems [31]. This opens up an entirely new field of dynamic studies. It is now highly desirable to study a variety of molecules, different model membranes, and the effects of chemical membrane modifiers upon these relaxation processes. Unfortunately, the experimental studies of cholesterol in GMO bilayers discussed above [22] were not sufficiently precise to perceive any relaxation present in $\tilde{\gamma}$.

An area that offers much fundamental interest and that is thus far quite untouched concerns the effects of electromagnetic fields upon the membrane modes of fluctuation. Theoretical approaches exist [6, 7], and various effects apparently having quite distinct observable signatures are predicted. These theories also predict limits to membrane stability, which could be changed by application of electric fields. While the theories seem not to be entirely complete, they provide sufficient guidance for sensible experiments to be undertaken.

ACKNOWLEDGMENTS

This work has been supported by the Science and Engineering Research Council. It is a pleasure to acknowledge the contributions and stimulation of past and present research students and associates, particularly Dr. J. F. Crilly and Dr. G. E. Crawford.

REFERENCES

1. R. N. Robertson, *The Lively Membrane,* Cambridge Univ. Press, Cambridge, 1983, pp. 59–79.
2. F. C. Goodrich, *J. Phys. Chem. 66:* 1858 (1962).
3. L. Kramer, *J. Chem. Phys. 55:* 2097 (1971).
4. C. P. Fan, *J. Colloid Interface Sci. 44:* 369 (1973).
5. F. C. Goodrich, *Proc. Roy. Soc. London A 260:* 503 (1961).
6. P. M. Bisch and H. Wendel, *J. Chem. Phys. 83:* 5953 (1985).
7. P. M. Bisch and H. Wendel, *J. Chem. Phys. 83:* 5962 (1985).
8. S. B. Hladky and D. W. R. Gruen, *Biophys. J. 38:* 251 (1982).
9. M. A. Bouchiat and D. Langevin, *J. Colloid Interface Sci. 63:* 193 (1978).

10. F. C. Goodrich, *Proc. Roy. Soc. London A 374:* 341 (1981).
11. M. Baus, *J. Chem. Phys. 76:* 2003 (1982).
12. J. D. Ferry, *Viscoelastic Properties of Polymers,* John Wiley, New York, 1980.
13. H. Engelhardt, H. P. Duwe, and E. Sackmann, *J. de Phys. Lett. 46:* L395 (1985).
14. E. F. Grabowski and J. A. Cowen, *Biophys. J. 18:* 23 (1977).
15. J. F. Crilly and J. C. Earnshaw, *Biophys. J. 41:* 197 (1983).
16. S. H. White, *Biophys. J. 15:* 95 (1975).
17. G. E. Crawford and J. C. Earnshaw in *The Application of Laser Light Scattering to the Study of Biological Motion* (J. C. Earnshaw and M. W. Steer, eds.), Plenum, New York, 1983, pp. 319–324.
18. G. E. Crawford and J. C. Earnshaw, *Eur. Biophys. J. 11:* 25 (1984).
19. J. C. Earnshaw and R. C. McGivern, *J. Colloid Interface Sci. 123:* 36 (1988).
20. G. E. Crawford and J. C. Earnshaw, *J. Phys. D: Appl. Phys. 18:* 1029 (1985).
21. D. A. Haydon and J. L. Taylor, *Nature 217:* 739 (1968).
22. J. F. Crilly and J. C. Earnshaw, *Biophys. J. 41:* 211 (1983).
23. G. E. Crawford and J. C. Earnshaw, *Biophys. J. 49:* 869 (1986).
24. F. Jahnig, *Biophys. J. 36:* 329 (1981).
25. R. E. Pagano, R. J. Cherry, and D. Chapman, *Science, 181:* 557 (1973).
26. J. P. Dilger, *Biochim. Biophys. Acta 645:* 357 (1981).
27. P. F. Fahey and W. W. Webb, *Biochemistry 17:* 3046 (1978).
28. D. R. Nelson and B. I. Halperin, *Phys. Rev. B. 19:* 2457 (1979).
29. P. J. Winch and J. C. Earnshaw, *J. Phys.: Condens. Matter 1:* 7187 (1989).
30. O. G. Mouritsen and M. J. Zuckermann, *Eur. Biophys. J. 12:* 75(1985).
31. G. E. Crawford and J. C. Earnshaw, *Biophys. J. 52:* 87 (1987).
32. J. C. Earnshaw and G. E. Crawford, *Biophys. J. 55:* 1017 (1989).
33. J. C. Earnshaw, R. C. McGivern, and P. J. Winch, *J. Phys. France 49:* 1271 (1988).
34. B. Owenson and L. R. Pratt, *J. Phys. Chem. 88:* 6048 (1984).
35. A. Genz and J. F. Holzwarth, *Coll. Polymer Sci. 263:* 484 (1985).
36. T. Lookman, D. A. Pink, E. W. Grundke, M. J. Zuckermann, and F. deVerteuil, *Biochemistry 21*: 5593 (1982).
37. D. Papahadjopoulos, K. Jacobsen, S. Nir, and T. Isac, *Biochim. Biophys. Acta 311:* 330 (1973).
38. L. Cruzeiro-Hansson and O. G. Mouritsen, *Biochim. Biophys. Acta 944*: 63 (1988).

13

Multiphase Microemulsion Systems

DOMINIQUE LANGEVIN Laboratoire de Physique Statistique, Ecole Normale Supérieure, Paris, France

I. INTRODUCTION

Microemulsions are thermodynamically stable mixtures of oil and water, stabilized by surfactant molecules [1, 2]. For a given set of conditions (temperature, composition) the system displays well defined structures. The most frequent consist of droplets, either of oil or of water, surrounded by a surfactant monolayer. This is because the surfactant layer has a preferred curvature C_0. By convention, C_0 is positive for oil in water (o/w) and negative for water in oil (w/o) systems. The droplet radius is given by

$$R = \frac{3\phi}{C_s\,\Sigma} \tag{1}$$

where ϕ is the dispersed phase volume fraction, C_s the number of surfactant molecules per unit volume, and Σ the area per surfactant molecule in the monolayer. It has been experimentally found that Σ is relatively constant for a given surfactant: this supports the assumption that the surfactant layer is compact and rather incompressible [1].

Generally, the curvature of the surfactant layer is different from C_0. The layer may even have two different principal curvatures C_1 and C_2. This is accompanied by a bending energy per unit area [3]:

$$F = \frac{1}{2}K(C_1 + C_2 - 2C_0)^2 + \overline{K}\,C_1C_2 \tag{2}$$

where K and $\overline{K}$ are respectively the splay and the Gaussian bending elastic moduli. In the following, $\overline{K}$ will be assumed to be zero unless stated otherwise.

In droplet microemulsions, $C_1 = C_2 = C = 1/R$. When ϕ is small, $R < R_0 = 1/C_0$. When the oil (resp. water) amount is increased, the droplet radius increases according to Eq. 1 and reaches the spontaneous radius of curvature $R_0 = |C_0^{-1}|$. At this point, a further increase of the oil volume will increase the bending energy. An alternative possibility is to reject the oil in an excess phase. This is called the emulsification failure [4]. In some cases, phase separation occurs earlier: when the droplet radius increases, attractive interactions between the droplets can become important, and a phase separation between two microemulsion phases can be observed. The behavior is analogous to the liquid-gas transition in simpler fluids.

When the spontaneous curvature C_0 is small, different types of structures are obtained. In some cases, the structure is lamellar: oil and water layers separated by surfactant monolayers. This structure possesses long-range order and is distinct from a microemulsion phase that is disordered. The lamellar and microemulsion phases are indeed separated by first-order transitions. In other cases, a microemulsion phase is also obtained for small $|C_0|$ values: as in the case of a droplet microemulsion, it is an isotropic phase. Unlike lamellar phases, it is continuously connected to droplet microemulsion phases, but it has a random, sponge-like, oil and water continuous structure, i.e., it is a bicontinuous phase. From random space-filling models, the mean size ξ of the oil and water microdomains can be calculated. With a cubic lattice,

$$\xi = \frac{6\,\phi_o\,\phi_w}{C_S\Sigma} \tag{3}$$

where ϕ_o and ϕ_w are respectively the oil and water volume fractions. ξ increases with ϕ_o and ϕ_w and reaches a maximum ξ_0, above which the oil and water are rejected as excess phases: a generalization of the emulsification failure concept.

Experiments show that bicontinuous microemulsions are not completely random: there is indeed a rather well defined characteristic distance between the oil and water microstructural elements. This distance corresponds to the mean size in the space filling models, as given by Eq. 3.

When C_o is large in absolute value, we have seen that the maximum droplet size is close to $1/C_o$. When $C_o \sim 0$ and K is of the order of kT, this is no longer true. The maximum swelling is controlled by the surfactant layer elasticity, and the maximum size is close to the persistence length of the layer [1]:

$$\xi_K = a \exp \frac{2\pi K}{k_B T} \tag{4}$$

a being a molecular length. ξ_K is the correlation length for the normals to the layer, i.e., the distance over which the layer remains flat. It describes the competition between thermal and bending energies. ξ_o is smaller than C_o^{-1}, and the swelling is limited by surfactant layer thermal undulations.

There are two types of theory that describe the phase equilibria associated with the microemulsions. In the first type of theory, the free energy of the system is calculated by incorporating the dispersion entropy, the interfacial energy, and the bending energy (Eq. 2). The three-phase equilibria are found by assuming that the surfactant film is not fully incompressible (the interfacial energy accounts for the Σ variations) [5, 6]. The relationship between ξ_o and ξ_K obtained in this way is rather indirect. A second approach can be made by assuming that the surfactant monolayer is incompressible but that the elastic constant varies with the scale, as recently demonstrated [7, 8]. In this way, ξ_o is found to be proportional to ξ_K [9].

In lattice theories, the treatment of the system free energy is more rigorous [10–12]. However, a phenomenological microscopic bending elasticity has to be used, and the connections between the results and the macroscopic notion of persistence length is still far from clear. In these models an interaction parameter J is related to the surfactant amphiphilicity, i.e., to its probability of sitting between oil and water sites. When J is large, lamellar phases are found. The structural evolution predicted by the phenomenological theories is similar. Indeed, K increases rapidly with surfactant chain length l ($K \sim l^3$) as does the surfactant amphiphilicity.

The two and three-phase equilibria involving microemulsions and excess phases are called Winsor equilibria [13]. The corresponding interfacial tensions γ are ultralow but can be conveniently measured with surface light scattering. As will be seen in the following, they are related to the microemulsion structure: $\gamma \sim k_B T/L^2$, L being the characteristic size in the microemulsion. Systems of this type are model systems for oil recovery [14]. The pressures required to eject the oil out of the reservoirs are proportional to the capillary pressures between the oil droplets trapped in porous rocks and the injected water. The capillary pressures are themselves proportional to the interfacial tensions γ. If γ is decreased below 10^{-2} mN/m, the oil trapped in the smallest pores can be recovered.

II. MEASUREMENT OF THE LOW TENSIONS: EXPERIMENTAL DETAILS

The samples are prepared by mixing the components and shaking them vigorously before allowing for phase separation, which can take hours or sometimes days. The phase separation is very slow because the tensions are low and because of the presence of surfactants that have a tendency to stabilize the emulsions. In order to ensure that the mutual solubilization of the components is satisfactory, it can be useful to shake the mixture from time to time at the beginning of the equilibration process, which should be carried out in a thermostatted environment (in a water bath, for instance) where the temperature is stable within at least ± 0.1°C. Equilibration periods of one week are usually sufficient.

The different phases can then be put into separate test tubes in order to perform the different characterization experiments. The separation process must be carried out carefully to prevent mixing of the phases. Syringes with long needles can be helpful in this respect. The phase separation should be done in the same thermostatted environment as the sample equilibration whenever this is possible: indeed, when the temperature is changed, the phases are no longer at equilibrium, and slow diffusion of the components starts producing an increasing turbidity of the phases.

Because the tensions are low, the capillary length is small and small liquid cells can be used (diameter $\sim$ 2 cm). When the cells are filled, there is frequently an emulsification around the interface, and enough time must be allowed to obtain sharp interfaces: this can be checked by inspection of the reflected beam. The temperature of the cell should be identical to that of the phase equilibration. For this purpose, the same water bath could be used to temperature control the samples both during phase equilibration and in the light scattering cell. The equilibration times being usually long, series of liquid cells can be used in the light scattering study: some of the cells can be filled while others filled earlier are studied. A considerable amount of time is saved in this way.

Because the tensions are low, the capillary waves are overdamped, and the spectra are centered at zero frequency. No instrumental broadening has to be corrected. The spectra are usually Lorentzian, of half-width (Chap. 11)

$$\Delta\omega = \frac{\gamma q + (\rho_2 - \rho_1)\, g/q}{2(\eta_1 + \eta_2) + \mu q} \tag{5}$$

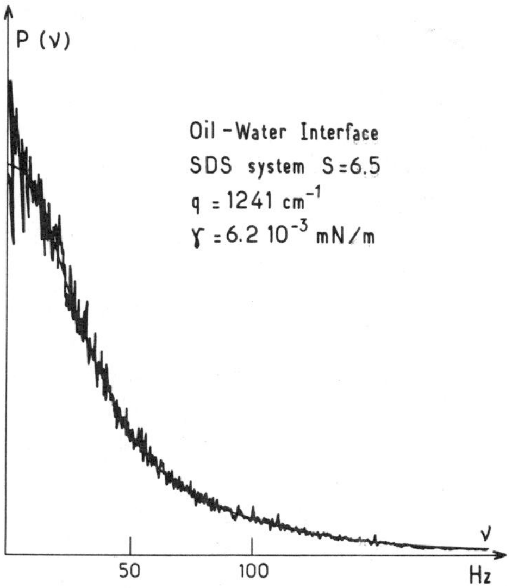

FIG. 1 Typical spectrum of the light scattered by an oil-water interface of low interfacial tension.

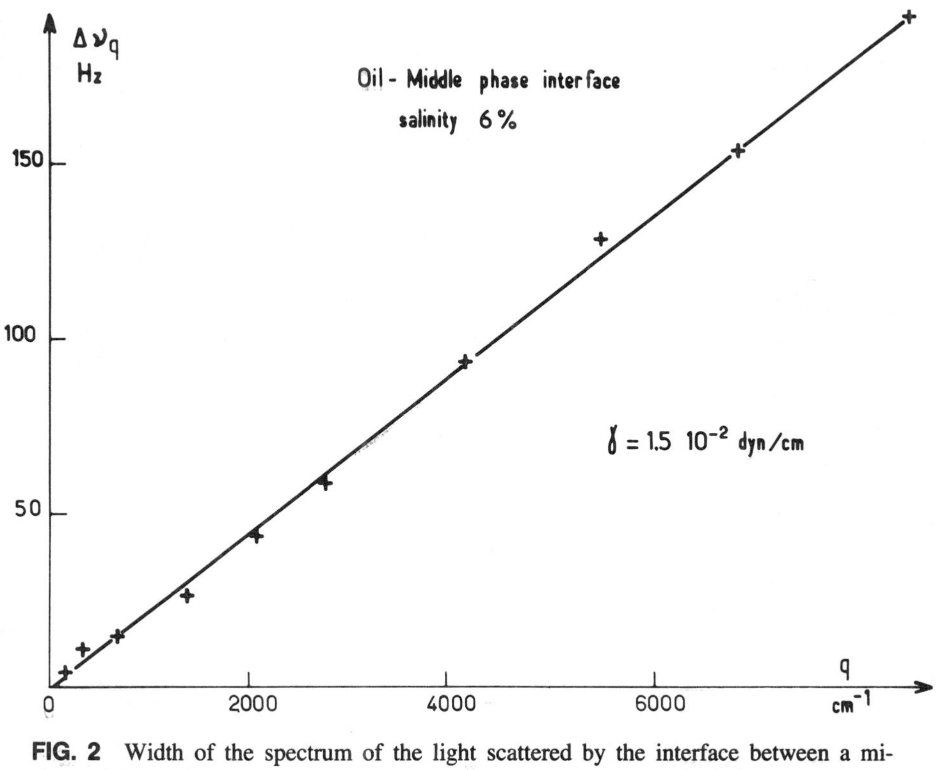

FIG. 2 Width of the spectrum of the light scattered by the interface between a microemulsion and excess oil (SDS system) versus the scattering wavevector q.

where γ is the interfacial tension, ρ_2 and ρ_1 the densities and η_2 and η_1 the viscosities of the coexisting phases, μ the transversal shear viscosity, and q the scattering wave vector. When γ is ultralow and q small, the gravity term $(\rho_2 - \rho_1)\, g/q$ might not be negligible. In order to determine γ and μ, ρ_1, ρ_2, η_1, and η_2 should be measured independently, at the same temperature. Because the surfactant layers are very flexible in microemulsion systems, μ is negligible: in all the published studies it was found to be zero within instrumental accuracy.

A typical spectrum is shown in Fig. 1. In order to improve the accuracy on γ, a large range of wavevectors q can be studied. The scattered intensity is large, and large scattering angles are accessible. An example is shown in Fig. 2 to illustrate the accuracy on γ that can be achieved. Rather than by instrumental problems, the accuracy is limited by the sample behavior: chemical evolution, small drifts of the interface due to residual temperature gradients, or slight evaporation. The accuracy can be improved by insuring good temperature control of the liquid cell, by removing the oxygen from the water to minimize surfactant oxidation, and by using appropriate sealing procedures for the cell (this is particularly important in systems containing alcohol, which evaporates quickly).

III. ORIGIN OF THE LOW TENSIONS

About ten years ago, there was a controversy about the physical origin of the ultralow tension involved in microemulsion systems. Schulman proposed early that these low tensions should be associated with the large surface pressures π of the surfactant monolayers, that almost exactly cancel the bare oil-water surface tension γ_0: $\gamma = \gamma_0 - \pi \sim 0$ [15]. In other models, microemulsion droplets were thought to accumulate close to the interface and to form a zone of thickness L; the phenomenological argument $\gamma \sim k_B T/L^2$ allowed the prediction of low tensions [16]. In still other models, the low tensions were associated with the vicinity of critical points [17, 18].

In order to clarify the problem, we have undertaken a detailed study of the microemulsion system of the following composition:

47 wt% water + sodium chloride (NaCl)
47 wt% toluene
4 wt% butanol
2 wt% sodium dodecyl sulfate (SDS)

According to the amount of salt in water S expressed in wt%, the system phase separates into

$S < 5.6$ o/w microemulsion in equilibrium with excess oil (Winsor I system)

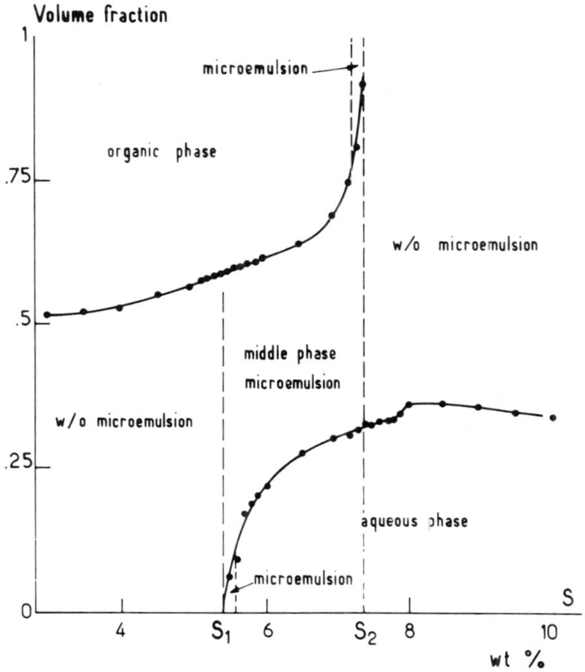

FIG. 3 Volume fraction of the different coexisting phases versus salinity in the SDS system.

$5.6. < S < 7.5$	a middle phase microemulsion in equilibrium with both excess oil and water (Winsor III system)
$S > 7.5$	w/o microemulsion in equilibrium with excess water (Winsor II system)

This evolution can be explained by the fact that the salt screens the electrostatic repulsion between the surfactant polar heads and thus decreases the spontaneous curvature of the surfactant layer from positive values at small salt concentration to negative values at large salt concentration. The relative volumes of the different phases are shown in Fig. 3 versus salinity.

A. Droplet Microemulsions

We have found that it was possible to dilute the microemulsion phase, i.e., to increase the volume of the continuous phase without changing the droplet size. The dilution procedure is quite complex: indeed, the continuous phase is not simply oil for w/o systems or water for o/w systems but contains also alcohol, traces of water in w/o systems, traces of oil in o/w systems, and traces of surfactant in both cases. The amount of surfactant ($\sim 10^{-5}$ by weight) corresponds to the critical micellar concentration in the particular aqueous phases studied here ($3.5 < S < 5.6$). The dilution procedure is the following in the case of a w/o system [19]: a given volume of oil v_o is added to 1 cm^3 of the microemulsion; this will define the dilution degree. The microemulsion becomes turbid, because it starts to phase separate (into a microemulsion and an excess phase); but if alcohol is added, the turbidity decreases and disappears above a certain volume of alcohol added v_A. The microemulsion has to be stirred during the procedure; otherwise the values of v_A are not well defined. Figure 4 shows a plot of v_o versus v_A: the plot is a straight line going through the origin as it should be if the composition of the continuous phase is well defined. Above $v_o = 20$ cm^3, however, traces of water and of SDS have to be added to conserve the same v_A/v_o ratio. This leads to the determination of the complete continuous phase composition. This method is a generalization of the Schulman method (who used only oil and alcohol) and the Graciaa et al. method (who used in addition trace water, but no surfactant) [20].

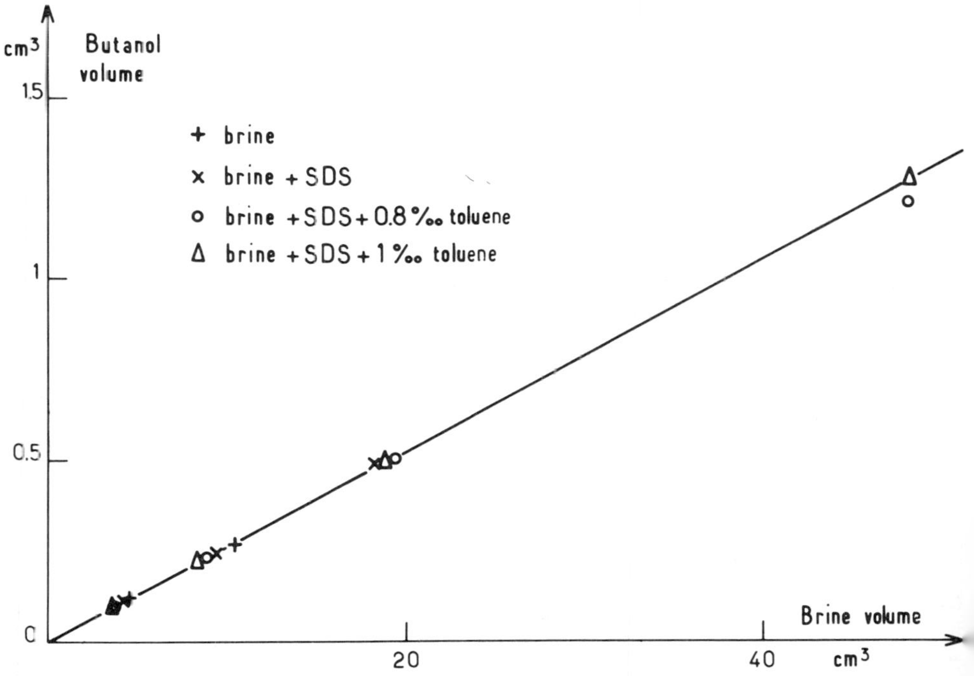

FIG. 4 Dilution line for an o/w microemulsion (SDS system).

TABLE 1 Interfacial Tensions Versus Micellar Concentration, $S = 8$ wt %

Upper phase	Lower phase	ϕ	γ (mN/m)
Microemulsion	Aqueous excess phase	0.18	0.023
Diluted microemulsion	Aqueous excess phase	0.036	0.014
Diluted microemulsion	Aqueous excess phase	0.018	0.013
Continuous phase, $C_s \sim$cmc	Aqueous excess phase	0	0.017
Continuous phase, $C_s \sim 2$cmc	Aqueous excess phase	~ 0	0.019
Continuous phase, $C_s = 0$	Aqueous excess phase	0	2.95

The volume of the dilute phases remains unchanged when they are put in contact with the corresponding excess phases. When a mixture of identical overall composition is prepared, it separates into two phases of the same aspect and volume. Furthermore, the interfacial tension between the two two-phase systems are about the same. This concurs to indicate that the dilute microemulsion phases are in thermodynamic equilibrium with the same excess phase as the initial microemulsion.

The measured interfacial tensions are shown in Table 1 versus surfactant concentration. These variations are very similar to the corresponding variations for aqueous surfactant solution: γ decreases when the surfactant concentration increases and saturates to a constant value above the critical micellar concentration (cmc). The value of the tension does not change appreciably between the cmc ($\sim 10^{-5}$) and the high surfactant concentration of the microemulsion. This proves unambiguously that the low value of the tension has nothing to do with the presence of droplets. As in aqueous surfactant systems, the value of the tension is determined by the surface pressure of the monolayer.

Similar results were found in o/w systems [19]. More recently, Overbeek and coworkers studied oil-water-alcohol-surfactant mixtures close to the cmc and showed that the ultralow tensions could be obtained as well in their systems [21]. They determined the monolayer composition from the $\gamma(c)$ curve below the cmc. Similar work on ternary oil-water-surfactant mixtures was also reported by Aveyard et al. [22]. In these recent studies, γ was measured using a spinning drop tensiometer (Sec. VII).

B. Middle-Phase Microemulsions

We found, as have others, that middle-phase microemulsions are not dilutable. However, one of the two interfacial tensions between the microemulsion and the excess phases is equal to the tension between the excess phases γ_{ow}:

$$\gamma_{ow} = \sup(\gamma_{om}, \gamma_{wm}) \tag{6}$$

o, w, m standing for organic, aqueous, and microemulsion phases (Fig. 5). The

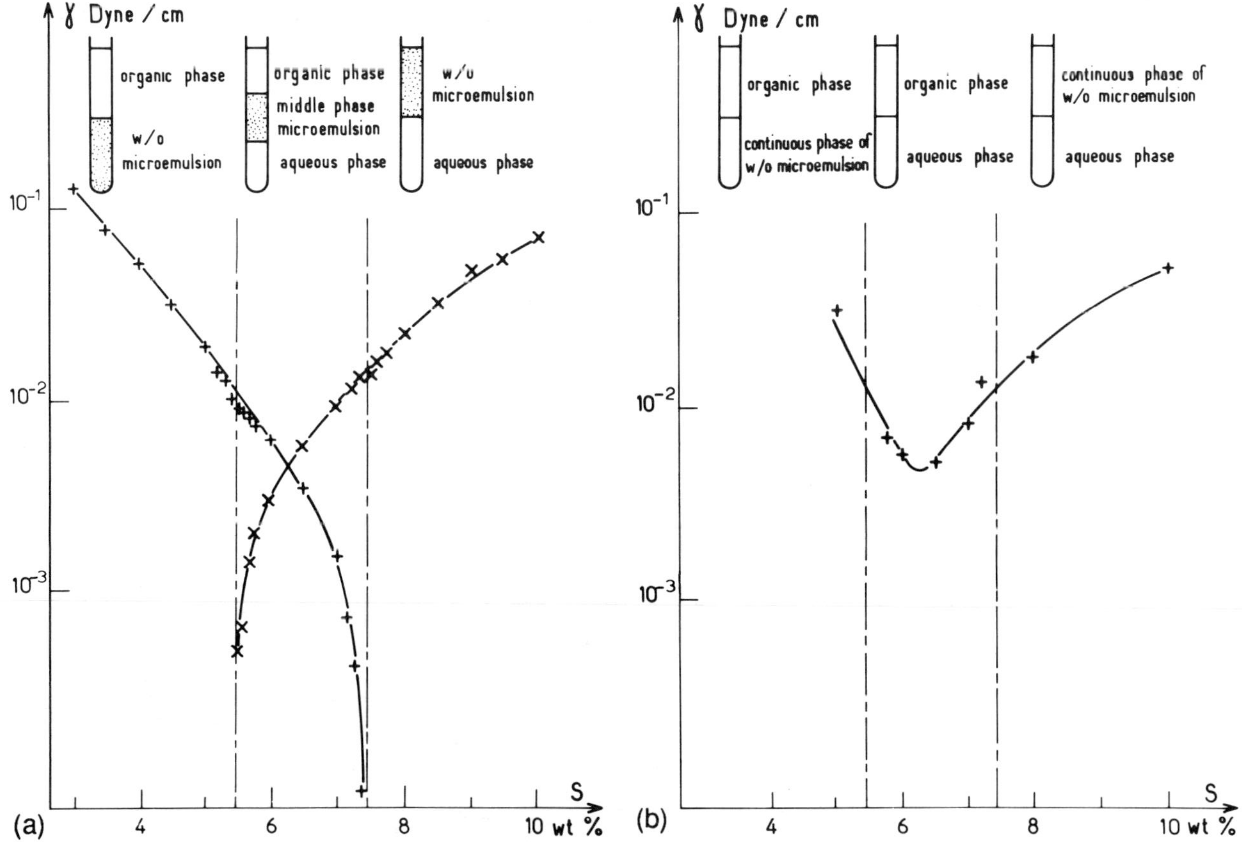

FIG. 5 Measured interfacial tensions in the SDS system versus salinity; (a) +, γ_{om}; x, γ_{wm}; (b) γ_{ow}.

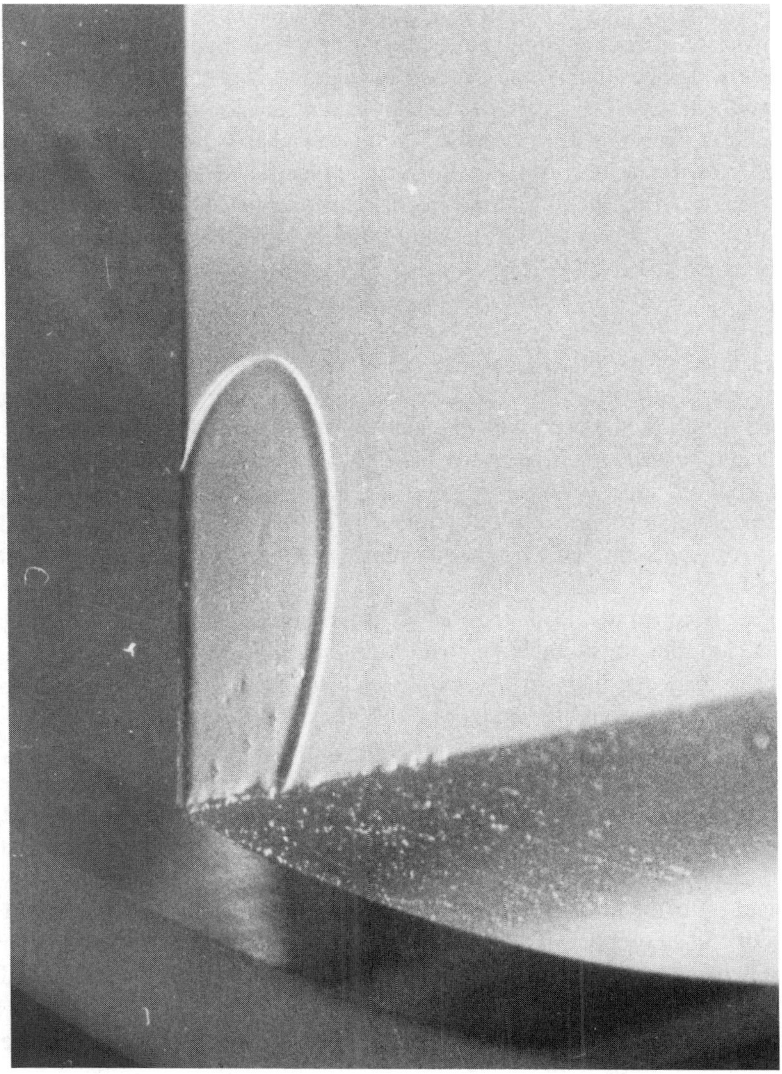

FIG. 6 Drop of middle-phase microemulsion at the oil-water interface (SDS system). Because of the low values of the three interfacial tensions, the drop is very flat. The picture has been taken after the drop has moved to a corner of the liquid cell where it becomes more visible. (After Ref. 62.)

tension between the excess phases is again ultralow, although these phases do not contain any surfactant aggregate (as seen from bulk light scattering). Furthermore, the composition of the excess phases at a given salinity corresponds to the extrapolation of the variation of the microemulsion continuous phases versus salinity determined previously. The excess phases likely contain trace surfactant, close to the cmc value ($\sim 10^{-5}$ g/cm^3). Indeed, when the excess phases are separated from the middle phase microemulsion and equilibrated together, large values of γ_{ow} are sometimes measured, much larger than those compatible with the Antonov inequality [23]:

$$\gamma_{ow} \leq \gamma_{om} + \gamma_{wm} \tag{7}$$

The explanation of the phenomenon is as follows: when the phases are equilibrated together, they reform a surfactant monolayer at their interface. If during the phase separation, the corresponding amount of surfactant is not added in the form of a minute volume of the microemulsion, the surfactant monolayer at the interface between the two excess phases will not be dense enough to have a large surface pressure, and γ_{ow} will be large. The Antonov inequality is valid only at thermodynamic equilibrium, meaning here that the three phases must always be present.

It might be argued that if the tension γ_{ow} is measured in the presence of trace microemulsion, this microemulsion could form a thin layer, invisible by eye, at the interface between the two excess phases. But in this case the measured tension would be the sum of γ_{om} and γ_{wm}. This is not the case, as shown in Eq. 7 and Fig. 5. At "optimal" salinity S^* (for oil recovery purposes), $\gamma_{ow} = \gamma_{om} = \gamma_{wm}$; the difference between γ_{ow} and $\gamma_{om} + \gamma_{wm}$ is maximal and well above the experimental accuracy. Furthermore, the microemulsion does not wet the interface between the excess phases (Fig. 6). This is in agreement with the fact that the strict Antonov inequality is verified: $\gamma_{ow} < \gamma_{om} + \gamma_{wm}$, the equality corresponding to the wetting case. It is also possible to measure γ_{ow} in the presence of a macroscopic middle phase drop if this drop stands at the border of the cell. In these cases, it was found that γ_{ow} is always ultralow and equal to the values of Fig. 5.

All these results concur to establish that γ_{ow} is low because of the presence of a large surface pressure monolayer at the interface, exactly as in the droplet microemulsion case. The tensions that are equal to γ_{ow}, γ_{om} for $S < S^*$ and γ_{wm} for $S > S^*$ are likely to be low for the same reason. The existence and properties of the monolayer at these interfaces can be studied in more detail by ellipsometry. This will be discussed in Chap. 17.

C. Critical Behavior

The lowest tensions have a different origin. They become still lower close to the phase boundaries: Winsor I → Winsor III → Winsor II systems. Close to the boundaries, the corresponding excess phase (aqueous close to S_1, organic close to S_2) becomes slightly turbid. This is reminiscent of the vicinity of a critical point. This particular type of critical point, where two phases become critical in the presence of a noncritical third phase, is called a critical end point [24]. S_1 and S_2 are not exactly the critical end points, as indicated by the phase volume behavior Fig. 3: the relative volume of the two critical phases should not vary upon approaching the critical end point, and the interface should vanish at the critical points. But S_1 and S_2 are likely to be close to the critical end points. The locus of these points in the phase diagram have been determined in a similar microemulsion system by Bellocq et al. [25]. The end points are approached here by changing the salinity. Like most other composition variables, S is not a "field" variable, i.e., a variable which is the same in both coexisting phases, like temperature or chemical potentials. Theory predicts that

$$\gamma = \gamma_0 \left(\frac{\varepsilon - \varepsilon_c}{\varepsilon_c} \right)^{\mu} \tag{8}$$

where μ is a critical exponent, ε a field variable, ε_c its critical value, and γ_0 a scale factor [24].

Instead of measuring chemical potentials, a very difficult task in these systems [26], it is easier to determine the critical behavior of a different property, the density difference between the phases, for instance:

$$\rho_2 - \rho_1 = \Delta\rho = \Delta\rho_0 \left(\frac{\varepsilon - \varepsilon_c}{\varepsilon_c} \right)^{\beta} \tag{9}$$

Eliminating $(\varepsilon - \varepsilon_c)/\varepsilon_c$ between Eqs. 8 and 9, one finds

$$\gamma = \gamma_0 \left(\frac{\Delta\rho}{\Delta\rho_0} \right)^{\mu/\beta} \tag{10}$$

This plot is shown in Fig. 7. Before S^* for γ_{wm} and after S^* for γ_{om}, the variation of γ with $\Delta\rho$ is of the same form as predicted by Eq. 10. It was found that $\mu/\beta = 4 \pm 0.4$ [27]. This result is in agreement with the three-dimensional Ising values of the exponents $\mu = 2\nu$, $\nu = 2\beta$, where ν is the critical exponent for the correlation length ξ. It disagrees with mean field exponents $\mu = 3\beta$. Similar results were found later by using a spinning drop technique [25] on other systems. The power law applies only to the smallest tensions. Above S^* for γ_{wm} and below S^* for γ_{om}, deviations clearly appear. These deviations are not surprising, because the larger tensions are associated with monolayer properties and having nothing to do with critical behavior. Several authors have extended the critical plots in the whole three-phase region and eventually in two-phase regions as well, and concluded that all the tensions were low because of the vicinity of the critical end points [28, 29]. This is because the accuracy of their measurements was not good enough to detect the deviations to critical behavior as in Fig. 7.

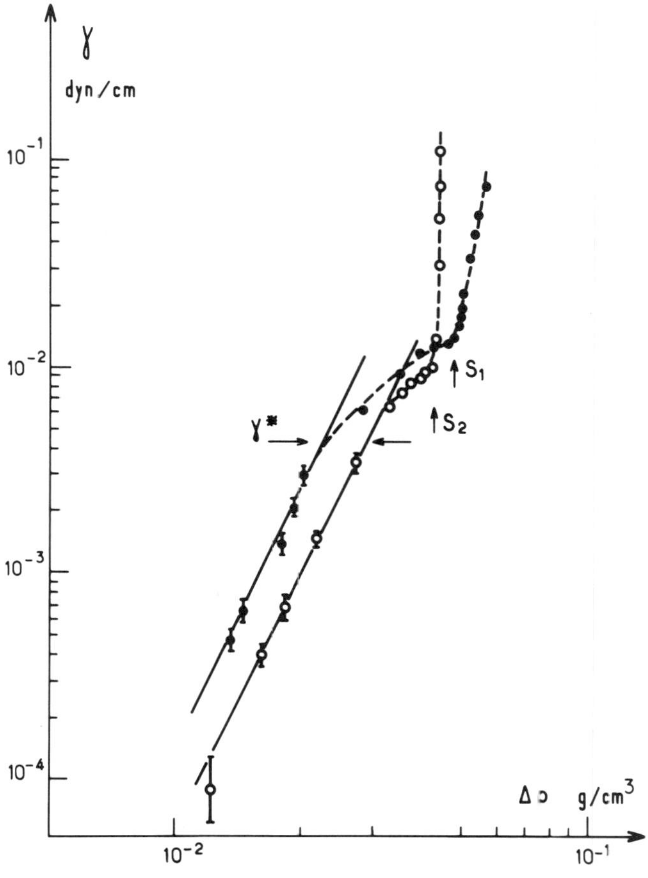

FIG. 7 Values of γ_{om} (open circles) and γ_{wm} (closed circles) versus the density difference between coexisting phases. SDS system.

Our own results were also reanalyzed in this way [18]. However, the plot of Fig. 7 as well as the discussions in Secs. 3.A and 3.B clearly demonstrate that most of the tensions are low because of the presence of high surface pressure monolayers. It must be stressed that deviations such as those of Fig. 7 can only be evidenced if the accuracy on the tension measurements is very good. The deviations can be much less visible if the data are plotted in an different way (see Ref. 18, for instance). When short chain amphiphiles are used, the critical plots extend in the whole three-phase region, but these surfactants do not form well-defined surfactant layers [30]. This point will be discussed further in Sec. 6.

Turning back to the lowest tensions, and in order fully to confirm the critical behavior, we have measured the correlation length in the bulk microemulsion by using elastic and inelastic bulk light scattering techniques. ξ diverges as

$$\xi = \xi_0 \left(\frac{\varepsilon - \varepsilon_c}{\varepsilon_c}\right)^{\nu}$$

with $\mu = 2\nu$. ξ was determined from the scattered intensity angular variation $I \sim (1 + q^2 \xi^2)^{-1}$, from the diffusion coefficient angular variation $D(q) = D_0 K(q\xi)/q^2\xi^2$, where K is the Kawasaki function

$$K(x) = \frac{3}{4}\left[x^2 + 1 + \left(x^3 - \frac{1}{x}\right) \text{Arctan } x\right]$$

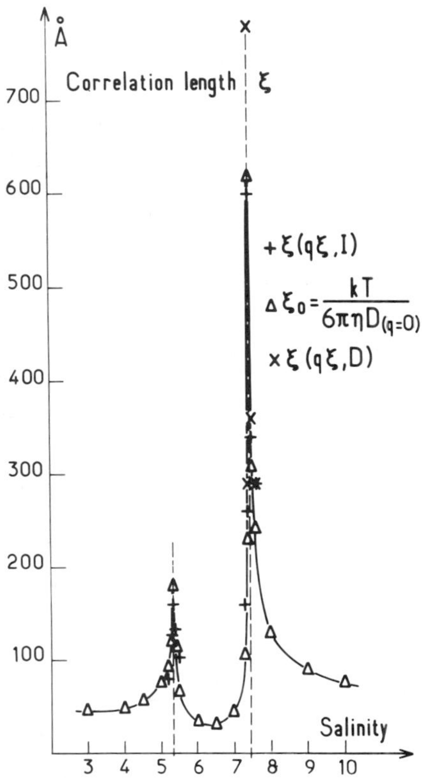

FIG. 8 Correlation length in the bulk microemulsion phase versus salinity as measured with bulk light scattering techniques; +, angular anisotropy of the scattered intensity; x, angular anisotropy of the diffusion coefficient; Δ, small angle value of the diffusion coefficient. SDS system.

TABLE 2 Radii of the Microemulsion Droplets that Have Been Diluted, as Deduced from Light Scattering Experiments, SDS System

	S (wt %)	R (Å)
o/w microemulsion	3.5	85
	5	130
	5.2	155
w/o microemulsion	8	180
	10	125

and from the $q = 0$ limit of D: $D_0 = k_B T/6\pi\eta\xi$, η being the microemulsion viscosity. The three ξ determinations are shown in Fig. 8. They are in agreement close to S_2, but close to S_1, I is q dependent whereas D is not. The description in terms of individual droplets is more appropriate, at least for $S < S_1$. The droplet radii determined by bulk light scattering experiments on microemulsions and their dilutions are given in Table 2. It is seen that $\xi \gg R$ only very close to S_2: the critical behavior is much less marked on bulk properties than on surface tension. This is specific of microemulsions, for which the universal combination of scale factors $\gamma_0\, \xi_0^2/k_B T$ is smaller than on simpler fluids: 0.012 ± 0.04, instead of 0.09 ± 0.02 [31]. This means that although the critical exponents found are compatible with those of three-dimensional Ising models with a scalar order parameter, microemulsions may not belong to the same universality class.

In the SDS system, ξ does not become large enough to determine the ratio μ/ν accurately. This was done by Kim et al. in another Winsor system where γ falls below 10^{-3} dyn/cm and ξ increases above 1000 Å; they found $\nu/\mu = 0.52 \pm 0.04$ [29].

Critical behavior was also observed in a different type of three-phase microemulsion system: a w/o microemulsion (lower phase), a sponge-like phase (middle phase), and excess oil [32]. The exponents measured for the interfacial tension were also in agreement with Ising exponents.

IV. INFLUENCE OF THE NATURE OF THE SURFACTANT MOLECULES

In Winsor systems, the shape of the surface tension curves is always similar to those of Fig. 5 whatever parameter is used to evolve from Winsor I to Winsor II phase equilibria: salinity or surfactant-to-alcohol ratio for ionic surfactants, temperature for nonionic surfactants.

Several examples are shown in Figs. 9 and 10 [33–35]. The corresponding

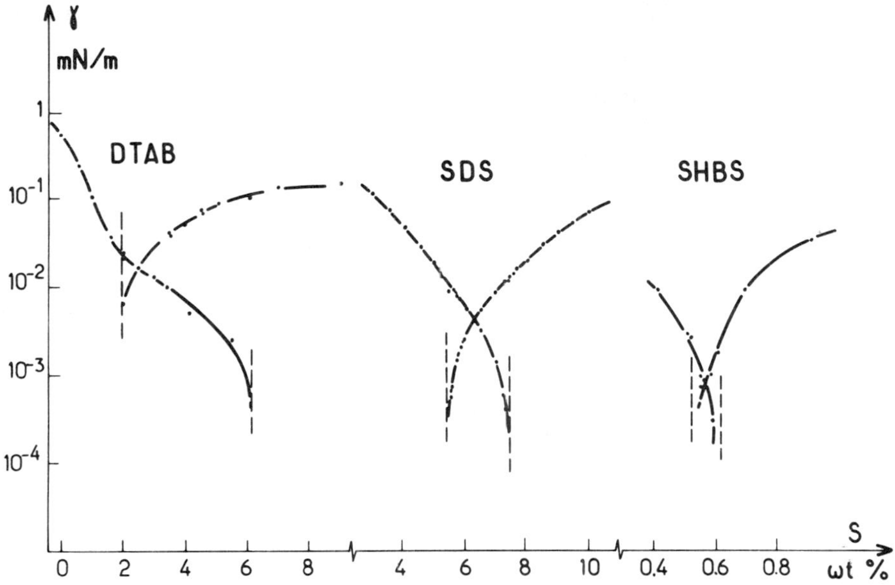

FIG. 9 Measured interfacial tensions γ_{om} and γ_{wm} in Winsor microemulsion systems made with different ionic surfactants (see Table 3) versus salinity.

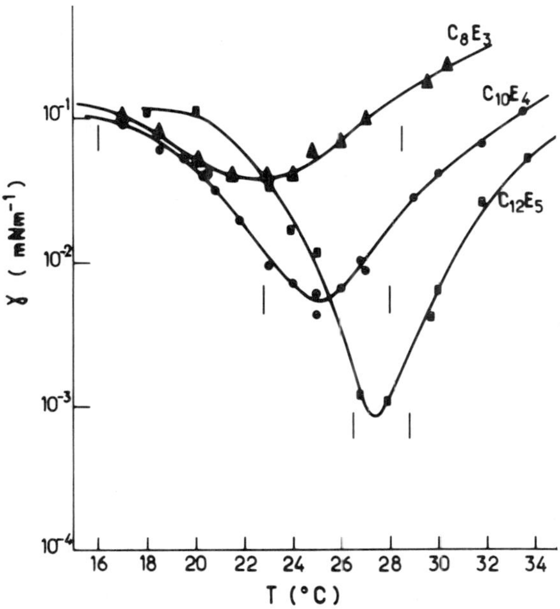

FIG. 10 Measured interfacial tensions γ_{ow} in Winsor microemulsion systems made with different nonionic surfactants (see Table 3) versus temperature.

TABLE 3 Surfactant Molecules

Sodium dodecylsulfate	SDS	CH_3—$(CH_2)_{11}$—SO_4 Na
Dodecyl trimethyl ammonium bromide	DTAB	CH_3—$(CH_2)_{11}$—N—$(CH_3)_3$Br
Hexadecyl trimethyl ammonium bromide	CTAB	CH_3—$(CH_2)_{15}$—N—$(CH_3)_3$Br
Sodium hexadecyl benzene sulfonate	SHBS	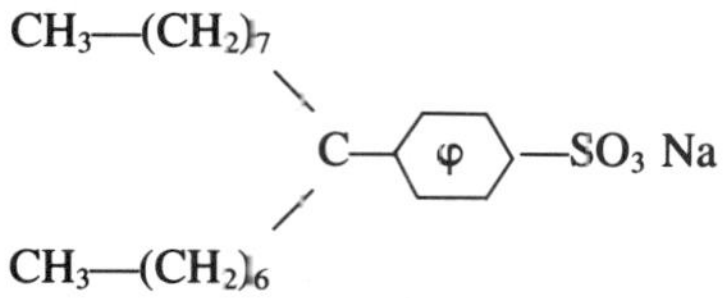
Sodium diethylhexyl sulfosuccinate	AOT	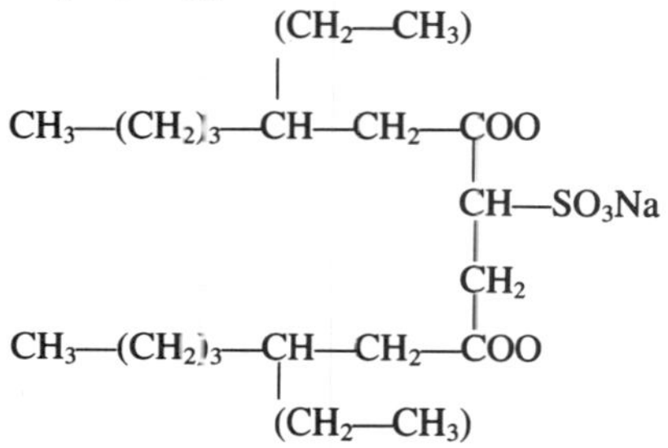
Pentaethylene glycol dodecylether	$C_{12}E_5$	CH_3—$(CH_2)_{11}$—O(CH_2—CH_2—O)$_5$—H
Tetraethylene glycol decylether	$C_{10}E_4$	CH_3—$(CH_2)_9$—O(CH_2—CH_2—O)$_4$—H
Triethylene glycol octylether	C_8E_3	CH_3—$(CH_2)_7$—O(CH_2—CH_2—O)$_3$—H
Diethylene glycol hexylether	C_6E_2	CH_3—$(CH_2)_5$—O(CH_2—CH_2—O)$_2$—H

surfactant molecules are described in Table 3. It is seen that there is a net correlation between the width of the three-phase region and the minimal value γ^* of γ_{ow} (optimal salinity for ionic surfactants): the smaller the width, the smaller γ^*. There is also a correlation between γ^* and the minimum concentration C^* required to solubilize a given amount of water and oil in the three-phase region: the longer the surfactant chain length, the smallest C^*, and according to Figs. 9 and 10 the smallest γ^*. This can be rationalized simply by using Eq. 3. The characteristic size in the middle-phase microemulsion at optimal conditions $\phi_o = \phi_w = 1/2$ is according to Eq. 3:

$$\xi = \frac{6\,\phi_o\,\phi_w}{C_s\,\Sigma} = \frac{3}{2C^*\,\Sigma} \qquad (11)$$

From dimensionality arguments, one expects

$$\gamma \sim \frac{k_B T}{\xi^2} \sim k_B T \, {C^*}^2 \, \Sigma^2 \tag{12}$$

The measurement of ξ is conveniently done by small angle x-ray or neutron scattering (the microemulsions cannot be diluted, so bulk light scattering tech-

TABLE 4 Measured Sizes and Interfacial Tensions for Various Systems. Dimensionless Group $\gamma L^2/k_B T$, where $L = R$ or ξ

	S (wt %)	R (Å)	ξ (Å)		γ(mN/m)	$\gamma L^2/k_B T$
	1	64		o/w	0.120	1.22
	2		120		0.026	0.93
	2.5		97		0.02	0.47
	3		103		0.033	0.87
	4		90		0.051	1.02
	5		78		0.091	1.37
DTAB	6	51			0.105	0.68
	7	54		w/o	0.138	1.0
	9	42			0.134	0.58
	3.5	85		o/w	0.075	1.35
	5	130			0.02	0.85
	5.2	155			0.014	0.85
SDS	S*		230		0.0045	0.65
	8	180		w/o	0.023	1.86
	10	125			0.075	2.93
	0.4	112			0.0089	0.28
	0.5	203		o/w	0.003	0.31
SHBS	0.52	250			0.0029	0.46
	S*		400		0.00087	0.35
	0.7	192		w/o	0.011	0.95
	0.9	129			0.035	1.46
	0.117	28			0.252	0.49
	0.146	58			0.088	0.74
	0.175	69		o/w	0.054	0.64
	0.234	118			0.012	0.42
AOT	0.357	176			0.0178	1.37
	0.415	132			0.0316	1.37
	0.474	113		w/o	0.0562	1.82
	0.532	95			0.0708	1.6
	0.591	89			0.0794	1.57

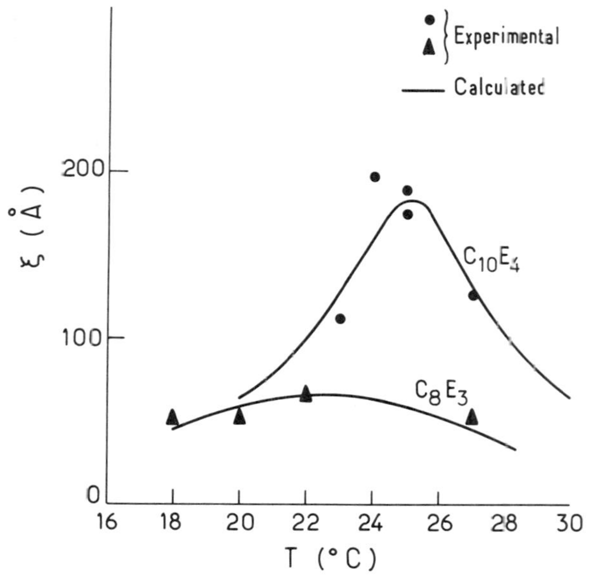

FIG. 11 Measured dispersion sizes in middle-phase microemulsions made with nonionic surfactants versus temperature. The line is calculated using the empirical relation $\xi^2 = 0.4\, k_BT/\gamma$.

niques cannot be used). The spectra exhibit broad peaks around $q = q_{max}$, which arise from position correlations between oil and water microdomains. It can be shown that $\xi = \pi/q_{max}$ [36]. Equation 12 applies satisfactorily to all the systems where γ and ξ have been measured (Table 4 and Fig. 11).

V. INTERFACIAL TENSION THEORIES

There are many different types of interfacial tension theories for Winsor microemulsion systems, most of which apply to droplet systems. Several theories focus on the role of the interaction forces between the droplets [16, 37] and ultimately on critical behavior [18]. The rest are somewhat equivalent to the simple description given below [38]. If μ_N is the chemical potential of a surfactant molecule in an aggregate of N molecules, then

$$\mu_N = \mu_N^o + \frac{k_BT}{N} \ln \frac{X_N}{N}$$

where μ_N^o is a standard chemical potential containing the interaction terms between the surfactant molecules in the droplet and X_N is the surfactant concentration fraction present in the droplet.

The interface between the microemulsion and the excess phase is covered by a surfactant layer that can be viewed as an infinite aggregate. The surfactant chemical potential at the interface is then

$$\mu_S = \mu_\infty - \gamma\Sigma = \mu^\circ_\infty - \gamma\Sigma$$

Since $\mu_S = \mu_N$, this leads to

$$\gamma = -\frac{1}{N\Sigma} k_B T \ln \frac{X_N}{N} + \frac{1}{\Sigma} (\mu^\circ_\infty - \mu^\circ_N) \tag{13}$$

The first term γ_e is an entropic contribution arising from mixing. It is analogous to the expressions derived by Ruckenstein [39] and Overbeek et al. [40] on different thermodynamic bases. It can indeed be rewritten as

$$\gamma_e = -\frac{k_B T}{4\pi R^2} \ln \alpha\phi \tag{14}$$

where ϕ is the droplet volume fraction and α a numerical constant depending upon the approximations made in calculating the dispersion entropy.

The second term γ_c is a curvature contribution and has been calculated explicitly by other authors [41, 42, 43]. The bending free energy of Eq. 2 can be used for this purpose. The total free energy for a system of n droplets of radius R is

$$F = n \left[2K \left(\frac{1}{R} - \frac{1}{R_o} \right)^2 + \frac{\overline{K}}{R^2} \right] 4\pi R^2$$

Minimization of F with respect to R leads to

$$R = \frac{2K}{2K + \overline{K}} R_o \tag{15}$$

On the other hand, $\mu^\circ_N = F/nN$ and $\mu^\circ_\infty = 2K\Sigma/R_o^2$. It follows that

$$\gamma_c = \frac{2K}{RR_o} \tag{16}$$

In this derivation, the interdroplet interactions have been neglected. The validity of this approximation is supported by experimental data like those of Table 1, where it is seen that the interfacial tension does not depend significantly upon droplet volume fraction.

The approximation fails, of course, close to the critical points where the interaction term becomes dominant.

The bending elastic constant K can be measured using ellipsometry [44]. Table 5 shows a comparison between measured and calculated interfacial tensions using Eqs. 14 and 16 and taking $\alpha = 1$, $\overline{K} = 0$ in various systems. Although discrepancies between theory and experiment can be as large as 100%, the

TABLE 5 Measured Radii and Interfacial Tensions for Droplet Microemulsion in Various Systems. Calculated γ_e and γ_c Contributions to the Tensions. For the Three First Surfactants, $K = 0.4\ k_BT$, for SDS, $K = 0.65\ k_BT$, and for AOT, $K = 1.1\ k_BT$

	S (wt %)	ϕ	R (Å)		γ(mN/m)	γ_c(mN/m)	γ_e(mN/m)
	2	0.091	52		0.348	0.165	0.029
	4	0.110	61	o/w	0.129	0.120	0.019
CTAB	24	0.075	46	w/o	0.111	0.210	0.039
	30	0.070	42		0.147	0.252	0.049
	1	0.1	64	o/w	0.120	0.108	0.018
DTAB	6	0.076	51		0.105	0.171	0.032
	7	0.065	54	w/o	0.138	0.152	0.030
	9	0.053	42		0.134	0.252	0.053
	0.4	0.17	112		0.0089	0.058	0.0047
	0.5	0.255	203	o/w	0.003	0.017	0.0011
SHBS	0.52	0.28	250		0.0029	0.011	0.00065
	0.7	0.29	192	w/o	0.01	0.019	0.0011
	0.9	0.18	129		0.035	0.041	0.0033
	3.5	0.15	85		0.075	0.073	0.008
	5	0.21	130	o/w	0.02	0.031	0.003
SDS	5.2	0.24	155		0.014	0.022	0.002
	8	0.18	180	w/o	0.023	0.016	0.002
	10	0.14	125		0.075	0.034	0.005
	0.117		28		0.252	1.70	0.096
	0.146		58		0.088	0.40	0.022
	0.175		69	o/w	0.054	0.28	0.016
	0.234		118		0.012	0.096	0.0054
AOT	0.357		176		0.0178	0.043	0.0024
	0.415		132		0.0316	0.077	0.0043
	0.474		114	w/o	0.0562	0.10	0.0058
	0.532		95		0.0708	0.148	0.0083
	0.591		89		0.0794	0.169	0.0095

agreement is not bad in view of the large uncertainty on measured radii and the approximation made on $\bar{K}$. If the assumption $\bar{K} = 0$ is released, $\bar{K}$ can be calculated from the known values of K, γ, and R. The data suggest that $|\bar{K}| < k_B T/2$, except for AOT where $\bar{K} \sim -2k_B T$. But these results should be considered only as indicative, in view of the large uncertainties on R.

It is seen from Table 5 that the entropy contribution is generally small compared to the curvature contribution, except when the droplets are very small. The differences between γ and γ_c ($\bar{K} = 0$) may be ascribed to the Gaussian bending elastic modulus. Indeed, $\bar{K}$ should become increasingly large when bicontinuous phases are approached. When K increases, microemulsion phases become close to lamellar phases, and $\bar{K}$ should decrease, as observed in the case of AOT.

The interfacial tension between the middle-phase microemulsion and the excess phases is much more difficult to calculate. The dispersion entropy contribution is easy to generalize, but the curvature contribution is more difficult to estimate, because of the complicated shape of the aggregates, which furthermore is not yet fully elucidated from the experimental point of view. However, since the mean curvature of the aggregates is about zero, γ_c is expected to be small. But if it is totally neglected, the approximations used to calculate γ_e, when extended to the droplet microemulsion phases, imply also that $\gamma \sim \gamma_e$ here and therefore give a much too large γ_e contribution. Let us recall that γ_e varies as $\log \alpha\phi$ (Eq. 14); if α is adjusted so that $\gamma \approx \gamma_e$ in the droplet microemulsions, γ should vary appreciably with the droplet volume fraction, much more than seen, for instance, in Table 1. It is therefore likely that γ_c still plays a role in the bicontinuous microemulsion case.

Let us note that Talmon and Prager made an attempt to calculate the interfacial tensions between the middle-phase microemulsion and the excess phase [45]. They used the model structure of Voronoi polyedra and used a mean field description for the interfacial profiles like the one used by Cahn and Hilliard close to a critical point. Their idea was that the tension being small, the interfacial thickness was much larger than a molecular thickness, making the mean field treatment valid. The result was that the interfacial region close to the excess water phase was richer in water than the bulk microemulsion. Similarly, the region close to the oil phase was richer in oil. This is in contradiction with the experimental investigations made using ellipsometry in the case of the interface of larger tension [46], (Chap. 17), i.e., the noncritical interface. The origin of the discrepancy is simple: the interface is covered by a monolayer, and the region close to water (resp. oil) excess phase is in fact richer in oil (resp. water) than the bulk microemulsion phase. Because γ is small, the interface is very rough, but its "intrinsic" thickness is small, in the molecular range. For this reason, the mean field calculation fails.

Microscopic Ising microemulsion models were developed to resolve this issue. The existence of the monolayer was clearly demonstrated [47, 48]. The interfacial tensions were also calculated. However, the microscopic parameters used in these models are not yet clearly connected to the measurable microemulsion properties.

VI. WETTING TRANSITION AND RELATIONSHIP WITH THE MICROEMULSION STRUCTURE

We have described in Sec. III.B the unusual behavior of the middle-phase microemulsion at the interface between oil and water. Even though the tensions are ultralow, the microemulsion does not wet the oil-water interface. In simple liquid systems, interfacial tensions are small only in the vicinity of a critical point. Let us call a, b, and c the three phases in equilibrium, and suppose that the critical interface is ab and that phase b does not wet the ac interface. Then γ_{ab} vanishes as $(\varepsilon - \varepsilon_c)^{\mu}$ when the critical point is approached. It can be shown that $\gamma_{ac} - \gamma_{bc}$ vanishes as $(\varepsilon - \varepsilon_c)^{\beta'}$ [49]. Since the Antonov inequality is

$$\gamma_{ac} - \gamma_{bc} < \gamma_{ab}$$

it is seen that it cannot be satisfied very close to the critical point, because γ_{ab} decreases much faster than $\gamma_{ac} - \gamma_{bc}$ when this point is approached ($\mu > \beta'$). Obviously, the distance at which wetting occurs depends on the system. In microemulsion systems, the critical points must be approached closer than was done in the reported experiments, although the interfacial tensions were already ultralow. Again this points out that the main cause of the low tensions is the high surface pressures of the surfactant monolayers at the interface.

Experiments done in three-phase systems made with small molecules showed that the middle phase always wets the interface between the upper and lower phases [50]. This behavior was exhibited in particular by oil-water short-chain non-ionic surfactant mixtures [51, 52]. It was therefore tempting to observe the behavior of intermediate chain surfactants. As expected, a wetting transition was observed in the temperature range of the existence of the middle phase for the system C_6E_2-hexadecane water [53].

The particular behavior of microemulsions is associated with the presence of monolayers, that do not form with small molecules. The microstructure of microemulsions is stabilized by similar monolayers. The wetting transition is then probably associated with the disappearance of the monolayers and of the microemulsion microstructure, leading to molecular mixtures of oil, water, and surfactant. As expected, this happens when the surfactant chain length is small, and when the amphiphilicity of this molecule has been reduced compared to that of the "good" surfactants of Table 3.

The microscopic microemulsion models allow the good description of this wetting transition when the surfactant amphiphilicity decreases [47]. In one of the models [54], the relationship between the wetting transition and the fading away of the microemulsion structure is made explicit. The function describing the correlation between oil and water domains in the microemulsion has the asymptotic form $g(r) = \cos(2\pi r/d)\exp(-r/\xi)/r$ for $r \to \infty$. In simple fluids $g(r) = \exp(-r/\xi)/r$. The cos term describes periodic correlations, i.e., a structuration in oil and water microdomains. The theory predicts that $d \to \infty$ at the wetting transition, i.e., that the correlation function does not differ from that of a simple fluid. The parameters d and ξ can be easily deduced from x-ray or neutron scattering experiments [55]. Indeed, the scattered intensity variation with the scattering wave vector is simply the Fourier transform of $g(r)$:

$$I(q) = \frac{A}{1 + aq^2 + bq^4} \tag{17}$$

A being a constant and

$$\text{en(18)}\xi = \frac{\sqrt{2\sqrt{b}}}{(1 + a/2\sqrt{b})^{1/2}}$$

$$d = \frac{2\pi\sqrt{2\sqrt{b}}}{(1 - a/2\sqrt{b})^{1/2}}$$

The values of ξ and d measured for a series of nonionic surfactants [35, 56] are reported in Table 6. In the table, we also give the values of the persistence length ξ_K calculated from independent determinations of K using ellipsometry.

Figure 12 shows the evolution of the scattering spectra for C_6 E_2 versus temperature. Contrary to what was predicted, the peak does not move towards $q = 0$ at the wetting transition. It rather moves toward larger values when the temperature is increased. This discrepancy with the theory could be due to the role of long range forces [57]. Indeed, the wetting transition seems to be first order rather than second order according to the behavior of the contact angle

TABLE 6

System	T (°C)	$\xi_0 = \pi/q_{max}$ (Å)	ξ (Å)	d (Å)	d/ξ	ξ_K (Å)
C_6E_2	17	31	17	53	3.1	
C_8E_3	20	54	38	99	2.6	69
$C_{10}E_4$	25	196	151	361	2.39	211

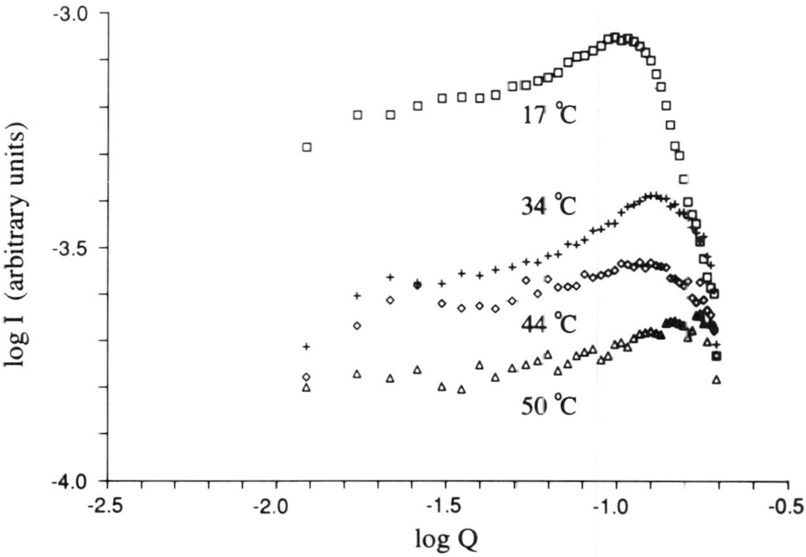

FIG. 12 Neutron scattered intensity versus scattering wavevector in microemulsions made with hexadecane and the nonionic surfactant C_6E_2 at different temperatures approaching the wetting transition.

[58]. In order to check this, we have measured the interfacial tensions between the different coexisting phases and calculated the contact angle using [23]:

$$\cos \theta = 1 - \frac{(\gamma_{om} + \gamma_{wm} - \gamma_{ow})(\gamma_{om} + \gamma_{wm} + \gamma_{ow})}{2\gamma_{om}\,\gamma_{wm}}$$

where γ_{om}, γ_{wm}, and γ_{ow} are respectively the tensions between the microemulsion phase and oil, the microemulsion phase and water, and oil and water. The interfacial tensions and the contact angles are shown in Fig. 13.

The accuracy on the contact angle determinations is rather poor close to T_w, because the difference $\gamma_{om} + \gamma_{wm} - \gamma_{ow}$ is small and comparable to the accuracy on γ measurements. However, it seems likely from Fig. 13 that the derivative of θ with respect to temperature is discontinuous at the wetting transition. The same trend is observed with more accurate direct θ measurements [58]. This would indicate that the transition is first order and that the role of long-range forces cannot be neglected. For this reason the temperature at which $d \rightarrow \infty$ would be shifted with respect to T_w. However, the scattered intensity decreases when the temperature increases (Fig. 12), indicating that the microemulsion structure is disappearing progressively around T_w. The wetting temperature and

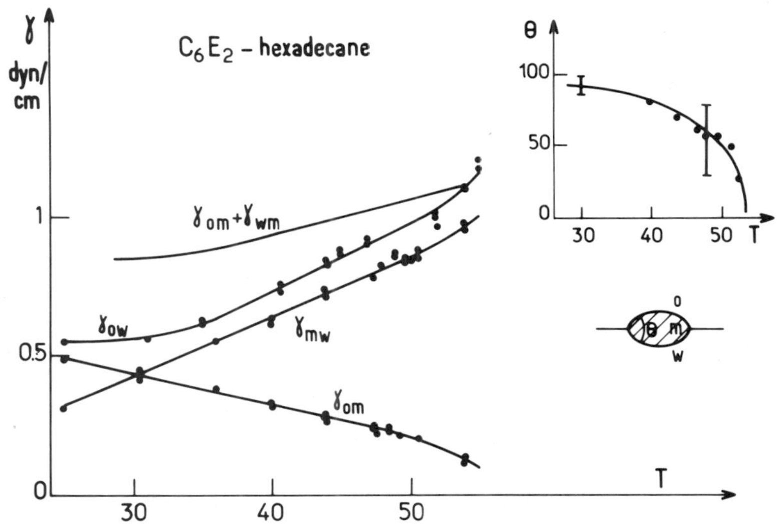

FIG. 13 Interfacial tensions versus temperature for the Winsor microemulsion system water-hexadecane-C_6E_2. The sum $\gamma_{om} + \gamma_{wm}$ has also been represented for comparison with γ_{ow}. Insert: calculated contact angles.

the contact angles are larger than in the experiments of Robert et al. [58]. This might be due to different impurities of the surfactant batches.

As expected, the characteristic lengths increase rapidly with surfactant chain length, while the ratio d/ξ decreases.

VII. COMPARISON WITH THE SPINNING DROP TECHNIQUE

A method based on the analysis of the shape of a spinning drop proposed by Vonnegut in 1942 was developed in the 1970s in view of the applications in tertiary oil recovery. It appeared rapidly that the method was perfectly suitable to low interfacial tension measurements. Several commercial instruments based on this principle are now available.

In the spinning drop technique, as in classical methods (sessile, pendent drops, etc.), interfacial tension is derived from shape and geometrical characteristics of a drop placed in a gravitational field. The system studied, consisting of a drop of liquid L_1 immersed in liquid L_2, is placed in a rotating tube. One assumes that the long axis of the drop coincides with the axis of rotation Ox. When the ratio between the length L and diameter d of the drop is larger than 4, one has the relation

$$\gamma = \frac{1}{4} d^3 \Delta\rho \, \omega^2$$

where ω is the rotation frequency. By considering the precision of measurement for ω, $\Delta\rho$, and d, one can estimate that the precision on γ is approximately 10%.

The method described so far places stringent requirements on experimental procedure if very precise measurements of interfacial tensions are needed. The most important concerns the necessity of the phases of a sample to be at thermodynamic equilibrium. Therefore, very precise control of the temperature and cell properties is required. The most serious disadvantages of this method are, on the one hand, the formation of artifacts (dumbell, dog-bone shapes, etc.), which most frequently appear when the interfacial tension level is low, and, on the other hand, the difficulty of specifying criteria characteristic of hydrodynamic equilibrium. This method, however, is perfectly suitable to measurement of low tensions, since instability of the meniscus does not occur, in contrast to what is observed in most classical methods, where a third solid phase is present. Considering vibration and bearing heating problems, such a method is less well suited for measurements of tensions requiring speeds either too low (< 1500 rpm) or too high (> 6000 rpm). Thus the measurable values of γ range from 5.10^{-4} mN/m to 5.10^{-1} mN/m.

The principles of the two techniques, spinning drop and surface light scattering, are very different. They give access to what can be called static and dynamic interfacial tension respectively. Indeed, in surface light scattering measurements, interfacial tensions are deduced from propagation characteristics of capillary waves. In low interfacial tension systems containing surfactant molecules, static and dynamic values could be different (Chap. 11). On the other hand, several authors have pointed out that the spinning drop technique does not always give perfectly reliable data [59].

We have made a comparison between the two techniques for several samples having interfacial tensions between 10^{-1} and 10^{-3} mN/m. The *same* samples were studied by both techniques in order to avoid possible discrepancies due to slight differences in chemical composition [60]. The systems studied were mixtures of sodium *p*-octylbenzene sulfonate (OBS)-pentanol-water-sodium chloride-dodecane. The agreement between the surface tension values obtained for the studied samples with the two techniques, spinning drop and surface light scattering, was satisfactory: the differences were lower than experimental accuracy.

In some few cases, the measurement with the spinning drop apparatus could not be made: the drop of the less dense liquid appeared to dissolve into the more dense one. This difficulty can arise from slight evaporation during the filling of the rotating tube. Indeed, the volume of the drops being small, the evaporation problems are likely to be more crucial than in the surface light scattering experiment, where the volumes of the two liquids are equal. More generally, if the composition of the phases put into contact does not correspond exactly to equilibrium compositions, the reequilibration processes can greatly modify the composition of the drop and lead to meaningless values of γ. This is, in our opinion, the probable origin of previously reported spinning-drop data dispersion in nonpreequilibrated samples.

This points out one advantage of the light scattering technique. In other aspects, the two techniques appear equivalent:

The lowest measurable γ are similar (about 10^{-5} mN/m).
They require similar total liquid volumes (a few cm^3).
The time of the measurement is a few hours to achieve an experimental accuracy of about 10%.
At least one liquid phase must be transparent (the denser one in the spinning-drop tensiometer).

VIII. SCATTERED INTENSITY MEASUREMENTS

Because the interfacial tensions are small in microemulsion systems, the intensity scattered by the surface is large. When the stray light coming from the cell windows and the other optical elements is carefully reduced, the intensity scattered by the surface is dominant and can be directly measured, and the interfacial tension determined after a careful calibration procedure. The method is similar to that used for soap films in Chap. 14.

It was applied to oil-microemulsion interfaces with success [61]. The intensity was optimized by working at the total reflection angle. In the case of the water-microemulsion interfaces, this condition is usually obtained when the beam travels in the microemulsion phase. The intensity scattered by the bulk microemulsion adds to the one scattered by the interface, and it is no longer possible to determine safely the interfacial tension.

IX. CONCLUSION

Surface light scattering is particularly well suited for the measurement of ultralow tensions found in multiphase microemulsion systems. It allows the investigation of systems at thermodynamic equilibrium and proved to be useful in clarifying the problem of the origin of the low tensions and in checking the different theories. It can also operate at high temperature and high pressure, conditions difficult to achieve with the only other existing method for low tension measurements, the spinning drop technique. This is the main reason why several oil companies are developing the scattering technique, to investigate the influence of extreme oil reservoir conditions on oil-water interfacial tensions.

REFERENCES

1. P. D. de Gennes and C. Taupin, *J. Phys. Chem. 86:* 2294 (1982).
2. D. Langevin, *Accounts Chem. Res 21:* 255 (1988).
3. W. Helfrich, *Z. Naturforsch. 28a:* 693 (1973).
4. S A. Safran and L. E. Turkevich, *Phys. Rev. Lett. 50:* 1930 (1983).
5. J. Jouffroy, P. Levinson, and P. G. de Gennes, *J. Phys. (France) 43:* 1241 (1982).
6. B. Widom, *J. Chem. Phys. 81:* 1030 (1984).
7. W. Helfrich, *J. Phys. (France) 46:* 1263 (1985).
8. L Peliti and S. Leibler, *Phys. Rev. Lett. 54:* 1690 (1985).
9. S A. Safran, D. Roux, M. E. Cates, and D. Andelman, *Phys. Rev. Lett. 57:* 491 (1986).
10. B. Widom, *J. Chem. Phys. 84:* 6943 (1986).
11. K. A. Dawson, *Phys. Rev. A 35:* 1766 (1987).
12. G. Gompper and M. Schick, *Phys. Rev. Lett. 62:* 1647 (1989).
13. P. A. Winsor, *Trans. Faraday Soc. 44:* 376 (1948).
14. R. N. Healy, R. L. Reed, and D. G. Stenmark, *Soc. Pet. Eng. J.: 147,* June 1976; D. O. Shah, ed., *Surface Phenomena in Enhanced Oil Recovery,* Plenum, New York, 1981.
15. J H. Schulman and J. B. Montagne, *Ann. N.Y Acad. Sci. 92:* 366 (1961).
16. C. A. Miller, R. Hwan, W. J. Benton, and T. Fort, *J. Colloid Interface Sci 61:* 554. (1977).
17. E. T. Davis and L. E. Scriven, SPE 9278, 55th Annual Fall Technical Conference of SPE-AIME in Dallas, September 21–24, 1980.
18. H. Kleinert, *J. Chem. Phys. 84:* 964 (1986).
19. A. Pouchelon, D. Chatenay, J. Meunier, and D. Langevin, *J. Colloid Interface Sci. 82:* 418 (1981).
20. A. Graciaa, J. Lachaise, A. Martinez, M. Bourrel, and C. Chambu, *C. R. Acad. Sci. 282B:* 547 (1976).
21. G. J. Verhoeckx, P. L. de Bruyn, and J.Th. Overbeek, *J. Colloid Interface Sci. 119:* 409 (1987).
22. R. Aveyard, B. P. Binks, S. Clark, and J. Mead, *J. Chem. Soc., Faraday Trans. 1, 82:* 125 (1986). R. Aveyard in *Solid Liquid Dispersion,* Academic Press, London, 1987.
23. J. S. Rowlinson and B. Widom, *Molecular Theory of Capillarity,* Clarendon Press, Oxford, 1982, Chap. 8.
24. B. Widom in *Fundamental Problems in Statistical Mechanics* (E. D. G. Cohen, ed.), 1975, Vol. 3.
25. A. M. Bellocq, D. Bourbon, B. Lemanceau, and G. Fourche, *J. Colloid Interface Sci. 89:* 427 (1982).

26. J. Biais, L. Odberg, and P. Stenius, *J. Colloid Interface Sci. 86:* 350 (1982).
27. A. M. Cazabat, D. Langevin, J. Meunier, and A. Pouchelon, *Adv. Colloid Interface Sci. 16:* 175 (1982).
28. P. D. Fleming, J. E. Vinatieri, and G. R. Glinsman, *J. Phys. Chem. 84:* 1526 (1980).
29. M. W. Kim, J. Bock, and J. S. Huang, in *Waves on Fluid Interfaces,* Academic Press, 1983, p. 151; *J. SPE/DOE* 10788 (1982).
30. K. Bonkhoff, A. Hirtz, and G. H. Findenegg, *Physica A, 172:* 174 (1991).
31. M. R. Moldover, *Phys Rev. A 31:* 1022 (1985).
32. G. H. Findenegg, A. Hirtz, R. Rasch, and F. Sowa, *J. Phys. Chem. 93:* 4580 (1989).
33. D. Guest and D. Langevin, *J. Colloid Interface Sci. 112:* 208 (1986).
34. B. P. Binks, J. Meunier, O. Abillon, and D. Langevin, *Langmuir 5:* 415 (1989).
35. L. T. Lee, D. Langevin, J. Meunier, K. Wong, and B. Cabane, *Prog. Colloid Polym. Sci. 81:* 209 (1990).
36. S. T. Milner, S. A. Safran, D. Andelman, M. E. Cates, and D. Roux, *J. Phys. (France) 49:* 1065 (1988).
37. Chun Huh, *J. Colloid Interface Sci. 71:* 408 (1979).
38. J. Israelachvili in *Surfactants in Solution* (K. L. Mittal and P. Bothorel, eds.), Plenum, New York, 1986, Vol. 4, 3.
39. E. Ruckenstein, *Chem. Phys. Lett. 118:* 435 (1985). E. Ruckenstein and I. V. Rao, *J. Colloid Interface Sci. 117:* 104 (1987).
40. J.Th.G. Overbeek, G. J. Verhoeckx, P. L. de Bruyn, and H. N. W. Lekkerkerker, *J. Colloid Interface Sci. 119:* 422 (1987).
41. M. Robbins in *Surfactants in Solution* (K. L. Mittal, ed.), Plenum, New York, 1977, Vol. 2, 713.
42. D. J. Mitchell and B. W. Ninham, *J. Phys. Chem. 87:* 2996 (1983).
43. Chun Huh, *J. Colloid Interface Sci. 97:* 201 (1984).
44. J. Meunier, *J. Physique, 48:* 1819 (1987).
45. Y. Talmon and S. Prager, *J. Chem. Phys. 69:* 2984 (1978).
46. J. Meunier, *J. Physique Lett. 46:* 1005 (1985).
47. C. Borzi, R. Lipowsky, and B. Widom, *J. Chem. Soc. Faraday Trans. 2, 82:* 1739 (1986).
48. K. A. Dawson, *Phys. Rev. A 35:* 1766 (1987).
49. J. Indekeu in *Lectures on Thermodynamics and Statistical Mechanics* (M. Lopez de Haro and C. Varea, eds.), World Scientific, 1991.
50. J. C. Lang, Jr. and B. Widom, *Physica 81A:* 190 (1975).
51. Y. Seeto, J. E. Puig, L. E. Scriven, and H. T. Davis, *J. Colloid Interface Sci. 96:* 360 (1983).
52. M. Kahlweit et al., *J. Colloid Interface Sci. 118:* 436 (1987).
53. M. Robert and J. F. Jeng, *J. Phys. (France) 49:* 1821 (1988).
54. G. Gompper and M. Schick, *Phys. Rev. Lett., 65:* 1116 (1990).
55. M. Teubner and R. Strey, *J. Chem. Phys. 87:* 3195 (1987).
56. O. Abillon, L. T. Lee, D. Langevin, and K. Wong, *Physica A, 172:* 209 (1991).
57. M. Schick, private communication.

58. L. J. Chen, J. F. Jeng, M. Robert, and K. Shikla, preprint.
59. J. E. Puig, E. I. Franses, H. T. Davis, W. G. Miller, and L. E. Scriven, *J. Soc. Pet. Eng. 71* (April 1979).
60. D. Chatenay, D. Langevin, J.Meunier, D. Bourbon, P. Lalanne, and A. M. Bellocq, *J. Dispersion Science and Technology 3:* 245 (1982).
61. J. Lachaise, A. Graciaa, A. Martinez, and A. Rousset, *J. Phys. Lett. 40:* L-599 (1979).
62. D. Chatenay, O. Abillon, J. Meunier, D. Langevin, and A. M. Cabazat in *Macro and Microemulsions* (D. O. Shah, ed.), ACS Publisher, 1985.

14

Thin Liquid Films

DOMINIQUE LANGEVIN Laboratoire de Physique Statistique, Ecole Normale Supérieure, Paris, France

I. INTRODUCTION

Thin liquid films have interesting characteristic properties that can be investigated with light scattering techniques. Soap films are water films stabilized by two surfactant monolayers at the air-water surfaces [1]. They can be formed on frames, dipped into and withdrawn from an aqueous surfactant solution. When the frame is vertical, the film thins because of gravity: this is the "drainage" process. When the film thickness h becomes comparable to the wavelength of light, beautiful interference colors are seen; these colors are also observed in soap bubbles for the same reason. In the last drainage stages, the film cannot be seen because it reaches a thickness of a few tens of nanometers or less, and it is called "black film."

In the early 1950s, Derjaguin proposed that the same interaction forces are operative in soap films and between colloidal particles suspended in an electrolyte solution [2]. These forces are the electrostatic repulsion between the surfactant monolayers and the van der Waals attraction between all molecules in the film [3]. Both forces are long range, and the total free energy can have a local minimum for a film thickness of a few tens of nanometers. This corresponds to the first type of black film, also called "common" black film. The free energy has a deeper minimum for a film thickness about equal to twice the monolayer thickness. This corresponds to the second or "Newton" black film. The height of the energy barrier between the two states depends on the range of the electrostatic forces, which in turn depends on the electrolyte concentration: the barrier is much larger than $k_B T$ at low electrolyte concentrations where very stable first black films are formed. At large electrolyte concentrations, the second black film is the only one stable; this film is similar to an inverted lipid bilayer (Chap. 12) for which the surfactant chains are in the interior of the bilayer, whereas in the case of the Newton black film the polar parts are in the interior, in association with residual hydration water. The above description of the free energy has been elaborated by Derjaguin, Landau, Verwey, and Overbeek; it applies to colloidal particles and soap films as well, and is known nowadays as the DLVO theory.

The interplay between electrostatic and van der Waals forces also affects the scattering of light by the soap films. Indeed, the range of the forces being comparable to the film thickness, the surface energy contains, in addition to the usual surface tension term, a contribution of the forces sometimes called "disjoining pressure." The scattered light intensity was studied in the 1960s by A. Vrij, who showed that this was a way to investigate interaction forces in soap films [4]. The method was implemented later by the study of the spectrum of the scattered light [5, 6, 7].

Freely suspended liquid crystal films have also been widely investigated [8]. These films can be formed with the help of frames, as with soap films. They exhibit a very rich phase behavior with symmetries and order intermediate between isotropic liquids and fully ordered crystals. New surface-specific types of phases like the hexatic phases have thus been evidenced. Surface order and bulk order being frequently different, they have been used to study the cross-over between two- and three-dimensional order. Indeed, surface-induced phases in the interior of the film can be several hundred layers thick; there are also surface induced tilt changes and surface-induced freezing transitions. Most of the present work in these systems is made by using x-ray scattering (Chap. 18). There have been several light scattering studies, most of which focus on orientation fluctuations involving the whole film [9, 10]. In this chapter we will deal with the scattering by the capillary waves at the boundaries of these films, for which nowadays there is only a limited amount of information [11, 12].

The behavior of thin films on solid substrates belongs to the field of "wetting" phenomena where the notion of disjoining pressure also plays an important role [13]. There exists preliminary information about the light scattering by the free surface of such a film [14]; it will be discussed later.

II. SOAP FILMS

Light scattering by soap films has been discussed in great detail by Joosten in a recent review [15]. In this chapter, we shall present only the basic concepts and describe the main experimental results obtained with these films.

A. Stability

It was early believed that soap films drawn from aqueous surfactant solutions consist of an aqueous core of thickness h_c sandwiched between two surface monolayers of adsorbed surfactant molecules of thickness h_s, the total film thickness being $h = h_c + 2h_s$. The picture was confirmed by recent neutron and x-ray reflectivity experiments [16, 17]. This structure is the basis of the model describing film stability. The intermolecular forces are conveniently expressed in terms of the disjoining pressure π, which is the outward force per unit area on the surface of the film attributable to long-range interactions across the film. The disjoining pressure is related to the interaction energy per unit area $V(h)$ by $\pi(h) = -\partial V(h)/\partial h$. By convention, π is positive when resisting thinning.

Electrostatic repulsion arises from adsorbed ions at the interfaces between the water films and the surfactant monolayers. This repulsion is screened by a diffuse layer of counterions solubilized in water, of thickness κ^{-1}, the "electric double layer." Its contribution to the disjoining pressure, in the limit of small overlap of the ionic diffuse layers $\kappa h \geqslant 2$, is given by [3]

$$\pi_e = B\phi^2 \kappa^2 \exp(-\kappa h_c) \tag{1}$$

where the inverse ionic diffuse layer thickness (Debye-Hückel length) κ is

$$\kappa^2 = \frac{2e^2 z^2 c_i}{\varepsilon k_B T} \tag{2}$$

and

$$B = 32\,\varepsilon\left(\frac{k_B T}{e}\right)^2 \tag{3}$$

$$\phi = \tanh\left(\frac{2e\Psi_0}{4k_B T}\right) \tag{4}$$

In these equations ε is the dielectric constant of the solution, e the elementary charge, z the valency of ions, Ψ_0 the surface potential, h_c the thickness of the aqueous core of the film, and c_i the number of ions per unit volume.

The van der Waals contribution to the disjoining pressure is

$$\pi_W = -\frac{A}{6\pi h^3} \tag{5}$$

for a uniform slab of material of thickness h. The Hamaker constant A can be calculated if the frequency dependence of the dielectric constant is known. In practice, Eq. 5 cannot be used here, because the three-layer nature of the film leads to significant corrections, as well as relativistic retardation effects also neglected in Eq. 5. Ninham and Parsegian have calculated a more rigorous expression for π_W [18], that can be numerically well approximated by

$$\pi_W = \frac{1}{h_c^3}\left(\frac{a + bh_c}{1 + dh_c + eh_c^2} + f\right) \tag{6}$$

where a, b, d, e, and f are tabulated constants that depend on the surfactant surface concentration Γ_s [19].

There is a final contribution to the disjoining pressure: the hydrostatic pressure

$$\pi_H = -\Delta\rho\, gH \tag{7}$$

where $\Delta\rho$ is the density difference between the liquid film and air, and H the distance between the film element that is considered and the surface of the bulk solution with which the film is in contact.

π_H and π_W both act to thin the film, and π_e resists thinning. Their interplay leads to the drainage of a freshly drawn film until it reaches the thickness where

$$\pi_e + \pi_W + \pi_H = 0 \tag{8}$$

This thickness is the equilibrium thickness h_{eq}.

B. Reflected Light Intensity

The thickness of the liquid film can easily be monitored by recording the intensity I_R of the light reflected by the film. In the case of a uniform slab of thickness h, and of a beam polarized normal to the plane of incidence, I_R is given by the Airy formula [20]:

$$I_R = \frac{I_0(4r^2 \sin^2 \beta)}{(1 - r^2)^2 + 4r^2 \sin^2 \beta} \tag{9}$$

where $r = \sin(\theta_i - \theta_t)/\sin(\theta_i + \theta_t)$ is the reflectivity of a single surface for this polarization, and $\beta = 2\pi nh \cos \theta_t/\lambda$, n being the refractive index of the slab.

I_R exhibits an oscillatory behavior from which h can be easily determined. In the case of a soap film, if one uses for n the refractive index of the surfactant solution n_w, Eq. 9 leads to an effective thickness $\tilde{h}$ related to the water core and surfactant monolayer thickness h_c and h_s by

$$\tilde{h} = h_c + 2h_s \frac{n_s^2 - 1}{n_w^2 - 1}$$

The surfactant refractive index in the film can be evaluated by using

$$n_s^2 - 1 = (n_s^2 + 2)\frac{\mathcal{R}\Gamma_s}{h_s}$$

where $\mathcal{R}$ is the surfactant molar refraction. Thus

$$\tilde{h} = h_c + \frac{2\mathcal{R}\Gamma_s(n_s^2 + 2)}{(n_W^2 - 1)} \tag{10}$$

$n_s^2 + 2$ being relatively independent of the surface concentration Γ_s [21]. Γ_s can be obtained independently from the variation of the surface tension of the surfactant solution with the bulk surfactant concentration by using the Gibbs relation $\Gamma_s = -d\gamma/d(\ln C_s)$.

In order to observe the whole drainage process, the film must be enclosed in a temperature and humidity controlled cell to avoid early film bursting. The drainage velocity can be decreased by using small circular frames (3 mm diameter in Ref. 7, 20 mm in Ref. 6) and adding glycerol to the surfactant solution [1]. Furthermore, without addition of the glycerol, it appeared impossible to reach equilibrium thicknesses that are in agreement with the DLVO theory [22]. Needless to say, the surfactant solution and the film must be kept clean and free of contamination.

Figure 1 shows some data for the equilibrium thickness of a hexadecyl trimethyl ammonium (CTAB) + potassium bromide (KBr) film for different heights above the bulk solution H [23]. H is directly related to the disjoining pressure $\pi = -\partial V/\partial h = -\pi_H = \Delta\rho\, gH$ (see Eq. 8). Table 1 shows data for films made with different chemicals, dodecyl trimethyl ammonium bromide (DTAB) and chloride (DTAC). In these systems, the ionic concentration c_i has been chosen in order to have comparable contributions of van der Waals and electrostatic forces: $c_i \sim 10$ mM. The solutions contain 1 M glycerol. π_H can be simply evaluated from Eq. 7, π_W from Eq. 6 and the data of Ref. 19. The equilibrium condition 8 can then be used to evaluate π_e and Ψ_0 [23]. The results are shown in Table 1. κ has been calculated by using for c_i the salt concentration plus the critical micellar concentration (cmc) in the case of CTAB (0.2 mM) and the surfactant concentration in the case of DTAB and DTAC; because these last two surfactants were more soluble in water and their cmc larger than 10 mM, it was

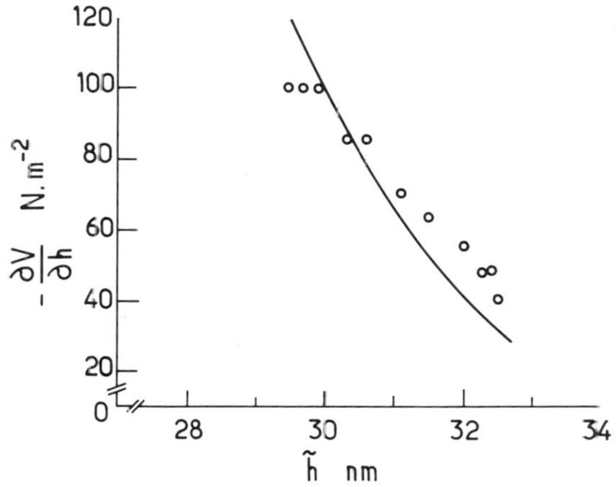

FIG. 1 Disjoining pressure $\pi = -\partial V/\partial h$ versus equilibrium optical thickness $\tilde{h}$, for a CTAB soap film, as deduced from the height above the bulk solution: $\pi = -\pi_H = \Delta\rho gH$ (see Eq. 8). (Data from Ref. 23.)

not necessary to add salt to reach $c_i \sim 10$ mM. The difference between the measured effective thickness and the core thickness calculated with Eq. 10 is also shown in Table 1. It is roughly about twice the surfactant chain length as expected.

It is thus easy to check directly the variation of the disjoining pressure with the thickness of the film. In the following we shall show that light scattering is sensitive to the derivative of the disjoining pressure with respect to h (second derivative of the energy), and that this quantity can be measured even during the drainage process.

TABLE 1 Equilibrium Soap Film Core Thicknesses $h_{c_{eq}}$ for Different Heights H above the Surfactant Solution. Optical Correction $\tilde{h}_{eq} - h_{c_{eq}}$ Calculated with Eq. 10. Comparison with the DLVO Theory.

Solution	c_i (mM)	H (mm)	$h_{c_{eq}}$ (nm)	$\tilde{h}_{eq} - h_{c_{eq}}$ (nm)	ψ_0 (mV)
0.8 mN CTAB + 10 mM KBr	10.2	4–10	29–33	3.4	130
10 mM DTAB	10.0	9	29.1	2.8	103
10 mM DTAC	10.0	8.7	29.4	2.7	113

Source: Data from Ref. 23.

C. Scattered Intensity

Light is scattered by thermal fluctuations on both sides of the film. At equilibrium, the two limit surfaces are parallel. Let us take Oz as their common normal. For points of the two limit surfaces having the same projection **r** in the Oxy plane, the displacements with respect to equilibrium will be called $\zeta(\mathbf{r})$ and $\zeta'(\mathbf{r})$. After a Fourier transform,

$$\zeta(\mathbf{r}) = \sum_q \zeta_{\mathrm{q}}\, e^{i\cdot\mathbf{q}\cdot\mathbf{r}} \qquad \zeta'(r) = \sum_q \zeta'_{\mathrm{q}}\, e^{i\mathbf{q}\cdot\mathbf{r}} \tag{11}$$

The excess free energy due to these fluctuations is

$$F = \int d^2r\left[\frac{\gamma}{2}\{(\nabla\zeta(\mathbf{r}))^2 + (\nabla\zeta'(\mathbf{r}))^2\} + \frac{1}{2}\frac{\partial^2 V}{\partial h^2}[\zeta(\mathbf{r}) - \zeta'(\mathbf{r})]^2\right] \tag{12}$$

which becomes with Eqs. 11

$$F = \sum_q \left\{\frac{\gamma q^2}{4}|\zeta_{\mathrm{q}} + \zeta'_{\mathrm{q}}|^2 + \left(\frac{\gamma q^2}{4} + \frac{1}{2}\frac{\partial^2 V}{\partial h^2}\right)|\zeta_{\mathrm{q}} - \zeta'_{\mathrm{q}}|^2\right\} \tag{13}$$

Equation 13 shows that there are two statistically independent fluctuation modes: the mode where ζ_{q} and ζ'_{q} are equal, called "undulation," "bending," or "stretching" mode, and the mode where ζ_{q} and ζ'_{q} are opposite, called "peristaltic" or "squeezing" mode (Fig. 2). Obviously, only the second mode is sensitive to interaction forces. By applying the equipartition theorem, one finds

$$\begin{aligned} \langle|\zeta_{\mathrm{q}} + \zeta'_{\mathrm{q}}|^2\rangle &= \frac{2k_{\mathrm{B}}T}{\gamma q^2} \\ \langle|\zeta_{\mathrm{q}} - \zeta'_{\mathrm{q}}|^2\rangle &= \frac{2k_{\mathrm{B}}T}{\gamma q^2 + 2\dfrac{\partial^2 V}{\partial h^2}} \end{aligned} \tag{14}$$

With this result, the relative scattered intensity per unit solid angle can be calculated following a geometrical calculation similar to that of Chap. 3. The result for a one-layer model is [4]

$$\frac{1}{I}\frac{dI}{d\Omega} = \frac{\pi^2(n^2 - 1)^2\, k_{\mathrm{B}}T}{\left(\gamma q^2 + 2\dfrac{\partial^2 V}{\partial h^2}\right)\lambda^4}\, G_0\left[G + \frac{\partial^2 V}{\partial h^2}\, G'/\gamma q^2\right] \tag{15}$$

G_0, G, and G' are geometrical factors

$$G_0 = \frac{t^2 t_{\mathrm{d}}^2}{R(r_{\mathrm{d}}^2, \beta_{\mathrm{d}})\, R(r^2, \beta)} \tag{16}$$

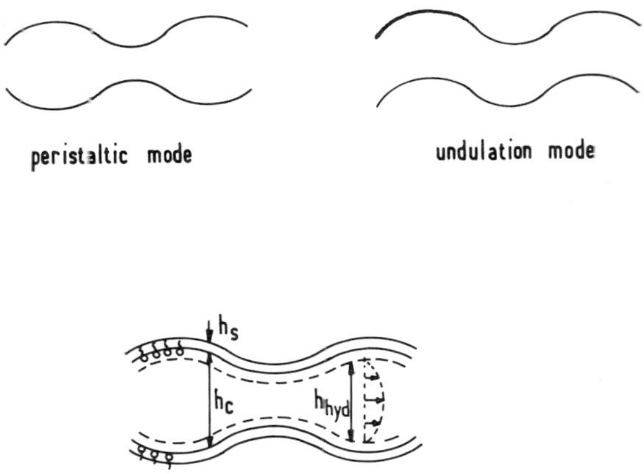

FIG. 2 Schematic representation of the undulation and peristaltic modes.

r being as in Sec. B the reflectivity coefficient for the light incident as an angle θ_i on the first film boundary and $\beta = 2\pi nh \cos \theta_t/\lambda$; r_d and β_d are the same quantities for the scattering angles θ and θ_d (around reflection and refraction respectively); $t = 1 + r$ and $t_d = 1 + r_d$. The factors G and G', for scattering around reflection, are

$$G_R = R(r, \beta)\, R(r_d, \beta_d) + t'^2\, t_d'^2 \qquad (17)$$
$$G_R' = t_d'^2 R(r, \beta) + t'^2\, R(r_d, \beta_d)$$

For scattering around transmission, they are

$$G_T = G_R - 2t'^2\, t_d'^2 \cos \beta \cos \beta_d + 2tt' t_d t_d' \sin \beta \sin \beta_d \qquad (18)$$
$$G_T' = G_R' - 2t'^2\, t_d'^2 \cos \beta \cos \beta_d - 2tt' t_d t_d' \sin \beta \sin \beta_d$$

using $t' = 1 - r$, $t_d' = 1 - r_d$ and

$$R(r, \beta) = 1 + r^2 - 2r \cos 2\beta$$
$$R(r^2, \beta) = 1 + r^4 - 2r^2 \cos 2\beta$$

and similar expressions for $R(r_d, \beta)$ and $R(r_d^2, \beta)$. Three layer models do not lead to significant differences [7].

The calculated relative scattered intensity is plotted in Fig. 3 for the two surface modes together with the reflected intensity for $n = 1.39$, $\theta = 47°$, $\theta_i = 43°$, $\lambda = 5\ 145$ Å, and $\partial^2 V/\partial h^2 = 0$. As expected, the undulation mode scatters an intensity proportional to the reflected intensity, like a single surface (Chap. 3), whereas the peristaltic mode scattering intensity maxima and minima correspond to minima and maxima respectively for the undulation mode.

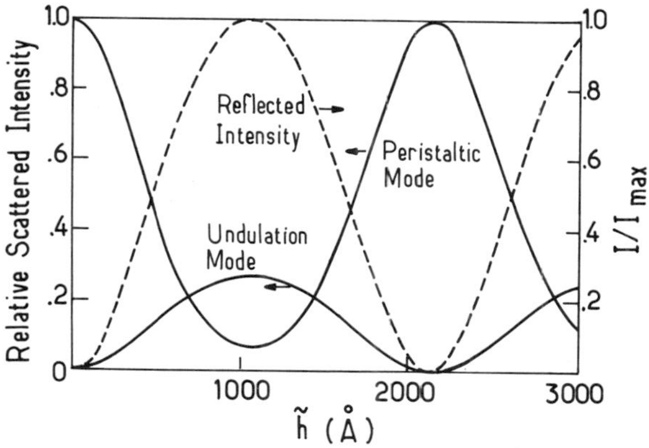

FIG. 3 Theoretical scattered intensity for the two modes (left scale) and reflected intensity (right scale) versus optical thickness; $n = 1.39$, $\theta = 47°$, $\lambda = 5145$ Å, $\partial^2 V/\partial h^2 = 0$. (Data from Ref. 7.)

The calculated total scattered intensity is shown in Fig. 4 for $\partial^2 V/\partial h^2 = 0$ together with experimental measurements on draining CTAB soap films (0.4 mM CTAB, 40 wt % glycerol, 10 mM NaCl) [7]. Only two adjustable parameters are used in obtaining the theoretical curve: the maximum amplitude and the refractive index of the solution, which determines the position of the extrema.

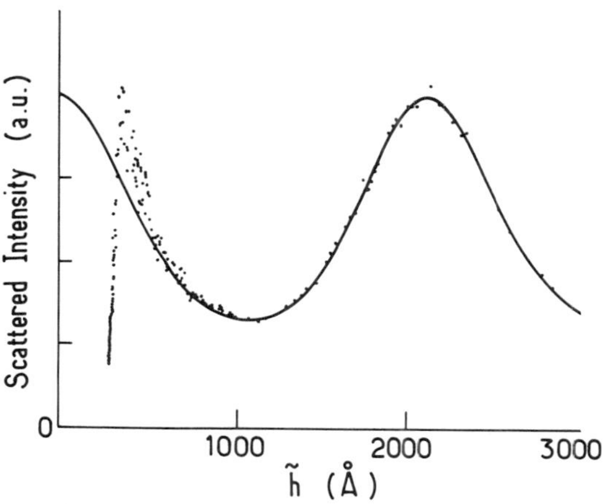

FIG. 4 Relative scattered intensity versus optical thickness for a CTAB soap film. The points are experimental. The line is the theoretical variation for $\partial^2 V/\partial h^2 = 0$ adjusted to fit the data at large h. (Data from Ref. 7.)

For small h, the deviations indicate that the experiment becomes sensitive to interaction forces. The corresponding values of $\partial^2V/\partial h^2$ are shown in Fig. 5. The fit with Eq. 1 and Eq. 6 with no adjustable parameters for π_W leads to $\kappa^{-1} = 28 \pm 3$ Å and $\Psi_0 = 43$ mV. κ is in good agreement with the calculated value using Eq. 2. Ψ_0 is smaller than for the films of Table 1, which were drawn from a more concentrated CTAB solution ($\gamma = 33.1$ mN/m instead of 54.4 mN/m).

More complete scattered intensity measurements were done for the films of Table 1, on both draining and equilibrium films [15, 23]. The results are similar in nature to those shown in Figs. 4 and 5. The fit of the $\partial^2V/\partial h^2$ data with Eqs. 1 and 6 gives better results for $f = 0$; this is indeed a zero frequency term in the van der Waals interaction potential that becomes negligible for large κh. Using the calculated values of the other parameters except for Ψ_0, it was found that the best fit corresponds to $\Psi_0 = 125 \pm 5$ mV for CTAB and $\Psi_0 = 109 \pm 1$ mV for DTAB, in good agreement with the value deduced from equilibrium thicknesses and Eq. 8 [23].

By changing the ionic concentration it is possible to change the relative weight of the electrostatic and the van der Waals forces. Earlier intensity measurements were done in CTAB soap films drawn from solution containing 0.8 mM CTAB, 8.4 wt % glycerol, and either no salt or 0.1 M KBr [24]. In the first case, the data were in agreement with electrostatic repulsion as given by Eq. 1 and $\Psi_0 = 90 \pm 20$ mV and $\kappa^{-1} = 10.3 \pm 0.7$ nm (the calculated value is 10.5 nm). In the second case, the van der Waals forces were found to be dominant, and in agreement with Eq. 6, with the constants $a,b,d,e,$ and f taken from Ref. 19.

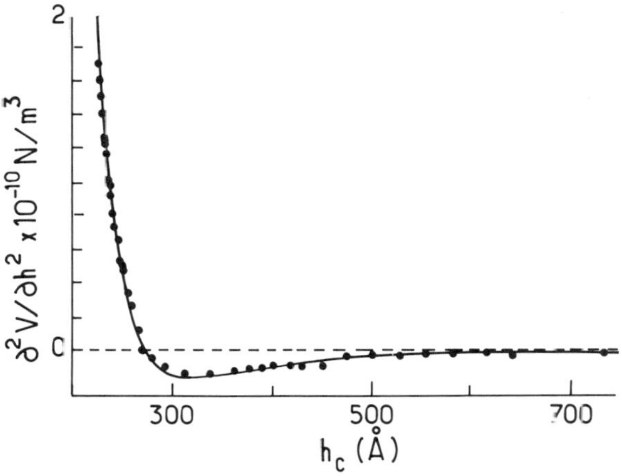

FIG. 5 Variation of $\partial^2V/\partial h^2$ with core thickness as deduced from the scattered intensity on CTAB films. (Data from Ref. 7.)

Although numerically the calculations of Ninham and Parsegian are better approached by $f = -5.4 \times 10^{-22}$, the fit of the light scattering data was also here better for $f = 0$. But as explained earlier, f being a zero frequency term, it should indeed become zero for large κh.

Results on a different system (SDS/NOP7) led to larger values for the Hamaker effective constant [25].

When equilibrium films are studied, the angular variation of the scattered intensity can be probed. This allows a better comparison with the theory; in particular, it was shown that Eqs. 15–18 are well obeyed. In a drainage experiment, intensity calibration can be made using the data in the thickness range where interaction is absent. In an equilibrium experiment, where the angular dependence of the scattered intensity is studied, the calibration factor can also be determined from the limit $q = 0$. These determinations, including the absolute calibration, all give consistent results [15]. Let us point out that in order to extract the $\partial^2 V/\partial h^2$ values from the scattered intensity measurements, experiments should be carried out preferentially at the transmission side of the film, where the undulation mode contribution is negligible (see Eqs. 15–18).

D. Inelastic Scattering—Peristaltic Mode

The dynamics of the fluctuations of a soap film may be calculated from the hydrodynamic equations

$$\rho \frac{d\mathbf{v}}{dt} = -\nabla p + \eta \nabla^2 \mathbf{v} \qquad \nabla \cdot \mathbf{v} = 0 \tag{19}$$

where $\mathbf{v}$ is the velocity of the fluid, ρ the density, η the viscosity, and p the local pressure. In the liquid phase p includes the disjoining pressure. Using the boundary conditions of vanishing velocities at infinity and equal velocity and stress at the liquid-air interfaces, the dispersion equation can be calculated [26].

Using $m^2 = q^2 - i\omega\rho/\eta$, $a = \tanh(mh/2)$, $b = \tanh(qh/2)$, and the superscript prime to denote variables of the gas phase, this dispersion equation for the peristaltic mode is

$$\begin{aligned}(mb - qa)\,[\tilde{\gamma}q^3 - i\omega\,\eta' m'(m' + q)]\,[i\omega\eta' q(m' + q) - \varepsilon q^3]\\ + ab\, i\omega\eta\, \tilde{\gamma}q^2(m^2 - q^2)q^2 - i\omega\eta\, m(m^2 - q^2)\,[i\omega\eta'(m' + q)q\\ - \varepsilon q^3] - ab\,\omega^2\,\eta' m' q(m' + q)(m^2 - q^2) - b\omega^2\, mq^2[2\eta q\\ + \eta'(m' - q)]^2 + a\omega^2\, q[(m^2 + q^2)\eta + \eta'(m'^2 - q^2)]^2 = 0\end{aligned} \tag{20}$$

where $\tilde{\gamma}$ is the effective surface tension

$$\tilde{\gamma} = \gamma + \frac{2}{q^2}\frac{\partial^2 V}{\partial h^2} \tag{21}$$

and ε is the surface compression modulus $\varepsilon = A\partial\gamma/\partial A$ (Chap. 11).

The film thickness is small compared to the light wavelength, itself much smaller than the wavelength of the thermal fluctuations if small scattering angles are studied. Therefore qh and mh are small compared to 1. If $\omega\rho'/\eta' q^2 \ll 1$, Eq. 20 simplifies to

$$\omega^3\left(1 + \frac{\rho'}{\rho q h}\right) + \frac{4i\eta q^2}{\rho}\left(1 + \frac{\eta'}{\eta q h}\right)\omega^2 - \frac{q^2}{\rho h}\left(2\varepsilon + \tilde{\gamma}\frac{q^2 h^2}{2}\right)\omega - \frac{i q^5 h^2 \varepsilon \tilde{\gamma}}{12\eta\rho} = 0 \tag{22}$$

For soap films, ε is generally large. Then the solution of Eq. 22 is

$$\omega_p \cong i\,\tilde{\gamma}\frac{q^4 h^3}{24\eta} \tag{23}$$

It corresponds to the relaxation by Poiseuille flow between two rigid surfaces. The capillary waves do not propagate, and the spectrum of the scattered light is centered around the frequency of the incident light. Its shape is Lorentzian and its half-width is $\Delta\omega = |\omega_p|$. Calculations show that the condition for the surfaces to be rigid is $\varepsilon \geqslant 10$ mN/m [27]. Experiments done on the free surface of the CTAB and DTAB solutions of Table 1 showed that $\varepsilon \geqslant \gamma/2$ [28], i.e., that the monolayer covered surfaces were rigid-like, as assumed to derive Eq. 23.

Because the bulk scattering is small, it is possible to investigate scattering up to large scattering angles: $|\theta - \theta_0| \leqslant 20°$ [20]. The detection is mainly homodyne at these large angles. A transition to heterodyne detection can be observed for scattering angles of a few degrees [22]. If the detection is homodyne, the inelastic scattering when analyzed with a correlator shows a single exponential decay. As for the scattered intensity, the thickness dependence can be probed with draining films, the angular dependence with equilibrium films. In their first paper, Fijnaut and Joosten [22] found that the variation of the characteristic relaxation time τ with the wave vector q was in agreement with

$$2\tau^{-1} = \Delta\omega = \frac{\gamma}{24\eta}h^3\left(q^4 + \frac{2q^2}{\gamma}\frac{\partial^2 V}{\partial h^2}\right) \tag{24}$$

but that the limit value for large q led to a value of γ/η 40% lower than the expected value. On the other hand, the limit value of $\partial^2 V/\partial h^2$ for large h was not zero as expected.

The origin of the discrepancy lies in the fact that the thickness h appearing in Eq. 24 is not the total film thickness but the thickness of the moving fluid [7]. This effective hydrodynamic thickness h_{hyd} can be determined by imposing $\partial^2 V/\partial h^2 \rightarrow 0$ when $h \rightarrow \infty$. The data obtained in this way for the film of Fig. 5 are shown in Fig. 6. In this case $h_{hyd} = \tilde{h} - (4.5 \pm 0.2)$ nm. The value of γ/η, with this choice of h, is now in agreement with the values of γ and η as

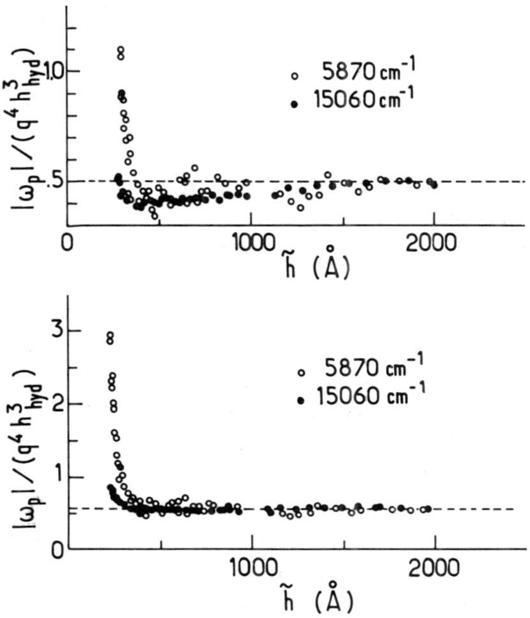

FIG. 6 Reduced half-width of the spectra for the peristaltic mode versus optical thickness for the CTAB films. Upper curve: $h_{hyd} = \tilde{h}$; lower curve: h_{hyd} adjusted in order to have $|\omega_p| \sim q^4$ above $\tilde{h} = 500$ Å. (Data from Ref. 7.)

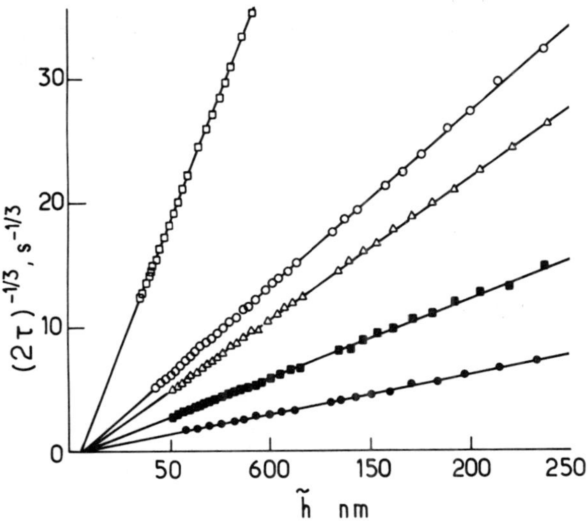

FIG. 7 Inverse cubic root of the relaxation time for the peristaltic mode of a CTAB film versus optical thickness; open squares, $q = 2.73 \times 10^4$ cm^{-1}; open circles, $q = 1.22 \times 10^4$ cm^{-1}; open triangles, $q = 1.04 \times 10^4$ cm^{-1}; filled squares, $q = 6.56 \times 10^3$ cm^{-1}; filled circles, $q = 4 \times 10^3$ cm^{-1}. (Data from Ref. 23.)

determined from the bulk solution. An alternative method to determine h_{hyd} is to plot $\tau^{-1/3}$ as a function of $\tilde{h}$: $\tau^{-1/3} = 0$ for $h_{hyd} = 0$ (Fig. 7) [15].

Detailed experiments on CTAB, DTAB, and DTAC soap films confirmed that the difference between $\tilde{h}$ and h_{hyd} was indeed of the order of a bilayer thickness, i.e., that the surfactant layers remained immobile during the film displacement. However, the difference between $\tilde{h}$ and h_{hyd} depends on the nature and the concentration of the salt added (Table 2).

The values of $\partial^2 V/\partial h^2$ obtained for the films of Fig. 5 are shown in Fig. 8. From the comparison between Figs. 5 and 8, it is seen that the accuracy on $\partial^2 V/\partial h^2$ as obtained from inelastic scattering is much poorer than from elastic scattering. In particular, the scatter of the data is too large to determine the van der Waals contribution. This is because the $\partial^2 V/\partial h^2$ term is only a small deviation from a dominant h^3 dependence. The data can however be analyzed to obtain information on the electrostatic forces. It was found that there is a systematic and yet unexplained difference betewen the static and dynamic values of $\partial^2 V/\partial h^2$ of about 20%. This difference is found even when the elastic and inelastic scattering experiments are performed simultaneously on the same film. It was found to be comparable for CTAB and DTAB films [23].

A possible origin of the discrepancy is that during the motion probed by light scattering, the ions do not have time to move in the diffuse layer to achieve local charge equilibrium, in which case the surface potential Ψ_0 could not be considered as constant. But the relaxation time τ_i for charges is of the order of $(D\kappa^2)^{-1}$ where D is the diffusion coefficient. Typically $D \sim 10^{-9}$ m^2/s, $\kappa^2 \sim 10^{17}$ m^{-2}, and $\tau_i \sim 10^{-8}$ s, much smaller than the relaxation time of the peristaltic mode (in the ms range).

TABLE 2 Difference Between Hydrodynamic and Water Core Thicknesses

Solution	Salt concentration (mM)	$h_c - h_{hyd}$ (nm)
0.8 mM CTAB + KBr [23]	—	2.4 ± 1
	1	2.0 ± 0.4
	10	1.1 ± 0.3
	100	0.8 ± 0.3
0.4 mM CTAB + NaCl [7]	10	−0.5 ± 0.4
0.8 mM CTAB + KF [23]	10	2.3 ± 0.3
10 mM DTAB [23]	—	0.7 ± 0.2
10 mM DTAC [23]	—	1.2 ± 0.2

Source: Data from Refs. 7 and 23.

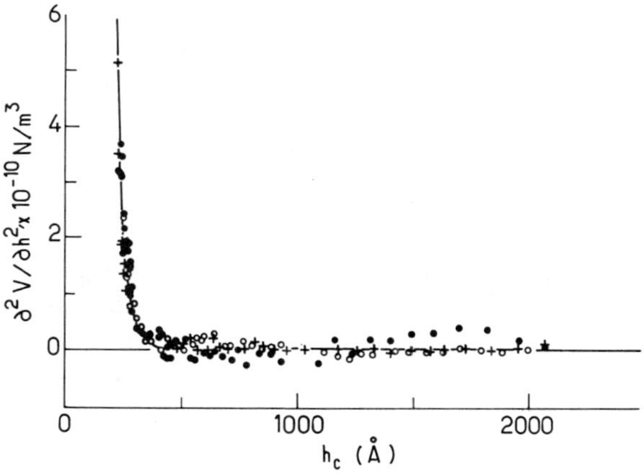

FIG. 8 Variation of $\partial^2V/\partial h^2$ with core thickness as deduced from the relaxation time of the peristaltic mode on CTAB films. (Data from Ref. 7.)

E. Undulation Mode

The dispersion equation for the undulation mode can be obtained by using the hydrodynamic equation Eq. (19) and limit conditions similar to those of Sec. D for the peristaltic mode except for the relation between the displacements of the two sides of the film: here $\zeta(\mathbf{r}) = \zeta'(\mathbf{r})$. The dispersion equation is then

$$\begin{aligned}&(bq - ma)\,[\gamma q^3 - i\omega\eta'\, m'(m' + q)]\,[\varepsilon q^2 - i\omega\eta'(m' + q)]\\&+ b\omega[\eta(m^2 + q^2) + \eta'(m' - q)]^2 + i\eta(m^2 - q^2)\,\{mab[\varepsilon q^2\\&- i\omega\eta'(m' + q)] + \gamma q^3 - i\omega\eta'\, m'(m' + q)\} - amq\omega[2\eta q\\&+ \eta'\,(m' - q)]^2 = 0\end{aligned} \tag{25}$$

Now $\omega\rho'/\eta' q^2$ is no longer small compared to one, and the same limit cannot be taken as for the peristaltic mode. However, assuming the damping terms to be small, the frequency of the mode can be calculated in the limit of small qh and $\eta' \rightarrow 0$. In this limit, one has

$$\begin{aligned}&\frac{qh}{2}\,(\gamma q^3 - \omega^2\rho')\varepsilon q^2 - i\eta q\left(m^2\,\frac{qh^2}{4}\,\varepsilon + \gamma q^3 - \omega^2\rho'\right)\\&+ i\eta q^2\,\frac{h}{2}\,\rho\omega^2 = 0\end{aligned}$$

For both small and large ε, it reduces to

$$\omega_u^2 \sim \frac{2\gamma q^2}{\rho h + 2\rho'/q} \tag{26}$$

This dispersion relation shows that the effective inertial mass in the equation of motion of the film is the mass per unit area of the film, plus the mass of air that is moved along with the film per unit area $2\rho'/q$. Because $h \ll q^{-1}$, this effective density is much smaller than the liquid density, and $\omega^2 \sim \gamma q^3/\rho'$. The characteristic frequencies are therefore much larger than the frequencies of capillary waves propagating at the surface of the bulk fluid $\omega^2 \sim \gamma q^3/\rho$: typically 1 MHz instead of 10 kHz.

We also see in Eq. 26 that the disjoining pressure plays no role in the frequency of the mode. This is understandable, since there are no thickness fluctuations during the propagation.

This mode has been studied either with a correlator [7] or with a wave analyzer [29], on both draining and equilibrium films [30]. In the earliest work (Ref. 29), differences between static and dynamic surface tensions were found, but this was due to accidental contamination [23].

Depending upon the relative intensity scattered by the undulation and the peristaltic modes, the detection can be homodyne or heterodyne, because the characteristic frequencies of the peristaltic mode are much smaller than those of the undulation mode. By fitting the correlation functions with a mixture of the corresponding damped sinusoids, Young and Clark were able to determine the homodyne-to-heterodyne ratio and to show that it was proportional to the square root of the intensity scattered by these modes [7] (Fig. 9). This showed unambiguously that the only source of scattered light was the soap film, i.e., that the detection was purely homodyne in the case of the peristaltic mode in their experiments.

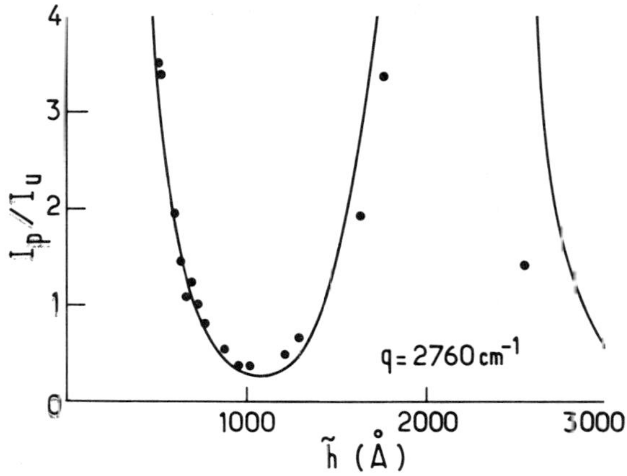

FIG. 9 Ratio of the intensities scattered for the peristaltic and undulation modes as obtained from the homodyne-to-heterodyne ratio in the inelastic scattering from the undulation mode. (The points are data from Ref. 7; the line is theoretical.)

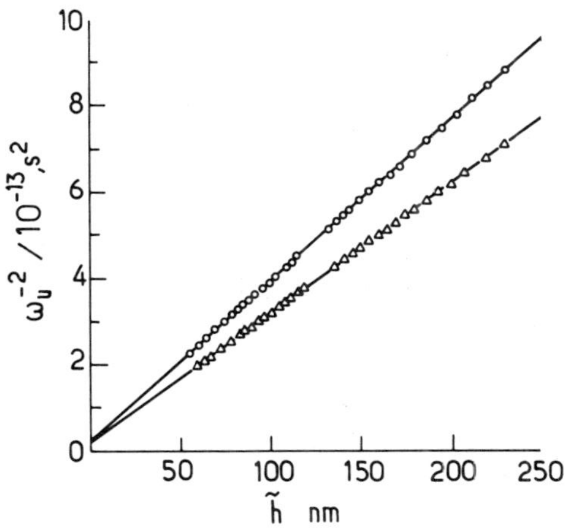

FIG. 10 Inverse squared frequency of the undulation mode versus optical thickness for CTAB films. Circles, $q = 4520\ \text{cm}^{-1}$; triangles, $q = 4000\ \text{cm}^{-1}$. (Data from Ref. 23.)

The width of the spectrum is mostly due to instrumental broadening. Indeed, in these experiments, the laser beam is slightly focused onto the film and the wavevector spread Δq is large (about 5% in Ref. 7). This width is roughly $\Delta\omega_u \sim \eta' q^2/\rho'$, so that $\omega_u/\Delta\omega_u \sim \gamma q/\eta'$, much larger than the corresponding ratio for the free surface of the bulk solution. Let us recall that the ratio of the instrumental width $\Delta\omega_u^i$ and $\Delta\omega_u$ is $3/2(\omega_u/\Delta\omega_u)$ (see Chap. 6).

The variation of ω_u with film thickness (Fig. 10) and wavevector q (studied with equilibrium films) was found to agree with Eq. 26. The frequency of the mode was in good agreement with the values calculated from Eq. 26 using the bulk solution values for γ and ρ and the air value for ρ'. In principle γ should be slightly different from the surface tension of the bulk solution γ_b:

$$\gamma = \gamma_b + \frac{1}{2}\left(V - h\frac{\partial V}{\partial h}\right)$$

Estimating this difference shows that it is always less than 0.1 mN/m, less than experimental error (± 0.3 mN/m). It is not therefore possible to find further information on the potential V in this way.

III. FREELY SUSPENDED LIQUID CRYSTAL FILMS

As mentioned earlier, most published light scattering experiments on this type of film have been concerned with the study of two-dimensional orientational fluctuations. The corresponding modes are damped and diffusive. The diffusion coefficient D is the ratio of an equivalent elastic constant K to a viscosity η. The viscosities were found similar to the bulk viscosities, but the elastic constants are sometimes surface-enhanced [9]. In the case of hexatic layers, K and η are specific to tilt-bond coupling [10]. The smectic layer displacements at the free surface are quenched by the surface tension and are smaller than in the bulk. This has been recently evidenced by x-ray scattering [31].

The types of modes observed in soap films, undulation and peristaltic modes, are also present. These modes were studied in films of octyl-cyanobiphenyl (8CB) in the smectic and in the nematic phase close to the nematic-smectic phase transition (t_{NS} = 33.6°C, t_{NI} = 40.6°C). Near t_{NS}, the film thickness changes in discrete steps during the final stages of the drainage $h \sim h_{eq} \sim 100$ nm. Preliminary measurements of the scattered light intensity showed that the peristaltic mode is weak, implying that the thickness dependence of the forces across the film is strong. The frequency of the undulation mode was linear in q, and the slope was in agreement with the expected value from bulk liquid crystal properties [11].

Scattered intensity measurements were performed later on nematic films of 8CB close to the nematic-smectic transition: $2° > t - t_{NS} > 1°$ [12]. The temperature range was very limited, first because it was not possible to form stable films at higher temperatures: indeed, the van der Waals forces cause a continued thinning until the film ruptures; second, at lower temperatures, the viscosity was such that virtually no drainage took place in the time scale of the experiments. Between 1 and 2° above t_{NS}, the film reaches a pseudoequilibrium thickness, probably determined by kinetic factors. However, once the film thickness has reached its final value, the film can be cooled at constant thickness down into the smectic phase.

For nematic films, Eq. 14 can be generalized into

$$\langle|\zeta_q + \zeta'_q|^2\rangle = \frac{2k_BT}{\gamma q^2 + K_1hq^4} \tag{27}$$

$$\langle|\zeta_q - \zeta'_q|^2\rangle = \frac{2k_BT}{\gamma q^2 + \frac{4}{3}K_1hq^4 + 8K_3q^2/h} \tag{28}$$

K_1 and K_3 being the splay and bending elastic constants respectively [12]. As for the free surface of nematic liquid crystals, the undulation deformation couples to splay elastic deformations. But the order of magnitude of K_1 is such that $\gamma q^2 \gg K_1hq^4$, and as for free surfaces, the elastic energy is negligible (Chap. 9). The case of the peristaltic mode, which involves both splay and bending elastic deformations, is more subtle; indeed, K_3 (unlike K_1) diverges at a second-order nematic-smectic transition, and for 8CB it becomes large for $t - t_{NS} \sim$ 1°.

The main difficulty of the experiment was the calibration. For this purpose, the surface tension was taken from the inelastic scattering data [11]. The calibration factor deduced from the angular variation of the scattered intensity proved to be unreliable because of large statistical errors in the data. The calibration at large thicknesses is complicated because for such thicknesses the film is wedge shaped. The accuracy on the measurement of $\partial^2V/\partial h^2$ is therefore poor (~100%). However, below a certain critical thickness, the interaction can be seen to rise precipitously, in an exponential fashion. This is incompatible with Eq. 28, even accounting for the K^3 increase close to t_{NS}. A more plausible interpretation attributes the onset of a strong interaction between the two film surfaces to the point where the range ξ of smectic-induced ordering becomes comparable to the film thickness. By analyzing the data within the framework of the de Gennes theory for smectic fluctuations in a nematic medium, Böttger and coworkers generalized Eq. 28 into

$$\langle|\zeta_q - \zeta_q'|^2\rangle = \frac{2k_BT}{\gamma q^2 + 2B/\xi \sinh(h/\xi)} \tag{29}$$

where B is the elastic modulus associated with the compression of smectic layers, and ξ the correlation length of the smectic fluctuations. This expression is in qualitative agreement with the data, and ξ was found to vary between 204 and 374 nm in the temperature range studied: $2° > t - t_{NS} > 1°$. B can be estimated from the data on the films frozen into the smectic phase (see above) where $\xi \gg h$, and the denominator of Eq. 29 becomes $\gamma q^2 + 2B/h$. It was thus found that $B \sim 2\ 10^6\ \mathrm{N/m^2}$.

IV. ASYMMETRIC SOAP FILMS ON VERTICAL SOLID SUBSTRATES

Preliminary experiments were done on silicon wafers dipped into and partially withdrawn from a Langmuir trough in which an arachidic monolayer was spread on water. A stable film thickness of 17 nm was reached at the end of the drainage process. In some cases a thickness of ~4 nm is also obtained. These two equilibrium states resemble the first and second black soap films [14].

Surface scattering was studied at scattering angles of a few degrees, during the drainage process. Toward the end of the thinning process, the reflected light oscillations (which allow the determination of the thickness) are damped out, whereas the scattered intensity starts to exhibit strong fluctuations (Fig. 11). The intensity of the scattered light is proportional to the amplitude of the surface fluctuations

$$\langle\zeta_q^2\rangle = \frac{k_BT}{\gamma q^2 + \partial^2V/\partial h^2}$$

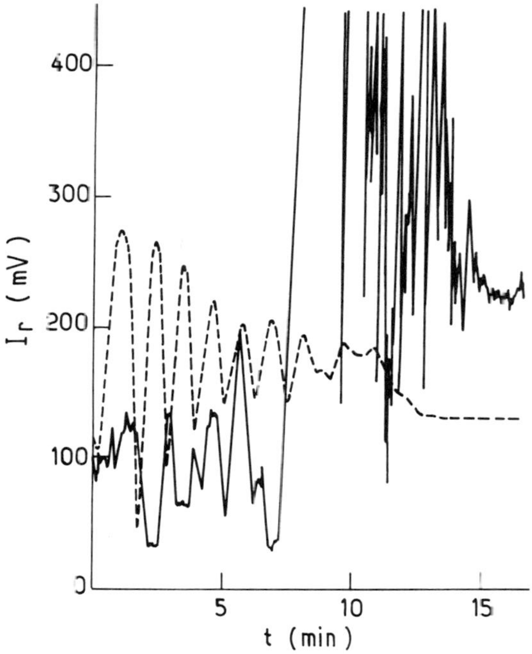

FIG. 11 Reflected light intensity (dotted line) and scattered line intensity (full line) versus time. Thinning water film covered by an arachidic monolayer on a silicon substrate. (Data from Ref. 14.)

It is possible that, at the end of the drainage process, one reaches a situation where for small q $\gamma q^2 \approx -\partial V/\partial h^2$. Further work is needed to clarify this point. Inelastic scattering should also be observable, although because of the boundary condition of vanishing velocity at the solid surface, the modes are strongly overdamped [32].

The possibility of obtaining also information about intermolecular forces from the light scattering study of wetting layers at a liquid-vapor interface has been recently discussed theoretically [33].

V. CONCLUSION

The study of light scattering by thin films is a rich subject that can provide good quality information about molecular forces across the film. In the case of soap films, contrary to bulk liquids, the information contained in the scattered intensity is of better quality than that contained in the inelastic scattering. This is because it is much easier to get rid of the other sources of scattered light. Furthermore, the inelastic scattering is dominated by hydrodynamic contributions independent of the forces.

To date, there is only preliminary information about the scattering by liquid crystal films and wetting films. But these subjects certainly deserve further work.

REFERENCES

1. K. J. Mysels, K. Shinoda, and S. Frankel, *Soap Films, Studies of Their Thinning,* Pergamon Press, New York, 1959.
2. B. V. Derjaguin and A. S. Titievskaya, *Discuss. Faraday Soc. 18:* 27, 1954.
3. E. J. W. Verwey and J. Th. G. Overbeek, *Theory of the Stability of Lyophobic Colloids,* Elsevier, Amsterdam, 1948.
4. A. Vrij, *J. Colloid Interface Sci. 19:* 1 (1964); *Adv. Colloid Interface Sci. 2:* 39 (1968).
5. H. M. Fijnaut and A. Vrij, *Nature, 246:* 118 (1973).
6. A. Vrij, J. G. H. Joosten, and H. M. Fijnaut, *Adv. Chem. Phys. 48:* 329 (1981).
7. C. Y. Young and N. A. Clark, *J. Chem. Phys. 74:* 4171 (1981).
8. P. S. Pershan, *Structure of Liquid Crystal Phases,* World Scientific, Singapore, 1988.
9. C. Rosenblat, R. Pindak, N. A. Clark, and R. B. Meyer, *Phys. Rev. Lett. 42:* 1220 (1979).
10. S. B. Dierker and R. Pindak, *Phys. Rev. Lett. 59:* 1002 (1987); S. Sprunt and J. D. Litster, *Phys. Rev. Lett. 59:* 2682 (1987).
11. A. Bötger and J. G. H. Joosten, *Europhys. Lett. 4:* 1297 (1987).
12. A. Böttger, D. Frenkel, J. G. H. Joosten, and G. Krooshof, *Phys. Rev. A, 38:* 6316 (1988).
13. P. G. de Gennes, *Rev. Mod. Phys. 57:* 827 (1985).
14. G. Schneider, W. Knoll, E. Sackmann, and J. G. H. Joosten, *Europhys. Lett. 1:* 449 (1986).
15. J. G. H. Joosten in *Thin Liquid Films* (I. B. Ivanov, ed.) Surfactant Science Series, Marcel Dekker, New York, 1988.
16. R. R. Highfield, R. P. Humes, R. K. Thomas, P. G. Cummins, D. P. Gregory, J. Mingins, J. B. Hayter, and O. Schaerpf, *J. Colloid Interface Sci. 97:* 367 (1984).
17. O. Belorgey and J. J. Benattar, *Phys. Rev. Lett. 66:* 313 (1991).
18. B. W. Ninham and V. A. Parsegian, *J. Chem. Phys. 52:* 4578 (1970).
19. W. A. B. Donners, J. B. Rijnbout, and A. Vrij, *J. Colloid Interface Sci. 60:* 540 (1977).
20. M. Born and E. Wolf, *Principles of Optics,* Pergamon, London, 1959.
21. J. B. Rijnbout, *J. Phys. Chem. 74:* 2001 (1970).
22. H. M. Fijnaut and J. G. H. Joosten, *J. Chem. Phys. 69:* 1022 (1978).
23. J. G. H. Joosten, thesis, University of Utrecht, 1982.
24. W. A. Donners, J. B. Rijnbout, and A. Vrij, *J. Colloid Interface Sci. 61:* 249, 535 (1977).

25. J. A. Mann, *J. Colloid Interface Sci. 25:* 437 (1967).
26. J. Lucassen, M. van den Tempel, A. Vrij, and F. Th. Hesselink, *Proc. K. Ned. Akad. Wet. B73:* 109 (1970); A. Vrij, F. Th. Hesselink, J. Lucassen, and M. van den Tempel, *Proc. K. Ned. Acad. Wet. B73:* 124 (1970).
27. J. G. H. Joosten, *J. Chem. Phys. 80:* 2363, 2383 (1984).
28. J. G. H. Joosten and D. Langevin, unpublished results.
29. J. G. H. Joosten and H. M. Fijnaut, *Chem. Phys.Lett. 60:* 483 (1979).
30. J. G. H. Joosten, *Ber. Bunsenges Phys. Chem. 88:* 1153 (1984).
31. R. Holyst, D. J. Tweet, and L. B. Sorensen, *Phys. Rev. Lett. 65:* 2153 (1990); D. J. Tweet, R. Holyst, B. D. Swanson, H. Stragier, and L. B. Sorensen, *Phys. Rev. Lett. 65:* 2157 (1990).
32. F. de Voeght and P. Joos, *J. Colloid Interface Sci. 98:* 20 (1984).
33. J. L. Harden and R. F. Kayser, *J. Colloid Interface Sci. 127:* 548 (1989).

15

The Nonequilibrium Crystal-Melt Interface

HERMAN Z. CUMMINS and LLOYD M. WILLIAMS Department of Physics, The City College, The City University of New York, New York, New York

I. INTRODUCTION

In 1978 Bilgram, Güttinger, and Känzig reported a new and unexpected interfacial light scattering phenomenon [1, 2]. A laser beam incident on the horizontal interface between an ice crystal and highly purified water in a zone-refining apparatus [2] initially produced virtually no scattering. But when growth was initiated by driving the sample tube downward at speeds exceeding 1.5 μm/s, intense quasi-elastic light scattering appeared at the interface after an initial onset time on the order of 1 hr. The intensity autocorrelation function of the scattered light was found to be exponential, with decay rate Γ (corresponding to the optical linewidth) that scaled with q^2, which is characteristic of relaxational (diffusive) processes; i.e., $\Gamma = Dq^2$, where D is an effective diffusion constant.

Since the initial discovery, this phenomenon (or phenomena) has been observed in a total of eight different materials as shown in Table 1 in experiments performed at the ETH in Zurich [1–19], the City College of New York (CCNY)

TABLE 1 Diffusion Coefficients and Equivalent Radii from Boundary Layer Light Scattering

Material	D Diffusion coefficient (cm^2/s)	R Radius (μm)
Benzophenone[a]	$\sim 1 \times 10^{-10}$	~2
Biphenyl[b]	4.9×10^{-10}–1.5×19^{-9}	1.0–3.6
Cyclohexane[c]	6×10^{-9}	0.26
Cyclohexanol[d]	2×10^{-11}–9×10^{-11}	0.34–1.5
Naphthalene[b]	1.6×10^{-10}–1.3×10^{-9}	1.7–8.2
Salol		
a axis[e]	7.5×10^{-10}–1.25×10^{-9}	0.22–0.36
a axis[f]	0.6×10^{-10}–1.0×10^{-10}	2.7–4.6
b axis[f]	1.0×10^{-10}–1.0×10^{-9}	0.27–2.7
b axis[g]	2.3×10^{-12}–7.8×10^{-10}	0.35–117.0
Succinonitrile[h]	1.5×10^{-9}–3.0×10^{-9}	0.6–1.2
degassed	2.9×10^{-11}–6.9×10^{-11}	31.0–52
Water		
(H_2O) a c axis[i]	1.5×10^{-11}–3×10^{-9}	.3–65
c axis[j]	1.4×10^{-8}–5.6×10^{-8}	0.014–0.079
c axis[i]	4.5×10^{-9}–6.3×10^{-9}	0.059–0.072
(D_2O)[k]	(close to H_2O values)	

[a]L. O. Ladeira and O. N. Mesquita (1989, unpublished).
[b]O. N. Mesquita, L. O. Ladeira, I. Gontijo, A. G. Oliveira, and G. A. Barbosa, *Phys. Rev. B 38*:1550 (1988).
[c]R. Steininger and J. Bilgram, *Helv. Phys. Acta 62*:215 (1989).
[d]G.Livescu, M. R. Srinivasan, H. Chou, H. Z. Cummins, and O. Mesquita, *Phys. Rev. A 36*:2293 (1987).
[e]U. Dürig, J. H. Bilgram, and W. Känzig, *Phys. Rev. B 30*:946 (1984).
[f]O. N. Mesquita and H. Z. Cummins, *Physico-Chemical Hydrodynamics 5*:389 (1984).
[g]L. Williams, M. R. Srinivasan, and H. Z. Cummins, *Phys. Rev. Lett. 64*:1526 (1990).
[h]J. M. Laherrere, H. Savary, R. Mellet, and J. C. Toledano, *Phys. Rev. A 41*:1142 (1990).
[i]J. P. Vesenka and Y. Yeh, *Phys. Rev. A 38*:5310 (1988).
[j]P. Böni, J. H. Bilgram, and W. Känzig, *Phys. Rev. A 28*:2953 (1983).
[k]B. Zysset, P. Böni, and J. H. Bilgram, *Helv. Phys. Acta 54*:265 (1981).

[20–24], the University of California at Davis [25–27], the University of Minas Gerais (Brazil) [28–31], and the CNET laboratory in France [32–34]. There are actually two quite different effects observed in materials that are strongly faceted. In Fig. 1 we show the intensity correlation function of light scattered from the crystal-melt interface of a salol crystal growing along its b axis. (Note: a, b,

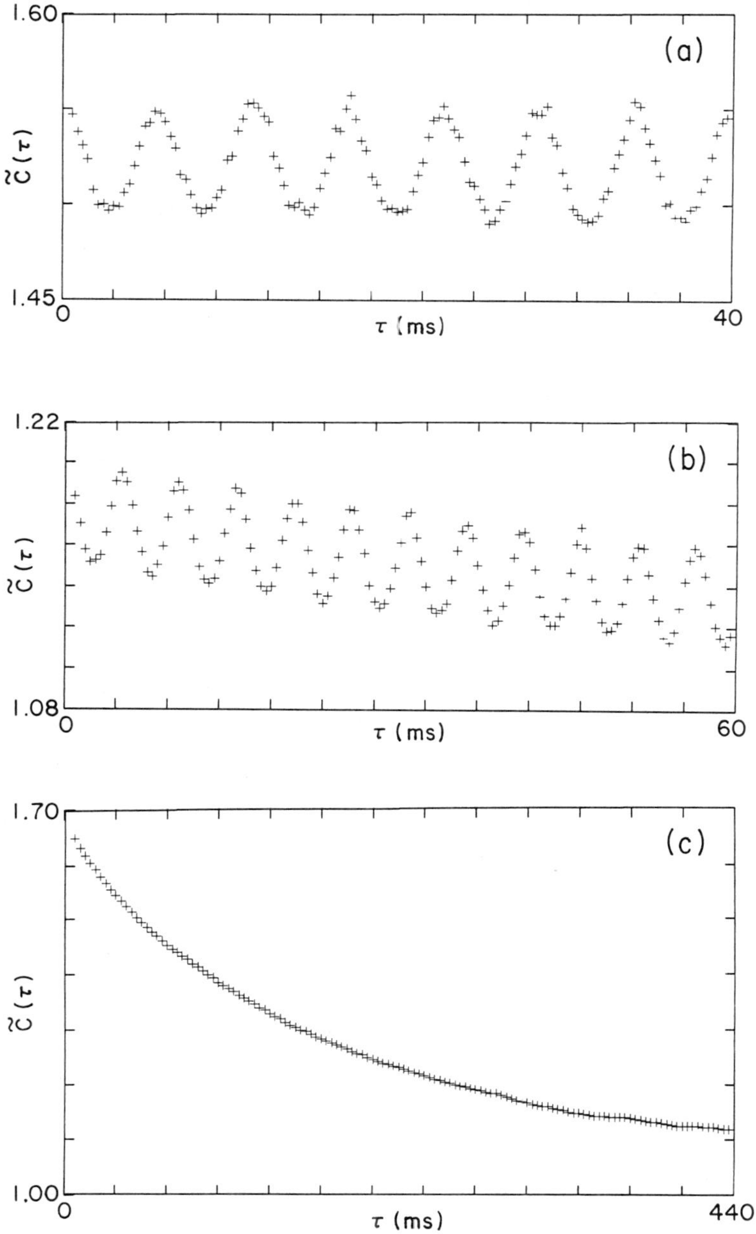

FIG. 1 Reduced intensity autocorrelation data $\tilde{C}(\tau) = [C(\tau) - C(\infty)]/C(\infty)$, where $C(\tau) = \langle I(t)I(t + \tau)\rangle$ for light scattering from the salol crystal-melt interface. (a) 30 min, (b) 3–4 hr, and (c) 24 hr after the initiation of growth at 0.212 μm/s.

and c refer to the three crystallographic axes. Salol is orthorhombic, and the b axis, which corresponds to the longest side of the unit cell, is the direction of slowest growth. For ice, the hexagonal c axis is the direction of slowest growth.) In 1a, 30 min after growth has begun, an oscillatory correlation function is seen indicative of a propagating excitation. Unlike the oscillatory correlation functions observed in scattering from capillary waves on fluid surfaces, these oscillations are not damped. They are caused by scattering from screw dislocations producing a crystal growth that generates a constant Doppler shift, and not from thermal fluctuations.

After 3 to 4 hr, an exponential relaxation builds up under the oscillatory component, as shown in Fig. 1b, and after 24 hr, the diffusive scattering, which is much more intense than the oscillatory scattering, completely dominates the correlation function, as shown in Fig. 1c. These times are extremely dependent on the sample history and growth conditions. Prolonged pumping on the sample lengthens the time for onset of the diffusive scattering. Experiments by Mesquita et al. [20] at CCNY showed that these two scattering processes occur simultaneously and can be separated by appropriate choices of wavevector and polarization directions.

In contrast to the inelastic light scattering observed on the crystal side of the interface in salol [7, 13, 20] and benzophenone [30], which is understood to result from Doppler-shifted scattering from the advancing ledges of polygonized screw dislocations, the origin of the diffusive light scattering observed on the fluid side of the interface in all the materials shown in Table 1 has been a matter of considerably controversy.

The basic properties of this scattering phenomenon, as revealed by the experiments to be described below, are

1. It occurs in a fluid boundary layer at least several microns in thickness, and in salol and benzophenone it coexists with the translational dynamics seen on the crystal surface.
2. It is much stronger than scattering from either the crystal surface or the bulk fluid.
3. It requires an onset time on the order of hours to begin.
4. The intensity correlation function of the scattered light is approximately exponential with a polydispersity that varies in some experiments from a small value at onset to a value on the order of unity after prolonged growth.
5. It occurs on both atomically smooth and atomically rough interfaces.
6. The effective diffusion constant D is typically 10^4 times smaller than the molecular self-diffusion coefficient and 10^6 times smaller than the thermal diffusion coefficient of normal liquids.

Several models have been proposed to explain the origin of this diffusive scattering:

1. Capillary waves on the interface (Bilgram et al.) [2]
2. Diffusion of defects in the surface (Keizer et al.) [35–37]
3. Partial orientation of molecules in the liquid layer adjacent to the interface (Mesquita et al.) [20]
4. Dynamic roughening (van Saarloos and Gilmer) [38]
5. A new mesophase (Böni et al.) [12]
6. Gaseous microbubbles (Cummins et al.) [22]

Models 1 through 4 are now generally regarded as inconsistent with various elements of the experimental data, particularly the observed thickness of the scattering layer, which increases with decreasing growth velocity [16]. Thus the mesophase model 5 [12] and the gaseous microbubble model 6 [22] are currently the only extant models consistent with the basic properties (1–6) given above.

In Sec. II we present a brief survey of the experimental literature on this phenomenon. In Sec. III we discuss the microbubble model, which was first proposed in 1986 [22]. Discussion and conclusions are given in Sec. IV.

II. SURVEY OF PAST EXPERIMENTS

Anomalous dynamic light scattering at the crystal-melt interface was first observed by Bilgram, Güttinger, and Känzig at the ETH in Zurich [2]. Their experiments were performed on ice crystals growing along the c axis in a zone refining apparatus placed in a cold room at $-18°C$. In this experiment it was found that after the initiation of growth and the formation of the basal plane on the top of the growing crystal, light scattering could be observed when a critical velocity v_c of 1.5 μm/s was exceeded, after an onset time of about 1 hr. The intensity correlation function of the scattered light was found to be accurately single exponential with a corresponding linewidth Γ of about 2,000 rad/s at a scattering angle of 90°. This would correspond to scattering from objects with a hydrodynamic radius R_h of about 25 nm. The scattering was found to exhibit hysteresis, that is, once the scattering had been initiated the scattering persisted even if the growth velocity dropped below v_c. The scattering was seen to disappear only sometime after the growth was completely stopped or when the interface was melted away. Γ was found to have the same linear dependence on q^2 (where q is the scattering vector) whatever the growth rate. The authors proposed that the scattering was caused by a fluctuating corrugated interface (i.e., overdamped capillary waves). In this model the decay time of a Fourier component with wavevector q could be calculated from the Gibbs-Thomson equation.

A preliminary version of these results had previously been reported in 1975 [1]. In this preliminary paper, the diffusive scattering exhibited two simultaneous decay rates, one fast and one slow. The process observed in Ref. 2 corresponded to the fast decay rate. The slow decay rate process was not seen in the later work of the ETH group, although it has been studied subsequently by Yeh and coworkers in California [26, 27], as described below. It is a matter of some controversy whether these two decay rates correspond to two distinct phenomena. Vesenka and Yeh [26] have suggested that the slow component may correspond to the gaseous microbubble mechanism [5], while the fast component is intrinsic, produced by crystal formation from the pure melt.

In 1979 the ETH group [3] measured the dependence of the scattering intensity on the azimuthal angle θ and found it essentially independent of angle. Also, from studying the coherence properties of the scattered light, they were able to estimate that the scattering layer was less than 6 μm thick.

In the first publications of the ETH group, it was assumed that the scattering process was two-dimensional, i.e., that it occurred in the planar interface between the crystal and the melt. Experimentally, two-dimensional scattering can be distinguished from three-dimensional scattering by studying the effect of varying the polar angle ψ; for two-dimensional processes, the projection of the scattering vector on the plane determines the spectrum, while for three-dimensional processes, it is it the total scattering vector that is important.

In a 1980 article Bilgram and Böni [4] measured the variation in the scattered intensity and linewidth with the polar angle ψ. Comparisons of linewidths at $\psi = 65°$ and $\psi = 90°$ revealed that the scattering process is three-dimensional. This conclusion was incompatible with the two-dimensional model of the corrugated interface, and a new model was introduced to explain the scattering. In this model an intermediate layer forms between the ice and water, and light is scattered by inhomogeneities in this layer.

In 1983 Böni, Bilgram and Känzig [12] reported a more complete study of the ice-water system that revealed several interesting properties of the scattering. A Brillouin scattering study revealed that the increased scattering is due to the central Rayleigh component and that the intensities of the Brillouin components are unaffected. It was also found that the critical growth velocity increased with temperature gradient in the ice. Measurements were made on the time dependence of the intensity and linewidth of the scattered light over a 3-hr period after the onset of scattering. The intensity I was seen to increase and the linewidth Γ to decrease. Γ was found to be related to I by a power law dependence: $\Gamma \propto I^{-\alpha}$, where $\alpha = 0.23 \pm 1\%$. It was stated that after this transient period both I and Γ are stationary although no data were presented to support this statement.

In the discussion of these results the boundary layer and corrugated interface models were compared. The corrugated boundary model was ruled out and the theory for the boundary layer model was discussed in more detail. From measurements of the intensity of the scattered light it was inferred that the boundary layer has an isothermal compressibility κ_T about 700 times larger than that of pure water. This large compressibility causes the entropy fluctuations responsible for the Rayleigh line in normal liquids to become larger and slower, somewhat resembling the characteristics of fluids near the critical point. From measurements of the linewidth it was inferred that the correlation length of the fluctuations in the boundary layer is in the range $20 < \xi < 80$ nm.

In their 1988 study of the ice-water system [16], Bilgram and coworkers reverted to a Czochralski growth apparatus [1] with the water below the ice crystal rather than above it. The essential features of the scattering were found to be the same as described previously [2, 3, 12]. However, several new features were measured. For growth rates less than 1.5 μm/s the scattering layer was thick enough to be imaged and measured. The thickness of the layer d was measured as a function of v_g, and it was found that $d = D/v_g$. For zero growth rate the layer reached a thickness of 20 μm, and upon melting the scattering layer "lifted off" the interface and stayed back in the water. The Raman spectra of the scattering from the interface did not differ significantly from water. Although some small differences in the intensity of the polarized Rayleigh wing and in the O—H stretching band were observed, the interface was described as "essentially water-like." From the different growth conditions relating to the previous zone-refining geometry it was concluded that the formation of the scattering layer does not depend on gravity.

The first confirmation of diffusive interfacial light scattering by investigators outside of Zurich came in the paper of Brown et al. [25] at the University of California, in which scattering was seen in ice growing along the a axis in a Bridgman growth apparatus. The linewidths reported in this work corresponded to a decay time 3 times slower than that reported by the ETH group [2, 3].

In the California group's work on ice growing along the c axis, both a fast and a slow component were reported. In this study several interesting features of the slow component were observed. It was noted that the slow component could be enhanced if high purity helium or argon was bubbled through the melt prior to growth to saturate it. It was also reported that there was a deviation from the linear dependence of Γ on q^2. This was explained in terms of a phenomenological equation that includes both diffusion and dissipation.

A "lift off layer" was observed in samples injected with helium, although not with argon. When growth was stopped after scattering had begun and the crystal was melted back (by reversing the motor drive), the scattering layer was observed to remain stationary in the melt as the interface moved away. This effect was first seen by Mcdonaugh (unpublished data, 1985) and reported in a preliminary form in 1987 [27]. It is interesting to note that helium so effectively enhances the scattering, since the ice crystals grown by the ETH group in the Czochralski apparatus [16] were subjected to an overpressure of about 100 mbar of helium. This was done because they asserted that helium is more soluble in ice than in water. The California group also reported that the polydispersity P ($P = \mu_2/\langle\Gamma^2\rangle$) increased with the hydrodynamic radius R_h. This was especially observable in the gas-injected samples. In the Ph.D. thesis of Vesenka it was noted that upon applying a significant overpressure to the melt, R_h was found to decrease.

In order to investigate whether this phenomenon was peculiar to the ice-water system, other systems were investigated. In a short article in 1981 [5] the ETH group reported that D_2O exhibited essentially the same scattering characteristics as H_2O. They then began an investigation of scattering in the highly faceted organic crystal salol (phenyl salicylate) [7]. This investigation revealed two different scattering mechanisms. One was isotropic, characterized by an exponentially decaying correlation function similar to the scattering seen in water; the other was highly anisotropic, characterized by an oscillatory correlation function. The oscillatory scattering had a critical growth velocity of 0.2 μm/s and began as soon as the crystal started growing. The diffusive scattering had a critical velocity of 0.4 μm/s and required about 2 hr for onset, at which time over the course of about 15 min the oscillatory correlation became damped and the exponential one emerged and dominated the scattering. The oscillatory scattering was interpreted to be Doppler shifted by a scatterer moving parallel to the scattering vector with a velocity about 150 times greater than the crystal growth velocity.

The diffusive scattering in salol exhibited the same hysteresis as the scattering in the ice-water system; however, the linewidth of the scattered light was 10 to 20 times smaller than the ice-water linewidth reported in Ref. 2, which corresponds to a hydrodynamic radius R_h of 0.25–0.5 μm. It was noted that this decay time agreed with the predictions of the Gibbs-Thomson equation of the model proposed in Ref. 2 even though that model did not take into account the three-dimensional nature of the scattering.

New observations on light scattering in salol were reported by the CCNY group in 1984 [20]. By exploiting the birefringence of the crystal, they were able to separate the diffusive and oscillatory scattering and show that the two phenomena coexist at the interface. The Doppler-shifted light causing the oscillatory scattering was suggested to be generated by moving growth steps on the faceted surface of the crystal. Two growth mechanisms were considered to generate the steps: nucleated layer growth and screw dislocation growth. The predictions of the dependence of the growth rate v_g on undercooling ΔT were tested for both mechanisms, and it was established that for low growth rates the screw dislocation mechanism was dominant, although at higher growth rates nucleated layer growth seemed to play an important role.

The observed properties of the diffusive scattering were essentially the same as those reported by the ETH group [7]. A tentative model for the scattering based on the existence of some orientational ordering of the molecules at the interface was proposed.

In a subsequent article by the CCNY group, the occurrence of scattering on both the a and the b face of salol was reported [21]. Scattering was also observed during melting, although at much higher velocities. The previously proposed model of orientational order was ruled out by a preliminary observation of scattering in cyclohexanol, which crystalizes in a plastic phase and therefore possesses no orientational order.

In 1984 Dürig, Bilgram, and Känzig [13] reported a comprehensive study of the salol system. Both the oscillatory and the diffusive scattering were investigated. They also interpreted the oscillatory scattering as due to growth steps originating from screw dislocation spirals. From measurements of the Bragg peaks in the scattering intensity, the spacing d_0 between steps was shown to be typically 0.4 μm. From the period of the oscillations in the correlation function a typical lateral step velocity was $v_l = 40$ μm/s. For growth along the b axis they found $d_0 = C/v_l$. The height h of the steps could also be determined. As $h = d_0(v_g/v_l)$, where v_g is the growth speed of the crystal, h was calculated to be about 6 nm or 2.5 lattice constants.

In their measurements of the diffusive scattering they reported seeing scattering only for growth along the a axis, in contrast to their earlier report [7] of scattering on both the a and b faces. The measurements of the linewidth Γ are in essential agreement with the earlier paper [7], and the interpretation of the origin of the scattering mechanism is along the same lines as for the ice-water system [12].

In 1986 the CCNY group [22] proposed the microbubble hypothesis for the scattering; this model will be outlined in the next section. Preliminary measurements of the angular dependence $I(\theta)$ were also reported. It was found that $I(\theta)$ could be well fitted by Mie theory but that the intensity fell off too rapidly with scattering angle to be fitted with Ornstein-Zernike theory, even in the limit of infinite correlation length ξ.

A full report of the scattering in cyclohexanol was given by the CCNY group [23]. These measurements gave a value for R_h in the range 0.6–3.0 μm. A "lift off layer" similar to that observed by Yeh [26] and Bilgram [16] was also observed. The persistence of scattering long after growth had stopped was also noted, although unlike salol the relaxation time for the cyclohexanol interface to return to its equilibrium position is very short, ~10 s compared to ~20 hr for salol.

In a major review of dynamics at the solid-liquid transition [14], Bilgram developed the theory of the boundary layer, in which it was postulated that the creation of the "mesophase" is brought about by a nonequilibrium wetting transition. In the same article, the microbubble hypothesis of Cummins et al. [22] was discussed and rejected, primarily on the ground that the energy barrier to bubble nucleation ($e^{-\Delta F/kT}$, where ΔF is the energy of formation of a gas bubble) is prohibitive.

Mesquita et al. [28] in Brazil observed scattering in two new materials, naphthalene and biphenyl. The R_h observed in these materials is also in the micron range. They too observed that smaller polydispersities corresponded to smaller radii. In this study the relationship between time for the onset of scattering t_i and the radius of onset R_i was investigated. It was observed that the product $t_i R_i v_g^2$ was a constant for each material.

Mesquita et al. [28] also analyzed the data on the onset of scattering in the ice-water system reported by Böni et al. [12]. They found that the experimental relationship between the linewidth Γ and the intensity I, $I \propto \Gamma^{-\alpha}$, could be modeled if the scatters were assumed to be bubbles and the intensity I was given by Mie theory. The agreement in the values of α was quite good: Bilgram $\alpha = 0.23$, Mesquita $\alpha = 0.22$.

The existence of a critical velocity for the onset of scattering has been reported by the ETH group [2, 3, 12] for the ice-water system. It appears that this may be a unique feature of this system. Mesquita [29] has suggested a mechanism based on the Rayleigh-Benard convective instability that might explain this phenomenon. While it is true that water is unique among the systems studied in having a negative thermal expansion in the temperature region around the freezing point, it is hard to see how this could explain the same critical velocity being observed by Bilgram in ice grown in a Czochralski apparatus [16], where the melt is stable against convective instabilities.

Mesquita has also observed oscillatory scattering in benzophenone [30] growing along the [110] direction. The scattering is assumed to come also from growth steps propagating out from screw dislocations. In the same work Mesquita [30] has also suggested that force balance considerations could play a role in the selection of the initial radius.

Laherrere et al. in Paris observed scattering in the plastic crystal succinonitrile [32, 33]. The values of R_h were in the range of 0.3–0.6 μm. The effect on the scattering of pumping on the crystal was observed. The crystal was melted and then recrystallized at room temperature. Any released gases were then pumped out. After 10 such cycles the scattered intensity was reduced by a factor of at least 4, and the onset radius was also seen to increase by a factor of 10. Laherrere et al. have recently repeated this experiment and carefully studied the dependence of the initial radius and onset time of residual gas pressure and growth velocity. They also reported a preliminary attempt to identify the gas pumped out of the sample by infrared absorption spectroscopy [33], which was subsequently shown to be methanol, a material used in the manufacturing of succinonitrile [34]. This group has also recently investigated the unstable crystal-melt interface in directional solidification of a binary system and has shown that the light scattering spectrum is modified by the occurrence of a cellular instability [34].

The first direct observation of the microbubbles was reported in 1990 by the CCNY group [24] in long-time evolution studies on salol. By utilizing slow growth rates, the evolution of the scattering was followed for a period of up to

14 days, which is much longer than the observation times in any previous experiment. The scattering was seen to onset at a radius of about 1 μm, which is typical for many experiments, and gradually increases to about 100 μm. On two occasions the scattering layer was trapped, and upon removal of the completely solidified sample from the thermostat and inspection under a microscope, a layer of bubbles could be seen included in the crystal. The size of the bubbles could then be measured independent of the light scattering measurements; it agreed with the light scattering values obtained after long growth periods, just before trapping occurred. These measurements were used to test the empirical scaling dependence of particle size versus growth velocity for the capture of inclusions found by Chernov and Temkin [39]. At a growth velocity of 0.212 μm/s trapped bubbles were observed with radii in the range 50–100 μm, which is consistent with the radius obtained from light scattering. The radius was also consistent with Chernov and Temkin's survey [39].

The crystal was also subjected to pumping, and after 24 hr the scattering was seen to disappear. A preliminary force balance model along the lines of that proposed by Mesquita [30] was discussed as a possible mechanism for selection of the initial radius R_i. Comparisons were made with data from the experiments on salol and water [12, 13, 21].

In 1989 the ETH group initiated a new study of diffusive scattering in cyclohexane [17]. Cyclohexane was chosen because of the absence of a permanent dipole moment. Thus the possibility that some long range dipole-dipole interaction causes the scattering phenomena could be ruled out. Measurements of the linewidth Γ showed no departure from the q^2 behavior seen in the other materials and corresponded to an R_h of about 0.26 μm. The angular dependence $I(\theta)$ of the scattering intensity was measured and found to fit the Ornstein-Zernike form $I(q) \sim (1 + q^2\xi^2)^{-1}$ with $\xi = 0.2$ μm. It was stated that $I(\theta)$ could only be fitted by Mie theory for a broad size distribution of the form $N(\alpha) \sim \alpha^{-5}$ with $1 < \alpha < 10$; such a distribution of sizes would have produced a large deviation from single exponential behavior in the correlation function, which was not observed.

A new geometry was also used to observe scattering. A temperature sensor was embedded in a crystal that had been zone refined several times, and a 3-mm liquid sphere was melted in the crystal. When the heat source was removed, the sphere grew back inwards with velocities in the range 0–30 μm/s, considerably faster than the velocities obtained in the zone refining apparatus, and scattering was seen. The linewidth of the scattering was similar to the previous results, corresponding to a slightly larger R_h. The onset time for this geometry was extremely short, on the order of several seconds. If the growth was stopped and the crystal melted back, the interface layer followed the interface and did not "lift off."

III. THE MICROBUBBLE MODEL

We proposed the microbubble model in 1986 after performing a preliminary study of the dependence of the scattered intensity $I(\theta)$ on the scattering angle θ in salol [22]. This study showed that $I(\theta)$ decreases too rapidly with increasing θ to be described by an Ornstein-Zernike form characteristic of thermodynamic density (or entropy) fluctuations, but *could* be fit with a Mie scattering function for spheres of the size indicated by the effective diffusion constant deduced from the correlation function. (In contrast to this result, Steininger and Bilgram [17] measured the angular dependence of the scattered intensity for cyclohexane and found that their data fit the Ornstein-Zernike form but *not* the Mie form!)

The microbubble model is based on the assumed existence of dissolved gaseous impurities in the melt. If the segregation coefficient k is sufficiently small, the resulting buildup of segregated dissolved gas in the melt ahead of the interface can reach a level sufficient to allow heterogeneous bubble nucleation to occur. Assume that the liquid initially contains a dissolved gaseous impurity with uniform concentration C_0. When crystal growth begins, the solidifying crystal will initially have impurity concentration kC_0, where k is the partition (or segregation) coefficient, as shown in Fig. 2a. We assume that $k < 1$, which is usually, though not always true.

Since the excess impurity is rejected by the solidifying crystal, the concentration will build up ahead of the advancing solidification front. In the frame of reference moving with the interface, the impurity concentration $C(z)$ must satisfy the diffusion equation

$$D\left(\frac{\partial^2 C}{\partial z^2}\right) + v_g\left(\frac{\partial C}{\partial Z}\right) = \frac{\partial C}{\partial t} \tag{1}$$

and the continuity equation

$$C_i(k - 1)v_g = D\left(\frac{\partial C}{\partial z}\right)_{z=0} + \left(\frac{\partial C}{\partial t}\right)_{z=0} \tag{2}$$

where z is distance into the fluid measured from the solidification front, $\mathbf{v}_g$ is the growth velocity, $C_i = C(z)_{z=0}$ is the concentration in the fluid in contact with the solidification front, and D is the impurity diffusion constant in the fluid.

At steady state, Eqs. 1 and 2 are solved by

$$C(z) = C_0\left[1 + \frac{1 - k}{k}\exp(-v_g z/D)\right] \tag{3}$$

so that $C(z)_{z=0} = C_0/k$, which can be much larger than C_0 if $k \ll 1$. The steady-state solution of Eq. 3 is shown in Fig. 2c. Note that the thickness of the high-

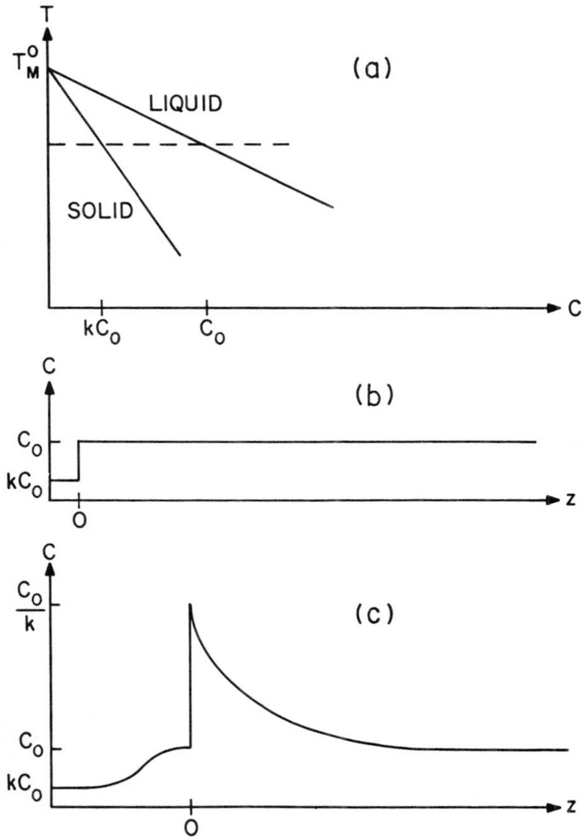

FIG. 2 (a) Phase diagram of a two-component system. At a given temperature T, the equilibrium concentration C of the (gaseous) solute in the liquid and solid are given by the intersections of a horizontal line at T with the upper (liquidus) and lower (solidus) lines respectively. (b) Concentration of the solute as a function of position relative to the interface ($z = 0$) at the beginning of solidification. (c) Concentration of solute as a function of position after steady state has been reached. The solute concentration in the crystal has increased from its initial value of kC_0 as the interface concentration builds up due to solute rejection. At steady state, the crystal grows with solute concentration C_0, equal to the original concentration in the liquid, while the concentration C_i in the fluid adjacent to the surface is C_0/k.

concentration boundary layer is given by Eq. 3 as

$$d = \frac{D}{v_g} \tag{4}$$

i.e., the thickness of the boundary layer increases as the inverse of the growth speed [40].

Next, consider a bubble of gas of radius R. If surface tension is neglected, then a bubble of *any* radius could exist in equilibrium with the fluid if the concentration C were exactly the saturation concentration C_s. When surface tension is included, however, bubbles can only exist in equilibrium if the liquid is supersaturated, i.e., if $f = C/C_s > 1$. If the fluid pressure is p and the surface tension is σ, then the effective pressure inside the bubble will be $p' = p + 2\sigma/R$, and bubbles can exist in (unstable) equilibrium only if $f = p'/p = 1 + (2\sigma/Rp)$, where Henry's law has been assumed to apply, or equivalently, if

$$R_0 = \frac{2\sigma}{[p(f - 1)]} \tag{5}$$

where p is assumed to be the vapor pressure of gas in equilibrium with liquid having dissolved gas concentration C_s. Bubbles with $R < R_0$ will collapse, while those with $R > R_0$ will grow, although the initial rate of change in R will be small [41].

As solidification proceeds, the solute concentration at the interface C_i will gradually increase from the initial C_0 towards C_0/k. (It may never reach the steady-state value because of bubble nucleation or convection.) Once it passes the saturation value C_s, however, the system may become unstable against the formation of bubbles. Initially, the critical radius R_0 given by Eq. 5 will be very large. As the concentration continues to increase, however, R_0 will eventually decrease to a value small enough to allow bubbles to nucleate as we discuss below.

The microbubble model immediately explains the long onset time for diffusive scattering, since the buildup of excess dissolved gas concentration in the boundary layer is a slow process at the low growth speeds employed. The narrowness of the scattering layer is then due to the exponential decay with distance from the interface predicted by Eq. 3 and shown in Fig. 2c. The prediction that the thickness of the scattering layer should scale with the inverse of the growth velocity v_g agrees with the results for ice reported in Ref. 16. Mazur and Keizer have shown that Eq. 4 is generally correct for any type of scattering object residing in the melt, while for scatterers residing in the crystal, the width of the boundary layer would increase with increasing growth velocity rather than decreasing [36, 37].

Further support for the model is provided by the observation that the scattering intensity is reduced or eliminated by extensive pumping on the sample [24, 32], while the scattering is enhanced by injecting gas into the melt [25]. Recently, Mesquita has prepared ultrapure samples of biphenyl and naphthalene by vacuum sublimation, which showed no evidence of light scattering; after bubbling argon through the samples, however, scattering was again observed [31].

Mesquita et al. [28] noted that the approximate time-dependent solution to Eqs. 1 and 2 for the solute concentration C_i at the interface is

$$\frac{C_i}{C_0} = 1 + \left(\frac{1 - k}{k}\right)\left(1 - \exp\left[-\frac{v_g^2(1 - k)kt}{D}\right]\right) \tag{6}$$

which, when linearized, becomes

$$\frac{C_i}{C_0} \approx 1 + \frac{v_g^2(1 - k)^2}{D}t \approx 1 + \frac{v_g^2}{D}t \tag{7}$$

Combining Eq. 7 with Eq. 5, and assuming that the melt is saturated ($C_0 = C_s$), the following scaling law was found [28]:

$$R_i t_i = \frac{2\sigma D}{p}\frac{1}{v_g^2} \tag{8}$$

Mesquita et al. measured the onset bubble size R_i and the onset time t_i for different growth speeds v_g in biphenyl and naphthalene and found for each material that the product $R_i \cdot t_i$ is proportional to v_g^{-2} as predicted by Eq. 8.

Additional evidence for microbubbles is provided by the observation of a "lift-off layer." This effect was seen in water by Vasenka and Yeh [26, 27] and by Halter et al. [16], and in cyclohexanol by Livescu et al. [23]. Note that the long persistence time of the "lift-off layer" and its observed gradual spreading with time [16] can be understood as a consequence of the very small diffusion constants of the bubbles. Finally, we note that the recent direct observation of bubbles trapped by the advancing crystal [24] provides the strongest evidence for the microbubble mechanism.

Although the experimental evidence in agreement with the predictions of the microbubble model is now rather extensive, there are two remaining unresolved problems with the model: (1) how does bubble nucleation occur, and (2) what determines the initial radius at the onset of scattering? These two problems, which are probably related, are addressed below.

A. The Nucleation Problem

The theory of homogeneous nucleation [42] predicts that J, the number of bubbles of radius R nucleated per s $\cdot$ cm^3, is given by

$$J = Ae^{-\Delta F/kT} \tag{9}$$

where the prefactor A depends on various material parameters and is typically $\sim 10^{32}$. The free energy barrier height ΔF for formation of a bubble of critical radius R_0 is given by

$$\Delta F = \frac{4}{3}\pi\sigma R_0^2 = \frac{16\pi\sigma^3}{3(P_g^* - P_l)^2} \tag{10}$$

where σ is the interfacial tension, P_g^* is the gas pressure inside the bubble, and P_l is the hydrostatic pressure.

In salol, for bubbles of radius $\sim 1\ \mu m$ with $\sigma = 20$ dyn/cm^2, $\Delta F/kT \sim 2 \times 10^7$, and $J \sim 10^{32} \times e^{-2\times 10^7} \sim 10^{-9\times 10^6}$. Wilcox and Kuo [42] estimate that an air bubble nucleation rate of 1/cm^3 · s in naphthalene would require a supersaturation of 380 atm, and in water, 1720 atm. Various models of heterogeneous nucleation in the presence of solid surfaces [42, 43] suggest that ΔF could be reduced by at most two orders of magnitude, leaving the theoretical nucleation rate still unobservably small at small supersaturation. It was this argument that Bilgram discussed in his 1987 review article [14] in which he rejected the microbubble model.

In spite of the depressingly small nucleation rates predicted by nucleation theory, however, the nucleation of gas bubbles during crystal growth is actually a well-established phenomenon that has been known for many years. In 1973, Wilcox and Kuo [42] summarized the extensive literature on the observation of the generation and trapping of gas bubbles in melt-grown crystals. They noted that bubble formation during the freezing of water is particularly well known, and that in 1870 Bunsen had obtained bubble-free ice for the first time by directional freezing downward of water boiled in the tube and then sealed to exclude air. The capture of gas bubbles during solidification of both water and salol has been observed and photographed extensively in the USSR by Geguzin and Dzuba [40]. Gas bubble trapping during crystal growth is also an important technical problem, since it leads to voids and imperfections in crystals.

An important clue to the effect of crystal surfaces on gas bubble nucleation was found by Hemmingsen and coworkers [44]. They studied bubble nucleation in water equilibrated with various gases at high pressure and then decompressed to atmospheric pressure. No nucleation was observed for initial pressures below 190 atm for N_2, or 300 atm for He, but massive nucleation rates were observed for 20–30 atm higher supersaturation.

When the water was first saturated with succinic acid or potassium nitrate and cooled to produce crystalline precipitates, it was found that the gaseous supersaturation threshold for bubble nucleation on the crystals was far lower than the value required for nucleation in the bulk fluid or on the glass-liquid interface. For succinic acid, the threshold was found to be less than 5 atm. It therefore seems that the free energy barrier required to nucleate gas bubbles on the surface of a crystal is enormously decreased relative to the bulk value $\Delta F = (4/3)\pi\sigma R_0^2$. The mechanism for this effect is unknown, although Hemmingsen discussed the possible role of pits or cracks on grain boundaries as nucleation centers.

B. The Initial Radius

Another problem with the microbubble model concerns the observation that for a given material and growth velocity, the bubble size at onset R_i is fairly repeatable. One possibility is that the critical bubble radius R_0 given by Eq. 5 must be small enough to avoid capture by the advancing interface. However, our recent observations of bubble capture in salol after prolonged growth suggest that the capture mechanism sets an upper limit to bubble side ~100 times larger than the observed onset value.

Another possibility is that bubbles, once nucleated, may be pushed out of the boundary layer if their size does not satisfy some criterion. We initially speculated that the buoyancy force would be sufficient to displace bubbles that exceeded some maximum size [22]. More recently, we performed an analysis of the forces on the bubbles as proposed by Mesquita [30], assuming that the three main forces are the viscous Stokes force $F_v = 4\pi\eta R v_g$, the thermocapillary force $F_t = -2\pi R^2(d\sigma/dT)G$, and the buoyancy force $F_b = (4/3)\pi R^3\rho g$, where G is the temperature gradient. If these three forces are in equilibrium ($F_t + F_v + F_b = 0$), the bubble will be carried along ahead of the interface. The buoyancy force turns out to be unimportant compared to the thermocapillary force for small bubbles. In water, for example, with a 1 K/cm temperature gradient, the buoyancy force for 0.1 μm bubbles is only 5% as big as the thermocapillary force. Therefore, for small bubbles the results of this calculation are nearly identical if the liquid is either above or below the interface.

This equilibrium bubble radius can then be plotted as a function of velocity for the relevant temperature gradients (i.e., estimates of those actually used in the experimental work). We have done this for ice and salol and have plotted the range of data points for the initial radius and growth velocity obtained in experiments at CCNY and at the ETH. The agreement was seen to be quite good [24]. One intriguing aspect of this analysis is that the predicted bubble radius for water is approximately ten times smaller than that for salol, primarily because of the difference in viscosity; this ratio is close to the ratio of the experimental values shown in Table 1.

The mechanism that is probably crucial for determining the onset bubble size is the requirement of a sufficiently large nucleation rate. Suppose that the (still unspecified) heterogeneous nucleation mechanism reduces the free energy of bubble formation of Eq. 10 from $\Delta F = (4/3)\pi R_0^2\sigma$ to $\Delta F = \phi \cdot (4/3)\pi R_0^2\sigma$ and that onset requires that the nucleation rate J reach a minimum observable value of, e.g., $J = 10^3$, or $J = 1/\text{s} \cdot \text{mm}^3$. Suppose that for salol $\phi = 3.3 \times 10^{-6}$ so that the nucleation rate of Eq. 9, for $R = 1\ \mu\text{m}$, would be 10^3.

As growth proceeds and the dissolved gas concentration at the interface increases, the minimum size for a noncollapsing bubble decreases following Eq. 5, and when r_0 reaches 1 μm, the nucleation rate would be observable. Note, however, that when R_0 is 2 μm, $J \sim 10^{-84}$. Therefore no bubbles will appear until R_0 is very close to 1 μm, so that the combined effects of Eqs. 5 and 9 can provide a very tight restriction on the initial bubble size. The nucleation rate equation would then provide the size selection mechanism. Of course the physical mechanism for such a small ϕ, and the way it varies with material properties, is completely unknown, but Gilmer has suggested that wetting effects might play a crucial role [45].

Indirect evidence supporting this explanation was provided by a recent nucleation study by Cumming, Wiltzius, and Bates [46]. Off-critical mixed polymer melts quenched into the two-phase coexistence region produced nucleation of highly monodisperse spherical droplets of the minority phase (polydispersity $<$ 3%), which subsequently grew with time. This remarkable initial monodispersity presumably reflects the extreme dependence of nucleation rate on droplet size.

IV. DISCUSSION AND CONCLUSIONS

In our review, we have concentrated on the diffusive scattering process because of the controversy surrounding its origin, and we have discussed the microbubble model in detail because it appears to explain much of the experimental data. Nevertheless, there may be other processes involved in some cases. Mazur and Keizer [36], and Vesenka and Yeh [26], for example, have suggested that there might be two different scattering phenomena involved, one for the c face of ice, and microbubbles for all other systems.

The only other explanation currently available for the diffusive scattering is the mesophase model of Bilgram. In his 1987 review [14], the theory of the formation of the mesophase boundary layer was formulated in terms of a wetting transition. It was suggested that the solid-liquid interface is not a two-dimensional plane but possesses a finite thickness that scales with the correlation length ξ of density fluctuations in the layer. It was proposed that at equilibrium this phase is restricted to a microscopically thin layer by its imperfect wetting properties. Several theoretical studies have indicated a thin preordered phase such as this [47]. Upon initiation of growth the system passes into a nonequilibrium state, where the liquid phase is supercooled and in contact with the crystal. After a certain growth rate has been attained, at a low enough thermal gradient a wetting transition is assumed to take place. The mesophase perfectly wets the interface, which enables it to grow from microscopic to macroscopic dimensions. From the hysteretic behavior of the scattering it was concluded that the transition is first order.

We note that Burke, Broughton, and Gilmer performed a molecular dynamics study of a Lennard-Jones system in which large undercoolings were introduced producing growth rates of up to 80 m/s [48]. In this study, the transition layer between the crystal and melt was *not* found to increase in thickness during growth [45]. Furthermore, since the isothermal compressibility of a crystal is typically much smaller than that of its melt, it is difficult to see how the transition layer could have a compressibility hundreds of times larger than that of the melt. Therefore, the mesophase concept must be considered an essentially ad hoc explanation until further additional evidence can be offered to support it.

Finally, we emphasize that the angular dependence of the scattered intensity is an important criterion, for which contradictory results have been reported in the past [17, 22]. Additional studies of $I(\theta)$ are currently in progress in our laboratory.

ACKNOWLEDGMENTS

We wish to acknowledge helpful discussions and correspondence with J. H. Bilgram, O. Mesquita, Y. Yeh, J. C. Toledano, W. H. van Saarloos, A. A. Chernov, Ya. E. Geguzin, P. Wiltzius, G. Gilmer, P. Mazur, J. Keizer, X. W. Qian, and M. Muschol. This work was supported by the U.S. Department of Energy under grant no. DE-FG02-84ER45132.

REFERENCES

1. H. Güttinger, J. H. Bilgram, E. Serrallach, and W. Känzig, *Helv. Physica Acta 48:* 399 (1975). "Lichtstreuung an der Phasengrenze fest-flüssig."
2. J. H. Bilgram, H. Güttinger, and W. Känzig, *Phys. Rev. Lett. 40:* 1394 (1978). "Fluctuations of the Ice-Water Interface During Solidification."
3. H. Gütinger, J. H. Bilgram, and W. Känzig, *J. Phys. Chem. Solids 40:* 55 (1979). "Dynamic Light Scattering at the Ice-Water Interface During Freezing."
4. J. H. Bilgram and P. Böni in *Light Scattering in Liquids and Macromolecular Solutions* (V. DeGiorgio, M. Corti, and M. Giglio, eds.), Plenum, New York, 1980, p. 203. "Interface Fluctuations of Growing Ice Crystals."
5. B. Zysset, P. Böni, and J. H. Bilgram, *Helv. Phys. Acta 54:* 265 (1981). "Lichtstreuung an der Phasengrenze fest/flüssig von D_2O."
6. J. H. Bilgram in *Nonlinear Phenomena at Phase Transitions and Instabilities* (T. Riste, ed.), Plenum, New York, 1982, p. 343. "Instabilities During Crystal Growth (Experiments)."
7. U. Dürig and J. H. Bilgram in *Nonlinear Phenomena at Phase Transitions and Instabilities* (T. Riste, ed.), Plenum, New York, 1982, p. 371. "Periodic Fluctuations at the Solid Liquid Interface of Salol."

8. J. H. Bilgram, *Ann. N.Y. Acad. Sci. 404:* 335 (1983) (Proceedings of the 4th International Conference on Physico-Chemical Hydrodynamics, R. Pfeffer, ed.). "Asymmetries of the Melting Transition."
9. J. H. Bilgram, *Naturwissenschaften 69:* 472 (1982). "Die Dynamic des Gefrierens."
10. J. H. Bilgram and P. Böni in *Chemical Instabilities* (G. Nicolis and F. Baras, eds.), D. Reidel, Dordrecht, 1984, p. 351. "Dynamical Processes During Solidification."
11. J. H. Bilgram, *Helv. Physica Acta 56:* 863 (1983). "Freezing Transition Phenomena far from Equilibrium."
12. P. Böni, J. H. Bilgram, and W. Känzig, *Phys. Rev. A 28:* 2953 (1983). "Properties of the Solid-Liquid Interface Layer of Growing Ice Crystals: A Dynamic Light Scattering Study."
13. U. Dürig, J. H. Bilgram, and W. Känzig, *Phys. Rev. B 30:* 946 (1984). "Properties of the Solid-Liquid Interface of Growing Salol Crystals: A Dynamic Light Scattering Investigation."
14. J. H. Bilgram, *Physics Reports 153:* 1 (1987). "Dynamics at the Solid Liquid Transition: Experiments at the Freezing Point."
15. P. Halter and J. Bilgram, *Helv. Phys. Acta 61:* 185 (1988). "Raman Scattering in the Interface Layer in Front of a Growing Ice Crystal."
16. P. U. Halter, J. H. Bilgram, and W. Känzig, *J. Chem. Phys. 89:* 2622 (1988). "Properties of the Solid-Liquid Interface Layer of Growing Ice Crystals: A Raman and Rayleigh Scattering Study."
17. R. Steininger and J. Bilgram, *Helv. Phys. Acta 62:* 215 (1989). *J. Cryst. Growth 99:* 98 (1990) (Proc. ICCG-9). "Light Scattering at the Solid-Liquid Interface of Cyclohexane."
18. J. H. Bilgram and R. Steininger, *J. Cryst. Growth 99:* 30 (1990) (Proc. ICCG-9). "Light Scattering in Crystal Growth."
19. R. Steininger and J. H. Bilgram, *Proceedings of the 4th National Seminar on Crystal Growth, Mysore, Aug. 1989,* World Scientific, Singapore (in press). "In Situ Studies by Light Scattering: The Solid-Liquid Interface During Solidification."
20. O. N. Mesquita, D. G. Neal, M. Copic, and H. Z. Cummins, *Phys. Rev. B 29:* 2846 (1984). "Assymmetry of Dynamics at the Growing Crystal-Melt Interface."
21. O. N. Mesquita and H. Z. Cummins, *Physico-Chemical Hydrodynamics 5:* 389 (1984). "Dynamics at the Nonequilibrium Crystal-Melt Interface."
22. H. Z. Cummins, G. Livescu, H. Chou, and M. R. Srinivasan, *Solid State Commun. 60:* 857 (1986). "Origin of Diffusional Light Scattering at the Non-Equilibrium Crystal-Melt Interface."
23. G. Livescu, M. R. Srinivasan, H. Chou, H. Z. Cummins, and O. Mesquita, *Phys. Rev. A 36:* 2293 (1987). "Dynamic Light Scattering at the Nonequilibrium Crystal-Melt Interface in Cyclohexanole."
24. L. Williams, M. R. Srinivasan, and H. Z. Cummins, *Phys. Rev. Lett. 64:* 1526 (1990). "Direct Observation of Microbubbles in Directional Solidification of Salol."
25. R. A. Brown, J. Keizer, U. Stelger, and Y. Yeh, *J. Phys. Chem. 87:* 4135 (1983). "Enhanced Light Scattering at the Ice-Water Interface During Freezing."
26. J. P. Vesenka and Y. Yeh, *Phys. Rev. A 38:* 5310 (1988). "Dynamic Light Scattering at a Growing Crystal Interface: Ice-Water System."

27. J. P. Vesenka and Y. Yeh, *Bull. Am. Phys. Soc. 32:* 711 (1987). "Dynamic Light Scattering at the Crystal Melt Interface, the Ice-Water System," *33:* 781 (1988). "Dynamic Light Scattering at the Crystal Melt Interface: The Ice-Water System" and private conversation.
28. O. N. Mesquita, L. O. Ladeira, I. Gontijo, A. G. Oliveira, and G. A. Barbosa, *Phys. Rev. B 38:* 1550 (1988). "Dynamic Light Scattering at the Nonequilibrium Crystal-Melt Interface in Biphenyl and Naphthalene."
29. O. N. Mesquita and L. O. Ladeira, "On the Nature of the Dynamic Light Scattering at the Growing Ice-Water Interface" (1988, unpublished).
30. L. O. Ladeira and O. N. Mesquita, "Laser Light Beating Spectroscopic Measurements at Growing Crystal-Melt Interfaces" (1989, unpublished).
31. O. N. Mesquita, private communication (May 1990).
32. J. M. Laherrere, H. Savary, R. Mellet, and J. C. Toledano, *Phys. Rev. A 42:* 1142 (1990). "Dynamic Light Scattering Study of the Solid-Liquid Interface in Succinonitrile During Growth."
33. J. M. Laherrere, R. Mellett, H. Savary, J. F. Scott, C. Licoppe, and J. C. Toledano, "Onset of Light-Scattering by a Solid-Liquid Interface During Growth" *Phys. Rev. A 42*: 3634 (1990).
34. J. M. Laherrere, H. Savary, R. Mellet, and J. C. Toledano, "In Situ Dynamic-Light-Scattering Study of a Cellular Interface Arising During Directional Growth" (unpublished). J. M. Laherrere, thesis, University of Paris VI (1990). "Etude de l'Interface Solid-Liquide du Succinonitrile en Cours de Croissance par Diffusion de la Lumiere."
35. J. Keizer, P. Mazur, and T. Morita, *Phys. Rev. A. 32:* 2944 (1985). "Theory for the Anomalous Light Scattering in Growing Ice Crystals."
36. P. Mazur and J. Keizer, *Phys. Rev. A 38:* 5267 (1988). "Anomalous Light Scattering at the Surface of Growing Crystals: A Theoretical Analysis Based on Irreversible Thermodynamics."
37. P. Mazur and J. Keizer, *J. Statistical Physics 57:* 497 (1989). "Anomalous Light Scattering at the Surface of Growing Ice Crystals and Thermodynamics of Irreversible Processes."
38. W. H. van Saarloos and G. H. Gilmer, *Phys. Rev. B 33:* 4927 (1986). "Dynamical Properties of Long-Wavelength Interface Fluctuations During Nucleation-Dominated Crystal Growth."
39. A. A. Chernov and D. E. Temkin in *Current Topics in Material Science,* Vol. 2 (E. Kaldis, ed.), North Holland, 1977, p. 4. "Capture of Inclusions in Crystal Growth."
40. Ya. E. Geguzin and A. S. Dzuba, *J. Cryst. Growth 52:* 337 (1981). "Crystallization of a Gas-Saturated Melt."
41. P. S. Epstein and M. S. Plesset, *J. Chem. Phys 18:* 1505 (1950). "On the Stability of Gas Bubbles in Liquid-Gas Solutions."
42. W. R. Wilcox and V. H. S. Kuo, *J. Cryst. Growth 19:* 221 (1973). "Gas Bubble Nucleation During Crystallization."
43. M. Blander, *Adv. Coll. Int. Sci. 10:* 1 (1979). "Bubble Nucleation in Liquids." M. Blander and J. L. Katz, *AIChE. J. 21:* 833 (1975). "Bubble Nucleation in Liquids."

44. E. A. Hemmingsen, *J. Appl. Phys. 46:* 213 (1975). "Cavitation in Gas-Supersaturated Solutions." *Nature 267:* 141 (1977). "Spontaneous Formation of Bubbles in Gas-Supersaturated Water." W. A. Gerth and E. A. Hemmingsen, *J. Coll. Interf. Sci. 74:* 80 (1980). "Heterogeneous Nucleation of Bubbles at Solid Surfaces in Gas-Supersaturated Aqueous Solutions."
45. G. H. Gilmer, private communication (1990).
46. A. Cumming, P. Wiltzius, and F. S. Bates, *Phys. Rev. Lett. 65:* 863 (1990). "Nucleation and Growth of Monodisperse Droplets in a Binary Fluid System."
47. E. T. Chen, R. N. Barnett, and U. Landman, *Phys. Rev. B 40:* 924 (1989). "Crystal-Melt and Melt-Vapor Interfaces of Nickel." See also Ref. 324–327 in J. H. Bilgram [14].
48. E. Burke, J. Q. Broughton, and G. H. Gilmer, *J. Chem. Phys. 89:* 1030 (1988). "Crystallization of fcc (111) and (100) Crystal-Melt Interfaces: A Comparison by Molecular Dynamics for the Lennard-Jones Systems."

III

Complementary Techniques

16
Externally Excited Surface Waves

KENJIRO MIYANO Department of Applied Physics, University of Tokyo, Tokyo, Japan

I. INTRODUCTION

The capillary waves discussed so far in this book are thermally generated. An alternative approach to capillary wave measurement is possible, in which capillary waves of desired frequency are externally generated through a mechanical or electrical means. The externally excited capillary wave approach is complementary to that of the spontaneously generated waves in many respects. In the latter technique, the frequency of the wave is above the kHz range, and the experimental conditions, especially in avoiding mechanical noise and in keeping the optical alignment, are stringent. In the former technique, on the other hand, the frequency is usually below the kHz range. It is technically much simpler and intuitively straightforward, hence historically much older.

A simple approach of the externally excited surface waves would be to touch the surface of a liquid with a vibrating bar and visually to observe the generated ripples or to use a second bar touching the liquid in order to monitor the surface movement. A variety of techniques along this line have appeared. The reader is referred to a review [1] and the references therein for the earlier development of the techniques. However, these techniques are unsatisfactory in two aspects. (1) Because of the lack of detection sensitivity, either in visual observation or in mechanical pick-up, the amplitude of the wave has to be relatively large, jeopardizing the validity of the use of the linearized Navier-Stokes equation. (2) In studying monolayers at the interface, the generator bar may cause detrimental effects to the integrity of the monolayers. An alternative technique is now available, in which the generator operates by drawing one of the two fluids (whose boundary forms the interface) with the higher dielectric constant into a strong electric field, and the detection is done by monitoring the light beam reflected from the surface. The above-mentioned difficulties caused by the bar in mechanical contact with the interface are averted. In this section, we will mostly deal with this technique.

II. THEORY

The hydrodynamic description of the surface waves does not, of course, depend on how the waves are generated, e.g., mechanically or thermally. Therefore, the dispersion relation is identical in both cases. Experimentally, however, there is one distinction between the two approaches. In the light scattering experiment from the thermally fluctuating surfaces, one studies the time evolution of a wave of a selected wavelength, i.e., one measures the complex frequency ($\omega = \omega_r - i\omega_i$) as a function of the real wavenumber ($k = k_r$). On the other hand, when the surface waves are externally generated, their frequency is fixed and real ($\omega = \omega_r$) but a complex wavenumber ($k = k_r + ik_i$) is measured. The dispersion relation is, therefore, seen in a different perspective.

Let us consider a plane surface wave at a liquid-gas interface. (The treatment can be easily generalized to a liquid-liquid interface.) The velocity of the liquid can be written in terms of a superposition of two types of waves,

$$v = [A \exp(kz) + B \exp(mz)] \exp(ikx - i\omega t) \tag{1}$$

where the coordinate is set such that the z axis is along the surface normal pointing upward and the wave propagates in the position x direction.[1] From the Navier-Stokes equation, one obtains $m^2 = k^2 - i\omega\rho/\eta$, where ρ is the density and η is the viscosity of the liquid. The sign of m must be taken so that its real part is positive and $k_i > 0$. The dispersion relation for a surface wave of the form of Eq. 1 on a liquid covered with a monolayer is given by [1]

$$(\rho\omega^2 - \gamma k^3 - \rho g k)(\rho\omega^2 - mk^2\varepsilon) - \varepsilon k^3(\gamma k^3 + \rho g k) + 4i\rho\eta\omega^3 k^2 + 4\eta^2\omega^2 k^3(m - k) = 0 \quad (2)$$

where g is the gravity acceleration, γ is the surface tension, and ε is the uniaxial dilational elastic modulus of the monolayer, which can be complex and should be considered as a phenomenological parameter at this stage.[2] The imaginary part of ε ($\varepsilon = \varepsilon_r - i\varepsilon_i$) is due to the dissipative process associated with the periodic uniaxial area change of the monolayer, which includes shear viscosity, structural relaxation, and desorption and adsorption.

Equation 2 relates the measured k_r and k_i to the unknown physical parameters γ, ε_r, and ε_i. Obviously, it is not possible to deduce three unknowns uniquely from two known values. Therefore, independent measurements of these quantities or a capability to cover a wide frequency range is highly desirable.

It is also quite important to understand the general behavior of k as a function of γ and ε so that one knows where to look once the data are at hand. In removing the square root in Eq. 2, one obtains a tenth-order polynomial for k. Of the ten roots, only two satisfy the initial assumptions, i.e., k_r and k_i are positive. The two roots correspond to the transverse (capillary) wave and the longitudinal wave [5] respectively. The capillary wave is the familiar ripples on a liquid surface. The longitudinal wave is a density wave in which the concentration of the surfactant molecules adsorbed at the interface oscillates with a very small vertical amplitude. Unfortunately, the polynomial cannot be reduced in a parameterized form. Thus Eq. 2 has to be solved for each experimental condition separately [5–7]: a tedious task but not difficult with the use of the personal computer today. As an example, results of numerical calculation of Eq. (2) are shown in Fig. 1.

Several features regarding Fig. 1 should be pointed out. (1) The damping constant, k_{iC} (C for capillary mode) goes over a maximum as ε_r increases. This phenomenon is often attributed to a "resonance" effect with the longitudinal

[1]A different convention can be adopted resulting in a sign change in the definition of m and the dispersion relation (see Ref. 3). We follow the convention of Refs. 1 and 2.

[2]Phenomenologically, γ can be complex too. However, its imaginary part, if any, is small at frequencies employed in the externally excited waves [4].

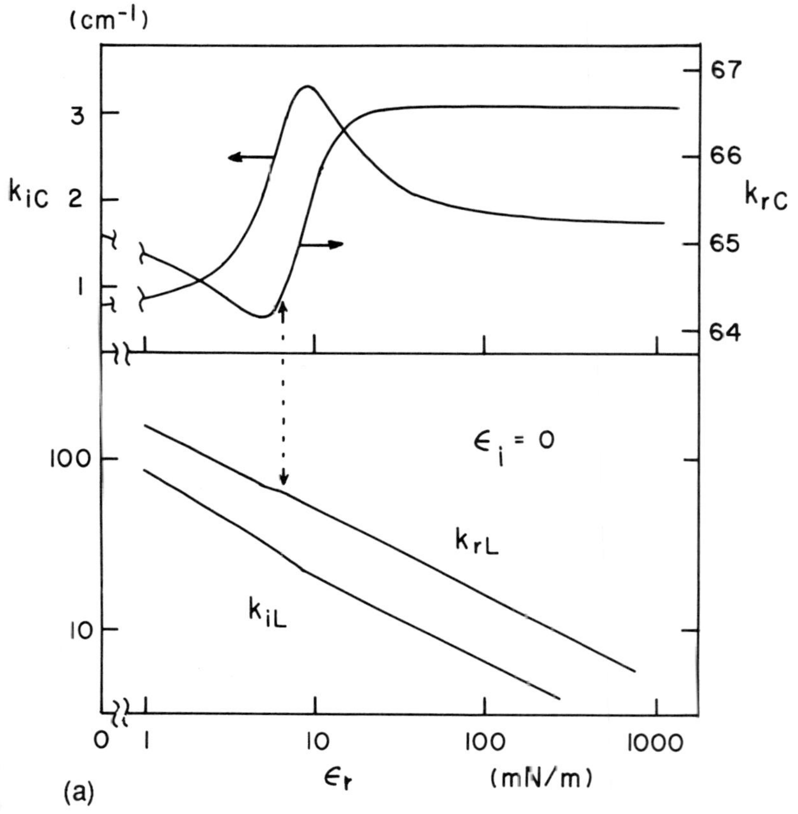

FIG. 1 Numerical solutions to Eq. 2 as a function of ε_r. k_{rC} and k_{iC} are for the capillary mode while k_{rL} and k_{iL} are for the longitudinal mode. The following parameters are used: $\rho = 1$ g/cm^3, $\eta = 0.01$ P, $\gamma = 70$ mN/m, and $\omega/2\pi = 700$ Hz. (a) $\varepsilon_i = 0$, (b) $\varepsilon_i = 3$ mN/m, and (c) $\varepsilon_i = 10$ mN/m. The dotted line in (a) indicates $k_{rC} = k_{rL}$.

mode and hence it is claimed that it occurs when the wavelength of the two modes match. It is easily noted, however, that even when $\varepsilon_i = 0$, the peak of the k_{iC} occurs not when $k_{rC} = k_{rL}$ (L for the longitudinal mode). At a moderately large ε_r, $k_{rC} = k_{rL}$ never happens, although the peak of k_{iC} appears. Discussions on the physical process causing this peak are found in Refs. 3 and 5.

(2) When ε_r is large, the longitudinal wave shows the limiting behavior $k_{iL}/k_{rL} = \tan(\pi/8) = 0.413$ and $k_{rL} \sim \sqrt{\varepsilon_r}$.

(3) Although the region $\varepsilon_r \lesssim \varepsilon_i$ is not accessible when adsorption-desorption is the sole mechanism that causes nonzero ε_i (see Sec. V), the longitudinal wave

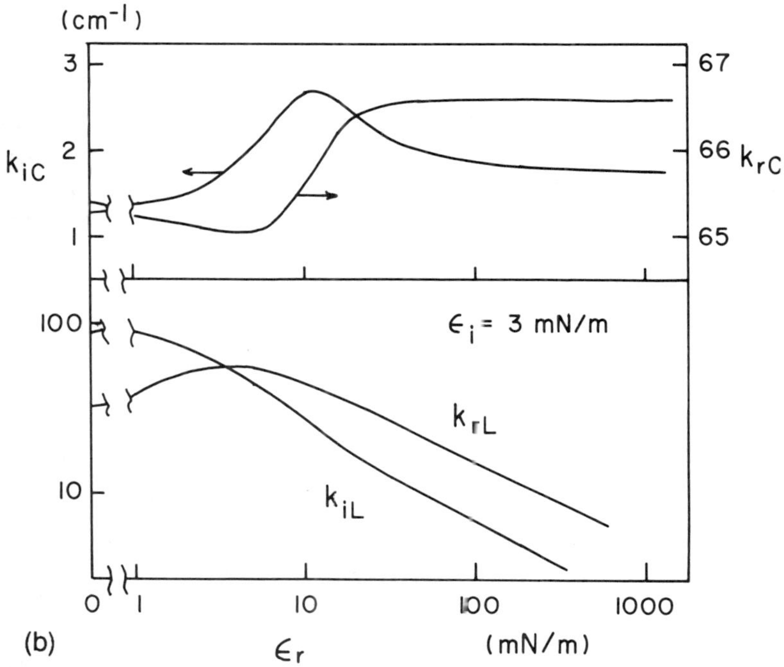

(cm⁻¹)
k_iC
k_rC
3
2
1
67
66
65
100
10
ε_i = 3 mN/m
k_rL
k_iL
0
1
10
100
1000
ε_r
(mN/m)

(b)

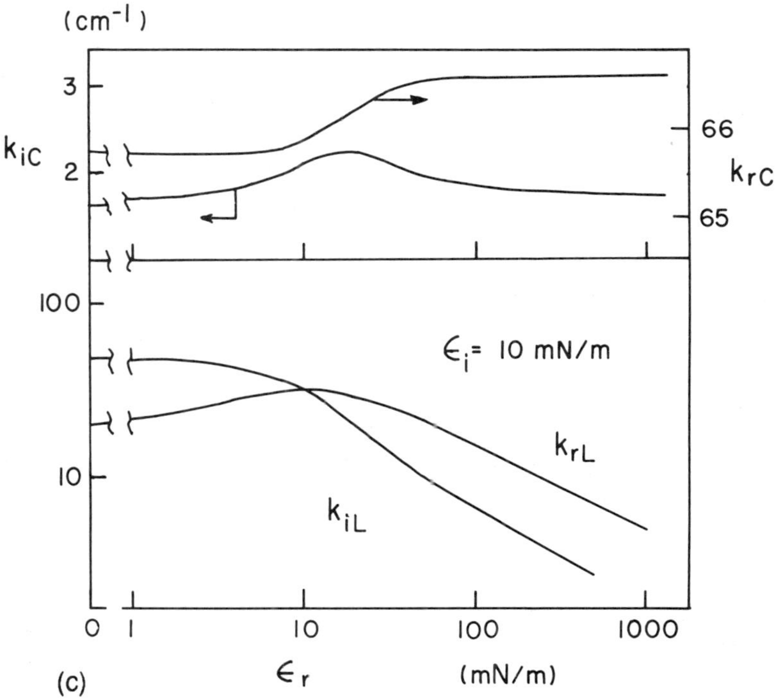

(cm⁻¹)
k_iC
k_rC
3
2
66
65
100
10
ε_i = 10 mN/m
k_rL
k_iL
0
1
10
100
1000
ε_r
(mN/m)

(c)

is overdamped if $\varepsilon_r \lesssim \varepsilon_i$ actually occurs. In this case, the wave propagates into the liquid instead of being localized near the surface. A similar situation has been observed in a highly viscous liquid without a spread film (see Sec. IV.A).

(4) The dependence of k_{rC} on ε_r is rather small compared with other quantities, and often not much is discussed about its variation. Experimentally, however, k_{rC} can be measured far more accurately than other quantities, and its variation, albeit small, is valuable information.

III. EXPERIMENTAL TECHNIQUE [8]

A. Generator

Let us imagine a sharp conducting object, e.g., a razor blade, located close to the surface of water (Fig. 2). When an electrical potential is applied between the bottom of the container and the blade, a strong electrical field results at the edge due to the small curvature (for the same reason that lightning rods have sharp pointed tips). Because the dielectric constant of water (~80) is much larger than that of air (~1), water is drawn into the region of high electric field. If the field alternates at a frequency f, the water surface oscillates, and surface waves are generated at the frequency, $2f$.[3]

The technique is not limited to air-water interfaces. It can readily be applied to any interface provided that the dielectric constants of the two adjacent fluids are different. Examples are liquid crystal surfaces [9] and oil-water interfaces [10, 11]. Actually, the difference in the dielectric constants can be extremely small. For example, capillary waves have been successfully generated electrostatically on liquid He, whose dielectric constant is only 1.05 [12].

The amplitude of the generated wave is proportional to the square of the electric field. Because the field increases rapidly toward the edge, the amplitude of the wave strongly depends on the edge-water distance. The distance can be easily set within a fraction of a mm. In such a case, several tens of volts applied to the blade is enough to generate waves of sufficient amplitude on water. In order to generate a clean plane wave, the edge of the blade has to be straight and parallel to the surface. By the use of a needle electrode instead, the trouble

[3] The word "electrocapillarity" has been coined in Ref. 8 to describe this phenomenon; it has gained some popularity in the subsequent literature. But the word has been historically used in relation to the change of the surface tension caused by an electric field. Thus the use of this term in the context of capillary wave generation should be avoided.

FIG. 2 Generation of capillary waves using alternating electrostatic attraction of the fluid.

of the alignment is reduced and the electrode can be set much closer to the surface [11]. In this case, the waves are circular and the amplitude decays much faster. This feature can be advantageous when one wants to avoid reflected waves from the container wall in working in a restricted geometry.

B. Detector

A simple detection scheme of the surface wave is illustrated in Fig. 3. A beam from a He-Ne laser is focused onto the surface under study. As the surface wave passes by, the direction of the light specularly reflected from the surface oscillates. The angular amplitude of the oscillation of the reflected beam is proportional to ak_{rC}, where a is the amplitude of the surface wave. The reflected beam is directed onto a position-sensitive photodiode (PSPD),

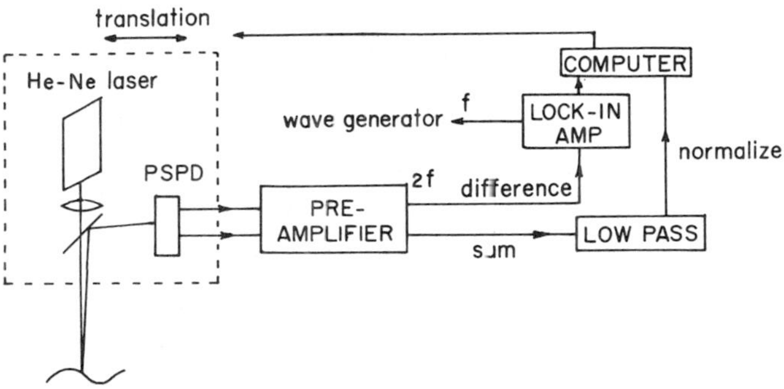

FIG. 3 A block diagram of the detector of the capillary waves.

which is essentially a photocurrent divider with two output leads. The difference between the two output currents normalized by their sum provides an absolute position of the light beam in the sense of the "center of mass" of the light intensity distribution over the photodiode face. Thus a submicrometer detection is possible even though the size of the light beam may be more than a mm.

The division (difference)/(sum) is necessary in order to offset the light intensity variation (of the laser beam itself or due to other causes) and the nonuniform sensitivity of the PSPD over the face. Because these changes occur slowly, the division is done numerically as follows. The difference (the signal) is fed to a lock-in amplifier directly, which measures the amplitude and the phase of the signal. The amplitude output of the lock-in amplifier is put into a computer and divided by the sum, which is smoothed through a low-pass filter. With the knowledge of the optical geometry and k_{rC}, the absolute amplitude of the wave can be calculated.

The lock-in amplifier supplies the AC voltage for generating the wave and is operated in a "$2f$ mode." The fact that the detection is done at $2f$ while the excitation is done at f is quite advantageous in reducing the leakage noise coupled into the detection side from the generator. As the laser beam is slowly scanned along the path of the surface wave, the amplitude and phase change. By knowing the distance of the scan, one can calculate k_r and k_i. A typical trace of such a scan is shown in Fig. 4.

The noise in the measurement is mainly mechanical oscillation of the liquid surface caused by the floor vibration. The amplitude of the noise is usually orders of magnitude larger than the signal. However, the noise frequency is mostly below 10 Hz, which is far away from the usual experimental frequencies of a few hundred Hz and can be easily filtered out by the phase-sensitive detection technique. Nevertheless, too large a noise can saturate the front end amplifier in the lock-in amplifier. Under the actual experimental conditions, measurements can be done comfortably with band-pass filtering and a dynamic reserve of more than 80 dB. Note that the amplitude of the noise oscillation *on the liquid surface* is a couple of orders of magnitude larger than that of the signal, but it affects less in the reflected beam, since it senses ak_{rC}, not the amplitude itself. The amplitude sensitivity of Å can be achieved without difficulty. The value ak_{rC} is thus easily kept below 10^{-4} all the time, well within the validity limit of the linearized Navier-Stokes equation.

The mechanical noise affects the phase stability also. Typically, the phase noise is 5° for an integration time of 0.3 ~ 1 s. If the wavelength is 1 mm and the scan distance is 2 cm, then the noise in k_{rC} is $5/(360 \times 20) = 0.1\%$. For water, this translates to the surface tension accuracy of 0.2 mN/m. The accuracy in k_{iC} is about 0.05 cm^{-1}. If the entire scan trace is least-square fitted, a factor of 5 improvement is possible.

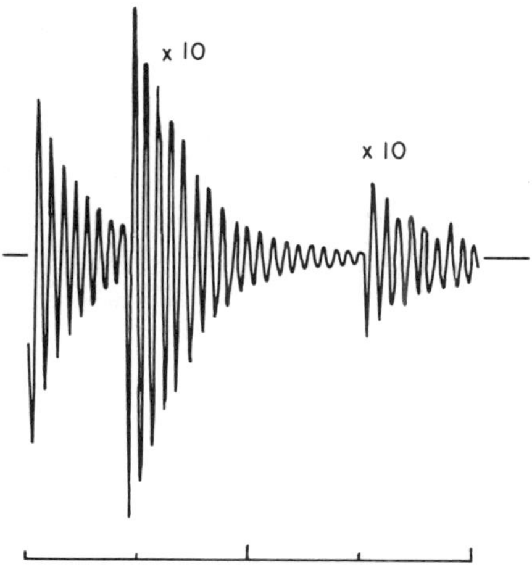

FIG. 4 A typical output from a lock-in amplifier when the water surface was scanned with the laser probe along the capillary wave path. The wave frequency was about 1.6 kHz. The sensitivity of the amplifier was increased by a factor of 10 twice. Note the wide dynamic range of the detector. The amplitude of the wave at the end of the trace was a fraction of a nm.

An alternative detection method is to use the diffraction of light from the corrugated surface caused by the electrostatically generated capillary waves. One may call it a forced Brillouin scattering. The angular deviation of the diffracted beam is measured and the wavelength calculated. Here, the information of the attenuation is thrown away, but one need not scan the laser beam. Hence it is much more suited for the measurement in a completely enclosed geometry. In a simple system without a spread monolayer, an accuracy of 0.002 mN/m has been claimed [13] in the determination of γ of about 20 mN/m.

C. Longitudinal Waves

The experimental considerations given so far are not applicable to the study of the longitudinal mode, for both the generation and the detection schemes do not couple well with this mode. Because the longitudinal mode is a density wave with its displacement almost entirely within the horizontal plane, it seems impossible to generate the longitudinal wave without a mechanical contact with the monolayer.[4] A bar or wire oscillating within the interface is a popular choice. Its movement, periodic compression and decompression of the monolayer, creates local density inhomogeneity, which then propagates as a longitudinal wave.

The detection of this mode can be done in two ways; either by looking at the movement of the surface or by monitoring the surface tension oscillation, which is the driving force of the waves. For the first approach, one can use a second bar in the path of the wave and detect the force acting on it by the wave motion [15]. Tracer particles may be placed on the surface whose movement is monitored with a video camera [16]. In the second approach, a Wilhelmy plate [17] or a capillary wave detection system [18] discussed above may be used.

IV. RESULTS WITHOUT MONOLAYERS

A. Simple Liquids

In the absence of the surface film, the dispersion equation Eq. 2 becomes simpler (a seventh-order polynomial). In the ordinary simple liquid of low viscosity, e.g., water, there is only one solution to the polynomial at low frequencies. The surface tension and the shear viscosity obtained from the capillary wave measurements agree quite well with those measured independently. Here, nothing is to be noted in particular.

When the viscosity is high, however, another mode comes into play. In order to facilitate the calculation and to see the two modes, let us ignore the gravity acceleration for the moment. Then the dispersion relation can be completely parameterized in terms of two variables,

$$Q = \frac{k}{(\omega\rho/\eta)^{1/2}} \qquad \text{and} \qquad P = \left(\frac{\omega\eta^3}{\rho\sigma^2}\right)^{1/2} \tag{3}$$

The result of a numerical calculation [19] is shown in Fig. 5. The effect of gravity does not change the overall mode structure. As has been mentioned in Sec. II, the second mode is overdamped. It is, in fact, a bulk viscous shear wave radiated from the surface into the liquid. Because the bulk viscous shear wave has a dispersion relation [20]

$$k = (i + 1)\left(\frac{\omega\rho}{2\eta}\right)^{1/2} \tag{4}$$

---

[4]One exception is to generate a surface compressional pulse using dyes, in which a dye-containing monolayer is irradiated with a flash of light. The sudden expansion of the occupying area of the dye molecules caused by the photoisomerization creates a pressure impulse in the surrounding film [14].

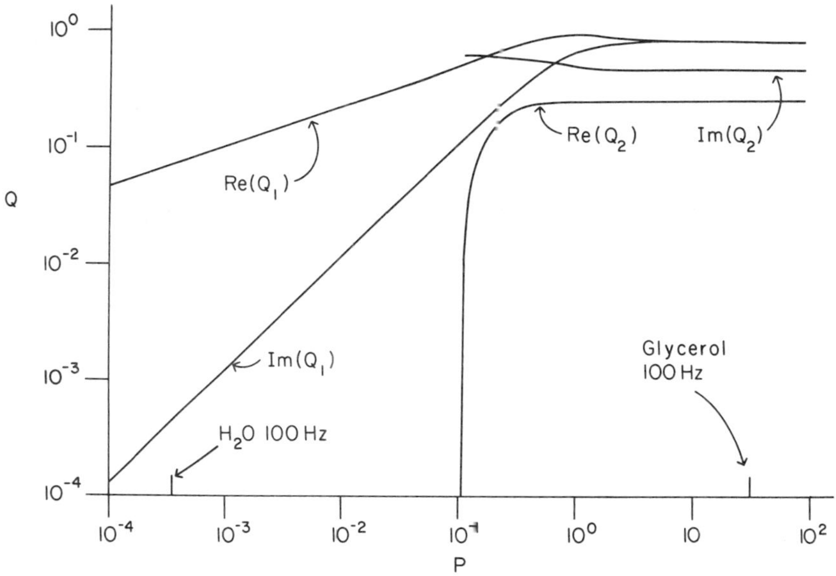

FIG. 5 Numerical solutions to Eq. 2 for a simple viscous liquid. $Q = k/(\omega\rho/\eta)^{1/2}$ and $P = (\omega\eta^3/\rho\gamma^2)^{1/2}$, so that $\mathrm{Re}(Q) \sim k_r$ and $\mathrm{Im}(Q) \sim k_i$. For a high P value, there are two solutions. (From Ref. 19.)

Q becomes independent of P. Although it is not a true surface wave, the second mode has to be taken into account in analyzing the experimental results. For example, results of a surface wave experiment on glycerol are shown in Fig. 6. Here, the lock-in amplifier output is shown as a function of the propagation distance. Two traces are in quadrature in order to improve the reliability of the data. Two modes have to be considered to fit both curves simultaneously [19].

B. Anisotropic Liquids (Liquid Crystals)

Liquid crystals show anisotropic viscosities when the molecules are aligned (see II.C). Particular combinations of the shear viscosities appear in the dispersion relation depending on the relative orientation of the director and the wave vector, but the rest of the calculation essentially remains the same as in the case of a simple liquid. Although there is some experimental complication in that the apparatus has to fit between the poles of a magnet [8], measurements are straightforward. In the smectic phase, liquid crystals are quite viscous and traces are just like those shown in Fig. 6. But no problem is experienced in analyzing the data.

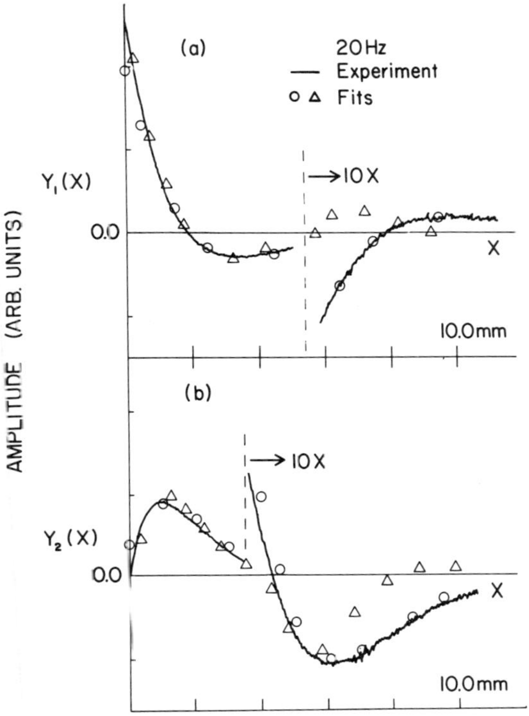

FIG. 6 Propagation of surface waves on glycerol. Triangles: the best fit to the data when only one propagating mode is assumed; circles: when an additional overdamped mode is also taken into account. Two traces are in quadrature with each other. (From Ref. 19.)

An example is shown in Fig. 7, in which the measured surface tension and the combinations of the shear viscosities relevant for the particular geometry are displayed. The expressions of the viscosities used here in terms of more conventional definitions of viscosity coefficients are given in Ref. 21. The viscosity η_3' diverges as the smectic-nematic transition is approached from the nematic side (a critical divergence) while others remain nearly constant. Note also that the surface tension *decreases* as the temperature is lowered around the transition point.

Thermodynamically $d\gamma/dT$ is related to the surface excess entropy over the bulk S_A as [22]

$$S_A = -\frac{d\gamma}{dT} \tag{5}$$

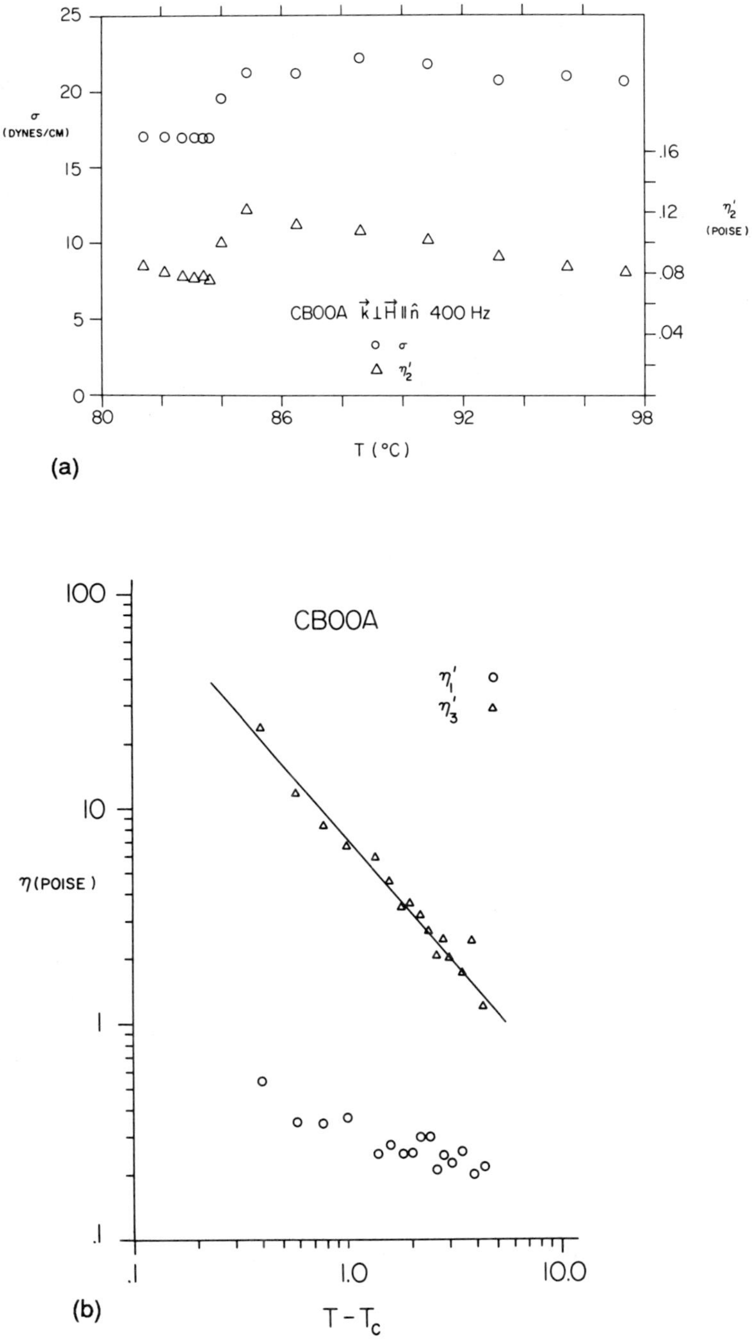

FIG. 7 Surface tension and the viscosities of a liquid crystal CBOOA around the nematic-to-smectic-A phase transition point ($T = T_c$).

Because the surface of a liquid is less ordered than the bulk, it is expected that $d\gamma/dT$ is negative, and indeed that is the case in ordinary liquids. The positive temperature coefficient, thus, signals the appearance of an ordered surface layer on a less ordered bulk liquid. This unusual behavior has been found at the smectic A to B transition also [23].

V. RESULTS WITH MONOLAYERS

A monolayer on the surface of a liquid affects the propagation of capillary waves in two ways. Firstly, it reduces the surface tension γ; secondly, an additional tangential force, the (complex) uniaxial dilational modulus ε, comes into play. In the absence of dissipative processes, ε is real and is given in terms of the area A by

$$\varepsilon = \varepsilon_r = \frac{d\gamma}{d \ln A} \tag{6}$$

which simply expresses the fact that a force has to be exerted to stretch or squeeze an elastic film. ε can be determined independently from equilibrium measurements such as the $\pi - A$ diagram obtained from the Langmuir trough experiment through the relation $\varepsilon = -d\pi/d \ln A$, because π is the surface tension of bare water surface minus that of the film-covered surface. But ε thus obtained can be different from that measured with the capillary wave technique.

It is often experienced that the $\pi - A$ curve is dependent on the rate of compression. Usually, the higher the rate, the larger ε becomes. Thus the modulus the capillary wave sees in the time scale short compared with that relevant for the $\pi - A$ measurement can be higher. The accompanying relaxation can introduce an imaginary part to Eq. 6.

One may also imagine a ubiquitous dissipation process due to the frictional motion of the molecules, the viscosity. In two dimensions, uniaxial compression is accompanied by the viscosity of the form $\varepsilon_i = \omega(\kappa + \mu)$, where κ is the surface dilational viscosity and μ is the surface shear viscosity. (For all possible types of motions in monolayers, see Ref. 24.) The shear viscosity μ can be separately measured with a deep channel viscometer, for example, and the value is small. On the other hand, the dilational viscosity κ is *not* usually measured as such but interchangeably used as a quantity representing the imaginary part of ε in Eq. 8 below, the adsorption-desorption process, although in this practice $\omega\kappa$ does not imply any frequency dependence. Two processes are physically distinct [25], and one should be careful in reading the literature.

A dissipative process of great importance is the adsorption-desorption of the molecules between the monolayer and the bulk liquid. If the molecules are soluble, γ depends only on the bulk concentration in equilibrium and hence Eq. 6 is always zero. However, it takes time to establish an equilibrium after a sudden area change, and this time lag contributes as a dissipation as well as a reactive response. Using the elasticity ε_0 in terms of the surface excess G,

$$\varepsilon_0 = -\frac{d\gamma}{d \ln G} \tag{7}$$

instead of Eq. 6, the effective dilational modulus is well approximated by [3]

$$\varepsilon = \frac{\varepsilon_0[(1 + r) - ir]}{1 + 2r + 2r^2} \tag{8}$$

Here, r is the rate of the molecular exchange ω_0 between the surface and the bulk relative to the wave frequency ω,

$$r = \left(\frac{\omega_0}{\omega}\right)^{1/2} \tag{9}$$

and ω_0 is given by

$$\omega_0 = \frac{D}{2(dc/dG)^2} \tag{10}$$

where c is the bulk concentration and D is the diffusion constant. One of the motivations in the externally generated surface wave experiments has been to understand this absorption-desorption process, which is hard to study by other means. Especially the low frequency obtainable in this technique is most suited for studying the slow diffusion process. But we shall start with the case of negligible dissolution.

A. Case of Negligible Dissolution

Spread monolayers all fall in this category. Even when the molecules are soluble into the liquid, sometimes it can be regarded as insoluble in the time scale of the capillary wave oscillation and thus belongs to this category. The elasticity modulus ε in this case is given by Eq. 6 with the caveat stated below the equation.

The most salient features in Fig. 1 are the peak in k_{iC} and the dip in k_{rC} at intermediate values of ε_r. Except for the gaseous state, however, spread monolayers usually exhibit some slope in the $\pi - A$ curve. By referring to Fig. 1, it is clear that, in most cases of spread monolayers, ε_r (Eq. 6) is significant and is above the range of values in which the peak and the dip are expected. Therefore, in the pioneering work by Lucassen and Hansen [26], the peak in k_{iC} was most clearly observed on *solutions* of octanoic acid and lauryl amine hydrochloride as a function of the concentration. Bendure and Hansen [27] subsequently studied *spread* monolayers in the intermediate range of ε_r and succeeded in observing both the dip in k_{rC} and the peak in k_{iC}. A quantitative agreement between the theory and the experiment was obtained for the case of distearyl dimethyl ammonium chloride, but for hexadecanol and tetradecanol the agreement was only qualitative. They attributed the lack of quantitative agreement in the latter substances to the "rigidity" of the films, although the precise meaning of the word was not offered. The result of distearyl dimethyl ammonium chloride is shown in Fig. 8.

The fact that some monolayers behave as if they are rigid at a very low pressure is now clearly understood. In a recent x-ray diffraction study from condensed monolayers [28, 29], the crystalline order was found in a number of monolayers at essentially zero pressure, although the details are still hazy at

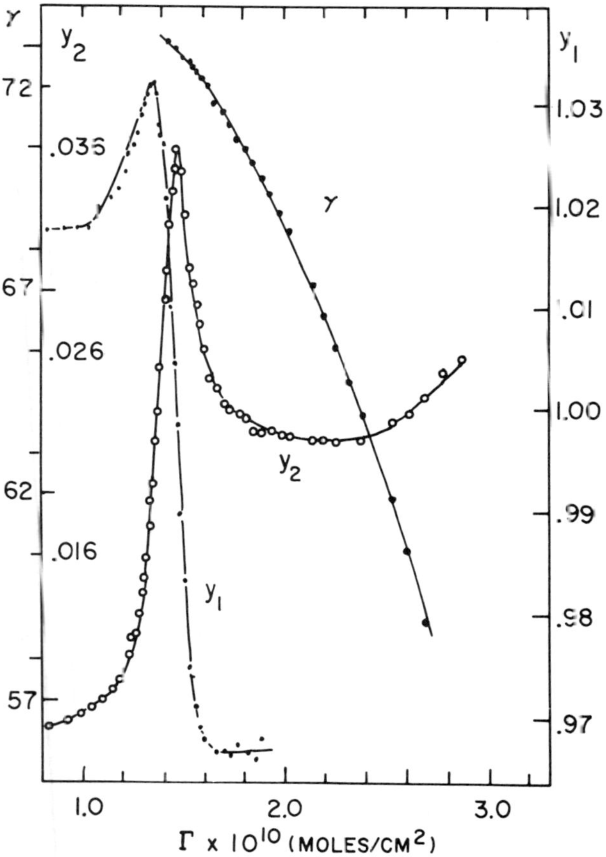

FIG. 8 Capillary wave propagation characteristics on distearyl dimethyl ammonium chloride monolayers. Compare with Fig. 1a. Note that γ = surface tension (mN/m), $y_1 = \rho\omega^2/\gamma k_r^3$, and $y_2 = k_i/k_r$, so that y_1 shows a peak rather than a dip. (Adapted from Ref. 27.)

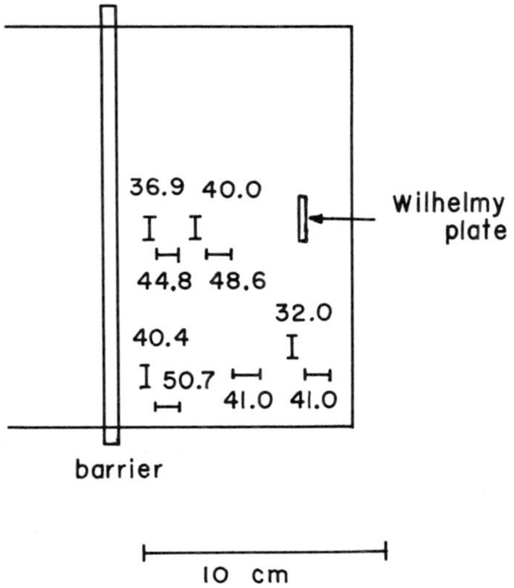

FIG. 9 The surface pressure distribution in a "rigid" monolayer (zinc tetra-*tert*-butylphthalocyanine) on water. The plan of the trough is shown, with a compression barrier and a Wilhelmy plate. The bars in the monolayer indicate the length and the direction of the paths of the capillary wave measurements. The accompanying numbers are the surface pressure calculated from Eq. 2 in mN/m. The pressure reading with the Wilhelmy plate was 26.6 mN/m, quite different from the capillary wave results.

extremely low density. An example of condensed monolayers is stearic acid at room temperature. In the zero-pressure region ($A > 25$ Å^2/molecule), the film is considered, from the above observation, to be in a solid-gas coexistence phase. In this case, although the surface tension γ may be constant throughout the inhomogeneous surface, ε is not. This is easily demonstrated by searching over the film covered surface [30]. Indeed, the capillary wave propagation characteristics vary between that of an almost bare surface and that of a film with a very large ε_r, i.e., almost the limiting case of $\varepsilon \to \infty$, but the same γ. Note that a large fluctuation in the light scattering data taken at a fixed location is also frequently encountered in a similar situation.

The search of inhomogeneity in the monolayers on water using capillary waves has more important consequences in the application of monolayers, i.e., in the Langmuir-Blodgett (LB) technique. In the LB technique, highly compressed monolayers are transferred onto a solid support. The transfer is accomplished by the flow of the material toward the solid support driven by the pressure gradient in the film. However, the monolayers are usually in the solid phase, which may not easily deform under an external stress. This seems to be the major cause of the failure of the transfer. In Fig. 9 it is shown that a solid film cannot deform easily and thus can support considerable surface pressure inhomogeneity [30]. It is to be noted that, since k is a vectorial quantity, the pressure anisotropy can be studied as well. The surface pressure π and hence the surface tension γ is inhomogeneous and anisotropic here.

Now, the usual definition of the surface tension being the energy per unit area seems rather strange, because surface energy should be a scalar. It is thus clear that the solid monolayer should be treated as a separate entity (an independent elastic slab), not just as something that modifies the liquid surface property. Although this notion is not new at all (Langmuir discussed that even a liquid monolayer must be considered in terms of hydrocarbon-tail/air interface and hydrophylic-head/water interface [31]), it has not been seriously pursued experimentally.

When the monolayer is solid, the uniaxial dilational modulus ε_r should be the sum of the dilational modulus $d\gamma/d \ln A$ and the shear modulus, just as in the case of the surface viscosity. Although the shear modulus can be fairly large in some cases [32], its contribution to the capillary wave has never been studied experimentally.

B. Case of Considerable Dissolution

When the adsorption-desorption of molecules to and from the monolayer occurs considerably during one cycle of surface wave oscillation, Eq. 8 has to be considered. Its imaginary part ε_i seriously reduces the ε_r dependence of k, as shown in Fig. 1. This is precisely what was observed experimentally [33], in which a series of fatty acids of chain length ranging from 5 carbons to 8 was investigated. As the chain length reduces, the rate of exchange gets faster, leading to the increase in ε_i in Eq. 8. The diffusion constant D for pentanoic acid and hexanoic acid was about 4×10^{-6} cm^2/s. Their behavior was considered to be normal, and no evidence for diffusion barrier between the surface and the bulk nor any anomalous dissipative process was found. The diffusion constant of octanoic acid was 6×10^{-5} cm^2/s, which was considered to be "wrong." The film behaves as insoluble at the frequency of the experiment (~200 Hz) and thus the analysis of the small effect of diffusion is subject to a large error.

An alternative approach to the diffusion problem is to generate the longitudinal wave and observe its propagation. Because the longitudinal wave is the density oscillation of the surfactant molecules at the interface, this mode is the natural choice for studying the diffusion problem. A qualitative feature of the results of a longitudinal wave study [10] is presented in Fig. 10.

A variation of the longitudinal wave technique is to cause homogeneous density change throughout the trough by oscillating barriers, which corresponds to the longitudinal waves in the long wavelength limit. Based on the earlier theoretical considerations [34], Lucassen and Giles found [35] that, in the nonionic surfactants below cmc, the data were consistent with the equation of state

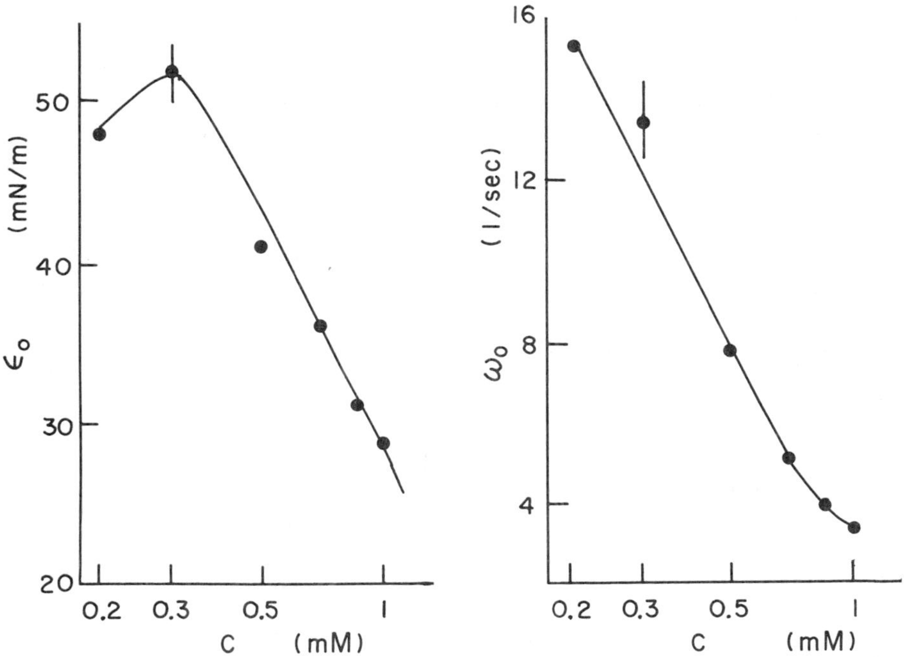

FIG. 10 The instantaneous elastic modulus ε_0 and the diffusion rate ω_0 of the adsorbed film from a sodium dodecyl sulfate solution at the decane/water interface. The maximum in ε_0 and the steady decline in ω_0 as a function of concentration c are general behaviors of adsorbed monolayers. (Adapted from Ref. 10.)

(the Frumkin isotherm) and a simple diffusion molecular exchange model between the surface and the bulk. The diffusion coefficients were reasonable.

The problem of soluble monolayers is, however, far from being settled. It was soon after argued [36] that the Frumkin isotherm is inadequate to describe the data and that too much correction needs to be added. By combining the results of the externally generated capillary waves and those of light scattering, which covers the frequency range of 100 Hz to more than 10 kHz, Stenvot and Langevin [37] reached a similar conclusion.

To determine the equation of state experimentally is still an outstanding problem. The time scale change that occurs from the low to the high concentration regime is many orders of magnitude, while the time scale covered with a single experimental arrangement is probably two orders of magnitude at most. What is lacking is the quantitative study in a wide frequency range, and for this purpose a simultaneous measurement employing the light scattering technique and the externally generated wave technique would be quite valuable.

ACKNOWLEDGMENTS

Work presented here was supported in part by a Grant-in-Aid for Scientific Research from the Ministry of Education, Science, and Culture and by a Grant for International Joint Research Project from the NEDO, Japan.

REFERENCES

1. R. S. Hansen and J. Ahmad in *Progress in Surface and Membrane Science* (J. F. Danielli, M. D. Rosenberg, and D. A. Cadenhead, eds.) Academic Press, New York, 1971, pp. 1–56.
2. R. S. Hansen and J. A. Mann, Jr., *J. Appl. Phys. 35:* 152 (1964).
3. E. H. Lucassen-Reynders and J. Lucassen, *Adv. Colloid Interface Sci. 2:* 347 (1969).
4. C. J. Earnshaw, R. C. McGivern, and P. J. Winch, *J. Phys. (Les Ulis, France 49:* 1271 (1988).
5. J. Lucassen, *Trans. Faraday Soc. 64:* 2221 (1968).
6. L. Ting, D. T. Wasan, K. Miyano, and S.-Q. Xu, *J. Colloid Interface Sci. 102:* 248 (1984).
7. E. Bock and J. A. Mann, *J. Colloid Interface Sci. 129:* 501 (1989).
8. C. H. Sohl, K. Miyano, and J. B. Ketterson, *Rev. Sci. Instrum. 49:* 1464 (1978).
9. C. H. Sohl, K. Miyano, J. B. Ketterson, and G. Wong, *Phys. Rev. A22:* 1256 (1980).
10. L. Ting, D. T. Wasan, and K. Miyano, *J. Colloid Interface Sci. 107:* 345 (1985).
11. K. Ito, B. B. Sauer, R. J. Skarlupka, M. Sano, and H. Yu, *Langmuir 6*: 1379 (1990).
12. S. T. Boldarev and V. P. Peshkov, *Soviet Phys. JETP 17:* 297 (1974).
13. N. Nagarajan, W. W. Webb, and B. Widom, *J. Chem. Phys. 77:* 5771 (1982).
14. M. Suzuki, D. Möbius, and R. Ahuja, *Thin Solid Films 138:* 151 (1986).
15. J. Lucassen, *Trans. Faraday Soc. 64:* 2230 (1968).
16. A. F. Snik, W. J. J. Kouijzer, J. C. M. Keltjens, and Z. Houkes, *J. Colloid Interface Sci. 93:* 301 (1983).
17. J. Lucassen and M. van den Tempel, *Chem. Eng. Sci. 27:* 1283 (1972).
18. K. Miyano, B. M. Abraham, L.Ting, and D. T. Wasan, *J. Colloid Interface Sci. 92:* 297 (1983).
19. C. H. Sohl and K. Miyano, *Phys. Rev. A20:* 616 (1979).
20. L. D. Landau and E. M. Lifshitz in *Fluid Mechanics,* Pergamon, London, 1959, Chap. 2.
21. M. J. Stephen and J. P. Straley, *Rev. Mod. Phys. 46:* 661 (1974).
22. A. W. Adamson in *Physical Chemistry of Surfaces,* Wiley, New York, 1976, Chap. 2.
23. K. Miyano, *Phys. Rev. A26:* 1820 (1982).
24. F. C. Goodrich, *J. Phys. Chem. 66:* 1858 (1962).
25. J. A. Mann and G. Du, *J. Colloid Interface Sci. 37:* 2 (1971).
26. J. Lucassen and R. S. Hansen, *J. Colloid Interface Sci. 22:* 32 (1966).
27. R. L. Bendure and R. S. Hansen, *J. Phys. Chem. 71:* 2889 (1967).

28. K. Kjaer, J. Als-Nielsen, C. A. Helm, P. Tippman-Krayer, and H. Möhwald, *J. Phys. Chem. 93:* 3200 (1989).
29. P. Dutta, J. B. Peng, B. Lin, J. B. Ketterson, M. Prakash, P. Georgopoulos, and S. Ehrlich, *Phys. Rev. Lett. 58:* 2228 (1987).
30. K. Miyano, *Langmuir 6*: 1254 (1990).
31. I. Langmuir, *J. Chem. Phys. 1:* 756 (1933).
32. B. M. Abraham, J. B. Ketterson, K. Miyano, and A. Kueny, *J. Chem. Phys. 75:* 3137 (1981).
33. J. Lucassen and R. S. Hansen, *J. Colloid Interface Sci. 23:* 319 (1967).
34. J. Lucassen and M. van den Tempel, *J. Colloid Interface Sci. 41:* 491 (1972).
35. J. Lucassen and D. Giles, *J. Chem. Soc., Faraday Trans. I 71:* 217 (1975).
36. E. H. Lucassen-Reynders, J. Lucassen, P. R. Garrett, D. Giles, and F. Hollway, *Adv. Chem. Ser. 144:* 272 (1975).
37. C. Stenvot and D. Langevin, *Langmuir 4:* 1179 (1988).

17

Light Reflectivity and Ellipsometry

JACQUES MEUNIER Laboratoire de Physique Statistique*, Ecole Normale Supérieure, Paris, France

Optical Techniques such as ellipsometry and reflected intensity measurement are well adapted to the study of flat liquid interfaces since they introduce very little perturbation. Light is an interesting choice of radiation because many liquids are reasonably optically transparent. X-rays or neutrons can take the place of

*URA 1306 du CNRS, associée aux Universités Paris VI et VIII.

light, giving more information at small scale because of their small wavelengths. However, x-rays have the disadvantage of being strongly absorbed by liquids. Neutron beams tend to have weak flux and a short scattering length, but they have one advantage: the contrast can be varied by isotopic substitution. The use of these two last types of radiation is limited to the free surface of liquids or interfaces covered with only a thin (about 0.1–10 μm) liquid layer.

Optical techniques give information about the variation of the refractive index $n(x, y, z)$ through the interface and of its anisotropy. A common model for plane liquid interfaces makes a clear distinction between the roughness, i.e., the vertical position $\zeta(\mathbf{r}, t)$ at time t of the point of the interface $[\mathbf{r}(x, y), \zeta]$, above its average value $\bar{\zeta} = 0$, and the structure or thickness of the interface, i.e., the variation of the refractive index along the vertical axis z at time t: $n[z - \zeta(t, \mathbf{r})]$ (Fig. 1a). The thickness h of the interface is the width of the domain where the derivative dn/dz does not vanish. The function n is supposed to be a characteristic of the interface, independent of the roughness ζ, the time t, and the point $\mathbf{r}$, at least if the slope of the roughness is small ($|\nabla\zeta| \ll 1$). This model

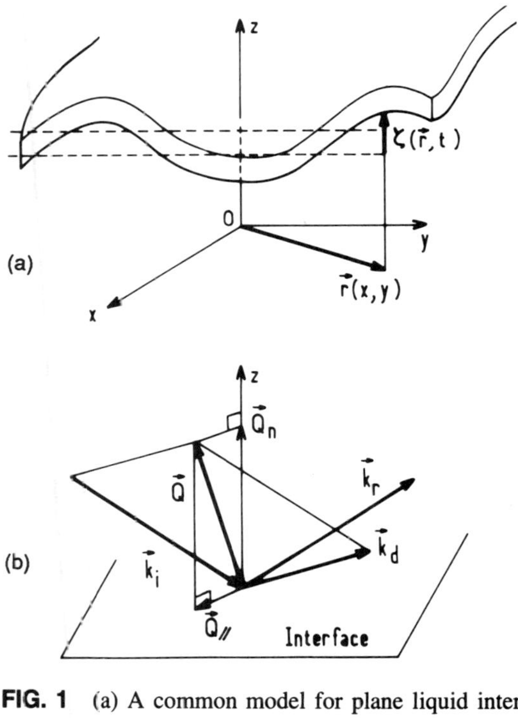

FIG. 1 (a) A common model for plane liquid interfaces: roughness and thickness are independent. (b) The different wavevectors in a surface scattering experiment.

clearly applies to amphiphilic monolayers at liquid interfaces. In this case, the origin of the roughness of the monolayer at a scale larger than the molecular scale is thermal fluctuations, while the main contribution to the interfacial thickness is the thickness of the monolayer of amphiphilic molecules. The thermal fluctuations of the surface density in the monolayer are neglected: they are too small to be observed by optical experiments [1] (Chap. 3). This model has been applied to the free surface of liquids and to critical interfaces: in these cases, the distinction between roughness and thickness is not clear and is more artificial than for monolayers.

In an optical experiment, the interface is illuminated with a beam of wavevector $\mathbf{k}_i$, and the light of wavevector $\mathbf{k}_d$, scattered by the interface, is analyzed in intensity, polarization, or frequency according to the experiment. The scattering vector $\mathbf{Q} = \mathbf{k}_d - \mathbf{k}_i$ is the sum of a normal component $\mathbf{Q}_n$, which probes the interface along its normal, and a component parallel to the interface $\mathbf{Q}_{\parallel}$, which probes the interface in a direction parallel to the interface (Fig. 1b).

In a surface light scattering experiment on a liquid interface (a technique developed since 1968 [2, 3]), the interfacial roughness is much smaller than the wavelength of the light ($\zeta/\lambda \ll 1$). Only first-order scattering processes are observed. Each mode (or Fourier component) of the roughness of wavevector $\mathbf{q}$ and amplitude ζ_q behaves like a phase grating of small amplitude ($\zeta_q/\lambda \ll 1$) at the interface and scatters the light. The component parallel to the interface of the scattering vector $\mathbf{Q}$ is given by

$$\mathbf{Q}_{\parallel} = \mathbf{q}$$

If $|(\mathbf{k}_d)_{\parallel}|\lambda < 2\pi$, the component normal to the interface of the wavevector of the scattered light, $(\mathbf{k}_d)_n$, is real. The scattered wave propagates and can be observed at a large distance from the interface (the index $\parallel$ indicates the component parallel to the interface and n indicates the component normal to the interface).

If $|(\mathbf{k}_d)_{\parallel}|\lambda > 2\pi$, the scattered wave is evanescent and cannot be observed far from the interface.

The roughness at small scale cannot be observed by this technique because it produces evanescent waves. The intensity of the light scattered by roughness at large scale ($q\lambda < 1$) contains information on the amplitude ζ_q of the scattering mode and on the thickness of the interface; it can be affected by multiple scattering (if $\zeta/\lambda \sim 1$) and bulk scattering. It is not easy to take advantage of the intensity measurement, and the technique is rarely used with light for liquid interfaces [4]. A spectral analysis of the scattered light is required most of the time: the spectra corresponding to the different scattering mechanisms are very different, and the first-order surface scattering can be easily distinguished. The spectrum corresponding to the time evolution of a surface mode ζ_q gives information on the parameters acting on this evolution: the restoring forces and the viscous forces [5]. Such scattering processes can be interpreted with two conservation laws:

$\mathbf{Q}_{\parallel} = (\mathbf{k}_d - \mathbf{k}_i)_{\parallel} = \pm\mathbf{q}$ (the total horizontal momentum is preserved when one surface phonon of momentum $\mathbf{q}$ interacts with an incident photon)

$\nu_d = \nu_i \pm \nu_q$ (energy conservation: the energy of the incident photon is increased or decreased from the energy of the annihilated or created surface photon)

Theoretically, modes such as $q < 10^5\ \text{cm}^{-1}$ can be studied with this technique, using a light beam. In practice, except for a difficult experiment performed with a Fabry-Perot [2], the heterodyne technique commonly used is limited to q smaller than $10^4\ \text{cm}^{-1}$, the scattered intensity being too weak for larger q. Modes of q higher than $10^5\ \text{cm}^{-1}$ are typically those that could be explored by x-ray or neutron scattering, but in practice the refractive indices for these two types of radiations are close to one, and the scattered intensity by the surface roughness of liquid interfaces is too low to be observed. A method to avoid this limitation is to use a multiple scattering experiment that can fulfill the equation $Q_{\parallel}\lambda < 1$ (required to obtain propagating scattered waves) with surface modes of wavelength smaller than the wavelength of the light. For instance, in the case of a scattering of order m, the light is successively scattered by the modes $\mathbf{q}_1, \mathbf{q}_2, \ldots, \mathbf{q}_m$:

$$\mathbf{Q}_{\parallel} = \mathbf{q}_1 + \mathbf{q}_2 + \cdots + \mathbf{q}_m$$

The horizontal component $\mathbf{Q}_{\parallel}$ of the scattering vector has a small modulus if the momenta $\mathbf{q}_1, \mathbf{q}_2, \ldots \mathbf{q}_m$ point in opposite directions. The simplest case, the subject of this chapter, is obtained by reflectivity measurements for which the scattering vector $\mathbf{Q}$ is normal to the interface ($\mathbf{Q}_{\parallel} = 0$ and $\mathbf{k}_d = \mathbf{k}_r$). The multiple scattering processes are observable in two cases. First, when the processes of zero and first order vanish. This is the case for ellipsometry or reflectivity measurements at the Brewster angle. Second, if the amplitude of the roughness is comparable with λ as is the case for x-rays. The experimental signal contains information about the interfacial thickness because $\mathbf{Q}$ is normal to the interface, and about the interfacial roughness at any scale because of the multiple scattering. It is more difficult to take advantage of the multiple scattering than the first-order scattering, because the roughness contribution to the multiple scattering results from a summation over all the surface modes.

The reflectivity of an interface is fully determined by a 2×2 matrix $r_{\alpha\beta}(\theta)$. The reflected field of polarization α, $\mathbf{E}_r^{\alpha\beta}$, for an incident field $\mathbf{E}_i^{\beta}$ of polarization β, is given by

$$E_r^{\alpha\beta} = r_{\alpha\beta}(\theta)\, E_i^{\beta} \tag{1}$$

where θ is the angle of incidence and α and β are equal to s or p depending on whether the electric field is perpendicular or parallel to the incident plane. $r_{\alpha\beta}(\theta)$ are the *reflectivity coefficients for the amplitude of the electric field*. They are complex numbers whose arguments indicate the phase retardation at the reflection. The matrix $r_{\alpha\beta}(\theta)$ is not fully determined in experiments; only some parameters related to this matrix are measured, according the required information.

The simplest experiments are the reflected intensity measurements allowing one to obtain the intensity reflectivities, $\mathrm{R}_{\alpha\beta}(\theta) = |r_{\alpha\beta}(\theta)|^2$. They are positive real numbers: the phase change at the reflection is lost. For a Fresnel interface, i.e., an interface without thickness and without roughness (the refractive index changes in one step, from n_1 to n_2 at $z = 0$), $\mathrm{R}_{\alpha\beta}(\theta) = 0$ if $\alpha \neq \beta$ and $R_{\alpha\alpha}(\theta) = \mathrm{R}^{\mathrm{F}}_{\alpha\alpha}(\theta)$, where $\mathrm{R}^{\mathrm{F}}_{\alpha\alpha}(\theta)$ are the Fresnel reflectivities for the s and p polarizations. Away from the Brewster angle, the parameters $R_{\alpha\alpha}(\theta) = \mathrm{R}_{\alpha\alpha}(\theta)/\mathrm{R}^{\mathrm{F}}_{\alpha\alpha}(\theta)$, obtained from the diagonal terms, are convenient. The deviation from one contains information about the structure and the roughness of the interface. It is significantly different from one only if the interfacial thickness or roughness is not too much smaller than the wavelength of the radiation. For this reason, except for very thick interfaces such as critical ones (Sec. II.F), this technique requires x-rays or neutron beams. The off-diagonal terms $\mathrm{R}_{\alpha\beta}$ ($\alpha \neq \beta$) are more sensitive to the interfacial properties than $R_{\alpha\alpha}(\theta)$ because they measure a deviation from zero. They can give information with light for interfaces as thin as monolayers (Sec. II.C) while the study of $R_{\alpha\alpha}(\theta)$ for the same systems requires x-rays.

An old but more sophisticated technique is ellipsometry, which allows the determination of the ratio $\rho_e = [r_{pp}(\theta) \cos \Phi + r_{ps}(\theta) \sin \Phi]/[r_{ss}(\theta) \sin \Phi + r_{sp}(\theta) \cos \Phi]$, where Φ is an angle introduced in the experimental setup. The relative phase between the matrix elements is measured by this technique; ρ_e is a complex number. This ratio is positive for $\theta = 0$ (at this angle the two polarizations are equivalent: $\rho_e = 1$) and negative for $\theta = \pi/2$ because of a phase change π between the two polarizations of the reflected field at grazing incidence. For a Fresnel interface, this ratio vanishes at the Brewster angle θ_B, where tg $\theta_B = n_2/n_1$: the dipoles induced by the incident field, of polarization p, point in the direction of observation and do not radiate in this direction. For a real interface this ratio ρ_e is complex; its real component vanishes at the Brewster angle but its imaginary component remains finite. This imaginary component is called the ellipticity $\bar{\rho}_B$ and is very sensitive to the interfacial structure and roughness because it is a deviation from zero. For thick interfaces (thickness + roughness $\sim \lambda$), the change in ρ_e with respect to Fresnel interface value is measurable over a large range of θ. Much more information can be obtained in this case, but most liquid interfaces are much thinner than the wavelength of the light. For this reason, the discussion of ellipsometry in this chapter is restricted to the measurement of $\bar{\rho}_B$. In this case, ellipsometry reduces to the measurement of the imaginary part of $\rho_e = [r_{pp}(\theta) \cos \Phi + r_{ps}(\theta) \sin \Phi]/r^F_{ss}(\theta) \sin \Phi$ close to the Brewster angle; in most cases $r_{ps}(\theta) = 0$ and ellipsometry is an easy method to measure the reflectivity coefficient $r_{pp}(\theta_B)$ at the Brewster angle. There is no equivalent technique for x-rays, for which the refractive indices are close to 1 and the Brewster angle is $\pi/4$. At this angle, the intensity of the reflected beam by surfaces is much weaker than the radiation scattered by the bulk and cannot be observed.

Many measurements of the reflectivity of liquid interfaces have been performed in the past, most of them at the Brewster angle (ellipsometry). In most cases, the measured difference from Fresnel reflectivity was interpreted as the result of a thickness h of the interface [6]. In some cases, the roughness was taken into account through an increase of thickness (7)] which becomes the sum of the intrinsic thickness h and the average roughness $(\bar{\zeta^2})^{1/2}$. This model fails for the reflectivity measurements at the Brewster angle, as was recently proved [8, 9]. The aim of Sec. I is to summarize the different parameters of the interface that affect the reflectivity. The difference between the roughness contribution to the reflectivity at and out of the Brewster angle is explained in detail. The roughness term is explicitly given for monolayers at liquid interfaces. In the following section, Sec. II, we give a small number of experimental examples where the different contributions to the reflectivity can be separated, while Sec. II.F concerns the critical points, where this separation is not possible.

I. THEORETICAL SECTION

A. Reflectivity of Interfaces

In order to simplify this section, we consider, first, the case of thick and flat interfaces (no roughness), second, the case of rough and infinitely thin interfaces (no thickness), and finally some cases of thick and rough interfaces. The calculations are limited to the parameters easily and commonly measured: the intensity reflectivity and the ellipticity $\bar{\rho}_B$ at the Brewster angle.

1. Thick Interfaces

The reflectivities $R_{\alpha\alpha}(\theta)$ of a planar, thick, isotropic interface, called $R^h_{\alpha\alpha}(\theta)$, are easy to deduce from $n(z)$, considering the interface as a stratified interface and using the Abelès 2×2 matrix method [7]. For most of the practical cases (critical interfaces with light, free surfaces with x-rays or neutron beams), the index variation is so small that, except in the vicinity of the critical total reflection angle, multiple reflections in the interface can be neglected, as can the change in direction of propagation of the incident beam and its attenuation into the interface; the reflectivity is

$$R^{h}_{\alpha\alpha}(\theta) = \left| \frac{f(\theta)}{2\bar{n}} \int_{-\infty}^{+\infty} \frac{dn}{dz} \exp[-i\bar{Q}z]\, dz \right|^2 \tag{2}$$

with $\bar{n} = (n_1 + n_2)/2$, $\bar{Q} = (4\pi/\lambda)\,\bar{n}\cos(\theta)$, and $f(\theta)$ the polarization factor: $[1 + \tan^2(\theta)]$ for the s polarization and $[1 - \tan^2(\theta)]$ for the p polarization.

This last approximation fails at the Brewster angle (tg $\theta_B = 1$), where the change in the direction of the beam as it passes through the interface has to be taken into account. The general 2×2 matrix method can be applied, but for thin interfaces a formula for the ellipticity $\bar{\rho}^{t}_{B}$ to first order in$1/\lambda$ was calculated by Drude [10]:

$$\bar{\rho}^{h}_{B} = \frac{\pi}{\lambda} \frac{\sqrt{n_1^2 + n_2^2}}{n_1^2 - n_2^2} \eta^{h}_{B} \qquad \text{with}$$

$$\eta^{h}_{B} = \int_{-\infty}^{+\infty} \frac{[n_1^2 - n(z)^2]\,[n_2^2 - n(z)^2]}{n(z)^2}\, dz \tag{3}$$

from which we deduce

$$r^{h}_{pp} = r^{F}_{ss}\, \bar{\rho}^{1}_{B} \tag{4}$$

The ellipsometric parameter η^{h}_{B} term can be negative or positive. A positive value is found when the refractive index $n(z)$ in the interfacial layer takes on values either smaller or greater than the indices, n_1 and n_2, of both mediums. The Drude formula supposes that the matter in the interface is optically isotropic, i.e., it is characterized by only one refractive index $n(z)$.

The case of an optically anisotropic interface is usual when there is a monolayer at the interface. For a Fresnel interface between optically isotropic media, $R^{F}_{sp} = R^{F}_{ps} = 0$. For an interface having an index anistropy in the interfacial layer, R_{sp} and R_{ps} do not vanish. The observations of the depolarized reflect light are very sensitive to the interfacial index anisotropy because of the deviation from zero. In this case, light can give information on interfacial index anisotropy of interfaces as thin as monolayers. The calculation can be made using a 4×4 matrix method [7], a generalization of the Abelès method. In this section, we only consider the case of a homogeneous biaxial interfacial layer of thickness h and principal axes OX, OY, OZ such that OY is in the plane of the interface and OZ is at an angle Θ from the normal Oz to the interface. The corresponding refractive indices are n_X, n_Y, n_Z. In the case where the interfacial layer is made of small domains and is observed by ellipsometry on a large surface, all the directions OY are equally probable and one finds an average value:

$$\eta_B^t = \frac{n_X^2 n_Z^2 + [n_Y^2 - 2(n_1^2 + n_2^2)][n_X^2 \sin^2(\Theta) + n_Z^2 \cos^2(\Theta)] + 2n_1^2 n_2^2}{2[n_X^2 \sin^2(\Theta) + n_Z^2 \cos^2(\Theta)]} h \tag{5}$$

If the interfacial layer is uniaxial, $n_X = n_Y = n_o$ and $n_Z = n_e$; n_o and n_e are the ordinary and extraordinary indices. In addition, if the optical symmetry axis is normal to the interface ($\Theta = 0$) [11],

$$\eta_B^h = \frac{(n_1^2 - n_e^2)(n_2^2 - n_e^2)}{n_e^2} h + (n_o^2 - n_e^2) h \tag{6}$$

and if the optical symmetry axis is in the plane of the interface ($\Theta = \pi/2$) [12],

$$\eta_B^h = \frac{(n_1^2 - n_o^2)(n_2^2 - n_o^2)}{n_e^2} h + \frac{(n_e^2 - n_o^2)}{2} h \tag{7}$$

In these last two cases, the ellipsometric parameter η_B^h can be seen as the sum of a thickness term calculated with the Drude formula, using the refractive index for an electric field normal to the interface, and a correction term due to the anisotropy:

$$\eta_B^h = (\eta_B^h)_D + (\eta_B^h)_A \tag{8}$$

These two terms can easily have the same order of magnitude.

2. Rough Interfaces

In the case of a surface involving only roughness at large scale ($q\lambda \ll 1$), with a small slope ($q|\zeta_q| \ll 1$), each element dS of the interface acts as a horizontal plane interface reflecting the light with a phase factor $\exp(-iQ_1\zeta)$:

$$r_{\alpha\alpha}^R(\theta) = r_{\alpha\alpha}^F(\theta) \frac{1}{A} \int_A \exp(-iQ_1\zeta)\, dA \tag{9}$$

Q_1 is the notation for the scattering vector Q in the incident medium 1. A is the area of the projection of the interface on the horizontal plane. If ζ is a Gaussian function [13],

$$|r_{\alpha\alpha}(\theta)|^2 = |r_{\alpha\alpha}^F(\theta)|^2 \exp(-Q_1^2 \overline{\zeta^2}) \tag{10}$$

In the case of roughness at small scale ($q\lambda \gg 1$), the electromagnetic wave propagates in a medium with a refractive index changing on a scale smaller than the wavelength λ and sees in first approximation an average index [14, 15].

$$n(z) = \langle n_1 [1 - Y(z - \zeta)] + n_2 Y(z - \zeta) \rangle_{x,y} \tag{11}$$

where $Y(z) = 1$ if $z > o$ and $Y(z) = 0$ if $z < o$. The reflectivity of the interface can be calculated as the reflectivity of a thick interface with a refractive index

$n(z)$. The reflected wave is a sum of plane waves reflected at different levels inside the interface. If ζ is a Gaussian function, the reflectivity is again given by Eq. 10.

The demonstration of the Eq. 10 neglects some effects limiting its validity. First, the local variation of the angle of incidence due to the slope of the roughness, $\nabla\zeta$, is not taken into account in the calculation of the interfacial reflectivity. It supposes that the variation of the Fresnel reflectivity $r^F_{\alpha\alpha}(\theta)$ with θ is small. This approximation is rough in the vicinity of the critical total refraction angle where the Fresnel reflectivity varies greatly for a small variation of the incidence angle. Second, the calculation supposes that the electromagnetic waves propagate as plane waves, without deformation, in the vicinity of the rough interface. This is valid when the field scattered by the rough interface can be neglected ($|n_1 - n_2| \ll n_1$), which is the case for interfaces close to a critical point [16–18] or for interfaces studied with x-rays [19–24].

Equations 9 and 10 fail for the p polarization in the vicinity of the Brewster angle. In this case, the incident plane wave induces dipolar moments in the medium, which are pointed in the direction of the specular reflection (at the Brewster angle, the direction of the transmitted light and the direction of the specular reflection are perpendicular). The Fresnel reflectivity vanishes: $r^F_{pp}(\theta_B) = 0$. The origin of the deviation from zero to this reflectivity for a rough interface is the deformation of the electromagnetic field, due to the scattered field, which induces dipolar moments, the directions of which are scattered around the direction of reflection. The reflected field results from a double scattering process: the field scattered at first order by a surface mode $\mathbf{q}$ is scattered a second time in the direction $\mathbf{k}$ by a surface mode $-\mathbf{q}$. Higher orders than two are neglected in the case $|\zeta_q|/\lambda \ll 1$ [25]. The scattering at first order is proportional to the component normal to the interface of the wavevector of the incident light $(\mathbf{k_i})_n$. For a double scattering corresponding to two first-order scatterings successively, the scattered field is proportional to $(\mathbf{k_i})_n$ $(\mathbf{k_d})_n$. If $q\lambda \ll 1$, the two components are real and electromagnetic waves scattered at second order have the same phase or the opposite phase as the Fresnel component (which vanishes at the Brewster angle): roughness at a scale larger than the wavelength only introduces a small change in the Brewster angle. If $q\lambda \gg 1$, the second component is imaginary, because the scattered wave at first order is evanescent and introduces a phase change of $\pi/2$: the reflected field has a minimum of amplitude at the Brewster angle but does not vanish when the interface has roughness at a scale smaller than the wavelength of the light. Ellipsometry at the Brewster angle measures this minimum of reflectivity and consequently is only sensitive to roughness at a scale smaller than the wavelength of the light.

The deformation of the field by roughness at very small scale ($q\lambda \gg 1$ and $|\zeta_q|/\lambda \ll 1$) is located close to the interface because the first-order scattered waves do not propagate. The phase retardation due to the finite propagation time of the light can be neglected; in this case, the solution for the deformation of the electromagnetic field is an electrostatic one with only one characteristic length: that introduced by the mode q [26]. The greater the index difference $(n_1 - n_2)$ and the interfacial slope $q\zeta_q$, the larger is the field deformation. The reflected field is proportional to the volume of the deformation(the volume of the space included between the rough interface and its average plane; this volume is proportional to ζ_q) and is a function of the deformation. At second and lowest order in ζ_q one obtains

$$\bar{\rho}_B^R \sim \frac{(n_1 - n_2)}{\lambda} q \, |\zeta_q|^2 \tag{12}$$

A summation of 12 over q gives the ellipticity for an isotropic rough interface with roughness at small scale:

$$\eta_B^R = -\frac{3}{2} \frac{(n_1^2 - n_2^2)^2}{n_1^2 + n_2^2} \sum_{q>2\pi/\lambda} q \, |\zeta_q|^2 \tag{13}$$

The ellipsometric coefficient for roughness is negative.

Equation 13 was recently given by Zielinska et al. [9] and Beaglehole [8]. It must be noted again that it is valid for small-scale roughness ($q\lambda \gg 1$) of small amplitude ($[\zeta/\lambda] \ll 1$) and weak slope ($[d\zeta/dx]^2 + [d\zeta/dy]^2 \ll 1$) and small wavelength on an isotropic interface. For any value of $q\lambda$, one can write

$$\eta_B^R = -\frac{3}{2} \frac{(n_1^2 - n_2^2)^2}{n_1^2 + n_2^2} \sum g(q) \, q \, |\zeta|_q^2 \tag{14}$$

Where $g(q)$ is a function such that $g(q) = 0$ if $q\lambda \ll 1$ and $g(q) = 1$ if $q\lambda \gg 1$. Its value for $q\lambda \sim 1$ can be deduced from the calculations of Croce [25]. An example is given Fig. 2 for the octane-water interface ($n_2 = 1.333$; $n_1 = 1.397$).

In conclusion, when $|n_1 - n_2| \ll n_1$, a reflected intensity measurement out of the Brewster angle is sensitive to the phase retardation, which depends only upon the depth of the interfacial roughness: $\overline{\zeta^2} = \Sigma_q |\zeta_q|^2$. Ellipsometry at the Brewster angle is sensitive to the field deformation in the vicinity of the interface, a function of the slope of the interfacial roughness: at lower order in ζ_q it is proportional to $\Sigma_q \, q|\zeta_q|^2$ if $q\lambda \gg 1$. Consequently, ellipsometry at the Brewster angle is more sensitive to small scale roughness than reflected intensity measurements out of the Brewster angle.

3. Rough and Thick Interfaces

For roughness of weak slope ($q\zeta_q \ll 1$), the function $n[z - \zeta(\mathbf{r})]$ is independent of x, y; i.e., the thickness and the roughness are independent. In this case the reflectivity of a rough and thick interface is given by Eq. 10, in which the

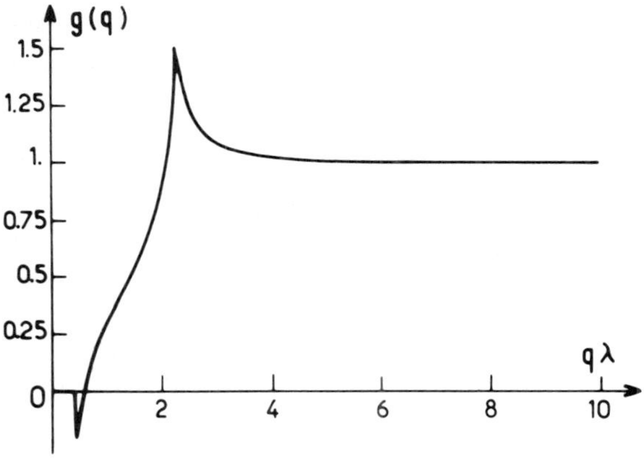

FIG. 2 The function $g(q)$ for the octane-water interface ($n_1 = 1.397$, $n_2 = 1.333$) and $\lambda = 6328$ Å.

reflectivity of the thick interface without roughness [$R^{h}_{\alpha\alpha}(\theta)$ Eq. 2] takes the place of the Fresnel reflectivity $|r^{F}_{\alpha\alpha}(\theta)|^2$:

$$R_{\alpha\alpha}(\theta) = R^{h}_{\alpha\alpha}(\theta)\exp(-Q_1^2\overline{\zeta^2}) \tag{15}$$

In the same way, for ellipsometry at the Brewster angle, the ellipticity due to the roughness and that due to the thickness are independent:

$$\bar{\rho}_B = \bar{\rho}_B^{h} + \bar{\rho}_B^{R} \tag{16a}$$

or, using the ellipsometric parameter $\eta = \dfrac{\lambda}{\pi}\dfrac{n_1^2 - n_2^2}{\sqrt{n_1^2 + n_2^2}}\bar{\rho}_B$:

$$\eta_B = \eta_B^{h} + \eta_B^{R} \tag{16b}$$

The coupling between the ellipsometric parameter due to the thickness (η_B^h) and that due to the roughness (η_B^R) is neglected.

B. Roughness of Liquid Interfaces

In the case of an interface between two structureless fluid phases, the roughness at a scale larger than the molecular scale originates entirely from thermal or spontaneous fluctuations. The creation of a sinusoidal deformation at a plane interface, i.e., a mode of wavevector **q** and amplitude ζ_q, requires energy to overcome the three restoring forces: gravity, surface tension, and mean bending elasticity. This is the sum of three terms:

$$F(\zeta_q) = \frac{A}{2}(\Delta\rho\, g + \gamma q^2 + K q^4)\zeta_q^2 \tag{17}$$

where $\Delta\rho$ is the density difference between the two bulk phases, g the acceleration of gravity, γ the interfacial tension, and K the mean bending elastic modulus of the interface.

At a macroscopic scale (>1 mm) i.e., at small q, gravitational energy dominates.
At a microscopic scale (10 to 100 μm), i.e., at $q \sim 1000\ \text{cm}^{-1}$, the capillary energy dominates.
At an ultramicroscopic scale (100 Å), i.e., at very large q, the curvature energy dominates.
At a molecular scale ($q \geqslant q_{\text{mol}}$), the continum model of the interface fails.

The intermediate (or microscopic) scale is studied by surface light scattering, as was described earlier, because this scale is close to but larger than the light wavelength $\lambda (q\lambda < 1)$. The ultramicroscopic scale can be studied by reflectivity measurements.

1. Approximation of Independent Modes

If the restoring forces are large enough, the average slope of the interfacial fluctuations is small: $(\nabla\zeta)^2 \ll 1$. The coupling between the thermal modes can be neglected; i.e., the energy of an interfacial configuration ζ can be written as the sum of the energy of each mode ζ_q of the configuration:

$$F(\zeta) = \sum_{\mathbf{q}} E(\zeta_q)$$

With this approximation, the average energy of each mode is $k_B T$ and

$$\overline{|\zeta_q|^2} = \frac{k_B T}{\Delta\rho\, g + \gamma q^2 + K q^4} \tag{18}$$

This equation allows the calculation of the roughness term in a reflected intensity measurement:

$$\overline{\zeta^2} = \sum_q \overline{|\zeta_q|^2} = \int_{q_0}^{q_{\text{mol}}} \frac{k_B T}{\Delta\rho\, g + \gamma q^2 + K q^4} \frac{q\, dq}{2\pi} \tag{19}$$

where q_0 is the smallest wavevector selected by the aperture of the experimental setup. One obtains [27]

$$\overline{\zeta^2} = \frac{k_B T}{2\pi\gamma} \ln\left(\frac{q_R}{q_{\text{min}}}\right) \tag{20}$$

with

$$q_{\text{min}} = \text{MAX}\left(\sqrt{\frac{\Delta\rho \cdot g}{\gamma}}\ ;\ q_0\right) \tag{21}$$

$$q_R = \text{MIN}\left(\sqrt{\frac{\gamma}{K}}\ ;\ q_{\text{mol}}\right) \tag{22}$$

where MAX and MIN indicate respectively the greatest and the smallest value. In ellipsometry at the Brewster angle, the roughness term is obtained by the summation (Eq. 14):

$$\sum g(q)\, q\, \overline{|\zeta_q|^2}$$

or

$$\sum q \overline{|\zeta_q|^2} = \int_{q0}^{q_{mol}} \frac{k_B T}{\Delta\rho\, g + \gamma q^2 + K q^4} \frac{q^2\, dq}{2\pi} \tag{23}$$

if $\lambda\, (\gamma/K)^{1/2} \gg 1$, because in this case $g(q) = 1$ in the domain where the terms under the summation are significant. One obtains [27–28]

$$\sum_q q\, \overline{|\zeta_q|^2} = \frac{k_B T}{2\pi\gamma}\, q_e \tag{24}$$

where

$$q_e = \text{MIN}\left(\frac{\pi}{2}\sqrt{\frac{\gamma}{K}}\ \ ;\ q_{mol}\right) \tag{25}$$

This calculation does not take into account the roughness at the molecular scale, which can be neglected if

$$q_{mol}^2\, \overline{\zeta^2} \gg 1$$

$$q_{mol} \sum_q q\, \overline{|\zeta_q|^2} \gg 1$$

2. Coupled Modes

The mean square slope of the interface, calculated with the independent mode approximation, is

$$\overline{|\Delta\zeta|^2} = \int_0^{q_{mol}} \frac{k_B T}{\Delta\rho\, g + \gamma q^2 + K q^4} \frac{q^3\, dq}{2\pi}$$

One obtains

$$\overline{|\nabla\zeta|^2} = \frac{k_B T}{4\pi K} \ln\left(1 + \frac{k}{\gamma}\, q_{mol}^2\right) \tag{26}$$

The approximation of independent modes is valid if the bending elastic modulus is large ($k \gg k_B T$) or if the interfacial tension is large and dominates at any scale ($\gamma \gg K \cdot q_{mol}^2$). For large slopes, this approximation fails because the coupling between modes has to be taken into account. This is possible for weak coupling ($\overline{|\nabla\zeta|^2} < 1$) through a renormalization of the bending elastic modulus. This renormalization was done for thermal fluctuations of an interface at constant area [29, 30]: K decreases as the logarithm of the scale. In the case of a flat interface between two liquids, the area fluctuates. There are two domains [27]:

$\gamma > Kq^2$; the capillary energy dominates. K is a constant.
$\gamma < Kq^2$; the bending elastic energy dominates. K decreases as the logarithm of the scale.

In this case, $\overline{\zeta^2}$ and $\Sigma_q\, q\, \overline{|\zeta_q|^2}$ are obtained by a numerical calculation [27]. The bending elastic constant which is used in Eqs. 24 and 25 is the bending elastic constant at the scale $q_e = (\pi/2)\,(\gamma/k)^{1/2}$, the scale at which the terms under the summation in Eq. 23 are significant.

II. EXPERIMENTAL SECTION

In this section, we give some examples of reflectivity experiments, especially experiments where the different origins of the signal (thickness, anisotropy, roughness) are distinguished.

A. Monolayer at the Free Surface of Water: Observation of First-Order Phase Transitions

The surface tension of water is large: $\gamma_w = 72$ mN/m. When the water surface is covered with a monolayer of amphiphilic molecules of surface pressure Π, this surface tension decreases by Π, but its minimum value is still large because this is approximately the surface energy of an interface between air and a liquid made of the hydrophobic part of the amphiphilic molecules. For hydrocarbon chains one obtains $\gamma \sim 50$ mN/m; for fluorocarbon chains $\gamma \sim 20$ mN/m. With so large a surface tension, the roughness of the interface is small: 3 Å for the free surface of water and 3.5 Å for a dense monolayer with hydrocarbon chains. The roughness is smaller than the thickness of the monolayer (10–20 Å) and varies weakly with the surface pressure in the monolayer if $\Pi \ll \gamma_w$. Reflectivity measurements at the Brewster angle are dominated by the surface density of amphiphilic molecules (thickness) at the interface. At a first-order phase transition in the monolayer, domains of different densities coexist in the monolayer. These domains are indicated by the fluctuations of the ellipsometric signal of the interface [31] but can be observed directly by microscopy, using reflectivity at the Brewster angle [32] (Fig. 3).

B. Monolayer at the Free Surface of Water: Measurement of the Thickness and of the Bending Elastic Modulus by X-Ray Reflectivity

The thickness and the roughness of monolayers at the free surface of water are small. Ellipsometry at the Brewster angle gives some information, but the in-

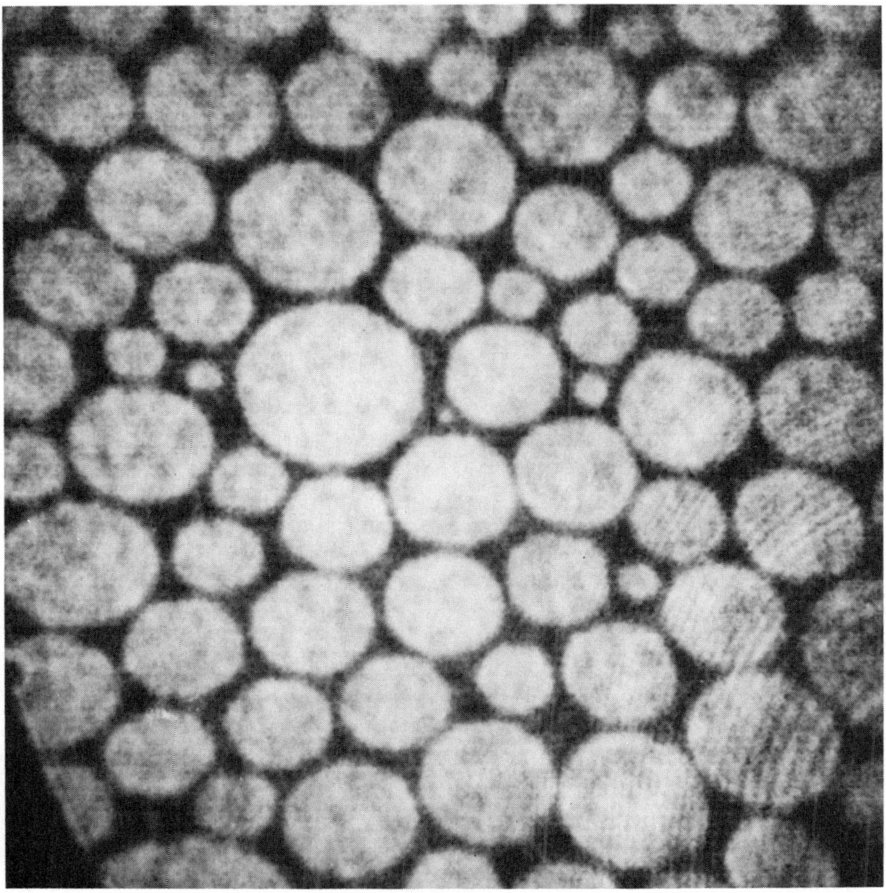

FIG. 3 Image for a myristic acid monolayer in the liquid expanded-liquid condensed coexistence region. It illustrates the sensitivity of a microscope using the Brewster angle [32]. The liquid condensed phase is more dense than the liquid expanded phase and appears bright. The size of the picture is 300 × 300 μm.

formation about thickness dominates and is mixed in the signal with the information about roughness. X-rays or other short wavelengths in radiation are suitable for separating these two pieces of information. The wavelength (λ = 1.5 Å) is smaller than the thickness of the monolayer: $R^{h}_{\alpha\alpha}(\theta)$ (Eq. 2) is a factor oscillating as a function of θ because of the constructive or destructive interferences between the beams reflected at different levels in the monolayer. This oscillating factor is damped by the roughness term (Eq. 15). The refractive indices for x-rays are close to one, and the reflected intensity is very weak as

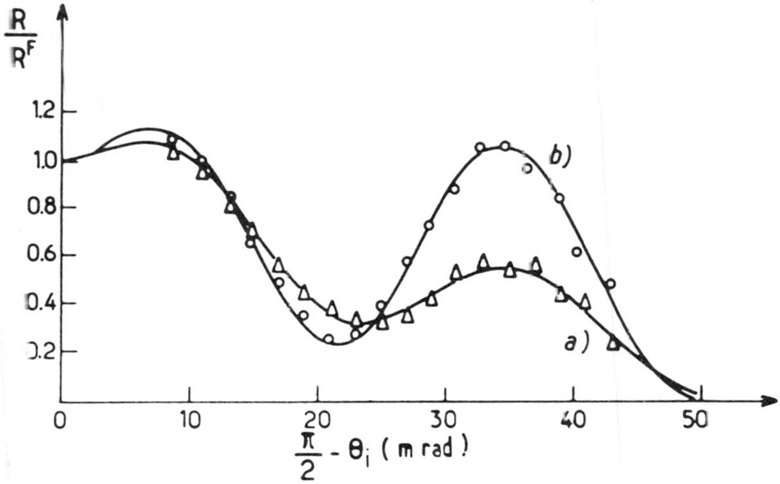

FIG. 4 The measured reflectivity of a monolayer of behenic acid at the free surface of water for two surface pressures, divided by the Fresnel reflectivity [24]. (a) In the liquid condensed 1 phase ($\Pi = 15$ mN/m). (b) In the liquid condensed 2 phase ($\Pi = 22$ mN/m). The curve is more damped than the curve b because the thermal fluctuations are larger in case a than in the second case b.

soon as the incident angle is smaller than the total reflection angle θ_L. For this reason, the incident beam is always grazing: $0.001 < (\pi/2 - \theta) < 0.1$ radian.

A monolayer of behenic acid on water has been studied in detail by this method [24] in the domain of surface pressure Π from 0 to 35 mN/m at 20°C. In this domain of surface pressure and temperature, three phase transitions have been observed [33]: the gas-liquid condensed 1 phase transition at very low surface pressure, the liquid condensed 1-liquid condensed 2 phase transition at $\Pi = 20$ mN/m (called liquid condensed-solid in Ref. 24), and the liquid condensed 2-solid phase transition at $\Pi \sim 30$ mN/m. Figure 4 gives two examples of reflected intensity measurements: one is obtained with a monolayer in the liquid condensed 1 phase for $\Pi = 15$ mN/m; the other one is obtained with a monolayer in the liquid condensed 2 phase for $\Pi = 22$ mN/m. The minima and maxima of reflectivity are not at exactly the same values of θ, indicating that the monolayer is thicker in the liquid condensed 2 phase than in the liquid condensed 1 phase (the tilt angle of the surfactant molecules at the surface, deduced from these thicknesses with a model, are respectively 18 and 28°). The damping of the reflectivity curve for the liquid condensed 1 phase is much larger than the damping of the reflectivity curve for the liquid condensed 2 phase: the roughness of the monolayer in the liquid condensed 1 state is larger than in the liquid condensed 2 state. The difference in roughness is explained by a large difference in the bending elastic modulus (Eqs. 20–22). Figure 5 shows the quantity $X = (\gamma\overline{\zeta^2})^{1/2}$ deduced from measurements versus the surface tension γ. X has a logarithmic, i.e., extremely weak dependence with γ (Eqs. 20–22). The experimental dependence of X with Π in the liquid condensed 1 domain and in the liquid condensed 2 domain is too weak to be observed. A large decrease of X at the phase transition indicates a large increase of the bending elastic modulus K.

In the liquid condensed 1 domain, $q_R = q_{mol}$; the bending elastic modulus is too small to be measured; it does not appear in Eq. 20.

In the liquid condensed 2 domain, $q_R = (\gamma/K)^{1/2}$; one found $120 < K < 300$ k_BT. The accuracy on K is low because it appears in a logarithm.

The tilt angle deduced from the measurement is a constant within each phase. The liquid condensed 2-solid phase transitions do not appear in these results.

C. Monolayer at the Free Surface of a Liquid: Thickness and Anisotropy

As said before, the part of the ellipsometric signal corresponding to the roughness of a monolayer at a free surface of water is smaller than the part of the signal

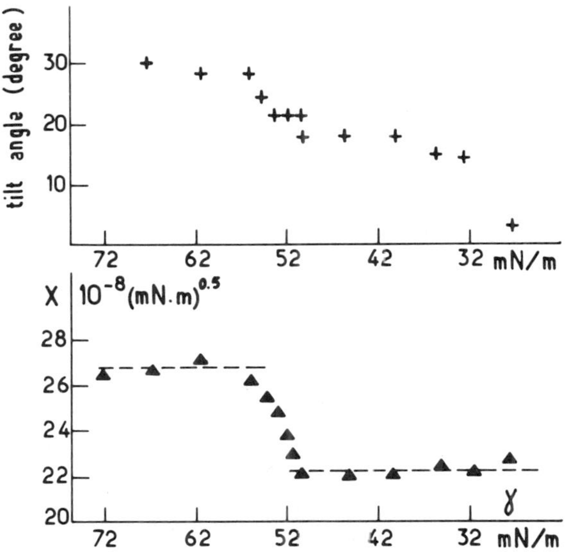

FIG. 5 The tilt angle and $X = (\gamma\overline{\zeta^2})^{1/2}$ versus the surface tension γ [24]. A large change in these quantities appears at the liquid condensed 1-liquid condensed 2 phase transition ($\gamma \simeq 52$ mN/m) illustrating the change in the structure of the monolayer and the effect of the bending elasticity.

corresponding to the thickness and varies slowly with the surface pressure Π if $\Pi \ll \gamma_w$. Consequently, in first approximation, the variation of $\bar{\rho}_B$ with the surface pressure can be attributed to the increase of the surface density of amphiphilic molecules, or to an increase of the thickness of the monolayer with the surface pressure. When the monolayer has optical anisotropy, whose origin is the anisotropy of the molecular arrangement in the monolayer, the anisotropy increases the ellipticity. In this case, $\bar{\rho}_B$ is a function of the surface density and of the anisotropy.

An optically anisotropic monolayer of NBD-stearic acid has been studied [12]. The phase diagram, Π versus the area per molecule Σ, exhibits a phase transition at Σ_c = 84 Å²/molecule (Fig. 6). The monolayer has been observed under a microscope by reflection, with yellow light (to avoid fluorescence), and between crossed polarizers. For $\Sigma > \Sigma_c$ = 84 Å²/molecule, no light is observed, showing no optical anisotropy in the monolayer. For $\Sigma <$ 84 Å²/molecule, some bright long domains looking like needles appear, indicating a first-order phase transition in the monolayer (Fig. 7). The total area of the needles increases when the surface concentration increases up to Σ = 32 Å²/molecule. At this surface area per molecule, the microscope field is almost entirely filled by needles or pieces of needles oriented at random. The needles oriented parallel to the polarizer or the analyzer appear to be in black while the brighter needles are oriented in

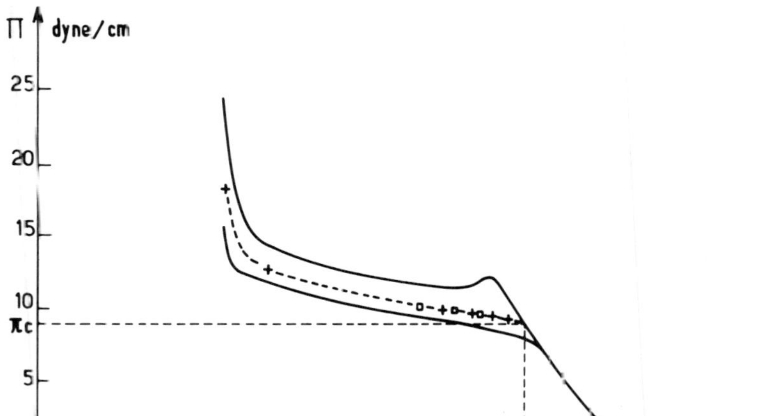

FIG. 6. Surface pressure curves for a monolayer of NBD-stearic acid [12]. Upper curve: continuous compression (2 Å/mol/min); lower curve: continuous decompression; dashed curve: points obtained after waiting about 30 mn. The small slope of the curves for 32 $< \Sigma <$ 84 Å² indicates the liquid-anisotropic phase coexistence region.

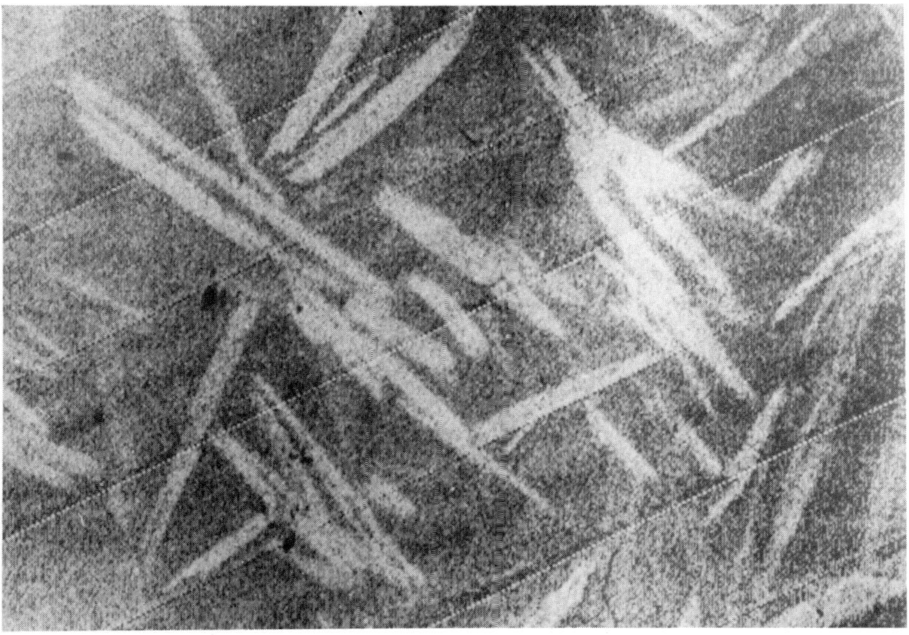

FIG. 7 Image of a monolayer of NBD stearic acid in the liquid-anisotropic phase coexistence region [12]. It was obtained by reflection with yellow light between crossed analyzer and polarizer. This image illustrates the sensitivity of a method that uses the nondiagonal terms of the matrix $R_{\alpha\beta}$.

the directions of the bisectors of analyzer and polarizer directions. This method of observation displays domains of the monolayer (needles) where the nondiagonal terms in the matrix $r_{\alpha\beta}(\theta)$ do not vanish. It is evidence of optical anisotropy for the needles. As has been said, this method of observation utilizes a deviation from zero (an isotropic monolayer appears in black) and consequently is very sensitive.

This monolayer was studied by ellipsometry at the Brewster angle. Figure 8 shows the ellipticity versus the surface concentration Γ. Three domains appear on the experimental curve.

(1) For a surface concentration Γ lower than $1/\Sigma_c$, the ellipticity is proportional to the surface concentration with large fluctuations of the signal around its mean value for $\Gamma < (1/110)$ molecule $/\text{Å}^2$. This corresponds to the gas-liquid coexistence region: the mean area of liquid domains under the light beam is proportional to the surface concentration, but its value varies from one point of the monolayer to another. This explains why the mean value of the ellipticity varies

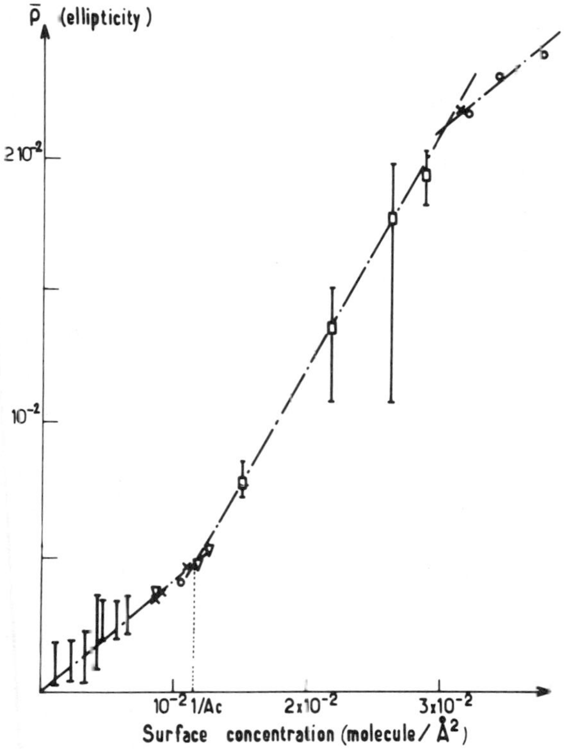

FIG. 8 The measured ellipticity versus the surface concentration for a monolayer of NBD-stearic acid on the free surface of water [12]. The open squares correspond to mean values. Large fluctuations appear in the gas-liquid and liquid-anisotropic phase coexistence regions. The slope of the curve is maximal in the liquid-anisotropic phase coexistence region illustrating the effect of the anisotropy.

linearly with Γ, from the value corresponding to the gas monolayer just before the phase transition to the value for the liquid monolayer just above the phase transition. In the concentration domain $(1/110) < \Gamma < 1/\Sigma_c$, the surface density increases as before, but without any fluctuation. The monolayer is homogeneous; its surface density has the same value at each point.

(2) For $1/\Sigma_c < \Gamma < 3.10^{-2}$ molecule/Å^2, fluctuations in the signal again occur and the mean ellipticity varies linearly with Γ, but with a larger slope than before. This domain of concentration corresponds to the liquid-anisotropic phase transition. The mean surface density of the anisotropic phase varies linearly with Γ. The slope for the ellipticity versus Γ is larger than for $\Gamma < 1/\Sigma_c$ because it has two origins: the increase in surface density of amphiphilic molecules as before

and the increase in the mean anisotropy of the monolayer. Supposing that the monolayer is a uniaxial medium with a horizontal optical axis, Eq. 7 shows the two effects and allows, with some assumptions, the deduction of the two refractive indices from the slopes of the two straight lines: $n_o = 1.564$ and $n_e = 1.72$, in good agreement with other experiments [12].

(3) For $\Gamma = 3.10^{-1}$ molecule/Å^2, all possible anisotropy of the monolayer is created, because all the surface is covered with anisotropic domains; the increase of the ellipticity with Γ, above this value, only originates in the increase of thickness (or density) of the monolayer. The slope of the increase of the ellipticity versus Γ in this domain is the same as for $\Gamma < 1/\Sigma_c$.

D. Monolayer at the Oil-Water Interface: Measurement of the Bending Elastic Modulus by Ellipsometry at the Brewster Angle

Some monolayers of soluble amphiphilic molecules at an oil-water interface have a very low surface tension [34, 35]: 10^{-2}–10^{-3} mN/m. The roughness (due to the thermal fluctuations) of such monolayers is about 100 times that of monolayers at the free surface of water: $(\overline{\zeta^2})^{1/2} \sim 300$ Å. They can easily be studied by ellipsometry at the Brewster angle, but the signal contains information about the roughness and about the structure: it is the sum of a roughness term and of a thickness term (Eq. 16). The thickness term being smaller than the roughness term, in a first approximation it can be estimated to deduce the roughness term and finally the bending elastic modulus [28] (Eqs. 24, 25). A more sophisticated method [36–38] is to vary one parameter that affects only one of the two terms of Eq. 16, for instance the surface tension. Such a system can be obtained with some soluble surfactants giving, when they are mixed with oil and water, a microemulsion in equilibrium with an excess of oil or water (Winsor equilibria [39]). The interfacial tension of a saturated monolayer of these surfactants at the oil-water interface is very low and exhibits a large variation with a small variation of the ionic strength of the water [34, 35] (for ionic surfactants) or the temperature [40] (non-ionic surfactants) (Fig. 9). The measured ellipticity for these monolayers has a pronounced peak at the same salinity or temperature as the minimum of interfacial tension (Fig. 10), indicating an increase of the thermal fluctuations for low γ. The ellipticity versus $1/\sqrt{\gamma}$ is a straight line (Fig. 11) as indicated by Eqs. 24 and 25 when the molecular cutoff is larger than the cutoff due to the bending elasticity: $(\pi/2)\,(\gamma/K)^{1/2} < q_{\mathrm{mol}}$ (for these monolayers, $K \sim k_B T$). The bending elastic modulus is deduced from the slope of this straight line. It must be remarked that the same surface tension is obtained for two different salinities or temperatures. The ellipticities for these two different salinities or temperatures are the same, indicating that only the interfacial tension is modified by the small variation in this parameter (ionic strength or temperature). The measured values

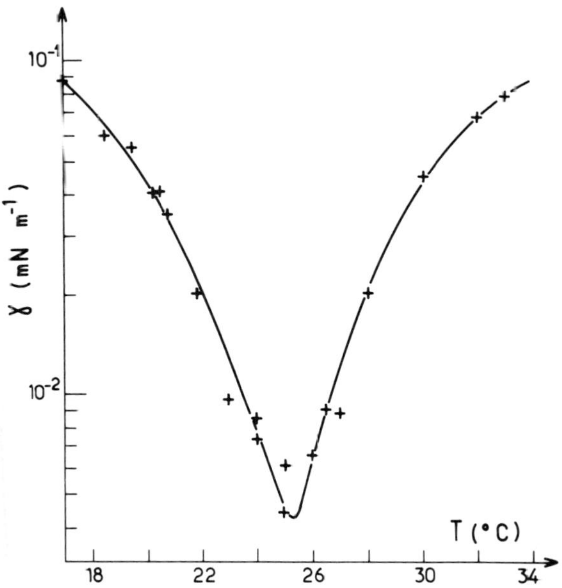

FIG. 9 The interfacial tension versus the temperature measured for a saturated monolayer of a nonionic surfactant ($C_{10}E_4$) at the octane-water interface [40].

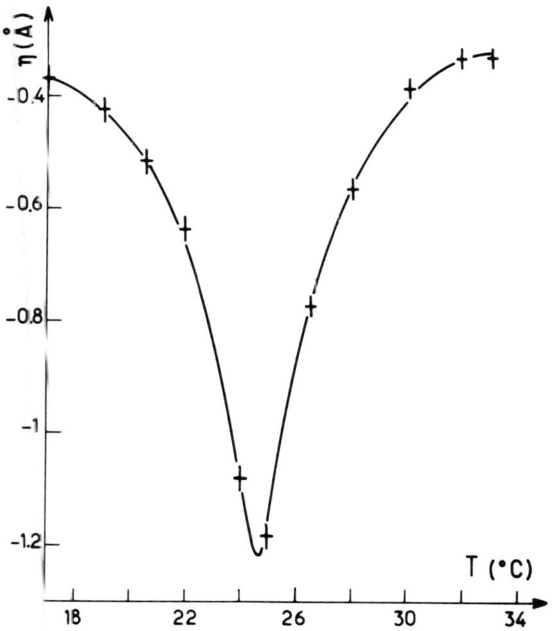

FIG. 10. The ellipsometric parameter η versus the temperature measured for a saturated monolayer of a nonionic surfactant ($C_{10}E_4$) at the octane-water interface. The minimum of η appears at the same temperature as the minimum of γ (Fig. 9) [40].

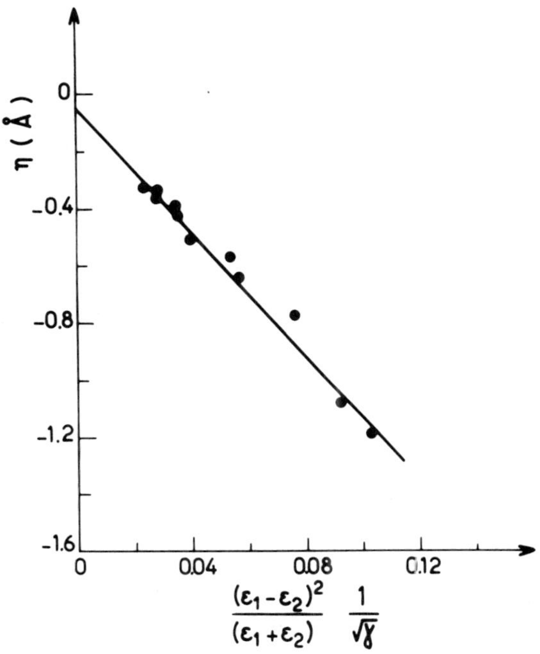

FIG. 11 The ellipsometric parameter η versus $1/\gamma^{1/2}$ for a saturated monolayer of a nonionic surfactant ($C_{10}E_4$) at the octane-water interface [40]. The bending elastic modulus of the monolayer is deduced from the slope of the straight line.

of the bending elastic modulus for these monolayers are very low $K \sim k_B T$ [28, 36, 38, 40]: the coupling between modes has to be taken into account [27]. Table 1 gives some results: the K values at the scale of the experiment, deduced from the measurements, and the K values at the molecular scale deduced from the same measurements and using the theory of Ref. 27. A correction due to the function $g(q)$ in Eq. 14 is important in the case where the $(\pi/2)\ (\gamma/K)^{1/2}$ is not much larger than $1/\lambda$: for instance if the interfacial tension γ is very small or if the bending elastic modulus is large. This correction has not been performed in the experimental values of K given in Table 1. Its calculation is in progress and must probably affect the value of K for the AOT at the heptane-brine interface because γ is small and K is large.

E. Depletion Layer: Structure of the Oil-Microemulsion and the Water-Microemulsion Interfaces in Winsor Equilibria

Winsor equilibria are observed when mixing oil, brine, and a small amount of surfactant and possibly cosurfactant [39]. When increasing the brine salinity S (with ionic surfactants), one obtains

TABLE 1 Bending Elastic Modulus of Monolayers at Oil/Water Interfaces

Monolayer	Scale of measurement (cm^{-1})	$K/k_B T$ measured	$K/k_B T$ at molecular scale
Ionic surfactants			
CTAB + butanol [38]	2.10^6	0.4	0.55
DTAB + butanol [38]	2.10^6	0.4	0.55
SDS + butanol [28]	6.10^5	0.65	1.00
SHBS + butanol [38]	5.10^5	0.4	0.86
AOT [36, 38]	5.10^5	1.10	1.65
Nonionic surfactants			
C_8E_3 [40]	3.10^6	~0.31	0.35
$C_{10}E_4$ [40]	8.10^5	0.51	0.76
$C_{12}E_5$ [40]	~2.10^6	~0.65	

Source: Refs. 28, 36, 38, 40.

$S < S_1$: 2-phase region with a lower microemulsion coexisting of an upper organic phase (Winsor I domain). The interface will be denoted by OM.

$S_1 < S < S_2$: 3-phase region with a coexisting lower aqueous phase, a middle-phase microemulsion and an upper organic phase (Winsor III domain). The two interfaces will be denoted by WM and OM respectively.

$S_2 < S$: 2-phase region with a coexisting lower aqueous phase and an upper microemulsion phase (Winsor II domain). The interface will be denoted by WM.

Microemulsions below S_1 (or above S_2) are dispersions of oil (or water) droplets in a continuous phase, aqueous or organic respectively [41–43]. These droplets are surrounded by a monolayer of surfactant molecules. The interface between the continuous phase and the organic phase in the Winsor I domain or the aqueous phase in the Winsor II domain will be denoted by OW.

The middle phase in the Winsor III domain is described by a bicontinuous structure of aqueous and organic phase separated by a surfactant monolayer [41, 43, 44]. In this domain, the interface obtained upon removing the microemulsion phase will also be denoted OW.

The OM and WM interfaces of the SDS system have been studied by ellipsometry [28]. Figure 12 shows the experimental ellipsometric parameter η_B versus brine salinity. Interfacial tension measurements reported earlier (Chap. 13) have shown that

For $\gamma > \gamma^*$ (γ^* is the minimum of surface tension of the OW interface) there is a flat surfactant monolayer at the OM and WM interfaces, the same as the one at the OW interface for the same brine salinity.

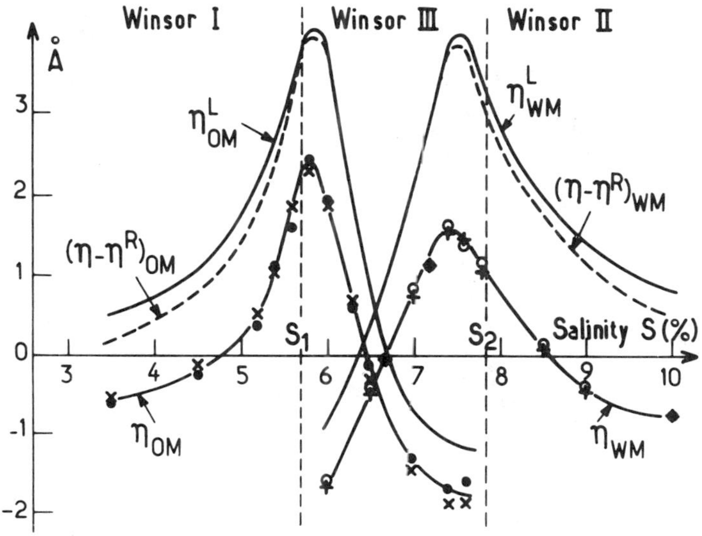

FIG. 12 η_{OM} at the organic phase-microemulsion phase (OM) interface and η_{WM} at the aqueous phase-microemulsion phase (WM) interface versus the salinity for a mixture of brine, toluene, SDS, and butanol [28]. $(\eta - \eta^R)_{OM}$, $(\eta - \eta^R)_{WM}$ (dashed lines); η^L_{OM}, η^L_{WM} (full lines) deduced from the experimental results η_{OM} and η_{WM}.

These OM (or WM) interfaces have the same interfacial tension as the OW interface at the same salinity.

It is therefore possible to calculate the contributions η^h_B and η^R_B to the ellipsometric parameter η_B of the thickness of the monolayer and of the interfacial roughness. But the sum of η^h_B and η^R_B is not equal to η_B: a third contribution $\eta^L_B = \eta_B - (\eta^h_B + \eta^R_B)$ must be added (Fig. 12). η^L_B being positive in the Winsor I and II and in a large part of the Winsor III, it cannot be a roughness contribution (always negative), and there is no reason to expect a larger optical anisotropy for the OM or WM interfaces than for the OW interface. A positive η^L_B is encountered when the refractive index in the interfacial region takes values smaller or larger than n_1 and n_2, the refractive indices of the two phases. This can be accounted for by the following model.

In the Winsor I (or II) domain, microemulsions are oil (or water) droplets in a continuous phase. A droplet cannot come closer to the interface than its radius R. This steric repulsion produces a decrease in droplet density in the vicinity of the interface, equivalent to a layer of continuous phase. The thickness L of this layer can be deduced from η^L_B and compared with R. L and R are in good agreement (Fig. 13).

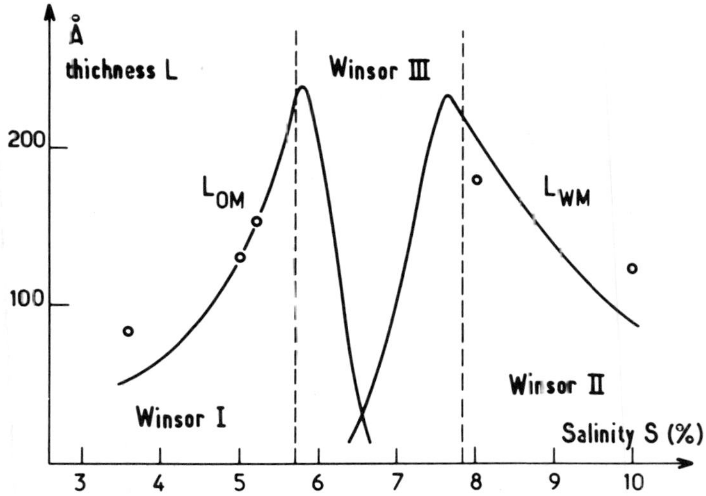

FIG. 13 The thickness L (full line) and the radii of the micelles in the Winsor I and II domains (open circles) versus the salinity [28].

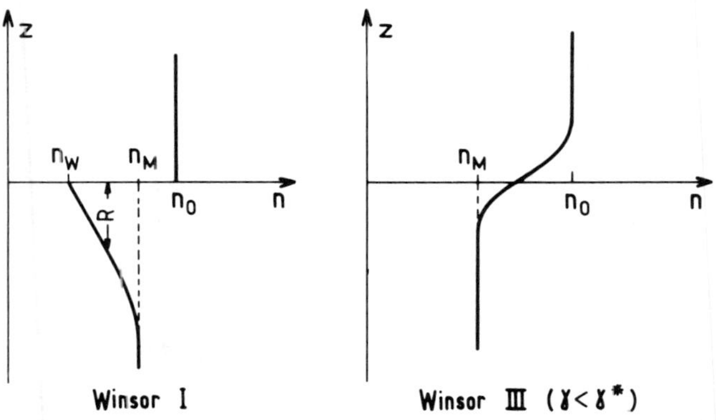

FIG. 14 Schematic representation of the variation of the refractive index through the microemulsion-oil interface for low and ultralow interfacial tensions. ($\gamma > \gamma^*$ and $\gamma < \gamma^*$, where γ is the minimum of the interfacial tension of the oil-water interface). n_o is the refractive index of the organic phase, n_M is the refractive index of the microemulsion, and n_W that of the continuous phase.

In the Winsor III domain, η_B reaches a maximum, which agrees with a similar picture, i.e., the microstructural elements in the bulk stay at a distance L away from the monolayer. It is found that $L = 240$ Å, which is close to the persistence length in the bulk ($\xi_K = 220$–250 Å), i.e., the size of the microstructural elements. As the interfacial tension decreases, η_B^L and consequently the calculated L value decrease and are close to zero for $\gamma = \gamma^*$, probably because attractive forces between the interfacial monolayer and the elements increase quickly. For $\gamma < \gamma^*$, η_B^L becomes negative, indicating that the refractive index in the interfacial layer has changed and lies between n_1 and n_2 (Fig. 14). This is explained by a model of open interfaces developed to account for the ultralow surface tensions in the vicinity of the critical points [45]. Indeed, since the two coexisting phases are becoming identical at the critical point, no monolayer should be present at the interface. As the interfacial tension decreases, the interface is expected to change from closed to open (Fig. 15).

F. Critical Interfaces

The theories of liquid-gas or binary mixture critical points predict a diffuse interface between the two phases with a thickness L increasing near the critical point as the density or concentration correlation function in bulk ξ. The van der Waals theory [46, 47] gives a hyperbolic tangent dependence of the profile of density, concentration or index of refraction. A more satisfying function has been proposed by Fisk and Widom [48]. These functions are close to the erf profil [17] proposed by Buff, Lovett, and Stillinger [49] and suggested by the capillary wave theory:

$$n(z) = \frac{1}{2}\left[(n_1 + n_2) + \frac{1}{\sqrt{\pi}}(n_1 - n_2)\int_0^{z/L\sqrt{2}} e^{-t^2}\,dt\right] \tag{27}$$

The reflectivity of this plane diffuse profile, given by Eq. 2, away from the Brewster angle, is [14, 15, 17]

$$R^1_{\alpha\alpha}(\theta) = R^F_{\alpha\alpha}(\theta)\exp[-Q_1^2 L^2]$$

Taking into account the thermal fluctuations (Eq. 15), one obtains

$$R_{\alpha\alpha}(\theta) = R^F_{\alpha\alpha}(\theta)\exp[-Q_1^2(L^2 + \overline{\zeta^2})] \tag{28}$$

γ and ξ or L in the vicinity of a critical point are given by scaling laws:

$$L \sim \xi = \xi_0\left[\frac{T - T_c}{T_c}\right]^{-\nu} \tag{29}$$

$$\gamma = \gamma_0\left[\frac{T - T_c}{T_c}\right]^{\mu} \tag{30}$$

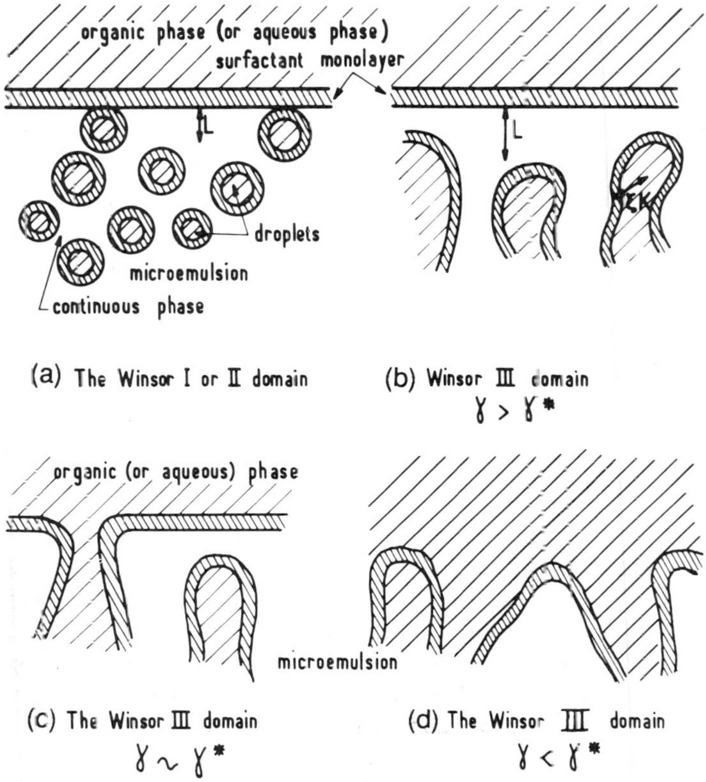

FIG. 15 The structure of the microemulsion-aqueous phase (or microemulsion-organic phase) interface and of the microemulsion versus the salinity S. The interface is passing gradually from a closed (a and b) to an open (c and d) interface [28].

The lattice gas model and experiments give $\mu = 2\nu$ ($\mu \sim 1.3$). $\overline{\zeta^2}$ is given by Eqs. 20, 21, and 22, in which the origin of the cutoff q_R for capillary waves is the thickness of the interface: $q_R \sim 1/\xi$. Neglecting the variation of the logarithm in Eq. 20:

$$\overline{\zeta^2} \sim \left[\frac{T - T_c}{T_c}\right]^{-\mu}$$
$$L^2 \sim \left[\frac{T - T_c}{T_c}\right]^{-2\nu} \qquad (31)$$

For ellipsometry at Brewster angle, Eqs. 3, 13, 24, and 25 with $q_e \sim 1/\xi$ give

$$
\begin{aligned}
\bar{\rho}_B^I &\sim L \sim \xi \sim \left[\frac{T - T_c}{T_c}\right]^{-\nu} \\
\bar{\rho}_B^R &\sim \frac{1}{\sqrt{\gamma}} \sim \left[\frac{T - T_c}{T_c}\right]^{-\frac{\mu}{2}}
\end{aligned}
\tag{32}
$$

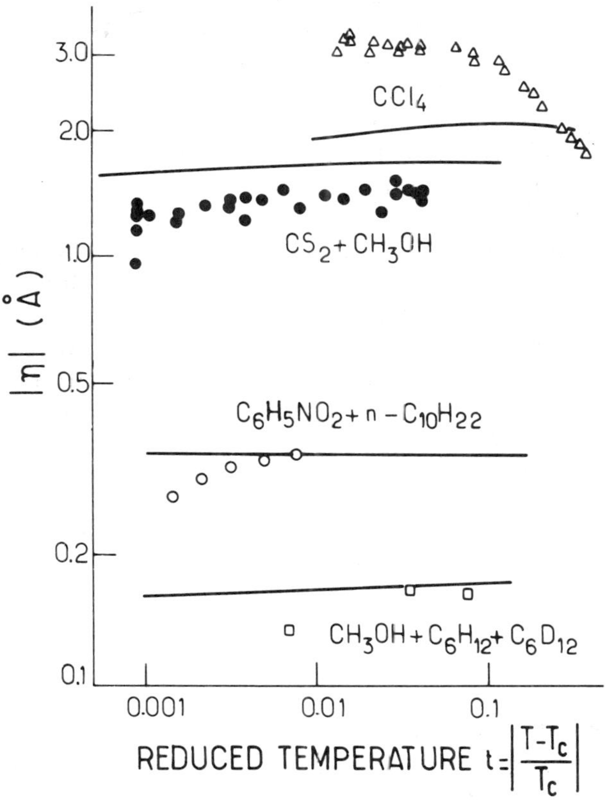

FIG. 16 The experimental ellipsometric parameter η of four different critical interfaces (Refs. 50 and 53) and theoretical fit (solid line) deduced from a coupled mode theory [27–37]. An unexplained discrepancy appears for CCl_4. No theory can explain this discrepancy; further experiments are needed for this critical system where ellipsometric measurements are very difficult because of the high pressure in the cell.

Thickness and roughness terms have the same variation with the temperature for reflectivity away from the Brewster angle (Eqs. 31) as for ellipsometry at the Brewster angle (Eq. 32).

Optical measurements have been performed with the intent of measuring the interfacial thickness L [16–18] (reflected intensity measurements) or to distinguish between the thickness and the roughness [50] (ellipsometry at Brewster angle). In fact, critical interfaces are typically the case where the interfacial thickness L and the interfacial roughness ζ cannot be distinguished with light reflectivity measurements or ellipsometric measurements [51]: they have the same effect on the reflectivity and the same variation with the temperature. The thickness can only be distinguished from the roughness at large scale ($q\lambda < 1$), because in this case, the roughness scatters the light. A way to eliminate this roughness at large scale in a reflectivity measurement is to include the greater part of the scattered light in the reflected light as in the Huang and Webb [17] and Wu and Webb [18] experiments. But these authors have not taken into account the roughness at small scale ($q\lambda > 1$), which decreases the reflectivity without scattering the light. They deduced an experimental thickness that is the quadratic average of the intrinsic thickness and of the roughness at small scale.

The distinction between thickness and roughness in the above semiphenomenological theory approach is somewhat artificial when $q \sim 1/\zeta$. More coherent descriptions of the interface have been proposed (see for instance Ref. 52 and references included in these papers), but each of them introduces at least one microscopic parameter the value of which is not exactly known. The interface can also be considered as formed by only capillary waves as in the original theory of Buff et al. [49]. A self-consistent treatment is based on the fact that the renormalized value of the surface tension at small scale increases strongly and leads to a natural cutoff [27]. This approach is rough because it supposes that the renormalization theory is valid for a roughness of large slope ($\overline{|\nabla\zeta|^2} \sim 1$), but it gives results close to the experimental measurements [27, 37] without an unknown microscopic parameter (Fig. 16).

III. CONCLUSION

Reflectivity measurements and ellipsometry at the Brewster angle are powerful tools to study liquid interfaces. They give information about the thickness and the roughness of these interfaces and about the optical anisotropy, if any; but these different pieces of information are mixed in the signal and are difficult to distinguish. This may be done by taking into account the particularities of each studied system, and it can be facilitated by the use of small wavelengths (x-rays or neutron beams). Another way could be a spectral analysis of the reflected light [54].

These techniques probe the thickness of plane interfaces because the scattering vector is perpendicular to the interface, and the roughness by multiple scattering. Reflectivity measurements out of the Brewster angle are equally sensitive to each surface mode $\Sigma_q \overline{|\zeta_q|^2}$, while reflectivity at the Brewster angle (ellipsometry at Brewster for most experiments) is only sensitive to surface modes at ultramicroscopic scale ($\Sigma\, q\zeta_q^2$ for $q \gg 2\pi/\lambda$).

REFERENCES

1. D. Langevin, J. Meunier, and D. Chatenay in *Surfactants in Solution* (K. L. Mittal and B. Lindman, eds) Plenum Press, New York, p. 1991.
2. R. H. Katyl and U. Ingard, *Phys. Rev. Lett. 19:* 64 (1967).
3. R. H. Katyl and U. Ingard, *Phys. Rev. Lett. 20:* 248 (1968). M. A. Bouchiat, J. Meunier, and J. Brossel, *C.R. Acad. Sci. 266B:* 255 (1968). M. A. Bouchiat and J. Meunier, *C.R. Acad. Sci. 266B:* 301 (1968).
4. Ch. Lalannes Cassou, thèse, Université de Pau, (1981).
5. M. A. Bouchiat and J. Meunier, *J. Phys. (France) 32:* 561 (1971).
6. See for instance Ch. Bouhet, *Ann. Phys.* Paris (1930).
7. R. M. A. Azzam and N. M. Bashara in *Ellipsometry and Polarized Light,* North Holland, 1977.
8. D. Beaglehole, *Physica 100B:* 163 (1980).
9. B. J. A. Zielinska, D. Bedeaux, and J. Vliegler, *Physica 107A:* 91 (1981).
10. P. Drude, *Ann. Phys. Chem.* (Leipzig) *43:* 126 (1891).
11. F. Abelès, *Ann. Phys.* (Paris) *5:* 596 (1950).
12. H. Bercegol, F. Gallet, D. Langevin, and J. Meunier, *J. Phys.* (France) *50:* 2277 (1989).
13. P. Beckmann and A. Spizzichino, in *The Scattering of Electromagnetic Waves from Rough Surfaces,* Pergamon/Macmillan, London/New York, 1963. P. Beckman, *Progress in Optics 6:* 55 (1967).
14. J. Meunier, *C.R. Acad. Sci. Paris 292:* 1469 (1981).
15. J. Meunier and D. Langevin,. *J. Phys. Lett.* (France) *43:* L-185 (1982).
16. G. H. Gilmer, W. Gilmore, J. Huang, and W. W. Webb, *Phys. Rev. Lett. 14:* 491 (1965).
17. J. S. Huang and W. W. Webb, *J. Chem. Physics 50:* 3677 (1969).
18. E. S. Wu and W. W. Webb, *Phys. Rev. 8A:* 2065 (1973).
19. A. Braslau, M. Deutsch, P. S. Pershan, A. H. Weiss, J. Als-Nielson, and J. Bohr, *Phys. Rev. Lett. 54:* 114 (1985).
20. C. A. Helm, H. Möhwald, K. Kjaer, and J. Als-Nielsen, *Europhys. Lett. 4:* 697 (1987).
21. L. Bosio, J. J. Benattar, and F. Rieutord, *Rev. Phys. Appl.* (Paris) *22:* 775 (1987).
22. R. M. Richardson and S. J. Roser, *Liquid Crystals 2:* 797 (1987).
23. A. Braslau, P. S. Pershan, G. Swislow, B. M. Ocko, and J. Als-Nielsen, *Phys. Rev. 38A:* 2457 (1988).
24. J. Daillant, L. Bosio, J. J. Benattar, and J. Meunier,*Europhys. Lett. 8:* 453 (1989).
25. P. Croce, *J. Optics* (Paris) *8:* 127 (1977).
26. P. Croce, *J. Optics* (Paris) *21:* 273 (1990).

27. J. Meunier, *J. Phys.* (Paris) *48:* 1819 (1987).
28. J. Meunier, *J. Phys. Lett.* (France) *46:* 1005 (1985).
29. W. Helfrich, *J. Phys.* (France) *46:* 1263 (1985).
30. L. Peliti and S. Leibler, *Phys. Rev. Lett. 54:* 1690 (1985).
31. Th. Rasing, H. Hsiung, Y. R. Shen, and N. W. Kim, *Phys. Rev. 37A:* 2732 (1988).
32. S. Henon and J. Meunier, *Rev. Sci. Inst. 62:* 936 (1991).
33. A. M. Bibo and I. R. Peterson, *Adv. Mat. 2:* 309 (1990).
34. A. Pouchelon, J. Meunier, D. Langevin, D. Chatenay, and A. M Cazabat, *Chem. Phys. Lett. 76:* 277 (1980).
35. A. Pouchelon, D. Chatenay, J. Meunier, and D. Langevin, *J. Coll. Interf. Sci. 82:* 418 (1981).
36. J. Meunier and B. Jerome, in *Surfactants in Solutions,* Vol 9 (K. L. Mittal, ed.), Plenum Press, New York, 1989, p. 463.
37. J. Meunier and B. P. Binks, *Progress Coll. Polym. Sci. 79:* 178 (1989).
38. B. P. Binks, J. Meunier, O. Abillon, and D. Langevin, *Langmuir 5:* 415 (1989).
39. P. A. Winsor, *Trans. Faraday Soc. 44:* 5685 (1948).
40. L. T. Lee, D. Langevin, J. Meunier, K. Wong, and B. Cabane, *Progress Coll. Polym. Sci. 81:* 209 (1990).
41. E. W. Kaler, H. T. Davis, and L. E. Scriven, *J. Chem. Phys. 79:* 5685 (1983).
42. A. M. Cazabat, *J. Phys. Lett.* (Paris) *44:* 593 (1983).
43. J. Tabony and A. de Geyer, in *Surfactants in Solution* Vol. 6, (K. L. Mittal and P. Bothorel, eds.), Plenum Press, New York, 1986, p. 1287.
44. L. E. Scriven, in *Micellization, Solubilization and Microemulsions* Vol. 2 (K. L. Mittal, ed.), Plenum Press, New York, 1977, p. 877.
45. L. Auvray, *J. Phys. Lett. 46:* L-163 (1985).
46. J. D. van der Waals, *Z. Phys. Chem. 13:* 657 (1894).
47. J. W. Cahn and J. E. Hilliard, *J. Chem. Phys. 28:* 258 (1958).
48. S. Fisk and B. Widom, *J. Chem. Phys. 50:* 3219 (1969).
49. F. P. Buff, R. A. Lovett, and F. H. Stillinger, Jr., *Phys. Rev. Lett. 15:* 621 (1963).
50. D. Beaglehole, *Phys. Rev. Lett. 58:* 1434 (1987).
51. D. Beysens and J. Meunier, *Phys. Rev. Lett. 61:* 2002 (1988).
52. J. W. Smith, *Physica A, 172:* 40 (1991). J. V. Sengers, J. M. J. van Leeuwen, and J. W. Schmidt, *172:* 20 (1991).
53. J. W. Smith, *Phys. Rev. 38A:* 567 (1988).
54. To be published.

18

X-Ray and Neutron Scattering from Liquid Surfaces

MARK L. SCHLOSSMAN* **and PETER S. PERSHAN** Division of Applied Sciences and Department of Physics, Harvard University, Cambridge, Massachusetts

**Current affiliation*: James Franck Institute, Center for Advanced Radiation Sources, University of Chicago, Chicago, Illinois.

I. INTRODUCTION

X-rays and neutrons are useful probes of liquid surfaces, primarily because the large penetration lengths and the 0.1 to 1 nm wavelengths used for scattering facilitate studies on atomic length scales. Recent developments in both scattering techniques and high-intensity sources, particularly synchrotron x-ray sources, have stimulated the extension of techniques, originally developed for studying bulk properties, to the study of surface or interfacial structures. Despite fundamental differences in the basic interactions between x-rays (or neutrons) and atoms, similarities in the wavelike properties of the two types of radiation have led to a widespread appreciation of the manner in which the two complement each other. In this chapter we shall review the applications of x-ray and neutron scattering to the study of the liquid-vapor interface of pure liquids and solutions, as well as to the study of insoluble monolayers at that interface. The three techniques of reflectivity, diffuse scattering, and surface diffraction will be reviewed; they will be illustrated by selected examples chosen from the wide variety of liquid interfacial phenomena that have recently been investigated.

Specular reflectivity is used to probe the interfacial profile of the interaction Hamiltonian density—between the wave and the material—normal to the interface. We shall discuss how specular reflectivity of x-rays can be used to measure the electron density profile with resolution of the order of a fraction of a nanometer. The interaction Hamiltonian for neutrons is slightly more complicated, perhaps richer, and below we shall discuss the type of profile that is measured by specular reflectivity of neutrons. Specular reflectivity has been used to measure structures of adsorbed interfacial phases as well as interfacial diffuseness, e.g., due to thermally excited capillary waves. Diffuse scattering is sensitive to in-plane inhomogeneities, and x-ray studies have extended previous measurements of thermally excited capillary waves from the smallest in-plane length scales typical of light scattering (10 μm) to several tens of nanometers. Surface-sensitive x-ray diffraction has been used to study in-plane ordered structures of surfactant monolayers adsorbed at the water-vapor interface.

The current synchrotron x-ray sources are several orders of magnitude more intense than neutron sources, and future x-ray sources currently under construction will be more brilliant by several more orders of magnitude. As will be shown below, the reflected intensity from ideally sharp and flat interfaces decays as θ^{-4} (where θ is a small reflection angle measured from the surface plane); the higher intensity allows the measurement of x-ray specular reflectivity over a much larger range of angles than is possible for neutrons. This can have a dramatic effect on the interpretation of interfacial profiles and is one of the primary advantages of x-rays. An advantage of neutrons is the seemingly random variation of neutron scattering lengths throughout the periodic table and between different isotopes of the same element; for example, the scattering lengths for hydrogen and deuterium are of different sign. This allows for the possibility of tuning the average neutron scattering amplitude for particular samples, e.g., appropriately deuterated water can be made invisible to coherently scattered neutrons.

In addition, for some neutron studies, isotopic substitution within the adsorbed molecules can highlight specific regions of the interfacial profile. X-rays regain some of this advantage for systems containing heavy atoms. Aside from the fact that heavy atoms, containing more electrons, scatter x-rays more strongly than light atoms, heavy atoms typically have absorption edges in the 7 to 15 keV range where x-ray scattering is most often performed. The x-ray wavelength, selected from the continuous distribution of the synchrotron, can be tuned through the absorption edge for a particular atomic species. This results in enhanced selectivity in both the scattering and x-ray induced fluorescence that can help to locate atomic positions relative to the interface.

The techniques of reflectivity, diffuse scattering, and surface diffraction will be further discussed in detail; however, there are other x-ray and neutron probes of liquid surfaces that are outside the scope of this article. These include neutron scattering from bulk colloidal solutions to probe polymer adsorption at the solid liquid interface [1, 2, 3] and the use of x-ray fluorescence [4, 5, 6] and x-ray standing waves [7, 8, 9] (induced by evanescent x-rays incident at variable angles $\theta \lesssim \theta_c$, where θ_c is the critical angle for total reflection) to probe the distribution of heavy atoms along the surface normal.

We shall first discuss the basic optics and kinematics of the different scattering techniques. The physics necessary to interpret measurements from real interfaces is discussed in Secs. II.B-C. The remainder of the article will review some of the important measurements, most having been made within the past five years. We shall cite recent review articles [10, 11, 12, 13] that discuss certain portions of this material in more detail than is done here.

II. EXPERIMENTAL TECHNIQUES

A. Optics and Kinematics

For any type of plane wave incident on a plane interface, the reflection and refraction conditions can be parameterized in terms of a macroscopic index of refraction that characterizes the average property of the radiation on either side of the interface. The general relation between scattering theory and refraction, as developed by Ewald and Oseen for classical optics, is elegantly reviewed by Born and Wolf [14]; this is a general result that applies to both neutrons scattering primarily from nuclei and x-rays scattering from electrons [15]. If k_0 and k are the respective wavevectors of the radiation in vacuum and in a material, then the index of refraction n is expressed as

$$\frac{k}{k_0} \equiv n = 1 - (\alpha_j - i\beta_j) \tag{1}$$

where the subscript j = x for x-rays or n for neutrons, and the imaginary part corresponds to absorption. In general it is necessary to take into account both the direction of propagation and the polarization of either the x-rays or the neutrons; however, for grazing incidence, which we shall show below to be appropriate to this discussion, polarization is not important. In addition, since this review is concerned with liquids, it is expedient to specialize the discussion to nonmagnetic materials that would otherwise affect neutron polarization.

Table 1 illustrates the relation between α, β and the more fundamental atomic constants where N is the atomic number density, λ is the x-ray or neutron wavelength, and $r_e = (e^2/mc^2) \cong 2.818$ fm is the classical electron radius. Following the standard conventions, the average x-ray atomic scattering factor is written in terms of f_0, f', and f'', where for small angle scattering $f_0 \cong Z$ (the atomic number) and f' and f'' are the real and imaginary parts of the anomalous dispersion correction, which is large near absorption edges; note that f'' determines the absorption [16, 17]. For neutrons, b_0 is the neutron coherent scattering length and b' and b'' are the real and imaginary parts of the anomalous dispersion correction, where b'' determines the absorption [18, 19, 20]. For most materials, b' and b'' are negligible, and for light materials (and x-ray energies greater than the material's electronic binding energies), f' can be neglected. A simplified expression for the index of refraction can then be written as

$$n = 1 - \lambda^2 A_j + i\lambda B_j \tag{2}$$

where $A_x = (\rho r_e/2\pi)$, ρ being the electron density in the material; $B_x = (\mu_a/4\pi)$, μ_a being the x-ray linear absorption coefficient for intensity; $A_n = (Nb_0/2\pi)$; $B_n = (N\sigma_a/4\pi)$, σ_a being the average neutron absorption cross section. For most materials, $\lambda^2 A \cong 10^{-5}$ and $\lambda B \lesssim 10^{-8}$ for x-rays and neutrons.

The kinematics of surface scattering is illustrated in Fig. 1; note that $\theta = 90°$ is normal to the surface. Specular reflectivity occurs when $\theta = \theta'$ and $\phi = 0$. Since polarization effects are not important at small θ, it is convenient to consider the special case of polarization parallel to the surface. Regardless of whether we are concerned about x-rays or neutrons, for radiation in a material of index of refraction n_1 incident on a plane surface at which the index of

TABLE 1 Expressions for α and β in Terms of Atomic Parameters

X-ray	Neutron
$\alpha_x = (N\lambda^2 r_e/2\pi)(f_0 + f')$	$\alpha_n = (N\lambda^2/2\pi)(b_0 + b')$
$\beta_x = (N\lambda^2 r_e/2\pi)\,\lvert f''\rvert$	$\beta_n = (N\lambda^2/2\pi)\,\lvert b''\rvert$

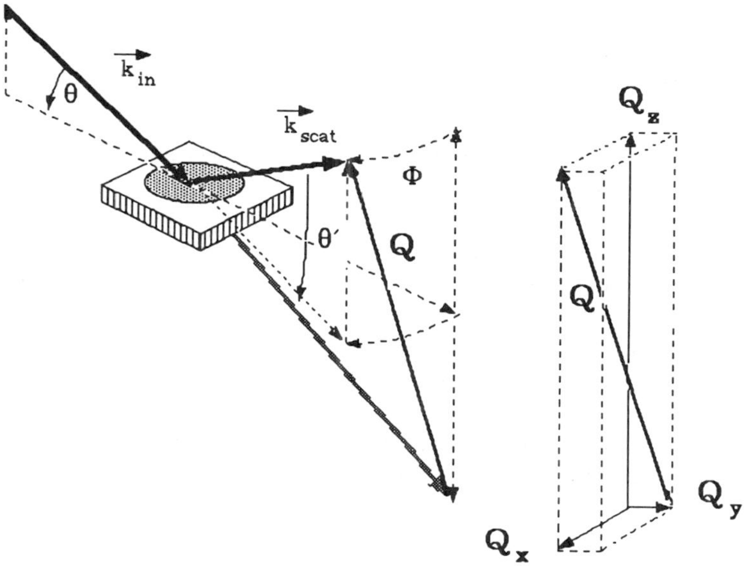

FIG. 1 Kinematics of surface scattering. The liquid surface is horizontal and the incident wavevector k_{in} makes an angle θ with the surface. The detector can be located at angles θ' from the horizontal and ϕ from the plane of incidence. Specular reflection occurs when $\theta = \theta'$ and $\phi = 0$. The drawing on the right indicates the components of the transferred momentum wavevector $Q = k_{scat} - k_{in}$, which is the difference between the scattered and incident wavevectors. For specular reflection, $Q_x = Q_y = 0$. For grazing incidence diffraction $\theta \cong \theta_c$ (the critical angle for total reflection), ϕ is adjusted to tune to the diffracted intensity at the Bragg rods, and θ' is adjusted to measure the intensity along the Bragg rods.

refraction abruptly changes to n_2, the specular reflectivity $R_F(\theta)$ is given by the Fresnel result of classical optics [14],

$$R_F(\theta) = \left| \frac{n_1 \sin\theta - \sqrt{n_2^2 - n_1^2 \cos^2\theta}}{n_1 \sin\theta + \sqrt{n_2^2 - n_1^2 \cos^2\theta}} \right|^2 \tag{3}$$

If material 1 is air ($n_1 \cong 1$) and $n_2 < 1$ (as it is for most materials), there is a critical angle for x-ray total reflection, $\theta_c = \cos^{-1}(n_2) \cong \sqrt{\rho r_e \lambda^2/\pi}$, and for neutron total reflection, $\theta_c \cong \sqrt{N b_0 \lambda^2/\pi}$. If absorption is neglected, then $R_F(\theta) = 1$ when $\theta < \theta_c$ and $R_F(\theta) \approx (\theta_c/2\theta)^4$ when $\theta \gg \theta_c$. For most materials, θ_c is approximately 2–5 mradian for an x-ray wavelength of $\lambda \cong 0.154$ nm. With real interfaces, for which the index of refraction profile normal to the interface is not a simple step function, the specularly reflected intensity $R(\theta)$ will deviate from $R_F(\theta)$ in a manner that allows characterization of the index of refraction profile normal to the surface. This will be discussed in detail in the next section.

Since $\lambda B \lesssim 10^{-8}$, both x-rays and neutrons, for incident angles $\theta \gg \theta_c$, penetrate into the bulk of most liquids large enough distances that scattering from the bulk is usually the dominant effect. The x-ray absorption length μ_a^{-1} at 8 keV ($\lambda \cong 0.15$ nm) is approximately 1 mm in water and in most organic fluids; for neutrons it is usually much longer. As a consequence, although specular reflection is the dominant scattering process in the direction of $\theta' = \theta$ and $\phi = 0$, at angles $\theta' \neq \theta$ and $\phi \neq 0$ far enough away from the specular condition scattering from the bulk is dominant. By simultaneously measuring the angular dependence of the specular reflectivity and the bulk scattering, it is possible to monitor the structure of the bulk liquid in the same experiment in which the surface is studied.

For arbitrary θ, θ', and ϕ (see Fig. 1), the three components of the transferred momentum of the scattered radiation, $\mathbf{Q} = \mathbf{k}_{scat} - \mathbf{k}_{in}$, are

$$\begin{aligned} Q_x &= \frac{2\pi}{\lambda} \cos(\theta') \sin(\phi) \\ Q_y &= \frac{2\pi}{\lambda} [\cos(\theta') \cos(\phi) - \cos(\theta)] \\ Q_z &= \frac{2\pi}{\lambda} [\sin(\theta') + \sin(\theta)] \end{aligned} \tag{4}$$

The specular condition corresponds to $Q_x = Q_y = 0$ and $Q_z = (4\pi/\lambda) \sin(\theta)$. Tuning off this condition results in a nonzero in-plane component $\mathbf{Q}_{xy} = (Q_x, Q_y)$. For example, if $|\theta - \theta'| \ll \theta \ll 1$ and $\phi = 0$, then $Q_z \cong (4\pi/\lambda)\theta$, $Q_x = 0$, and $Q_y \cong (\pi/\lambda)(\theta^2 - \theta'^2) \cong (Q_z/2)(\theta - \theta')$, resulting in values of Q_y that are factors of 10^{-4} to 10^{-5} smaller than $2\pi/\lambda$. For $\theta = 0.5°$, $\theta - \theta' \cong 0.1°$, and $\lambda = 0.15$ nm, $2\pi/Q_y \cong 10$ μm. This is comparable to the smallest length scales usually probed in scattering by visible light; however, with x-rays the measurements can be extended to length scales that are orders of magnitude smaller.

Scattering from the surface can also dominate when, for $\theta \leq \theta_c$, the incident wave is totally reflected [21]. Neglecting absorption, the incident intensity penetrates the surface evanescently, as $\exp(-z/\Lambda_i)$, where $\Lambda_i \cong [(4\pi/\lambda)\sqrt{\theta_c^2 - \theta^2}]^{-1}$. For $\theta_c \cong 0.2°$, $\theta = 0.8\,\theta_c$, and $\lambda = 0.15$ nm, the intensity is reduced to $1/e$th of its initial value in approximately 5 nm. For this value of θ and almost any value of $\theta' \lesssim 10°$, scattering at $Q_{xy} \equiv \sqrt{Q_x^2 + Q_y^2} \cong (4\pi/\lambda) \sin(\phi/2)$ will reflect structure parallel to the surface and within approximately 10 nm of the interface. This geometry has been widely applied to study the structure of insoluble Langmuir monolayers on the surface of aqueous solutions.

There are several different methods for experimentally achieving the kinematics just discussed [22–28]. One example illustrating the configuration at beamline X22B of the National Synchrotron Light Source at Brookhaven National Laboratory is shown in Figure 2 [10, 29]. The x-ray synchrotron radiation emitted from the electron storage ring is horizontally and vertically focused approximately 2:1 by a bent cylindrical mirror. This focuses the beam incident on the sample to approximately the same size as the source (0.3 mm vertically with divergence $\cong 2 \times 10^{-4}$ radian and approximately 1 mm horizontally with divergence $\cong 2 \times 10^{-3}$ radian). The wavelength can be varied from approximately 1.2 to 1.7 Å and is monochromated by Bragg diffraction from a Ge(111) crystal outside the experimental hutch. The beam diffracted from that crystal is approximately horizontal when entering the hutch. Another Ge(111) crystal, mounted in a 4-circle goniometer, is used to tilt the beam downward by an angle θ onto the liquid surface [10, 28]. To remain on the Bragg condition for the second crystal, the reciprocal lattice vector **G** of the crystal must be rotated along a cone so that the component of **G** parallel to the x-ray beam incident on the crystal remains constant. This means that the exit wavevector from the second crystal rotates downward along a cone, and this motion, unless compensated by a rotation of the goniometer, would move the beam off the central axis of the surface reflectometer, where the central axis is defined by a vertical axis through the center of the sample. The compensating horizontal sample motions are achieved by small rotations of the "two-theta" arm of the goniometer on which the second Ge(111) monochromator is mounted. Since the incoming angle θ is usually very small (approximately 1–100 mradian), the footprint of the beam on the sample can be very large (approximately 1–100 mm), necessitating the use of a slit before the sample to define the vertical size of the beam. The sample is moved

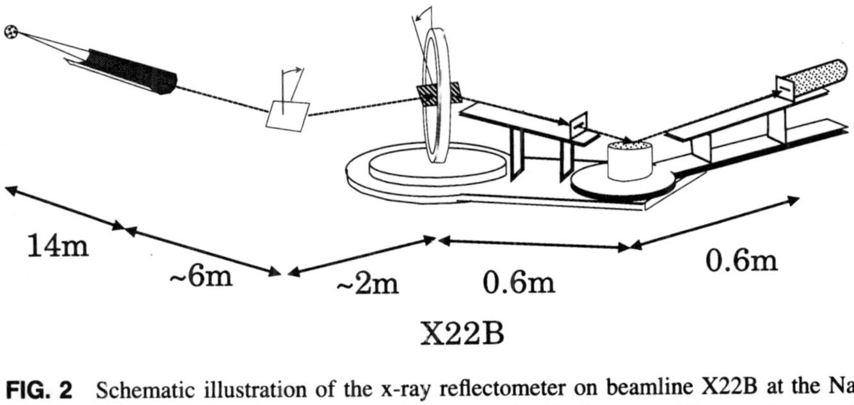

FIG. 2 Schematic illustration of the x-ray reflectometer on beamline X22B at the National Synchrotron Light Source, Brookhaven National Laboratory. The synchrotron is represented by the small circle at the far left. The optics are described in the text.

vertically so that the beam remains centered on the sample as θ changes and the position of the detector is appropriately adjusted for the desired transferred momentum **Q**. Another slit before the detector is set to accept fully the specular beam or to determine a particular resolution when measuring surface diffuse scattering or surface diffraction. Figure 2 illustrates a configuration to study nonspecular scattering with the detector rotated azimuthally along the angle ϕ (Eq. 4).

B. Optical Theories of Reflection from Interfacial Profiles

As discussed elsewhere [10, 12], the specularly reflected wave can be attributed to scattering from the interfacial profile averaged over the x-y plane of the surface. Neglecting deviations from the average profile, the reflected and transmitted waves have the same component of in-plane wavevector $k_{x,y}$ as the incident wave. For the special case of a wave incident from vacuum, the general solution for the reflected amplitude and phase (e.g., of the electric field E), can be obtained by solving the one-dimensional wave equation (derived from Maxwell's equations and Eq. 2)

$$\frac{d^2E}{dz^2} + \frac{1}{4}\left[Q_z^2 - Q_c^2\frac{\langle\rho(z)\rangle}{\rho}\right]E = 0 \tag{5}$$

where $Q_z = (4\pi/\lambda)\sin\theta$ is the normal component of the wavevector transfer between the specularly reflected and incident waves in vacuum, and $Q_c = (4\pi/\lambda)\sin\theta_c$. The essential boundary condition is that as $z \rightarrow \infty$, inside the material,

$$E \propto \exp\left[-i\frac{z}{2}\sqrt{Q_z^2 - Q_c^2}\right] \tag{6}$$

This approach has the advantage of being correct for all values of θ and $R(\theta)$ if the substitution of the average profile for the exact surface is justified. The principal disadvantage of this approach is that it is usually not possible to integrate Eq. 5 analytically. It is typically solved digitally using one of several matrix methods that are computationally complex. Profiles of varying complexity often require different matrix methods [11] or possibly more sophisticated integration routines [30]. Most of the recent neutron reflection measurements have been analyzed in this way [11].

Since the atomic scattering amplitudes for neutrons and x-rays are relatively small, when $\theta \gg \theta_c$ refraction and multiple scattering are not important. The intensity of the specularly reflected wave can then be simply calculated from the basic scattering formulae using the first Born approximation. Although this approach is not accurate for $0 \lesssim 4\theta_c$, it is quantitatively reliable for larger θ. Its analytic simplicity often facilitates an intuitive understanding of the effects to be expected from different types of surface profiles. The principal result of this approximation for x-rays is [10]

$$\frac{R(\theta)}{R_F(\theta)} \cong \left| \frac{1}{\rho_{\text{bulk}}} \int dz \frac{d\langle\rho(z)\rangle}{dz} \exp(iQ_z z) \right|^2 \tag{7}$$

where $d\langle\rho(z)\rangle/dz$ is the electron density gradient averaged over the x-y plane (parallel to the surface at constant z) and ρ_{bulk} is the electron density far below the surface. A similar result is applicable to neutron reflectivity [11] if $\rho = \Sigma N_i b_{0,i}$, where N_i is the number density of the ith nucleus with scattering length $b_{0,i}$. For interfaces in which $\langle\rho(z)\rangle$ varies slowly compared to $1/Q_c$, it is possible to demonstrate that $R(\theta)/R_F(\theta) \cong 1$ for all Q_z that are not too much larger than Q_c—the precise condition depending on details of the profile. In these cases, if Q_z is interpreted as the value inside the material, i.e., $Q_z \cong (4\pi/\lambda)\sqrt{\theta^2 - \theta_c^2}$, Eq. 7 can be accurate for all $\theta \geq \theta_c$.

To illustrate the application of Eq. 7, consider a profile in which $\langle\rho(z)\rangle$ first rises from $\langle\rho(z)\rangle = 0$ to ρ_1 in some small distance σ_1, and then, following a distance L in which it is nearly constant, rises from $\langle\rho(z)\rangle = \rho_1$ to ρ_{bulk} in a distance σ^2 (see Fig. 3). For small incident angles, at which $Q_z\sigma_1$ and $Q_z\sigma_2 \ll 1$ (i.e., the interfacial widths can be neglected), the integrand in Eq. 7 can be written as

$$\frac{1}{\rho_{\text{bulk}}} \frac{d\langle\rho(z)\rangle}{dz} \cong \frac{\rho_1}{\rho_{\text{bulk}}} \delta(z) + \frac{(\rho_{\text{bulk}} - \rho_1)}{\rho_{\text{bulk}}} \delta(z - L) \tag{8}$$

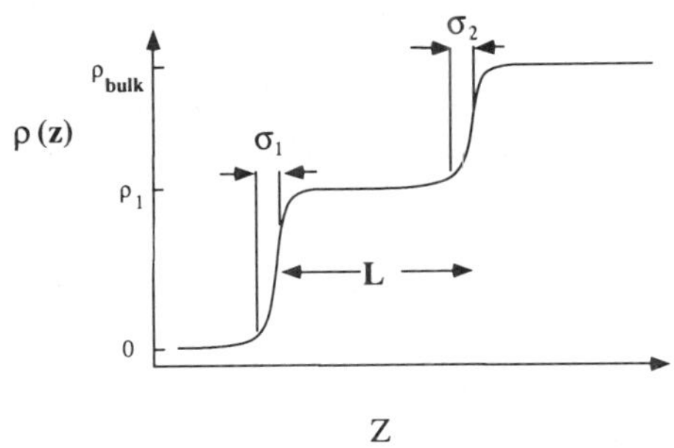

FIG. 3 Model real space electron density profile, $\rho(z)$ versus z, where z is the distance along the surface normal going from the vapor to the liquid. The electron densities of the bulk liquid, of the surface layer, and of the vapor are respectively, ρ_{bulk}, ρ_1, and 0. The width of each step-like change in $\rho(z)$, on either side of the surface layer, is given by σ_1 and σ_2. The thickness of the surface layer is given by L.

and substitution into Eq. 7 yields

$$\frac{R(\theta)}{R_F(\theta)} \cong \left| \frac{\rho_1}{\rho_{bulk}} + \frac{(\rho_{bulk} - \rho_1)}{\rho_{bulk}} e^{[iQ_z L]} \right|^2 \tag{9}$$

The immediate result of this analysis is to reveal the expected effect: the reflectivity from a surface layer of thickness L oscillates with a period $\Delta Q_z = 2\pi/L$. If the widths $\sigma_{1,2}$ are not sufficiently small to be neglected, then the δ-functions might be replaced by Gaussian functions with widths $\sigma_{1,2}$ and areas normalized to one. The integral of a Gaussian is another Gaussian and

$$\frac{R(\theta)}{R_F(\theta)} \cong \left| \frac{\rho_1}{\rho_{bulk}} e^{-Q_z^2\sigma_1^2/2} + \frac{(\rho_{bulk} - \rho_1)}{\rho_{bulk}} e^{-Q_z^2\sigma_2^2/2} e^{[iQ_z L]} \right|^2 \tag{10}$$

This reveals a second expected effect for interfaces of finite widths: the ratio of $R(\theta)/R_F(\theta)$ decreases with increasing angle of reflection.

Another advantage of the analytic approach of Eq. 7 is that its Fourier transform is directly related to the autocorrelation function [31];

$$P(s) = \frac{1}{2\pi} \int_{-\infty}^{\infty} dQ_z \left[\frac{R(Q_z)}{R_F(Q_z)} \right] e^{-iQ_z s}$$
$$= \frac{1}{\rho_{bulk}^2} \int_{-\infty}^{\infty} \frac{d\langle\rho(z)\rangle}{dz} \frac{d\langle\rho(s+z)\rangle}{dz} dz \tag{11}$$

Although the function $P(s)$ is slightly different from the Patterson function used in standard diffraction analysis, we do not think that the differences are particularly important and we shall refer to $P(s)$ as "the Patterson function."† The primary deficiency in all Patterson function analyses is the absence of phase information about the various Fourier components of $d\langle\rho(z)\rangle/dz$. The problem is minimized for surface profiles in which the dominant part of $d\langle\rho(z)\rangle/dz$ arises from the interface with the vacuum, for example,

$$\frac{1}{\rho_{bulk}} \frac{d\langle\rho(z)\rangle}{dz} \cong \frac{1}{\rho_{bulk}} \frac{d\langle\rho_0(z)\rangle}{dz} + f(z) \tag{12}$$

where $d\langle\rho_0(z)\rangle/dz$ represents the vacuum interface and $f(z)$ represents variations in $\rho(z)$ on length scales longer than those at the vacuum interface. The Patterson function can then be written as

$$P(s) \cong P_0(s) + 2f(s) + \cdots \tag{13}$$

†The conventional Patterson function is the autocorrelation function of $\rho(\mathbf{r})$; the function in Eq. 11 is the autocorrelation function of $d\langle\rho(z)\rangle/dz$.

where $P_0(s)$ is the autocorrelation function of $\rho_{\rm bulk}^{-1}[d\langle\rho_0(z)\rangle/dz]$. In this particular case, as well as others in which the connection is less direct, calculation of $P(s)$ can be used as a model independent method to visualize directly features of the real-space surface density profile.

A subtler issue in the interpretation of specular reflectivity data concerns the precise meaning of the "average" involved in $\langle\rho(z)\rangle$. For example, in the case of two coexisting surface phases, the meaning of both $\langle\rho(z)\rangle$ and the observed reflectivity will depend on the size of the single phase domains relative to the resolution of the spectrometer. If the domains are large compared to the spatial resolution of the spectrometer, each domain will reflect independently; therefore, the observed reflectivity will be a weighted average of the reflectivity from the two different phases. If, however, the domains are sufficiently small, the reflectivity will be determined by $\langle\rho(z)\rangle$ averaged over both phases. The in-plane coherence length—defined by the resolution of the spectrometer—that separates these two conditions is determined in the same way as in conventional optics, i.e., as the longest distance between two surface points for which interference can be observed. In practice this depends on the spectral purity, $\Delta\lambda/\lambda$, the angular divergence of the incident wave, and the angular acceptance of the detector. For scattering within the plane of incidence and near to the specular condition ($\phi = 0$ and $\theta \cong \theta' \ll 1$ in the geometry illustrated by Fig. 1), the reciprocal of the in-plane coherence length is given by

$$\Delta Q_{xy} \cong \sqrt{(\Delta Q_x)^2 + (\Delta Q_y)^2} \tag{14}$$

where $(\Delta Q_x)^2 \cong \langle((2\pi/\lambda)\ \Delta\phi)^2\rangle$ and $(\Delta Q_y)^2 \cong \langle[Q_y\ (\Delta\lambda/\lambda) + (Q_z/2)\ (\Delta\theta - \Delta\theta')]^2\rangle$ (see Eq. 4). As in all other wave problems, when the coherence length of the waves—x-rays or neutrons—becomes comparable to, or smaller than, the size of the domains, a halo of small angle diffuse scattering surrounds the specularly reflected signal.

Regardless of the method of analysis, all scattering measurements suffer from the same problem of extracting a *real-space* model of the scatterer from scattering information confined to *a finite region of reciprocal space*. Most reflectivity measurements have been analyzed by comparing the measured reflectivity to the reflectivity calculated from various model surface profiles. Unfortunately, agreement between the two, when the data are only over a finite range of Q_z, does not prove that other model profiles might not have agreed as well.

C. Grazing Incidence Diffraction and Diffuse Scattering

As mentioned in Sec. A, when $\theta \leq \theta_c$ the incident wave penetrates into the sample as an evanescent wave confined to a distance $\Lambda_i \lesssim 5$ nm from the interface. Scattering of this wave is ideally suited for studying surfaces in which the interfacial width is less than Λ_i and the surface inhomogeneities are within Λ_i

of the surface. The simplest approach to scattering by this wave, known as the distorted wave approximation [32, 33], uses the exact solution to Eq. 5 for the interface with a homogeneous bulk (that is, the Fresnel solution) and then calculates the scattering of that wave from surface inhomogeneities. The scattering of the "distorted wave" of the refracted field is in contrast to the first Born approximation for which the scattering is of the incident plane wave.

For scattering from inhomogeneities within Λ_i of the surface, the scattered intensity can be written as [34–37]

$$I(\mathbf{Q}) \propto |T(\theta)|^2 \, |S(\mathbf{Q})|^2 \, |T(\theta')|^2 \tag{15}$$

where $\mathbf{Q}$ is the wavevector transfer, $S(\mathbf{Q})$ is the scattering structure factor, and $T(\theta)$ is the Fresnel transmission coefficient for the amplitude of the total wave field at the surface, polarized perpendicularly to the plane of incidence, written as [14]

$$T(\theta) = \frac{2n_1 \sin\theta}{n_1 \sin\theta + [n_2^2 - n_1^2 \cos^2\theta]^{1/2}} \tag{16}$$

For small angles and $n_1 = 1$, $T(\theta) \cong 2\theta/[\theta + \sqrt{\theta^2 - \theta_c^2}]$ and $|T(\theta)|^2 \cong (2\theta/\theta_c)^2\sqrt{R_F(\theta)}$. The factor $T(\theta)$ enhances the surface scattering because the scattered intensity is proportional to the square of the total surface field, not the square of the incident field. For $\theta \leqslant \theta_c$, the amplitude of the reflected and incident fields are equal, and because the phase of the reflected wave varies from π when $\theta \cong 0$ to 0 when $\theta \cong \theta_c$, the total field at the surface varies from zero to twice the incident field for $\theta \cong \theta_c$. The function $|T(\theta)|^2$, which is proportional to the square of the field, varies from 0 to 4, as θ increases from 0 to θ_c, and then falls to unity for $\theta \gg \theta_c$. The factor $T(\theta')$ appears because of a similar effect [34–39] in the coupling between the surface currents and the scattered fields. The angular dependence of the two factors $T(\theta)$ and $T(\theta')$ enables the surface scattering to be distinguished from other diffuse scattering processes.

For the idealized case of a monolayer of point particles at an interface between vacuum and a bulk material, the distorted wave approximation yields the following simple form for the differential cross section for x-rays [33, 35, 40]:

$$\frac{d\sigma}{d\Omega} = r_e^2 A_{xy} \, |T(\theta)|^2 \, |T(\theta')|^2 \int dx\, dy \, \langle \rho_s \, \rho_s(x, y) \rangle \, e^{i\mathbf{Q}_{xy}\cdot\mathbf{r}} \tag{17}$$

where $\rho_s(x, y)$ describes the surface electron density of the point particles, that is, $[\rho(x, y, z) - \langle \rho(z) \rangle] \cong \rho_s(x, y)\delta(z)$, and A_{xy} is the illuminated area of the sample. If the particles form a two-dimensional ordered phase, the evanescent wave will diffract from the in-plane order when the in-plane transferred momentum $\mathbf{Q}_{xy}$ corresponds to an in-plane reciprocal lattice vector $\mathbf{G}_{xy}$, regardless of the value of Q_z. As a consequence, there are Bragg rods of scattering for surface order (instead of Bragg peaks as for bulk diffraction) [41]. For point particles the intensity along the Bragg rod will be determined only by the classical surface field enhancement, $T(\theta)$ and $T(\theta')$, described previously, and by polarization effects. If the particles in the monolayer are molecules, then the intensity $I(Q_z)$ along the Bragg rod is modulated by the molecular structure factor $F(\mathbf{G}_{xy}, Q_z)$ normal to the surface and is represented as

$$I(Q_z) \propto |T(\theta)|^2 \; |(F(\mathbf{G}_{xy}, Q_z)|^2 \; |T(\theta')|^2 \qquad (18)$$

For monolayers of long molecules that exhibit ordering of the molecular tilts, the factor $F(\mathbf{G}_{xy}, Q_z)$ is a function of both the molecular tilt and the intrinsic molecular structure factor. The intensity distribution along the Bragg rod provides information analogous to that contained in reflectivity measurements, but only for the ordered part of the monolayer.

Although in general it is difficult to observe the relatively weak surface diffuse scattering in the presence of much stronger diffuse scattering from the bulk, small-angle scattering away from the specular condition can be an exception. In particular, if the interface is not perfectly flat and if the amplitude of the height fluctuations, $\zeta(x, y)$, are small in comparison with $1/Q_z$, the diffuse scattering can be calculated by setting $\langle \rho_s \rho_s(x, y) \rangle = \rho^2 \langle \zeta\zeta(x, y) \rangle$ in Eq. 17 [33, 40, 42].

III. PURE LIQUIDS

The interfacial profile along the surface normal between a pure liquid and its vapor is characterized by a combination of an intrinsic profile that describes the molecular distribution of an otherwise flat interface and thermally induced fluctuations that roughen the interface [43]. The interplay between these aspects of the interfacial profile has generated considerable theoretical interest, but experimental information has been meager. Although it is not clear that there is an unambiguous way to separate the two, thermally induced roughening increases the average width of the interfacial profile. At temperatures much lower than the liquid-vapor critical point, the particle-particle statistical correlation length in the bulk will be on the order of the molecular size. According to standard theoretical ideas—neglecting surface roughening—the decay of surface-induced perturbations will be on the same length scale as the decay length of bulk correlations. Far from the critical point, roughening is simply modeled by thermal capillary waves; however, near the liquid-vapor critical point, divergence of the particle-particle correlation length will induce divergences in both the intrinsic width and thermal fluctuation, or capillary, roughening. These effects have been seen in light scattering studies [44–46]. The most extensive x-ray and neutron scattering studies on pure liquids have been on the water-vapor interface at room temperature, far from the critical point. A few experiments have been performed on other simple solvents [26, 33].

A. Water—Reflectivity

X-ray specular reflectivity data from the free surface of H_2O, as shown in Fig. 4a, is typical of the reflectivity from pure liquid surfaces [33]. These data were taken from the surface of H_2O in a vapor- and temperature-controlled Langmuir trough in which the surface tension was monitored in situ to be 72.5 ± 0.4 mN/m [40]. There is a small region for $\theta < \theta_c$ where the reflectivity is essentially 100% followed by a rapid fall, shown here over eight orders of magnitude. As shown in Fig. 4b, where the same data are normalized to $R_F(\theta)$ and plotted as $R(\theta)/R_F(\theta)$ versus Q_z^2 on a semilog scale, the data are reasonably well represented by the form

$$\frac{R(\theta)}{R_F(\theta)} = \exp(-Q_z^2\sigma^2) \tag{19}$$

where $\sigma^2 = (2.70 \pm 0.03\ \text{Å})^2$. When the effects of thermal capillary waves are taken into account, there is a weak logarithmic dependence of σ on Q_z [33,

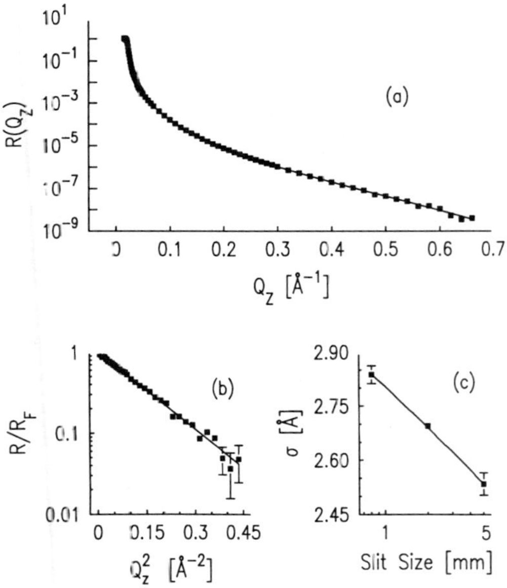

FIG. 4 (a) Measured reflectivity from a water-vapor interface with a detector slit height of 2 mm at 600 mm from the sample. The solid line is the best fit of the theoretical form given by Eqs. 19 and 20 to the data. (b) The same data as in (a) plotted versus Q_z^2 on a semilog plot. The slope of the solid line indicates a value of $\sigma^2 = (2.70 \pm 0.03\ \text{Å})^2$. (c) Comparison between the slope of data similar to that shown in (b) for detector slit sizes (heights) of 0.8, 2.0, and 5.0 mm and the theoretical result for capillary wave roughness (Eq. 20) indicated by the line [40].

40]. It is straightforward to demonstrate that the form in Eq. 19—with a constant σ—is consistent with the profile $(d\langle\rho(z)\rangle/dz)/\rho_{water} = (1/\sqrt{2\pi\sigma^2})\exp(-z^2/2\sigma^2)$ either by substitution into Eq. 7 or by integrating the data using Eq. 11 to obtain the Patterson function $P(s) \cong (1/\sqrt{4\pi\sigma^2})\exp(-s^2/4\sigma^2)$.

The interfacial width σ has previously been represented as a combination of a local microscopic electron density gradient σ_0 and the capillary wave contribution [33, 40, 47]

$$\sigma^2 = \sigma_0^2 + \frac{k_B T}{4\pi^2\gamma}\int_{A_s}\frac{d\mathbf{q}}{q^2 + k_g^2} \tag{20}$$

where γ is the measured interfacial tension, $k_g^2 = \rho_m g/\gamma$ where ρ_m is the *mass* density, and the area A_s, in reciprocal space, is a circular region of outer radius q_{max} and an inner boundary defined by a rectangular cutout. The inner boundary of A_s is determined by the known resolution of the x-ray reflectometer, such that x-rays scattered by capillary waves with wavevectors $\mathbf{q}$ outside of the cutout, and thus contained in A_s, are deflected beyond the angular acceptance of a detector centered on the position for specular reflection. Since the longest wavelength capillary waves scatter at very small wavevectors, i.e., within the rectangular cutout of A_s, they do not have an effect on the value of σ^2 as determined from the reflectivity measured by a detector with finite resolution. Consequently, once the resolution of the reflectometer and the incident angle have been set, σ^2 is specified in terms of only two adjustable parameters: the intrinsic width of the electron density profile σ_0 and the radius q_{max} of the outer perimeter of A_s. Although the integration in Eq. 20 actually results in a logarithmic dependence of σ^2 on Q_z, the variation is slow and, as mentioned above, the data are well represented by a value of σ^2 that is independent of Q_z but logarithmically dependent on the reflectometer resolution.

The origin of both σ_0 and q_{max} are microscopic (on the scale of angstroms), and therefore these two parameters cannot be distinguished by this set of measurements. In our most recent experiment on the water surface, three sets of data, taken with three different detector heights (resolutions), were simultaneously fit by numerical integration of Eq. 20 over the measured resolution function—the "intrinsic width" σ_0 was set to be zero, and the fit had only one adjustable parameter, q_{max}. The lines in Figs. 4a and 4b display this fit, with $q_{max} = (\pi/1.4 \pm 0.1$ Å), for the data set with a detector height of 2 mm. Figure 4c displays the comparison of the measured values of σ^2, as obtained from the best fits like that represented by the solid line in Fig. 4b, and those calculated for $Q_z = 0.4$ Å^{-1}. X-ray measurements over the accessible range of angles cannot distinguish between this approach of using one particular value of q_{max} for an interface that is locally sharp [40] and alternative hypotheses about the nature of the surface on atomic length scales [48].

Recent neutron reflectivity measurements on H_2O-D_2O mixtures at room temperature reported an interfacial width σ of approximately 2.8 Å, but without commenting on its dependence on Q_z [11]. Full details of these experiments have yet to be published; therefore a detailed comparison of this value with those quoted above for the various resolutions cannot be made.

B. Water—Diffuse Scattering

In the capillary wave model for the interfacial width, the reduction of $R(Q_z)/R_F(Q_z)$ to less than unity occurs because capillary waves scatter radiation away from the specular condition. Measurements of this diffuse scattering from the surfaces of ethanol [50] and of water [40] have been recently reported. To examine this diffuse scattering from the water surface, x-ray intensity was measured for two fixed incident angles θ by scanning the outgoing angle θ' in the plane of incidence ($\phi = 0$, see Fig. 1) [40]. The peaks at $\theta = \theta'$ in Fig. 5 are the specular signals, and the weaker peaks at $\theta' \cong \theta_c = 0.14°$, due to the surface field enhancement $T(\theta')$, are sensitive to the scattering cross-section from the capillary waves. The solid lines through the data are calculated by averaging the cross-section, Eq. 17, over the angular distribution of the incident beam and

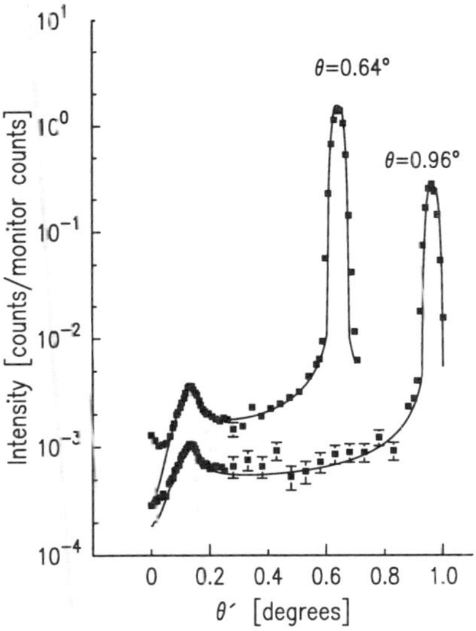

FIG. 5 Scattered intensity, in the plane of incidence, as a function of detector angle θ' for incident angles $\theta = 0.64°$ and $\theta = 0.96°$. The weak peaks at $\theta \cong \theta_c = 0.14°$ are due to the surface scattering enhancement factor; the strong peaks at $\theta = \theta'$ are the specular reflectivity signals. The solid line is the theoretical prediction calculated with no significant adjustable parameters. (From Ref. 40.)

integrating over the detector resolution. The scattered intensity I_D is written as [40]

$$I_D = I_0 \frac{(r_e\rho)^2}{\theta\,\Delta\theta} \frac{k_B T}{\gamma} \int_{(\theta-\Delta\theta/2)}^{(\theta+\Delta\theta/2)} d\theta \int_{(-\Delta\phi/2)}^{(\Delta\phi/2)} d\phi \int_{(\theta'-\Delta\theta'/2)}^{(\theta'+\Delta\theta'/2)} d\theta' \frac{|T(\theta)|^2\,|T(\theta')|^2}{q^2 + k_g^2} \tag{21}$$

where I_0 is the total incident flux and $\Delta\theta$, $\Delta\theta'$, and $\Delta\phi$ are the widths of the distributions of the respective angles.‡ The only adjustable parameter is a small constant background, of the order of 10% of the peak at $\theta' = \theta_c$. The high quality of the fit indicates that the capillary wave model is a good representation of the height-height correlation function over a region in reciprocal space corresponding to capillary waves with wavelengths from 40 to 8000 nm [40]. This extends the range of the smallest measured capillary wavelength from approximately 10 μm, as measured by optical light scattering [49], to approximately 40 nm, as measured by x-ray scattering. Agreement between data and theory for both the diffuse scattering and the resolution dependence of $R(Q_z)/R_F(Q_z)$ confirms the role of thermally excited capillary waves and demonstrates the quantitative reliability of the experimental technique.

IV. LIQUID CRYSTALS

This section has been prepared under the assumption that the reader is familiar with the general properties of liquid crystals; those for whom this is not the case might find the various reviews of liquid crystals useful [51–53]. Specular reflectivity is the only one of the three types of scattering techniques mentioned above that has been widely applied to liquid crystals. For liquid crystals, in-plane surface order that might be studied using grazing incidence diffraction from the surface of bulk samples can be more easily studied in transmission from freely suspended thin smectic films, for which the ratio of surface-to-bulk scattering is enhanced by the relatively high surface-to-volume ratio [54]. In this section we shall review both x-ray specular reflectivity from the surfaces of bulk samples and diffraction studies on thin films, and neutron specular reflectivity from the surfaces of solutions of lyotropic liquid crystals.

A. X-Ray Scattering

One of the properties that has contributed to general interest in the physics of liquid crystals has been the manner in which they form lamellae-like smectic phases [55, 56]. For example, on lowering the temperature, molecules in the

‡We have chosen constant, or square wave, angular intensity distributions. For example, for any incident angle within the range $\theta \pm (\Delta\theta/2)$ the incident intensity is constant. This is a good approximation in this case; however, the results are insensitive to the exact form of the distribution.

homologous cyanobiphenol series, nCB, with $n \leqslant 9$, exhibit second-order nematic-to-smectic-A transitions; for $9 < n \lesssim 9\ 5$ the transitions are first-order, while for $9.5 \lesssim n$, there is no nematic phase and there are direct first-order transitions from the isotropic to the smectic-A phase [57–60]. One would intuitively expect smectic layers to form at the surfaces of these materials [61–64], even at temperatures for which the bulk is either isotropic or nematic. Figure 6 shows x-ray data for $R(\theta)/R_F(\theta)$ that illustrate this effect for four different liquid crystals at temperatures approximately 0.05K above the transition temperature to the smectic phase. For 8OCB (a) [29, 65] and 9CB (b) the transition from the nematic to smectic phase is second-order; for 9.15CB (c)—a mixture of 9CB and 10CB—the transition from the nematic to smectic-A phase is first-order; and for 10CB (d) the transition from the isotropic to smectic-A phase is first-order [10, 66, 67]. The abscissa for each of these plots is the ratio of the wavevector transfer Q_z to that of the peak in the ordered smectic-A phase (i.e., Q/Q_0). In large part, the structure at $Q/Q_0 \approx 1$ arises from constructive interference between the waves that are back-scattered from surface-induced smectic layers and waves reflected directly from the free surface; however, details of the structure are influenced by the number of surface layers, their position relative to the surface, and their relative amplitude, which, among others things, describes the nature of the interface between the surface-induced smectic and the nematic or isotropic phase below the surface.

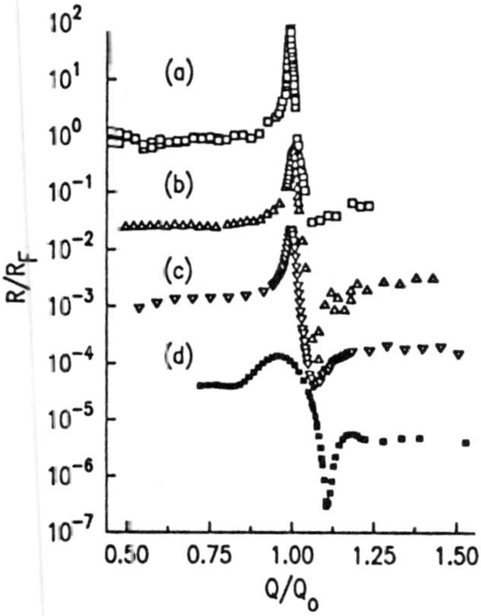

FIG. 6 The ratio of $R(Q_z)/R_F(Q_z)$ for (a) 8OCB (b) 9CB (c) 9.15CB and (d) 10CB as a function of Q_z/Q_0 at temperatures approximately 0.05 to 0.06°C above the smectic-A transition temperature. The values of Q are respectively (a) 0.0218 Å^{-1}, (b) 0.0212 Å^{-1}, (c) 0.0206 Å^{-1}, and (d) 0.0213 Å^{-1}. (From Ref. 29.)

From these and other experiments on similar materials, the surface appears always to induce some smectic layering; however, the number of surface layers always decreases with increasing $|T - T^*|$, where T^* is the transition temperature to the smectic phase [67–72]. When the transition from the nematic to the smectic phase is second-order, the number of layers diverges proportionally to $\xi_{\parallel} \sim (T - T_{NA})^{-\nu_{\parallel}}$ [59, 60, 73–76] in a manner that is appropriately termed *critical adsorption* in the language that is used for physical wetting [77–80]. When the transition is first-order, or when the transition is from the isotropic to the smectic-A (also first-order), the number of layers appears to increase to a finite maximum at the transition, corresponding to incomplete wetting [10, 29, 66, 70]. It is not clear whether or not the phenomena of critical wetting has yet been observed.

The geometry of reflectivity studies of liquid crystals provides an ancillary benefit for the intensely studied phenomena of critical bulk scattering from smectic fluctuations near the second-order nematic-to-smectic-A transition [51a, 81]. A technical difficulty for these measurements is that as the critical correlation lengths diverge, the half-width of the peak becomes comparable to the width due to the finite angular spread in the orientation of different domains; this is one of the factors that limits the maximum correlation lengths that can be measured [59, 60, 73–76, 82–84]. Domain alignment induced by the free surface appears to be at least one order of magnitude better than alignment that had previously been possible using either external magnetic fields or the surfaces of confined samples, and by tuning of the specular condition it is possible to measure the critical bulk scattering in the reflection geometry [67, 71]. Although this technique has been demonstrated, it has not been fully developed.

Similar reflectivity and bulk critical scattering effects have been studied for the lyotropic liquid crystal formed by mixtures of cesium perfluorooctanoate (CsPFO) and water [85]. For $T \gtrsim 50$–60°C, mixtures containing approximately equal weight fractions of the two form micellar solutions, which on cooling exhibit a nematic phase—approximately 7°C wide—followed by a smectic-A phase on further cooling. Specular reflectivity studies indicate that even in the isotropic phase there is a relatively dense monolayer of CsPFO on the surface, oriented with the fluorocarbon toward the vapor; this layer does not appear to be affected by the bulk transition to the nematic phase. On cooling deeper into the nematic phase, the number of surface-induced smectic layers increases, as with observations for thermotropic systems; however, because of difficulties in maintaining the composition of the solution, it was not possible to say that this was truly a case of critical adsorption. In any event, within a few degrees of the transition to the smectic phase, the surface layers appear to be relatively intact bilayers, as distinct from the other possible case in which the same anisotropic micelles that form the bulk nematic phase are somehow aligned in layers. In contrast to this example, specular reflectivity studies show that the spherical micelles that form the microemulsion system made up of AOT (sodium di-2-

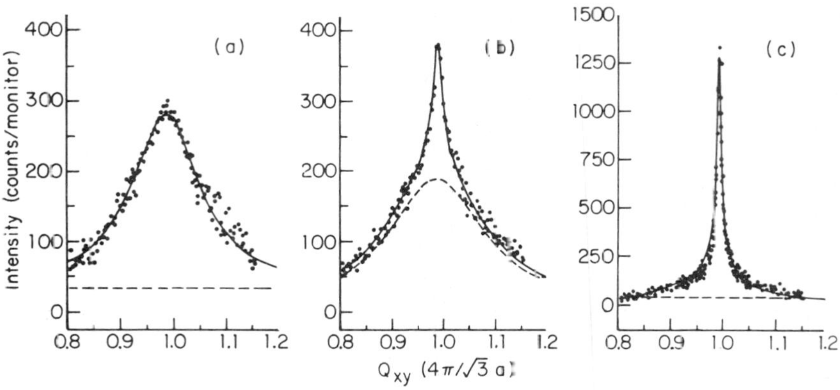

FIG. 7 Comparison of the measured scattering with the corresponding fits for the $Q_z = 0$ spots for a five-layer film of 7O.7. The data have been corrected for the no-film background and normalized to a constant incident flux. The fits are indicated by a solid line. (a) Scans taken at 78.14°C where the entire film is smectic-C. As shown, the data is well fit by a single Lorentzian with $\xi = 1$ nm plus constant diffuse background (dashed line). (b) Scans taken at 77.67°C where the two surface monolayers are smectic-I and the three interior layers are smectic-C. The fit shown is the sum of a broad smectic-C Lorentzian (dashed line) and a narrow smectic-I square-root Lorentzian. The square-root Lorentzian correlation length is $\xi = 9$ nm. (c) Scans taken at 69.09°C where the entire film is smectic-I. The fit shown is the sum of a single sharp square-root Lorentzian with $\xi = 26$ nm plus a small constant diffuse level (dashed line). (From Ref. 96).

ethylsulfosuccinate), D_2O, and decane come right up to the surface, without any intervening surfactant monolayer [86].

The second type of surface x-ray study concerns in-plane positional order within the smectic layers, or parallel to the surface. For most liquids, diffuse scattering from the bulk precludes studying this any other way than by the grazing incidence technique; however, for liquid crystals it is possible to form metastable freely suspended films of smectic phases that are only two layers thick and that are all surface [54]. By carrying out scattering studies as a function of film thickness it is possible to separate surface and bulk effects. Experiments of this type have been reviewed in a number of places [51b, 54, 87–98]. As an example, Figure 7 shows x-ray scattering spectra, taken in transmission, from a five-layer-thick film of *N*-[4′-(*h*-heptyl)benzylidene]-4-(*n*-heptyl)aniline (7O.7) at temperatures where the entire film is in the smectic-C phase (a), where the two surface layers are smectic-I and the interior three layers are smectic-C (b), and where all five layers are smectic-I (c) [96].

B. Neutron Scattering

Like the lyotropic liquid crystal CsPFO discussed above, the compound decyl-trimethyl-ammonium bromide (DTAB) will form liquid crystalline phases when mixed with water. The lowest concentration of DTAB in water at which organized structures appear in the bulk is the critical micelle concentration (cmc), at which micelles form. Below the cmc, isolated molecules of DTAB will be found in the bulk; in addition, there is the possibility of an adsorbed monolayer at the surface. Recent neutron reflectivity measurements [11, 99, 100] have explored the structure of the surface of a DTAB-in-water solution at concentrations below and slightly above the cmc. This study reveals some of the strengths and weaknesses of neutron reflectivity and will be discussed in some detail.

The neutron scattering lengths for hydrogen and deuterium are of opposite sign, allowing for a contrast-matched H_2O-D_2O mixture (0.912 molar ratio) which does not coherently scatter the neutrons; there is no specular reflection from the surface [11, 101]. A low-concentration solution, in contrast-matched water,of a soluble surfactant with a large enough scattering length, such as partially or fully deuterated DTAB, will only specularly reflect neutrons if the surfactant is adsorbed at the surface.

Neutron reflectivity was measured from the liquid-vapor interface of contrast-matched water solutions of partially and fully deuterated DTAB for values of $Q_z \leq 0.2$ Å^{-1} [99]. Because of the restricted range of Q_z, the data were usually featureless—without interference minima and maxima—and exhibited only a rapid decrease of intensity with increasing Q_z. As previously discussed, Eqs. 8 and 9 indicate that when interference features are observed their positions determine the thickness of a film. In the absence of identifiable interference minima or maxima it would seem difficult to infer the thickness of the film from just the shape of the initial decay in the reflected intensity. As was shown by Eq. 10, the decay is determined both by the interference features and by the roughness of the interfaces. Lee et al. [99] reported that the thicknesses derived from fits to their data, for the fully deuterated DTAB solution in contrast-matched water, varied from 1.5 nm to 2.1 nm depending on whether or not they included interfacial roughness in the functional form for the profile; the largest thickness is the result of no roughness. In general, they chose a model profile without any interfacial roughness, which tends to give larger thicknesses, and relied on the use of data from other isotopic compositions of DTAB to constrain their fits. For example, according to Lee et al., a protonated portion of DTAB has approximately the same scattering length density as air; therefore, selective deuteration of either the head group ($C_{10}H_{21}N(CD_3)_3^+Br^-$) or the tail group ($C_{10}D_{21}N(CH_3)_3^+Br^-$) allowed a determination—subject to the caveats mentioned above—of the thickness and neutron scattering length density of those regions. At a concentration of 0.05 M, the surface density, calculated from the model of the reflectivity, was 58 $\pm$ 5 Å^2/molecule, indicative of a closely packed monolayer. For this concentration a two-slab model was fit to three sets of reflectivity data for the three isotopic compositions just mentioned (partially or fully deuterated) and yielded an overall thickness of 2.1 nm. This indicates that the molecules are staggered or interdigitated in some manner, because the longest length for an all-trans single molecule is approximately 1.7 nm. One slab consisted of 90% of the tail region scattering length density, and the other slab contained the remaining 10% of the tail density as well as the head group scattering length density. The mixture of tail and head groups in one of the slabs introduced an effective disorder in that region of the profile. Because it assumes a sharp interface between the regions of different density, this representation of the disorder is not equivalent to an effective roughness, and it will not reduce the reflected intensity for large Q_z, as would be expected from a rough interface (see Eq. 10). Although their data and modeling are suggestive of staggered head groups, it is unclear how this conclusion would be affected if roughness were included in the model. At lower concentrations (0.002 to 0.01 M), the monolayer was less dense, and staggering of the head groups was not required by the fitted thickness of the surface layer.

Lee et al. [99] also studied a solution of fully protonated DTAB in pure D_2O. They were able to show (based on considerations of molecular volumes in addition to analysis of the reflectivity) that there is some D_2O in the head group region; however, there is little or no D_2O in the tail region. Although this is not a surprising result, it would be difficult to demonstrate this from x-ray reflectivity because the electron density contrast between closely packed tail groups and H_2O is very small.

Interference maxima in the reflectivity ratio were observed in a later study of fully protonated $C_{10}TAB$ (DTAB) at concentrations slightly above the cmc in D_2O, and also for the longer chain length isomorphs, $C_{12}TAB$ and $C_{14}TAB$, where the maxima occur at progressively smaller Q_z [102]. Although a quantitative analysis of the data was not presented, the authors argued that the surfaces consisted of a monolayer, followed by an alternating series of bilayers and regions of D_2O with counterions, with the thickness of the D_2O layer progressively increasing further from the surface until the bulk concentration of D_2O is attained [102].

Several other water-soluble surfactant systems, including tetramethylammonium dodecyl-sulphate and dodecanol-sodium dodecyl-sulphate, have recently been studied using techniques very similar to those just discussed for DTAB [11, 103].

V. INSOLUBLE MONOLAYERS ON WATER

A. Monomeric Surfactants

In the past few years, several research groups have performed x-ray and neutron studies of insoluble monolayers of surfactants such as fatty acids and lipids at the water-vapor interface. In this section we shall discuss representative examples of reflectivity and grazing incidence diffraction studies of these monolayers [4, 12, 13, 26, 48, 104–131].

1. X-Ray Reflectivity

The normalized x-ray reflectivity $R(Q_z)/R_F(Q_z)$ for a monolayer of tetracosanoic acid [$CH_3(CH_2)_{22}COOH$] on a pH 2 solution of H_2O and HCl, as shown in Fig. 8a [124], is typical of the reflectivity from insoluble monomeric surfactants at comparable surface densities [10, 12, 13, 48, 104, 107, 108, 111, 112, 117, 123, 124, 126–129]. The solid line is a fit of the normalized reflectivity to $R(Q_z)/R_F(Q_z)$ calculated for the electron density profile illustrated in Fig. 8b, which is written as [124]

$$\frac{\langle\rho(z)\rangle}{\rho_w} = \frac{\rho_t}{2\rho_w}\left[1 + \operatorname{erf}\left(\frac{z}{\sqrt{2\sigma^2}}\right)\right] + \frac{1 - (\rho_t/\rho_w)}{2}\left[1 + \operatorname{erf}\left(\frac{z - D}{\sqrt{2\sigma_2^2}}\right)\right] + \frac{A_G}{\sqrt{2\pi\sigma_G^2}}\exp\left[\frac{-(z - D)^2}{2\sigma_G^2}\right] \tag{22}$$

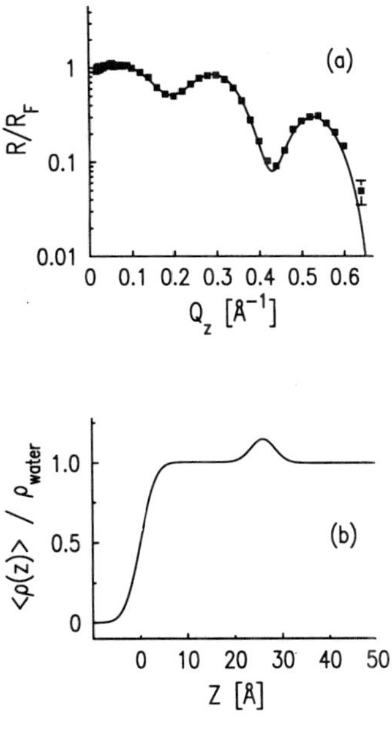

FIG. 8 (a) Measured $R(Q_z)/R_F(Q_z)$ for a tetracosanoic acid monolayer on the water-vapor interface at an area/molecule $A = 22.2$ Å^2 and a surface pressure $\Pi = 3.5$ mN/m (refer to Fig. 10 for an isotherm). The line represents the best fit of the data to the R/R_F calculated from the model electron density profile given by Eq. 22. (b) An illustration of the real-space electron density profile used to fit the data in (a) [124].

The first term describes the interface between the vapor and the hydrocarbon tail region [$CH_3(CH_2)_{22}$-] of the monolayer, where ρ_t is the electron density of the tail, ρ_w is the electron density of the bulk water, and σ is the RMS interfacial width. The third term is a Gaussian describing the excess electron density due to the head group (the carboxyl group—COOH), where A_G is the area under the Gaussian, σ_G is the Gaussian width, and D is the distance from the vapor-tail interface to the center of the Gaussian. The parameter D is a measure of the thickness of the monolayer. The second term is introduced to allow for the possibility that the water and tail regions could have different electron densities. Using a nonlinear least-squares fitting procedure, allowing ρ_t, σ, D, A_G, and σ_G as free parameters, the fits for many reflectivity curves yield $(\rho_t/\rho_w) = 1.0 \pm 0.02$, rendering σ_2 irrelevant. This value for $\rho_t \cong \rho_w$, which is typical of the literature, demonstrates the small electron density contrast between closely packed hydrocarbon chains and water and is very similar to the densities found in hydrocarbon three-dimensional crystals [132]. In addition, both the value of ρ_t/ρ_w and its constancy for all acceptable reflectivity measurements indicate a dense local packing of the molecules. Although the usual assumption of a molecular conformation with all-trans C—C bonds in the alkane chain is consistent with this result, the agreement is not a strong argument for either the presence or the absence of a small percentage of gauche C—C bonds, as has recently been predicted by numerical simulations for condensed monolayer phases [133–137]. Assuming the all-trans conformation, the thickness D can be interpreted in terms of a molecular tilt angle θ_T from the surface normal. The results of several reflectivity studies have been interpreted in terms of a tilt angle that changes from approximately 30° in the low-density phase to approximately 0° in the highest density region [104, 112, 114, 117, 124, 127, 129].

Two aspects of the head group structure can be extracted: the excess number of electrons and the effective width. The excess is the number of electrons due to the excess electron density above that of the surrounding regions (water and tail group). In this model (Eq. 22) the excess is directly related to the area A_G of the Gaussian piece of $\langle\rho(z)\rangle$. The excess number of electrons for many data sets for this molecule (tetracosanoic acid, a carboxylic acid) is 7 ± 0.1, which is the value calculated from the expected electron densities of the head group and surrounding regions. Constancy of this number implies, as expected, that the chemical structure of the head group is unchanged throughout the isotherm. The small error in determining the excess number of electrons indicates the potential of the reflectivity measurement to detect small changes in the chemical structure.

The primary difference between the present model and the slab models that have been used to represent much of the published reflectivity studies is the superposition of an independent slab, representing *excess electrons due to the head group,* on top of a broader slab that represents the average electron density near to the head group. The Gaussian in Eq. 22 represents this *extra slab*. The usual slab models are interpreted such that the *total* number of electrons in the head group is determined by the area of the head group slab (the product of the thickness of the head group and its electron density in the real-space model). Unfortunately, if the roughnesses of the two interfaces on either side of the slab (i.e., the tail–head-group and head-group–water interfaces) are appropriately adjusted, different thicknesses of the head-group slab, implying different numbers of electrons in the head group, can yield the same real-space interfacial profile. An unambiguous determination of these interfacial roughnesses, and therefore the *total* number of head-group electrons, would require higher spatial resolution than can be achieved by reflectivity taken only to $Q_z \leqslant 0.6$ Å^{-1}, as has been reported. This issue highlights the subtlety involved in determining a real-space model from reflectivity taken over a finite range of Q_z.

The effective head-group width σ_G is not the size of a single molecular head group but corresponds to the average width of the region occupied by the head groups along the surface normal. In one measurement on arachidic acid monolayers, the head group width was reported to decrease as the pressure increased; however, the authors did not provide an explanation for this effect [117]. In other measurements, on tetracosanoic acid and on an amino acid fluorinated surfactant, the head group width was reported to increase as the monolayer was compressed from low to high density [124, 127]. The latter would imply interdigitation or some other movement along the surface normal of the head groups relative to each other.

2. X-Ray Diffraction

X-ray grazing incidence Bragg diffraction has recently been used to investigate the in-plane ordered structures of monolayers of long-chain fatty acids and lipids on water [12, 106, 109, 110, 113–121, 129, 138–140]. The scattered intensity is very weak and, except in a few cases, only the lowest-order Bragg peaks have been measured [106, 126]. The weak scattering is also the explanation for the lack of similar neutron measurements [141]. Nevertheless, with the simple assumption of local triangular packing, a rich variety of in-plane structures, that often mimic well-known liquid crystalline bulk phases, have been discovered. In particular, transitions between phases with hexagonal and distorted hexagonal symmetries, as well as between phases with ordering in the azimuthal in-plane orientation of the molecular tilt direction, have been measured [113, 121, 140]. The detailed structural information provided by these studies indicates that the previously used phase nomenclature, such as liquid-expanded or liquid-condensed, is misleading, if not incorrect [142]. The weak scattering has precluded detailed line shape measurements; therefore, certain features of the phases, such as the distinction between two-dimensional crystallites or hexatic domains, have yet to be made in these systems [113, 116, 143–145].

Typical diffraction peaks corresponding to three different phases are illustrated in Figs. 9a–e for typical values of surface density and pressure, as illustrated in Table 2 and Fig. 10. These were measured along an approximately 20.5°C isotherm (Fig. 10) of tetracosanoic acid on a pH 2 water subphase [140]. The positions of the peaks along Q_{xy} determine the lattice spacings and, assuming the all-trans configuration, the positions of the peaks along Q_z determine the tilt from the surface normal θ_T and its azimuthal orientation within the lattice (see Eq. 18). There is some surface-pressure Π dependence of the peak positions, two-dimensional lattice spacings (a, b), tilt angles θ_T, and unit cell areas A_x; nevertheless, typical numerical values are listed in Table 2 for the lattices illustrated in Fig. 9f. The azimuthal orientations of θ_T, for the I and F phases, are indicated by the major axis of the ellipsoids in Fig. 9f; the U phase is untilted.

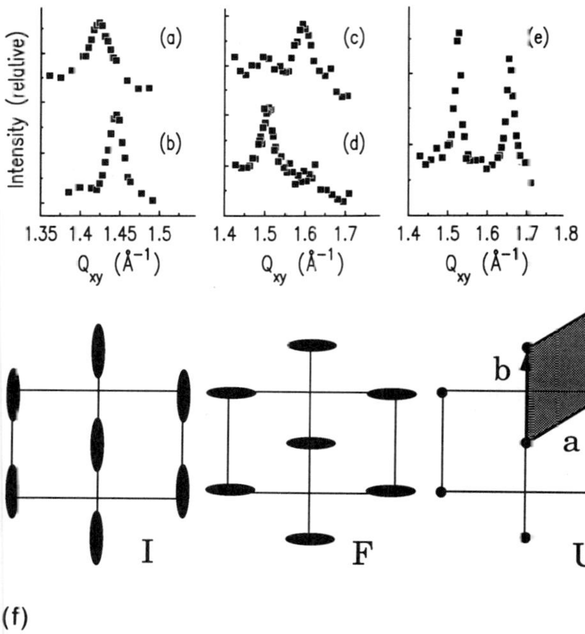

FIG. 9 Scans illustrating diffraction results for the I phase: (*a*) $Q_z = 0.78\text{Å}^{-1}$, (b) $Q_z \cong 0$); the F phase: (c) $Q_z = 0.50$ Å^{-1}, (d) $Q_z = 0.24$ Å^{-1}; and the U phase: (e) $Q_z \cong 0$ for surface densities indicated in Table 2. The peaks centered at different Q_z values are vertically offset for display. (f) Schematic illustration of the two-dimensional distorted hexagonal (e.g., rectangular) lattice corresponding to the diffraction results illustrated in (a)–(e). The primitive unit cell area A_x is indicated by the shaded parallelogram. The peak positions and lattice structure for these three data sets are listed in Table 2. (From Ref. 140.)

TABLE 2 The Phases (Ph), Peak Positions (Q_{xy}, Q_z), Two-Dimensional Lattice Spacings, (a, b), Unit Cell Areas (A_x), and Tilt Angles from the Surface Normal (θ_T) for Three Sets of Diffraction Data Shown in Fig. 9.

Ph	A_x	Q^a_{xy}, Q^a_z	Q^b_{xy}, Q^b_z	a	b	θ_T
	(Å²)	(Å⁻¹)		Å		°
I	22.3	1.445, 0.0	1.42, 078	5.05	5.14	30
F	19.4	1.60, 0.50	1.50, 0.24	4.64	4.92	15
U	18.5	1.66, 0.00	1.526, 0.0	4.50	4.89	0
	±0.1	$\Delta Q_{xy} \cong \pm 0.005$, $\Delta Q_z \cong \pm 0.05$		±0.02	±0.02	±3

Source: Ref. 140.

Some similar structures have been seen for monolayers of different fatty acids [113, 121] and also for smectic-F and -I liquid crystalline phases [29]. In addition, the in-plane unit cell of the U (and F) phase is similar to that of the three-dimensional untilted crystalline "B" (and tilted "C") phase for pure alkanes [132].

As is true for much of the literature for fatty acid monolayers, the in-plane peak widths ΔQ_{xy} are usually resolution limited, or nearly so, corresponding to two-dimensional positional correlation lengths $\geqslant$ 25 nm. Statistical analysis of intensities [140] at $\pm\phi$ and fluorescence microscopy of other similar systems indicate that some domains may be much larger, on the order of 10 μm [110, 116, 122]. The minimum width along the Bragg rod is approximately (monolayer thickness)$^{-1}$, which is much larger than the minimum instrumental resolution. The data shown in Fig. 9 were averaged over $\Delta Q_z \cong 0.06$ Å^{-1}. For higher resolution in Q_z, the intensity along the Bragg rod $I(Q_z)$ (see Eq. 18) can be analyzed to extract the electron density normal to the surface of the ordered

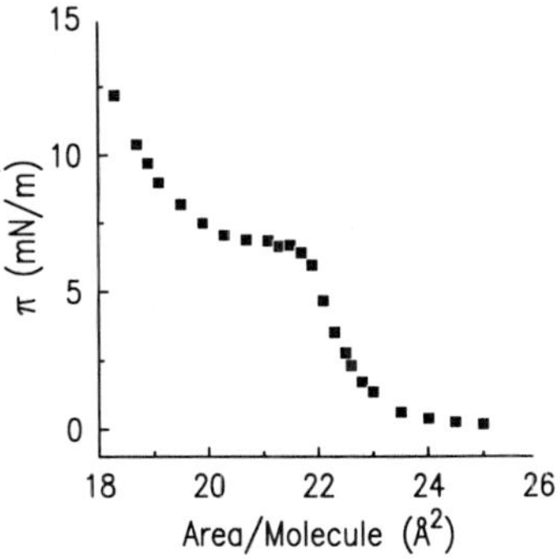

FIG. 10 Isotherm for a tetracosanoic acid monolayer, in the Π-A plane, on the surface of pH 2 water (with HCl) at 20.5°C.

regions of the monolayer. However, the weak scattering severely limits the counting statistics for high resolution in Q_z. Recent measurements on a fluorocarbon surfactant, in which the scattering is greater than that from a typical hydrocarbon surfactant, allowed for the analysis of the Bragg rod intensity and extracted information analogous to that from reflectivity measurements [112, 129].

An unanticipated discovery of some recent diffraction studies is the existence of *island phases* of crystalline domains (e.g., the I phase) in the low-density region of the isotherm. For tetracosanoic acid, this region is from 22 Å^2/molecule to at least 35 Å^2/molecule (Fig. 10 illustrates part of this region of the isotherm; for even lower densities the pressure remains at similarly low values [112, 121, 128, 131, 140, 146]. Even at these low pressures (approximately 0.1 mN/m) and densities the molecules self-assemble into crystalline or hexatic domains whose lattice unit cell area (typically 22 Å^2/molecule) does not change significantly throughout the low-density region, indicating that large regions of the monolayer are occupied by a very-low-density phase undetected by the diffraction measurements. One simple hypothesis regarding this *island phase* is coexistence of two-dimensional crystalline and gas phases. Coexistence alone cannot explain the $d\Pi/dA$ in this region and, as discussed elsewhere, impurities in the gas phase may play a role in producing the surface pressure in this region [140]. In addition, long-range electric dipole effects, like those discussed by Andelman et al., might also be important [147].

Detailed x-ray diffraction studies at various points in the Π-T phase diagram have recently been reported for two monolayer systems (heneicosanoic and behenic acids) [113, 121]. Both of these studies identified phases similar to those mentioned above and correlated their appearance with certain features (low- or high-compressibility regions or kinks) of the Π-A isotherms. In particular, in the highest-density region of the phase diagram, the molecules are untilted with the in-plane packing limited by the cross-sectional area of the alkane chain (in contrast to the larger cross-sectional area of the head group) [113, 121, 140]. Between the highest-density region and the low-density island phase, a rich variety of transitions has been seen for different temperatures or systems.

A recent study in which the diffraction measurements were taken only after the monolayer was allowed to relax until $d\Pi/dt \leqslant 0.01$ mN/m-hr has revealed dramatic structural changes (not due to x-ray damage) for small changes $\Delta\Pi \cong 0.1$ mN/m [140]. A reentrant phase was observed that did not appear under the faster compression rates typically reported in the literature. This raises the issue of equilibrium in studies of monolayers, especially those of long chain molecules that form solid phases near room temperature and may also have slow relaxation kinetics.

X-ray reflectivity and diffraction and neutron reflectivity (as well as x-ray fluorescence) have been used to investigate the effect of pH on the binding of salts in the subphase to the head groups in the monolayer [4, 105, 123, 131, 148]. The reflectivity studies utilize the sensitivity of x-rays to the heavy salt ions to determine their position relative to the monolayer and allow the determination of a pK for binding of the ions to the head groups [123, 131] (see Patil for an example of a similar thermodynamic study [149]). One combined x-ray and neutron-reflection study of the low-density phase of a docosanoic acid monolayer reported a reorientation of the molecules from an approximately 30° tilt on an ion-free subphase to an approximately 0° tilt with bound Cd^{2+} ions [131]. A recent x-ray diffraction study of a heneicosanoic acid monolayer has found pH-dependent structural changes in the monolayer lattice due to the binding of Ca^{2+} and Cu^{2+} ions and related these changes to qualitative features of the isotherm [148].

B. Polymeric Surfactants

In this section we discuss a recent x-ray reflectivity measurement on a monolayer of hydroxylated 1,2-polybutadiene (50% random substitution of butyl alcohol) with emphasis on illustrating a simple use of the Patterson function (Eq. 11) [124, 150]. X-ray reflectivity was measured as the monolayer approached and entered the high-density region in which Π reaches a plateau at approximately 28 mN/m. For three surface pressures (20.3, 23.7, 28.2 mN/m), the reflectivity data are graphed as $R(Q_z)/R_F(Q_z)$ versus $Q_z{}^2$ on a semilog plot in Fig. 11. Two aspects of the data allow qualitative features of the electron density profile to

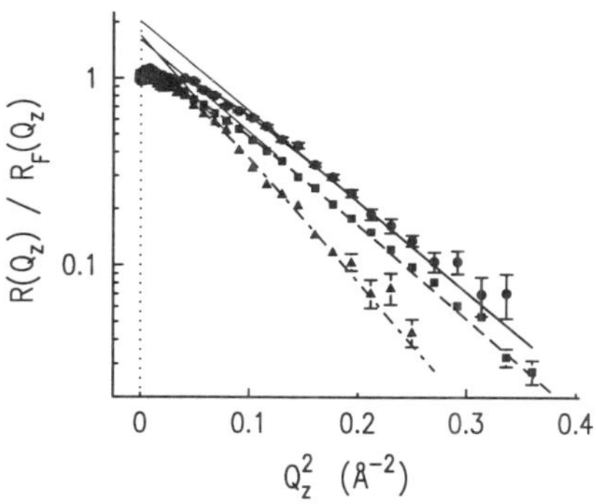

FIG. 11 The measured reflectivity from a monolayer of hydroxylated 1,2-polybutadiene (50% random substitution of butyl alcohol) on water, normalized to the theoretical Fresnel reflectivity, is plotted versus Q_z^2 for three surface pressures: 20.3 (solid circles, solid line), 23.7 (squares, dashed line), and 28.2 (triangles, long-short dashed line) mN/m. The lines indicate fits to a model for the electron density discussed in Ref. 150. The three thin solid lines indicate an extrapolation of the best fit from the linear, large Q_z^2 region back to $Q_z^2 = 0$ for the three data sets. (From Ref. 150).

be determined: (1) for large Q_z, $\log_e[R(Q_z)/R_F(Q_z)]$ decreases linearly with Q_z^2 and is well represented by the form $R(Q_z)/R_F(Q_z) \approx A^2 \exp(-Q_z^2\sigma^2)$; and (2) for all three data sets, extrapolation of the linear dependence back to $Q_z = 0$ rises above 1. Without reference to any model for the electron-density interfacial profile, it can be shown that these aspects indicate that the vapor-polymer interface is the sharpest feature of the interfacial profile, and the electron-density change associated with the sharp step is larger than the change from vapor to bulk water [150]. Again, without reference to a specific profile, a more quantitative determination of both of these features can be made by computing the Patterson function $P(s)$ of the data (Eq. 11).

$P(s)$ can be calculated with a fast Fourier transform (FFT) routine by expressing the data in terms of Q_z inside the bulk and making use of $|(\Phi(Q_z)|^2 = |(\Phi(-Q_z)|^2$. As with all FFT methods, it is necessary to make some assumptions about what the data would be like for Q_z larger than the maximum measured value. In the present case we assumed that the observed Gaussian falloff continues indefinitely past $Q_z \cong 0.6$ Å^{-1}. In fact the calculated form of $P(s)$ is not sensitive to the details of the extrapolation as long as $R(Q_z)$ continues to fall. Although Fourier analysis is often interpreted so that the spatial resolution $\Delta z \sim (\pi/Q_{max}) \approx 6$ Å, the accuracy in extracting widths for Gaussian type profiles is more like $\Delta\sigma \sim (1/Q_{max}) \approx 2$ Å. In the present case, since it is *very likely* that the amplitude of the reflectivity for $Q_z \geqslant 0.6$ Å^{-1} is small, the $P(s)$ calculated in this way probably represents the true Patterson function on length scales somewhat smaller than 2 Å. The Patterson function is plotted in Fig. 12 for one pressure (the other pressures are qualitatively similar) and for the pure water-vapor interface for comparison.

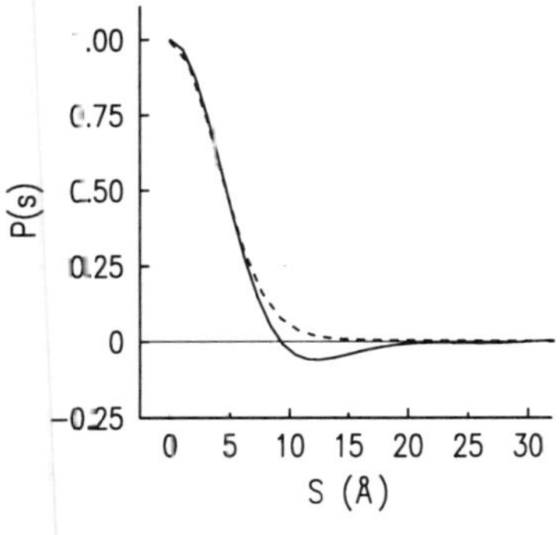

FIG. 12 Patterson function for the polymer monolayer (co-poly 1,2-butadiene/butyl alcohol—50% substitution) on water at a surface pressure of 20.3 mN/m (solid line) and for a pure water-vapor interface (dashed line). (From Ref. 150.)

The physical significance of this curve can be seen by referring to Eqs. 12 and 13. The pure water-vapor interface would be represented by the first term, $P_0(s)$, in Eq. 13. The form for $P_0(s)$ would then be a Gaussian,

$$P_0(s) = \frac{A^2}{\sqrt{2\pi\sigma^2}} e^{-s^2/2\sigma^2} \tag{23}$$

and is illustrated in Fig. 12. For values of $s \gg \sigma$, $f(s)$ in Eq. 13 is given by $f(s) \cong (1/\rho_{\text{bulk}})(d\langle\rho\rangle/ds)$. Since $P(s)$ is below zero for large s, the electron density for this polymer monolayer is *decreasing far from the surface* (into the bulk). This demonstrates that near the surface $\langle\rho(z)\rangle$ exceeds the value for bulk water. Although the density of the bulk polymer is not known, it can be compared with that of the similar polymer, pure 1,2 polybutadiene. Since the electron density of pure crystalline 1,2 polybutadiene is $0.96\rho_{\text{water}}$, this is a surprising result [151].

Analysis of the large Q_z data, using methods similar to those outlined in the section on water, allows the determination of the width of the vapor-monolayer interface. For lower pressures the experimental and predicted widths agree; however, for higher pressures there is some discrepancy. In particular, *for the highest pressure* the measured width is significantly larger than the capillary wave model prediction. One possible origin for this effect could be buckling of the monolayer; however, any molecular rearrangement that either (1) made the surface locally more diffuse or (2) made the surface rougher on a length scale that was small in comparison with our in-plane resolution of approximately 50 nm would result in the discrepancy.

Fitting the entire reflectivity curve (for $Q_z \geqslant 5\ Q_c$, where the Born approximation will be very accurate) to $R(Q_z)/R_F(Q_z)$ computed from a model of the real space profile determines both that the maximum electron density is typically 10% greater than the water density and that its position moves further from the vapor-polymer interface as the pressure increases. In addition, the high-density region of the monolayer remains quite thin (0.5–1 nm) even at these very high pressures.

A theoretically interesting property of this system is the way in which the excess electron density near the surface decays with increasing distance from the surface. Unfortunately, information on that decay would be contained in reflectivity measurements at $|Q_z - Q_c| \lesssim Q_c$ which is obscured by the rapid dependence of $R_F(Q_z)$ on Q_z. Neutron reflectivity has an advantage in this region since, as was mentioned above, suitable mixtures of H_2O and D_2O have Q_c that are essentially zero. In this case, the smaller neutron flux is not a limitation because $R(Q_z)$ becomes small only at large Q_z. A combined x-ray/neutron study of the near surface rise in the electron density and the more distant decay of the neutron-scattering-length density should result in a complete picture of polymer adsorption at the water-vapor interface [152].

VI. CONCLUSION

Although the x-ray and neutron scattering techniques have only recently been applied to the study of liquid surfaces, as has been seen in this chapter a large and ever-increasing variety of systems are being explored. For lack of space, only a few systems have been discussed. In particular, the study of liquid metal surfaces, one of the earliest applications of x-ray reflectivity, has been reviewed by Rice et al. [153] and references are indicated here [154–158]. As mentioned in the section on insoluble polymeric surfactants, neutron reflectivity is ideally suited to the study of polymer adsorption, and several measurements of adsorption at the liquid-vapor interface of polymer solutions have been made [101, 152, 159–162] as well as a recent measurement of adsorption at the quartz-water interface [163]. X-ray and neutron reflection measurements on soap films, systems similar to the freely suspended liquid crystal films previously discussed, have recently been reported [104, 164, 165]. In addition to experiments at the liquid-vapor interface, the x-ray techniques of reflection, diffraction, and EXAFS have been used to study the liquid-solid interface [166–169]. Recently, interesting new studies have been performed on the liquid-solid interfacial structure inside of an electrochemical cell in which the surface chemical potential can be altered. The technological importance of a better understanding of electrochemistry is self-evident; however, if the electrochemical cell can be used to tune electrically the surface structure, then these techniques may provide a convenient tool to study solid surfaces [167–169].

In conclusion, we have demonstrated both the versatility and the power of the x-ray and neutron surface scattering techniques to explore the microscopic structure of liquid surfaces. In the future we expect these techniques to play a crucial role in the continued exploration of novel aspects of two-dimensional physics and in the elucidation of the physical chemistry and biochemistry of phenomena occurring at liquid interfaces.

ACKNOWLEDGMENTS

We should like to acknowledge help from Dan Schwartz in the preparation of this manuscript and from Dr. Binhua Lin in proofreading. In addition, we should like to thank Professor T. C. Chung, Eric Kawamoto, Greg Kellogg, Dr. Mahn Won Kim, Dr. Soonil Lee, and Dan Schwartz for permission to report results that have not yet been published. This work was supported by the National Science Foundation through grants NSF-DMR-88-12855 and NSF-DMR-89-20490. Research carried out at the NSLS, Brookhaven National Laboratory, is supported by the Department of Energy, Material Sciences and Division of Chemical Sciences under contract DE-AC02-76CH00016.

REFERENCES

1. L. Auvray and J. P. Cotton, *Macromolecules 20:* 202 (1987).
2. M. A. Cohen Stuart, T. Cosgrove, and B. Vincent, *Adv. Coll. Int. Sci. 24:* 143 (1986).
3. T. Cosgrove, *J. Chem. Soc. Faraday Trans. 86:* 1323 (1990).
4. J. M. Bloch, W. B. Yun, X. Yang, M. Ramanathan, P. A. Montano, and C. Capasso, *Phys. Rev. Lett. 61:* 2941 (1988).
5. J. M. Bloch, M. Sansone, F. Rondelez, D. G. Peiffer, P. Pincus, M. W. Kim, and P. M. Eisenberger, *Phys. Rev. Lett. 54:* 1039 (1985).
6. J. M. Bloch, W. B. Yun, and K. M. Mohanty, *Phys. Rev. B 40:* 6529 (1989).
7. P. L. Cowan, J. A. Golovchenko, and M. F. Robbins, *Phys. Rev. Lett. 44:* 1680 (1980).
8. M. J. Bedzyk, D. H. Bilderback, G. M. Bommarito, M. Caffrey, and J. S. Schildkraut, *Science 241:* 1788 (1988).
9. M. J. Bedzyk, G. M. Bommarito, M. Caffrey, and T. L. Penner, *Science 248:* 52 (1990).
10. P. S. Pershan, *Faraday Discuss.Chem. Soc. 89:* 231 (1990).
11. J. Penfold and R. K. Thomas, *J. Phys. Condens. Matter 2:* 1369 (1990).
12. J. Als-Nielsen and K. Kjaer in *Phase Transitions in Soft Condensed Matter* (T. Riste and D. Sherrington, eds.), Plenum Press, New York, 1989, p. 145.
13. N. Dent, M. J. Grundy, R. M. Richardson, S. J. Roser, N. M. McKeown, and M. J. Cook, *J. Chim. Phys. 85:* 1003 (1988).
14. M. Born and E. Wolf, *Principles of Optics,* Pergamon Press, Oxford, 1980.
15. D. W. Oxtoby, F. Novak, and S. A. Rice, *J. Chem. Phys. 76:* 5278 (1982).
16. R. W. James, *The Optical Principles of the Diffraction of X-Rays: The Crystalline State, Volume 2,* Cornell Univ. Press, Ithaca, New York, 1965.
17. *The International Tables for X-Ray Crystallography,* Kynoch Press, 1962 and 1974.
18. J. B. Hayter and H. A. Mook, *J. Appl. Cryst. 22:* 35 (1989).
19. V. F. Sears, *Thermal Neutron Scattering Lengths and Cross Sections for Condensed Matter Research,* Atomic Energy of Canada Ltd., Chalk River, Canada, 1984.
20. V. F. Sears, *Neutron Optics,* Oxford Univ. Press, New York, 1989.
21. W. C. Marra, P. Eisenberger, and A. Y. Cho, *J. Appl. Phys. 50:* 6927 (1979).
22. A. H. Weiss, M. Deutsch, A. Braslau, B. M. Ocko, and P. S. Pershan, *Rev. Sci. Instrum. 57:* 2554 (1986).
23. M. Stamm, G. Reiter, and S. Huttenbach, *Physica B 156 and 157:* 564 (1989).
24. J. Penfold, *J. de Phys. (Paris) Colloq. C7 50:* C7-99 (1989).
25. G. P. Felcher, R. O. Hilleke, R. K. Crawford, J. Haumann, R. Kleb, and G. Ostrowski, *Rev. Sci. Instrum. 58:* 609 (1987).
26. L. Bosio, R. Cortes, G. Folcher, and M. Oumezine, *Revue Phys. Appl. 20:* 437 (1985).
27. J. M. Bloch and P. Eisenberger, *Nucl. Inst. & Meth.* (1988).
28. J. Als-Nielsen and P. S. Pershan, *Nucl. Inst. Meth. 208:* 545 (1983).

29. P. S. Pershan, *J. Phys. (Paris) Colloq. C-7 50:* C7-1 (1989).
30. A. J. Liu and M. E. Fisher, *Phys. Rev. A 40:* 7202 (1989).
31. I. M. Tidswell, B. M. Ocko, P. S. Pershan, S. R. Wasserman, G. M. Whitesides, and J. D. Axe, *Phys. Rev. B 41:* 1111 (1990).
32. G. H. Vineyard, *Phys. Rev. B 26:* 4146 (1982).
33. A. Braslau, P. S. Pershan, G. Swislow, B. M. Ocko, and J. Als-Nielsen, *Phys. Rev. A 38:* 2457 (1988).
34. R. S. Becker, J. A. Golovchenko, and J. R. Patel, *Phys. Rev. Lett. 50:* 153 (1983).
35. S. Dietrich and H. Wagner, *Z. Phys. B 56:* 207 (1984).
36. S. Dietrich and H. Wagner, *Z. Phys. B 59:* 35 (1985).
37. H. Dosch, *Phys. Rev. B 35:* 2137 (1987).
38. H. Dosch, B. W. Batterman, and D. C. Wack, *Phys. Rev. Lett. 56:* 1144 (1986).
39. R. Feidenhans'l, *Surf. Sci. Rep. 10:* 105 (1989).
40. D. K. Schwartz, M. L. Schlossman, E. H. Kawamoto, G. J. Kellogg, P. S. Pershan, and B. M. Ocko, *Phys. Rev. A 41:* 5687 (1990).
41. A. Zangwill, *Physics at Surfaces,* Cambridge Univ. Press, Cambridge, 1988.
42. S. K. Sinha, E. B. Sirota, S. Garoff, and H. B. Stanley, *Phys. Rev. B 38:* 2297 (1988).
43. a) B. Widom, *Faraday Symp. Chem. Soc. 16:* 7 (1981), and b) J. K. Percus and G. O. Williams, in *Fluid Interfacial Phenomena* (C. A. Croxton, ed.) John Wiley & Sons, Chichester, 1986, p. 1.
44. J. S. Huang and W. W. Webb, *J. Chem. Phys. 50:* 3677 (1969).
45. D. Beaglehole, *Phys. Rev. Lett. 58:* 1434 (1987).
46. D. Beaglehole in *Fluid Interfacial Phenomena* (C. A. Croxton, ed), John Wiley, Chichester, 1986, p. 523.
47. A. Braslau, M. Deutsch, P. S. Pershan, A. H. Weiss, J. Als-Nielsen, and J. Bohr, *Phys. Rev. Lett. 54:* 114 (1985).
48. J. Daillant, L. Bosio, J. J. Benattar, and J. Meunier, *Europhys. Lett. 8:* 453 (1989).
49. R. Loudon in *Surface Excitations* (V. M. Agranovich and R. Loudon, eds.), Elsevier, New York, 1984, pp. 589–638.
50. M. K. Sanyal, S. K. Sinha, K. G. Huang, and B. M. Ocko, *X-Ray Scattering Study of Capillary Wave Fluctuations at a Liquid Surface,* preprint, 1991.
51. a) J. D. Litster, and R. J. Birgeneau, *Physics Today, 35*, May: 26 (1982), and b) J. B. Brock, R. J. Birgeneau, J. D. Litster, and A. Aharony, *Physics Today, 42,* July: 52 (1989).
52. P. S. Pershan, *Structure of Liquid Crystal Phases,* World Scientific, Singapore, 1988.
53. P. G. de Gennes, *The Physics of Liquid Crystals,* Clarendon Press, Oxford, 1974.
54. R. Pindak and D. Moncton, *Physics Today 35,* May: 56 (1982).
55. P. G. de Gennes, *Solid State Comm. 10:* 753–756 (1972).
56. W. McMillan, *Phys. Rev. Lett. 7:* 1419–1422 (1973).
57. D. Brisbin, R. D. Hoff, T. E. Lockhart, and D. L. Johnson, *Phys. Rev. Lett. 43:* 1171 (1979).
58. J. Thoen, H. Marynissen, and W. V. Dael, *Phys. Rev. Lett. 52:* 204 (1984).

59. B. M. Ocko, R. J. Birgeneau, J. D. Litster, and M. E. Neubert, *Phys. Rev. Lett. 52:* 208–211 (1984).
60. B. M. Ocko, R. J. Birgeneau, and J. D. Litster, *Z. Physik B 62:* 487 (1986).
61. A. A. Chernov and L. V. Mikheev, *Phys. Rev. Lett. 60:* 2488 (1988).
62. P. Sheng, *Phys. Rev. 26:* 1610 (1982).
63. Z. Pawlowska, G. F. Kventsel, and T. J. Sluckin, *Phys. Rev. A36:* 992 (1987).
64. J. V. Selinger, *Phys. Rev. A37:* 1736 (1988).
65. J. Als-Nielsen, F. Christensen, and P. S. Pershan, *Phys. Rev. Lett. 48:* 1107 (1982).
66. A. Brasslau, *The Free Surface of Liquids and Liquid Crystals Studied by X-Ray Reflectivity* Ph.D. thesis, Harvard University, 1988.
67. P. S. Pershan, A. Braslau, A. H. Weiss, and J. Als-Nielsen, *Phys. Rev. A 35:* 4800 (1987).
68. E. F. Gramsbergen and W. H. d. Jeu, *Phys. France 49:* 363 (1988).
69. E. F. Gramsbergen, W. H. d. Jeu, and J. Als-Nielsen, *J. Physique 47:* 711 (1986).
70. B. M. Ocko, A. Braslau, P. S. Pershan, J. Als-Nielsen, and M. Deutsch, *Phys. Rev. Lett. 57:* 94 (1986).
71. P. S. Pershan, and J. Als-Nielsen, *Phys. Rev. Lett. 52:* 759 (1984).
72. B. M. Ocko, P. S. Pershan, C. R. Safinya, and L. Y. Chiang, *Phys. Rev. A 35:* 1868 (1987).
73. J. Als-Nielsen, R. J. Birgeneau, M. Kaplan, J. D. Litster, and C. R. Safinya, *Phys. Rev. Lett. 39:* 352 (1977).
74. C. W. Garland, M. Meichle, B. M. Ocko, A. R. Kortan, C. R. Safinya, L. J. Yu, J. D. Litster, and R. J. Birgeneau, *Phys. Rev. 27:* 3234 (1983).
75. K. K. Chan, P. S. Pershan, L. B. Sorensen, and F. Hardouin, *Phys. Rev. A 34:* 1420 (1986).
76. K. K. Chan, P. S. Pershan, L. B. Sorensen, and F. Hardouin, *Phys. Rev. Lett. 54:* 1694 (1985).
77. M. E. Fisher, *J. Chem. Soc., Faraday Trans. 82:* 1569 (1986).
78. P. G. de Gennes, *Rev. of Mod. Phys. 57:* 827 (1985).
79. J. G. Dash in *Solvay Conference on Surface Science* Univ. of Texas, 1988.
80. M. Schlossman, X.-L. Wu, and C. Franck, *Phys. Rev. B 31:* 1478 (1985).
81. T. C. Lubensky, *J. Chem. Physique 80:* 31 (1983).
82. E. Fontes, P. A. Heiney, J. H. Haseltine, I. Amos, and B. Smith, *J. de Physique 47:* 1533 (1986).
83. E. Fontes, P. A. Heiney, P. Barois, and A. M. Levelut, *Phys. Rev. Lett. 60:* 1138 (1988).
84. C. R. Safinya, W. A. Varady, L. Y. Chiang, and P. Dimon, *Phys. Rev. Lett. 57:* 432 (1986).
85. G. Swislow, D. Schwartz, B. M. Ocko, P. S. Pershan, and J. D. Litster, *Phys. Rev. A, 43:* 6815 (1991).
86. D. Schwartz, A. Braslau, B. Ocko, P. S. Pershan, J. Als-Nielsen, and J. S. Huang, *Phys. Rev. A 38:* 5817 (1988).
87. A. Aharony, R. J. Birgeneau, J. D. Brock, and J. D. Litster, *Phys. Rev. Lett. 57:* 1012 (1986).

88. J. D. Brock, A. Aharony, R. J. Birgeneau, K. W. Evans-Lutterodt, J. D. Litster, P. M. Horn, G. B. Stephenson, and A. R. Tajbakhsh, *Phys. Rev. Lett. 57:* 98 (1986).
89. D. E. Moncton and R. Pindak, *Phys. Rev. Lett 43:* 701 (1979).
90. D. E. Moncton, R. Pindak, S. C. Davey, and G. S. Brown, *Phys. Rev. Lett. 49:* 1865 (1982).
91. S. C. Davey, J. Budai, J. W. Goodby, R. Pindak, and D. E. Moncton, *Phys. Rev. Lett. 53:* 2129 (1984).
92. J. Collett, P. S. Pershan, E. B. Sirota, and L. B. Sorensen, *Phys. Rev. Lett. 52:* 356 (1984).
93. E. B. Sirota, P. S. Pershan, L. B. Sorensen, and J. Collett, *Phys. Rev. A 36:* 2890 (1987).
94. E. B. Sirota. *X-Ray Scattering Study of the Thickness Dependent Phase Diagram of Liquid Crystal Films,* Ph.D. thesis, Harvard University, 1986.
95. E. B. Sirota, P. S. Pershan, S. Amador, and L. B. Sorensen, *Phys. Rev. A 35:* 2283 (1987).
96. S. Amador, P. S. Pershan, H. Stragier, B. D. Swanson, D. J. Tweet, L. B. Sorensen, E. B. Sirota, G. E. Ice, and A. Habenschuss, *Phys. Rev. A 39:* 2703 (1989).
97. B. D. Swanson, H. Stragier, D. J. Tweet, and L. B. Sorensen, *Phys. Rev. Let. 62:* 909 (1989).
98. D. B. Swanson, H. Stragier, D. J. Tweet, and L. B. Sorensen, *Bull. Am. Phys. Soc. 33:* 311 (1988).
99. E. M. Lee, R. K. Thomas, J. Penfold, and R. C. Ward, *J. Phys. Chem. 93:* 381 (1989).
100. E. M. Lee and R. K. Thomas, *Physica B 156 and 157:* 525 (1989).
101. A. R. Rennie, R. J. Crawford, E. M. Lee, R. K. Thomas, T. L. Crowley, S. Roberts, M. S. Qureshi, and R. W. Richards, *Macromolecules 22:* 3466 (1989).
102. E. M. Lee, E. A. Simister, R. K. Thomas, and J. Penfold, *J. Phys. (Paris) Colloq., C7 50:* C7-75 (1989).
103. J. Penfold, E. M. Lee, and R. K. Thomas, *Mol. Phys. 68:* 33 (1989).
104. J. J. Benattar, J. Daillant, O. Belorgey, and L. Bosio, Annual Meeting of the American Chemical Society, Boston, 1990.
105. J. M. Bloch, and W. Yun, *Phys. Rev. A, 41:* 844 (1990).
106. T. M. Bohanon, B. Lin, M. C. Shih, G. E. Ice, and P. Dutta, *Phys. Rev. B. 41:* 4846 (1990).
107. L. Bosio, J. J. Benattar, and F. Rieutord, *Revue Phys. Appl. 22:* 775 (1987).
108. J. Daillant, L. Bosio, and J. J. Benattar, *Europhys. Lett. 12:* 715 (1990).
109. P. Dutta, J. B. Peng, B. Lin, J. B. Ketterson, M. Prakash, P. Georgopoulos, and S. Ehrlich, *Phys. Rev. Lett. 58:* 2228 (1987).
110. C. A. Helm, H. Möhwald, K. Kjaer, and J. Als-Nielsen, *Biophys. J. 52:* 381 (1987).
111. C. A. Helm, H. Möhwald, K. Kjaer, and J. Als-Nielsen, *Europhys. Lett. 4:* 697 (1987).
112. D. Jacquemain, S. G. Wolf, F. Leveiller, M. Lahav, L. Leiserowitz, J. Deutsch, K. Kjaer, and J. Als-Nielsen, *J. Phys. (Paris) Colloq. C-7 50:* C7-1 (1989).

113. R. M. Kenn, C. Bohm, A. M. Bibo, I. R. Peterson, H. Möhwald, K. Kjaer, and J. Als-Nielsen, *J. Phys. Chem 95:* 2092 (1991).
114. K. Kjaer, J. Als-Nielsen, C. A. Helm, P. Tippmann-Krayer, and H. Möhwald, *Thin Solid Films 159:* 17 (1988).
115. K. Kjaer, J. Als-Nielsen, C. A., Helm, L. A. Laxhuber, and H. Möhwald, *Phys. Rev. Lett. 58:* 2224 (1987).
116. K. Kjaer, J. Als-Nielsen, C. A. Helm, L. A. Laxhuber, and H. Möhwald, *Phys. Rev. Lett. 58:* 2224 (1987).
117. K. Kjaer, J. Als-Nielsen, C. A. Helm, P. Tippman-Krayer, and H. Möhwald, *J. Phys. Chem. 93:* 3200 (1989).
118. E. M. Landau, M. Levanon, L. Leiserowitz, M. Lahav, and J. Sagiv, *Nature 318:* 353 (1985).
119. E. M. Landau, R. Popovitz-Biro, M. Levanon, L. Leiserowitz, M. Lahav, and J. Sagiv, *Mol. Cryst. Liq. Cryst. 134:* 323 (1986).
120. B. Lin, J. B. Peng, J. B. Ketterson, P. Dutta, B. N. Thomas, J. Buontempo, and S. A. Rice, *J. Chem. Phys. 90:* 2393 (1989).
121. B. Lin, M. C. Shih, T. M. Bohanon, G. E. Ice, and P. Dutta, *Phys. Rev. Lett. 65:* 191 (1990).
122. H. Möhwald, in *Phase Transitions in Soft Condensed Matter* (T. Riste and D. Sherrington, eds.) Plenum Press, New York, 1989, p. 145.
123. R. M. Richardson and S. J. Roser, *Liquid Crystals 2:* 797 (1987).
124. M. L. Schlossman, D. K. Schwartz, E. H. Kawamoto, G. J. Kellogg, P. S. Pershan, B. M. Ocko, M. W. Kim, and T. C. Chung in *Materials Research Society Proceedings—Macromolecular Liquids* (C. R. Safinya, S. A. Safran, and P. A. Pincus, eds.) Materials Research Society, Boston, 1989, p. 351.
125. M. Seul, S. Subramaniam, and H. M. McConnell, *J. Phys. Chem. 89:* 3592 (1985).
126. S. G. Wolf, L. Leiserowitz, M. Lahav, M. Deutsch, K. Kjaer, and J. Als-Nielsen, *Nature 328:* 63 (1987).
127. S. G. Wolf, M. Deutsch, E. M. Landau, M. Lahav, L. Leiserowitz, K. Kjaer, and J. Als-Nielsen, *Science 242:* 1286 (1988).
128. S. G. Wolf, E. M. Landau, M. Lahav, L. Leiserowitz, M. Deutsch, K. Kjaer, and J. Als-Nielsen, *Thin Solid Films 159:* 29 (1988).
129. S. G. Wolf, D. Jacquemain, M., Deutsch, K. Kjaer, L. Leiserowitz, and J. Als-Nielsen, *Structural Characterization of the 2nd Crystalline Phases in a Langmuir Monolayer by X-Ray Bragg Rod and Reflectivity Measurments,* preprint, 1989.
130. T. M. Bayerl, R. K. Thomas, J. Penfold, A. Rennie, and E. Sackmann, *Biophys. J. 57:* 1095 (1990).
131. M. J. Grundy, R. M. Richardson, S. J. Roser, J. Penfold, and R. C. Ward, *Thin Solid Films 159:* 43 (1988).
132. B. Ewen, G. R. Strobl, and D. Richter, *Faraday Disc. Chem. Soc. 68:* 19 (1980).
133. J. P. Bareman, G. Cardini, and M. L. Klein, *Phys. Rev. Lett. 60:* 2152 (1988).
134. G. Cardini, J. P. Bareman, and M. L. Klein, *Chem. Phys. Lett. 145:* 493 (1988).
135. J. Harris and S. A. Rice, *J. Chem. Phys. 88:* 1298 (1988).
136. J. Popielawski and S. A. Rice, *J. Chem. Phys. 88:* 1279 (1988).
137. Z.-G. Wang and S. A. Rice, *J. Chem. Phys. 88:* 1290 (1988).

138. B. Lin, J. B. Peng, J. B. Ketterson, and P. Dutta, *Thin Solid Films 159:* 111 (1988).
139. H. Möhwald, *Thin Solid Films 159:* 1 (1988).
140. M. L. Schlossman, D. K. Schwartz, P. S. Pershan, E. H. Kawamoto, G. J. Kellogg, and S. Lee, *Phys. Rev. Lett. 66:* 1599 (1991).
141. J. F. Ankner, H. Zabel, D. A. Neumann, C. F. Majkrzak, J. A. Dura, and C. P. Flynn, *J. de Phys. (Paris) Colloq. C7 50:* 189 (1989).
142. G. L. Gaines, *Insoluble Monolayers at a Liquid Gas Interface,* John Wiley, New York, 1966.
143. J. M. Kosterlitz and D. J. Thouless, *J. Phys. C 6:* 1181 (1973).
144. D. R. Nelson and B. J. Halperin, *Phys. Rev. B 19:* 2457 (1979).
145. D. R. Nelson, *Phys. Rev. B 27:* 2902 (1983).
146. Z.-H. Cai and S. A. Rice, *Faraday Discuss. Chem. Soc. 89:* 211 (1990).
147. D. Andelman, F. Brochard, and J.-F. Joanny, *J. Chem. Phys. 86:* 3673 (1987).
148. B. Lin, *X-Ray Diffraction Studies of Fatty Acid and Alcohol Monolayers on the Surface of Water,* Ph.D. thesis, Northwestern University, 1990.
149. G. S. Patil, R. H. Matthews, and D. G. Cornwell in *Monolayers* (E. D. Goddard, ed.), American Chemical Society, Washington, D.C., 1975, p. 44.
150. M. L. Schlossman, D. K. Schwartz, E. H. Kawamoto, G. J. Kellogg, P. S. Pershan, M. W. Kim, and T. C. Chung, *J. Phys. Chem.,* to be published, 1991.
151. G. Nutta, *Science 147:* 261 (1965).
152. E. M. Lee, E. A. Simister, and R. K. Thomas, *Mol. Cryst. Liq. Cryst. 179:* 151 (1990).
153. S. A. Rice, J. Gryko, and U. Mohanty, in *Fluid Interfacial Phenomena* (C. A. Croxton, ed.), John Wiley, Chichester, 1986, p. 255.
154. S. W. Barton, B. N. Thomas, E. B. Flom, F. Novak, and S. A. Rice, *Langmuir 4:* 233 (1988).
155. S. W. Barton, B. N. Thomas, F. Novak, J. Harris, P. Dolmer, J. M. Bloch, and S. A. Rice, *Nature 321:* 685 (1986).
156. L. Bosio and M. Oumezine, *J. Chem. Phys. 80:* 959 (1984).
157. L. Bosio, R. Cortes, A. Defrain, and M. Oumezine, *J. of Non. Cryst. Solids 61/62:* 697 (1984).
158. L. Bosio, R. Cortes, M. Denoziere, and G. Folcher, *J. de Phys. (Paris) Colloq. C7 50:* 23 (1989).
159. X. Sun, E. Bouchaud, A. Lapp, B. Farnoux, M. Daoud, and G. Jannink, *Europhys. Lett. 6:* 207 (1988).
160. X. Sun, M. Daoud, B. Farnoux, and G. Jannink, *Physica B 156 and 157:* 407 (1989).
161. E. Bouchard, B. Farnous, X. Sun, M. Daoud, and G. Jannink, *Europhys. Lett. 2:* 315 (1986).
162. S. Dietrich and R. Schack, *Phys. Rev. Lett. 58:* 140 (1987).
163. E. M. Lee, R. K. Thomas, and A. R. Rennie, *Europhys. Lett. 13:* 135 (1990).
164. J. J. Benattar, J. Daillant, L. Bosio, and L. Leger, *J. de Phys. (Paris) Colloq. C7 50:* 39 (1989).

165. R. R. Highfield, R. P. Humes, R. K. Thomas, P. G. Cummins, D. P. Gregory, J. Mingins, J. B. Hayter, and O. Schaerpf, *J. Colloid Interface Sci. 97:* 367 (1984).
166. B. M. Ocko, *Phys. Rev. Lett. 64:* 2160 (1990).
167. H. D. Abruna, in *Modern Aspects of Electrochemistry* (J. O. Bockris, R. E. White, and B. E. Conway, eds.) Plenum Press, New York and London, 1989, p. 265.
168. B. M. Ocko, J. Wang, A. Davenport, and H. Isaacs, *Phys. Rev. Lett. 65:* 1466 (1990).
169. M. G. Samant, M. F. Toney, G. L. Borges, L. Blum, and O. R. Melroy, *Surface Sci. 193:* L29 (1987).

19

Fluorescence Methods at the Liquid-Vapor Interface

F. RONDELEZ Laboratoire de Structure et Reactivité aux Interfaces, Université Pierre et Marie Curie, Paris, France

Fluorescence is probably one of the most sensitive techniques in the panoply of experimental tools available to physicists. It is thus particularly suited to the study of interfaces where the layer thickness to be probed is in the 1–100 nm range. It is generally admitted that neutron magnetic resonance (NMR) requires a minimum of 10^{16} nuclear spins in the NMR cavity to produce a detectable signal. Electron paramagnetic resonance is much more sensitive since it can detect 10^{13} paramagnetic radicals. But fluorescence beats them all with a detection level of 10^4 molecules in the illuminated volume. This allows us to answer one frequent criticism when the system under investigation is not naturally fluorescent and it is necessary to add deliberately extrinsic fluorescent markers. Because of the extreme sensitivity of the technique, the amount of induced contamination is usually negligible in materials like polymers, surfactants, liquid crystals, and phospholipids, which in any case can never reach a purity level higher than 99 or 99.9%, contrary to the semiconductor industry in which figures with five or six 9's are quite common. Fluorescence is generally thought of as an optical technique, applicable mainly in the ultraviolet and visible range. However, it also extends to much shorter wavelengths, and we shall see in the following that x-ray fluorescence provides a powerful means to probe liquid interfaces over very small thicknesses comparable to the wavelength of the incident x-ray wavelength.

In this chapter, we have selected three examples, two in which fluorescence is used to study static structures, either in the plane of the interface or perpendicular to it, and one in which fluorescence gives access to dynamic phenomena such as lateral diffusion. We shall start by discussing the fluorescence observations of phase transitions in monolayers of insoluble surfactants at the air-water interface by optical microscopy in the reflection mode. Next, we shall present measurements of the concentration profile of flexible polymer chains absorbed onto the air-solution interface by x-ray fluorescence in the evanescent mode regime. Finally, we shall examine the possibilities of fluorescence photobleaching to monitor the two-dimensional Brownian diffusion of fluorescent markers along an interface that can be a Langmuir monolayer as well as a biological cell membrane.

I. EPIFLUORESCENCE MICROSCOPY APPLIED TO LANGMUIR MONOLAYERS

It is known since the pioneering works of Pockels [1] and Langmuir [2] that insoluble surfactants spread in the form of single monolayers at the air-liquid interface. The amphiphilic molecules have their hydrophilic polar head groups (a carboxylic acid in the case of fatty acids, and a glycerol phosphate in the case of phospholipids) in contact with the water subphase, whereas their hydrophobic chains (generally aliphatic hydrocarbon chains with 13 to 32 methylene units, or sometimes a fluorocarbon chain with 8 to 10 CF_2 units) point away from the water. Over the years it has been established that the physical state of the spread monolayer could be solid, liquid, or gaseous depending on the average molecular surface density [3]. The purely gaseous state is encountered at area per molecule in excess of 10 to 20 nm^2 per molecule. Each molecule behaves independently from the others, and the chains are fully disordered, lying flat on the interface. The homogeneous liquid state, so-called liquid expanded, is observed at areas of two to three times the cross-sectional area A_0 calculated from molecular models for vertical chains. The chains are still disordered, with a few gauche defects per chain, but they are now standing up and in strong interaction with each other. At areas of the order of A_0 ($A_0 = 0.19$ nm^2 per molecule for fatty acids), the chains are tightly packed next to each other and all the gauche defects have been squeezed out. This state can be liquid condensed, with no long-range positional ordering of the center of mass (the correlation length is at most a few molecular diameters), or solid (the correlation length is then a few tens of molecular diameters) [4]. In almost all known cases so far the transitions between these various states, when they exist, are first order. That is to say, they occur at fixed pressure [5] and there is a coexistence region where the two phases are in equilibrium with each other. As the surface density is increased (or decreased), the densest (most dilute) phase grows at the expense of the other until the whole monolayer is in its new, homogeneous, state. The phase transition involves two successive stages: first, nucleation of seeds of the terminal phase; second, growth of the domains that occupy a fraction of the interface given by the level rule [6].

Epifluorescence microscopy is ideally suited to monitor this nucleation and growth process and to study the shapes of the growing domains. We shall see in the following that in several instances, unexpected features, fractal growth, periodic ordering of the domains, stripe patterns, foams, etc., have also been detected by this method. In this technique the monolayer is observed with an optical microscope in the reflection mode [7]. Typical optical magnification is 200× with a numerical aperture of 0.4–0.6 for the objective. This allows a spatial lateral resolution of 1 μm, whereas the field of view is a few hundred μm. Since surfactant molecules are not fluorescent in general, a small amount of a fluorescent dye (0.01–1% by weight) is deliberately added. Its molecular structure is chosen as close as possible to that of the matrix so that the perturbation effect is minimized. An example is 4-hexadecylamino-7-nitrobenz-2-oxa-1,3-diazole (NBD-HDA for short). Compared to pentadecanoic acid (CH_3-$(CH_2)_{13}$-COOH), the chain length is only two methylene units longer and the carboxylic acid group is replaced by a conjugated ring with an amino group that provides the polar anchor to the water subphase. Maximum surface pressure variations are 0.2 $mN \cdot m^{-1}$ per mole % of fluorescent marker [8]. Moreover, NBD-HDA has an excitation band conveniently located in the green part of the visible spectrum, around 470 nm, and an emission band centered at 530 nm. When the monolayer is illuminated with a 100 W high-pressure mercury lamp through a dichroic narrow band filter and observed in the fluorescence band of the dye, two situations may arise depending on whether the monolayer is in a homogeneous phase or within a two-phase region. In the first case, a uniform light intensity is observed on the detector screen (a sensitive low-light level vidicon camera connected to a video monitor). In the second case, the field of view is no longer uniform, and characteristic spatial structures are easily discernable, evidencing the presence of liquid condensed domains in a continuous liquid expanded phase (or vice versa) or the presence of liquid expanded domains in a continuous gas phase (or vice versa). The optical contrast between these regions may come from several factors: (1) partitioning of the dye molecules between the two phases present according to the surface molecular density, e.g., a gas phase, being of lower density, will contain a smaller percentage of fluorescent dye than the liquid phase and will consequently appear darker. (2) solubility differences of the dye molecules between the various phases; these impurity molecules will have a tendency to be excluded from the more organized and

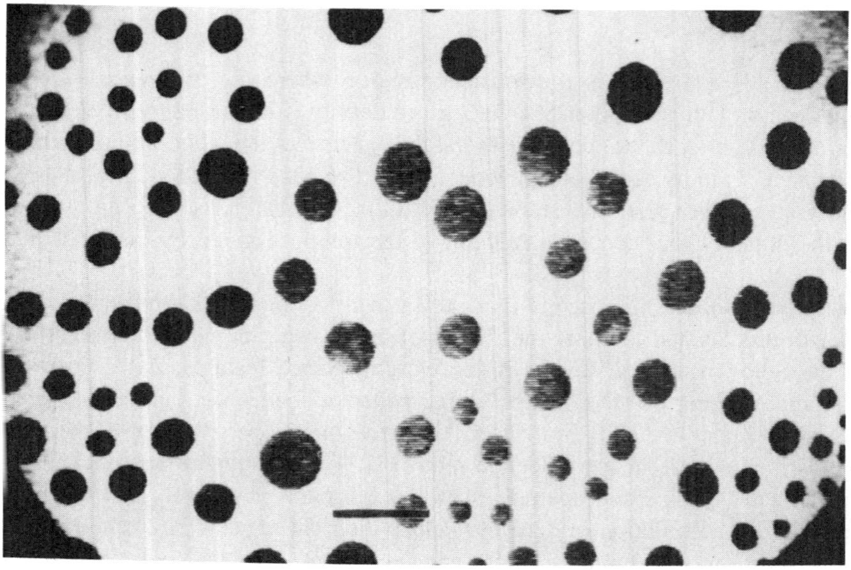

FIG. 1a Circular islands of the liquid condensed phase observed in the liquid condensed-liquid expanded coexistence region of myristic acid at 20°C. Sizes are between 20 and 50 μm.

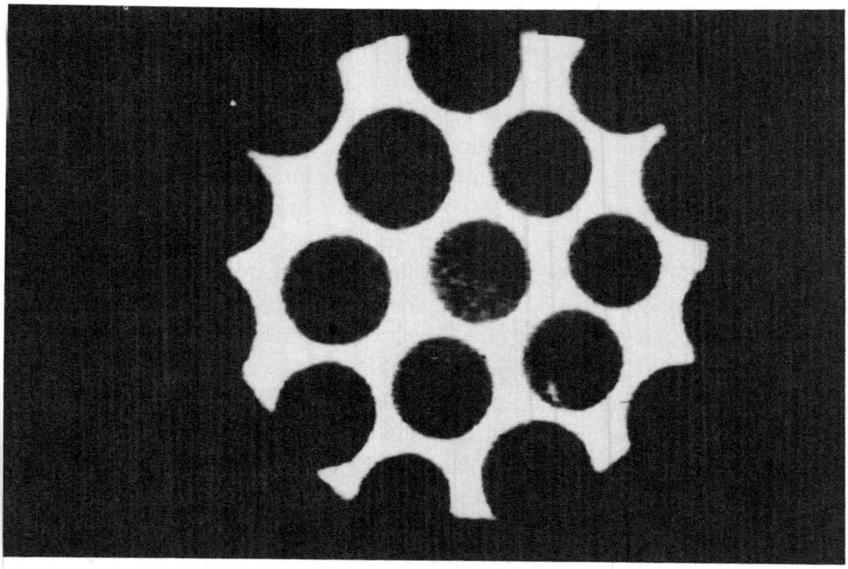

FIG. 1b Hexagonal ordering of the liquid condensed islands observed when their size becomes comparable to their spacing.

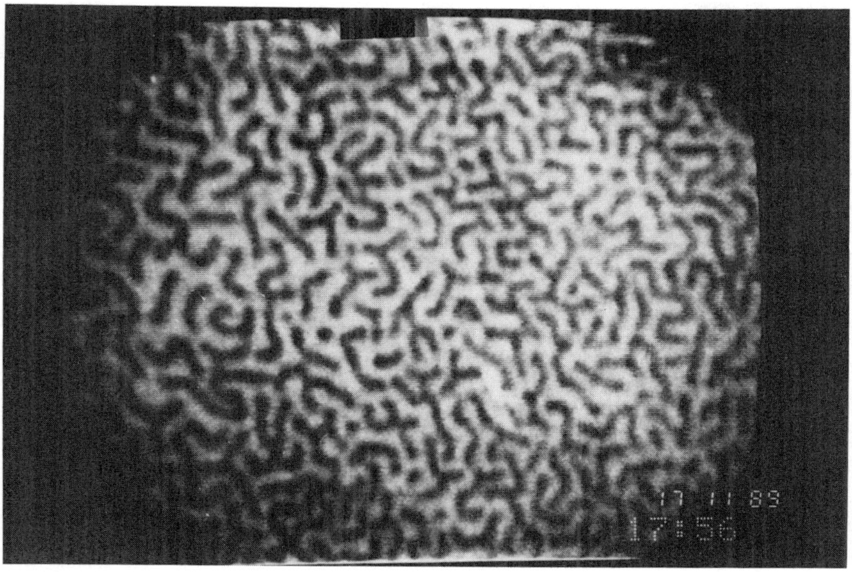

FIG. 1c Stripe phase observed in the liquid expanded-gas coexistence region of *d*-myristoyl alanine at room temperature. The white regions are liquid; the dark ones are gaseous. The periodicity is ≅ 20 μm.

FIG. 1d Unstable growth of liquid condensed domains of myristic acid at 30°C. The fractal geometry is characterized by Hausdorf exponent of 1.77 ± 0.09. The ramified pattern will eventually relax to a circle as imposed thermodynamically by the condition of minimization of the line tension.

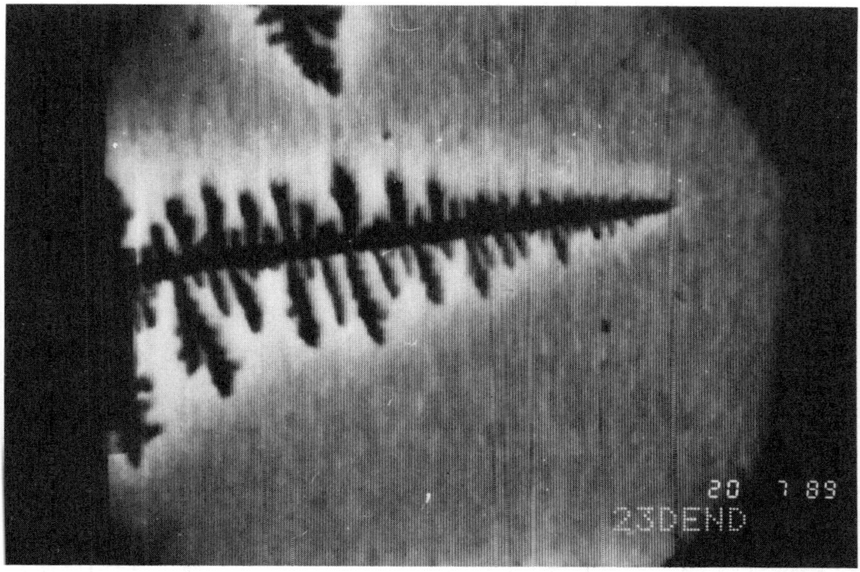

FIG. 1e Dendritic growth of liquid condensed domains when the monolayer is composed of chiral molecules. Monolayer of *d*-myristoyl alanine at room temperature. The dendrite size along its main axis is 210 μm.

FIG. 1f Spiral structure observed in the solid-liquid coexistence region of a chiral phospholipid monolayer. The sense of rotation is directly related to the molecular chirality. (Taken from R. M. Weis, and H. M. McConnell, *J. Phys. Chem. 89:* 4453 (1985).

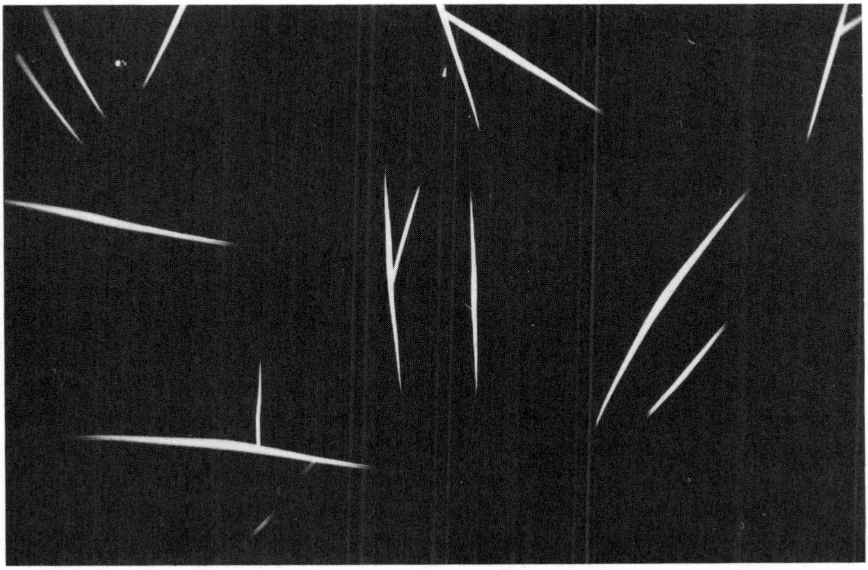

FIG. 1g Needle crystals in a monolayer of NBD-stearic acid at room temperature. (Courtesy of F. Gallet and P. Muller.)

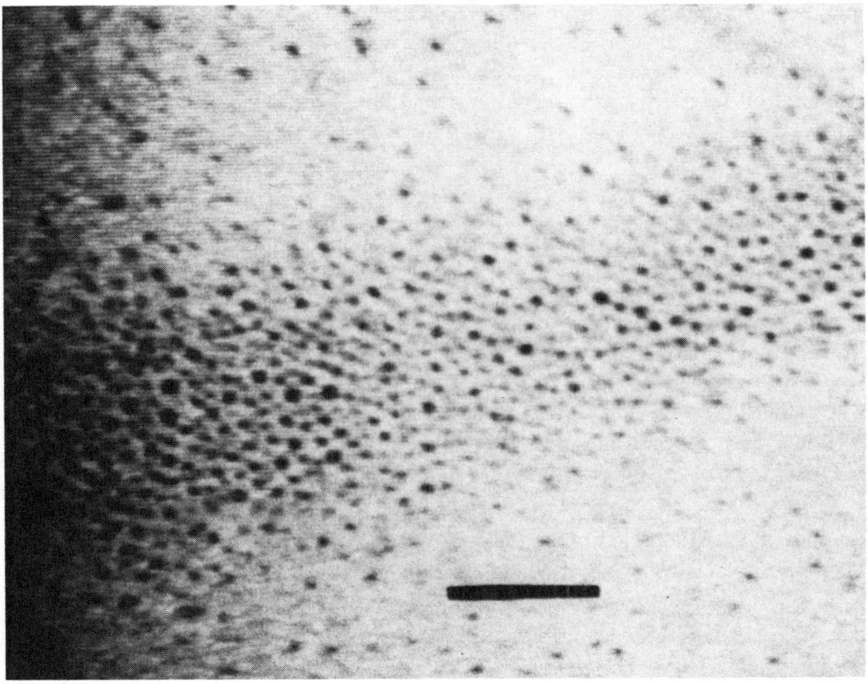

FIG. 1h Gas bubbles observed in the liquid expanded-gas coexistence region of pentadecanoic acid at room temperature. The bubble size is 10 μm.

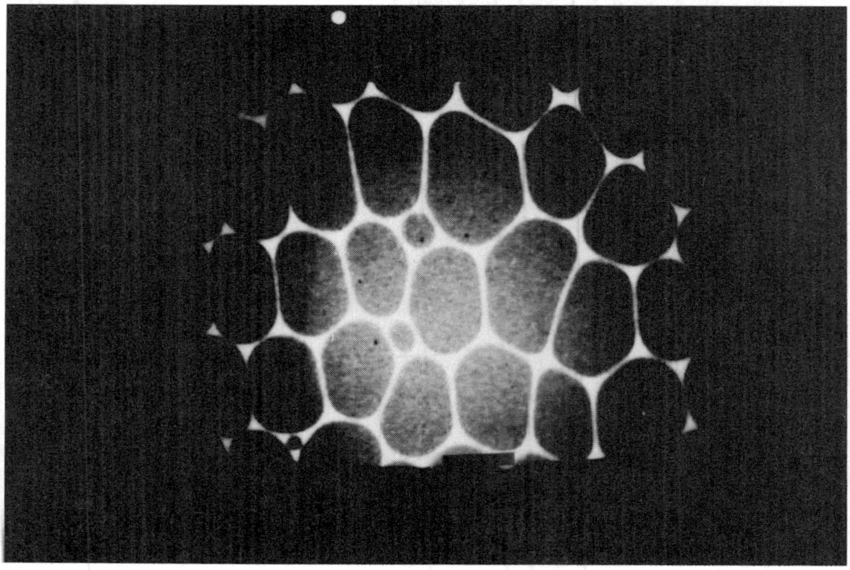

FIG. 1i Two-dimensional foams observed in the liquid expanded-gas coexistence region of pentadecanoic acid when the gas-to-liquid surface fraction exceeds 70%. The polyhedral structure with narrow liquid strips meeting at 120° angles and triangular plateau borders is typical of the dry foams observed in many three-dimensional gas-liquid emulsions. The gas cell size is 200 μm.

dense phases such as the solid and the liquid condensed phases, which then appear darker than the liquid expanded phase; (3) quenching of the dye fluorescence; this is a case of less frequent occurrence but it may happen with dye molecules that are very sensitive to the polarity of their environment. Upon lateral compression, the active part of the dye may come in closer contact with water and its fluorescence yield will increase (or decrease) [9].

In general, the gas and the solid, or liquid condensed, domains appear dark on the monitor screen, whereas the liquid expanded phase is much brighter. Figures 1a to 1i show a variety of structures observed by fluorescence microscopy in monolayers of fatty acids and phospholipids. Figure 1a shows liquid condensed islands in the liquid condensed-liquid expanded coexistence region of myristic acid at temperatures below 24.5°C. They appear as circular islands, with characteristic sizes ranging from 5 to 500 μm. Such domains are obtained by slow compression of the monolayer from the one-phase region into the two-phase region. They nucleate at random on the interface as the phase boundary is crossed and grow steadily upon further compression. Nucleation rate depends on the local supersaturation and can be made much faster than the growth rate. Consequently, if all the domains can be generated at once, their size distribution at any moment will be quite narrow. Such monodisperse domains will tend to form a supermolecular lattice when their size becomes on the order of the average interdomain distance [10, 11]. Figure 1b shows such a striking hexagonal pattern. The ordering mechanism is due to dipolar repulsion between the domains that act as macroscopic dipoles. Within a single domain all the molecules are aligned parallel to each other and they conjugate the effect of their molecular dipole component perpendicular to the interface. Above a certain domain size, the corresponding macrodipoles are large enough to overcome thermal agitation. Within a simple domain, the same dipolar effect between neighboring molecules may induce the deformation of the circular island into elongated stripes (Fig. 1c). Dipole energy is reduced by putting the molecules further apart, but at the expense of line tension [12, 13]. An equilibrium shape may result from the competition of these two effects. Stripe domains have been observed both in the liquid expanded-liquid condensed coexistence region [14, 15] and in the liquid expanded-gas coexistence region [16]. Figures 1d and 1e show that the growth of liquid condensed domains may be unstable under certain conditions. Fractal patterns, with circular symmetry [17, 18], as well as anisotropic dendrites [19, 20], can be generated. In most cases, the effect seems to be due to the expulsion of the dye molecules from the growing domains [21], but another diffusion mechanism that does not involve impurities has also been suggested [18]. When the dye exodiffusion is slow compared to the relaxation of a perturbation of the domain boundary by line tension, shape fluctuations get amplified. This occurs either when the line tension between the two coexisting phases is decreased by the addition of a line-active molecule [22] or when the viscosity of the growing domains is large [23]. Both effects tend to slow down the shape relaxation process, which makes the nonequilibrium ramified shapes easily observable.

Two last examples of striking liquid condensed and solid structures are shown in Figs. 1f and 1g respectively. The observation of spiral crystals is peculiar to the case where the monolayer is formed with chiral molecules. Here again the explanation for these elongated patterns is the balance between dipolar energy and line tension [24, 25]. The observation of solid needle-like crystals is a typical example of highly anisotropic molecular arrangement [26].

When the monolayer is expanded from its liquid expanded state into the liquid expanded-gas coexistence region, gas bubbles appear as black circles, as shown in Fig. 1h. As they grow, they transform into polyhedral domains separated by narrow liquid strips (Fig. 1i). The observed structure is highly reminiscent of the so-called dry foams [27] and has the same topology, with plateau regions where three strips meet at (almost) 120° angles and hexagonal cells. An important difference with three-dimensional foams is that there is no drainage by gravity and consequently their time evolution is very slow. Such two-dimensional foams can be examined for days and their characteristics can be compared in great detail with recent computer simulations and analytical predictions [28–30].

Epifluorescence microscopy observations also provide a way to probe the viscosity or the elasticity of the various phases. For instance, it is possible to deform the liquid condensed domains into ellipses by blowing an air-stream above the liquid surface [23]. Monitoring the relaxation toward the equilibrium circular shape gives access to the viscosity.

II. X-RAY FLUORESCENCE EXCITED BY EVANESCENT WAVES

So far we have discussed optical observations at the microscopic level of various structures in the plane of the interface. We shall now see how excitation of fluorescence by evanescent x-ray waves can probe the structure of the interface in the perpendicular direction. We discuss two examples, one in which one wants to study the monomer concentration profile within a polymer adsorption layer and the other in which one wants to access the spatial localization of counterions for a monolayer of ionizable surfactants. In both cases the basic principle is to illuminate the liquid-vapor interface with a collimated x-ray beam under grazing incidence. If the incident electromagnetic radiation propagates in the air, it will suffer total external reflection if the angle of incidence is less than the critical angle α_c given by $\alpha_c = \arcsin n_2/n_1$, where n_2 is the refractive index of the solution and n_1 is the refractive index of the air [31]. Contrary to classical optics, where the denser medium has a higher refractive index than air, x-ray optics shares with neutrons that $n_2 = 1 - \varepsilon$ (with $\varepsilon \geqslant 0$) and $n_1 = 1$. This unexpected phenomenon is due to the fact that the frequencies associated with the very short x-ray wavelengths exceed the characteristic frequencies of the electrons around their atomic orbitals [32]. X-ray reflectivity is thus equal to unity for $\alpha < \alpha_c$ [33]. Nevertheless, a part of the x-ray beam will penetrate the solution medium in the form of an evanescent wave. This wave is not propagating, however, and will decay over a characteristic penetration distance Λ, which depends on the incident wavelength λ_0, the refractive index mismatch $n_2 - n_1$, and the incidence α. Figure 2 shows a typical variation for Λ versus α in the case of bromocyclohexane, where $n_2 = 1 - \varepsilon$ with $\varepsilon = \varepsilon' + i\varepsilon'' = 9.06 \times 10^{-7} + i\, 2.95 \times 10^{-8}$ and the incident wavelength $\lambda_0 = 0.7093$ nm. The imaginary part corresponds to the unavoidable x-ray absorption. The critical angle for total external reflection is $\alpha_c = 1.35$ mrd. One observes that Λ stays practically constant around 4 nm for $\alpha_c/2 > \alpha > 0$ but increases rapidly up to 40 nm between $\alpha_c/2$ and α_c. For $\alpha > \alpha_c$, the incident beam penetrates into the solution and gets attenuated in the classical sense. The projection of this attenuation length d along the vertical axis z relative to the interface gives a penetration distance that increases linearly as $\Lambda = \alpha d$. The strong variation of Λ with α

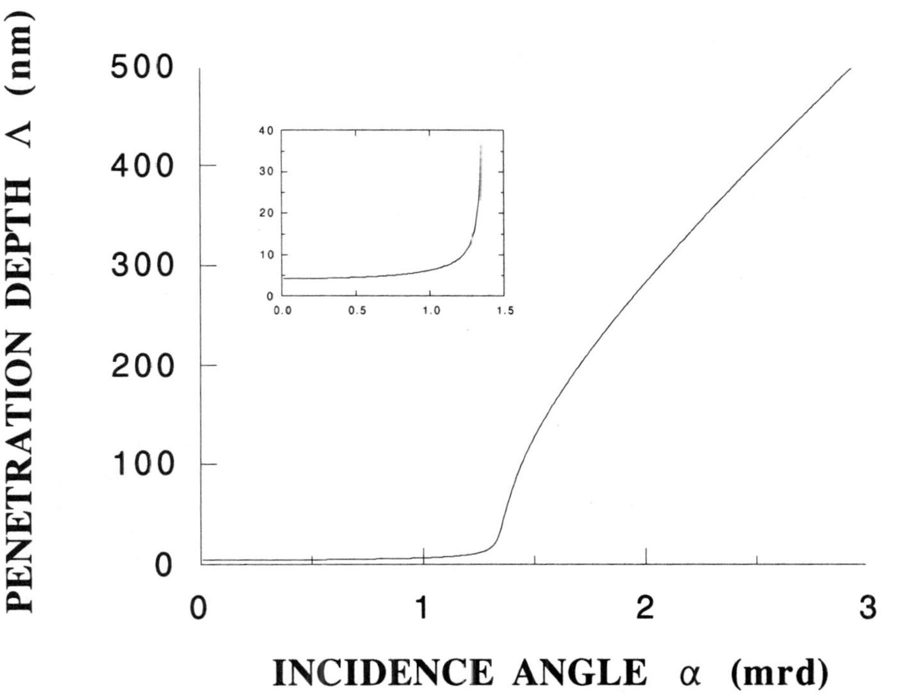

FIG. 2 Penetration depth of the x-ray evanescent wave at the bromocyclohexane-air interface as a function of the grazing incidence angle. The critical angle for total reflection α_c is 1.35 mrd. For $\alpha > \alpha_c$, the x-ray beam penetrates into the liquid but adsorption limits the vertical penetration to a distance $\Lambda = \alpha d$, where d is the characteristic bulk adsorption length. For $\alpha > \alpha_c$, the x-ray beam is totally reflected at the interface, except for an evanescent wave. The penetration depth of this evanescent wave decays very rapidly to a minimum value of 4 nm.

provides an easy way to probe the liquid-vapor interface over various distances. If fluorescent molecules are distributed in the solution with a concentration profile $\phi(z)$, the fluorescent intensity $I_F(\alpha)$ emitted under evanescent wave excitation is, in a simplified form,

$$I_F(\alpha) \propto \int_0^\infty \phi(z) \exp\left(\frac{-z}{\Lambda(\alpha)}\right) dz$$

Measuring I_F for various α yields information on $\phi(z)$ either by using an inverse Laplace transform, which is a difficult mathematical procedure and requires an infinite number of experimental data points, or by fitting the fluorescence intensity curve with a model function for $\phi(z)$. The latter method is by far the simplest and has been put to use to study the concentration profile of an absorbed polymer layer at the liquid-vapor interface [34]. Such an interface is well suited for adsorption studies for at least three reasons: (1) the surface roughness due to the inevitable thermal fluctuation is a few tenths of a nm at most [35] if the experimental setup is carefully isolated from external mechanical disturbances such as building vibrations; (2) the absorbing surface is easy to clean by sucking the top layers of the solution with an aspirator just prior to running the experiment; (3) the absorbing surface is chemically, and structurally, homogeneous. There are neither specific adsorption sites as in a solid surface nor crystallographic defects such as grain boundaries, atomic terraces, etc.

X-ray fluorescence is an atomic property: when an electron from a core orbital is knocked off by the high-energy incident x-ray photon, the vacant electron hole is filled by an electron coming from a higher orbital on the same atom and an x-ray photon is emitted [32]. Fluorescence yield is maximum for the K_α, K_β transitions where an electron from an L orbital falls into the empty space in the K shell. The signal is also maximized by selecting the incident energy to be next to the absorption edge where adsorption is the largest. Working with a manganese anticathode is particularly suited to excite fluorescence from atoms like bromine, rubidium, or manganese. In the first experiment of this sort [34], the fluorescent atoms were attached at random to the polymer backbone as counterions. More specifically, partially sulfonated polystyrene was dissolved into dimethylsulfoxide, which is a good solvent, and the classical sodium counterions were replaced by manganese ions, which have a characteristic K_α fluorescent emission at 5.898 keV. By comparing the fluorescent signal at each angle with the intensity from a solution containing manganese dichloride salt, it was readily observed that the manganese concentration next to the interface was much higher in the polymer case, evidencing a strong polymer adsorption. However, quantitative information was not easy to extract, because the polymer used exhibits intermediate properties between neutral and polyelectrolyte chains, and there is no firm theoretical prediction for this. In addition, the manganese ions are divalent and can thus couple electrostatically to two different sulfonate groups along the chain, creating internal cross-linking, which perturbs the natural polymer conformation.

In a second series of experiments [36], these difficulties have been solved by using a neutral, highly flexible polymer chain like polydimethylsiloxane in bromocyclohexane, which is a theta solvent at 29°C, or in bromoheptane, which is a good solvent. The use of a brominated solvent suppresses the necessity of chemically labeling the polymer, which is always a tedious procedure. If the polymer absorbs at the interface, this creates a depletion layer for the solvent, and the fluorescence intensity is decreased compared to the pure solvent case. Figure 3 shows the fluorescence intensity ratio R between polymer solution (volume fraction $\phi_{\text{bulk}} = 0.0015$, molecular weight $= 130{,}000$ daltons) and

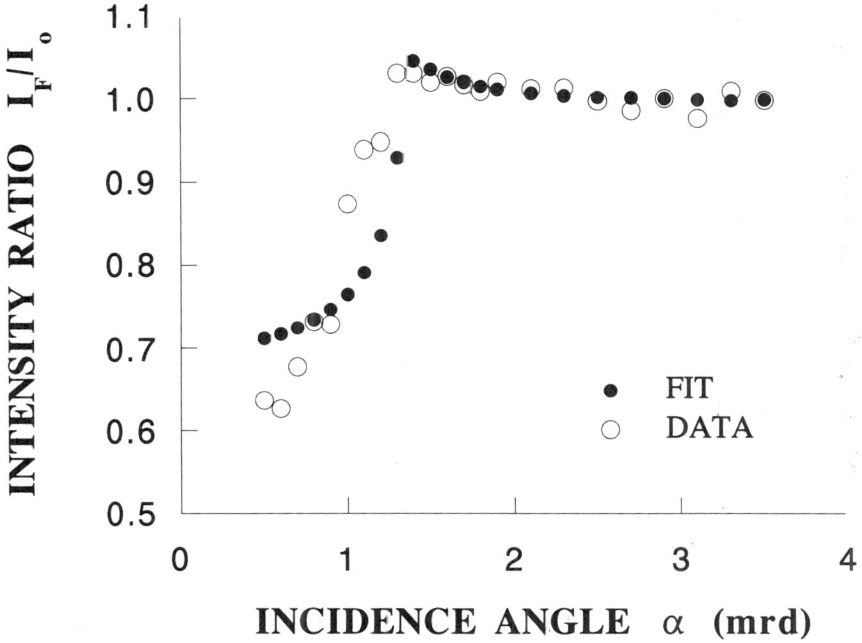

FIG. 3 Ratio of the fluorescence intensities emitted by pure bromocyclohexane and a polydimethylsiloxane solution respectively. Ratios less than unity prove the existence of an adsorbed polymer layer at the vapcr-solution interface. The experimental data have been fitted to a step profile of thickness 9.6 nm and 0.39 volume fraction. Polymer molecular weight is 130,000 and its volume fraction in the bulk is 0.0015.

pure solvent. The fact that R is less than unity for $\alpha < \alpha_c$ is an irrefutable proof of polymer adsorption. The solid line shows a fit with a step profile for the polymer concentration: $\phi = \phi_s$ for $0 < z < a$ and $\phi = 0$ for $z > a$. The best fit gives $\phi_s = 0.39$ and $a = 9.6$ nm, i.e., comparable to the chain radius of gyration in bulk solution [37]. Figure 4 displays three simulations corresponding to a step concentration profile with the same $\phi_s = 0.4$, but three different thicknesses of $a = 10$, 20, an 30 nm. We observe the interesting feature that as a increases, a strong peak appears around α_c. The maximum is a consequence of the surface having a lower critical angle than the bulk solution. Its increase is due to an interference effect between the reflections at the liquid-vapor interface and at the sharp boundary between the dense polymer layer and the dilute bulk solution. Such nonmonotonous behavior has indeed been observed experimentally [36] when the polymer adscrption process is coupled to a surface-induced demixtion. The polymer-rich phase preferentially wets the liquid-vapor interface and forms a layer of mesoscopic thickness, much larger than the chain radius of gyration [38].

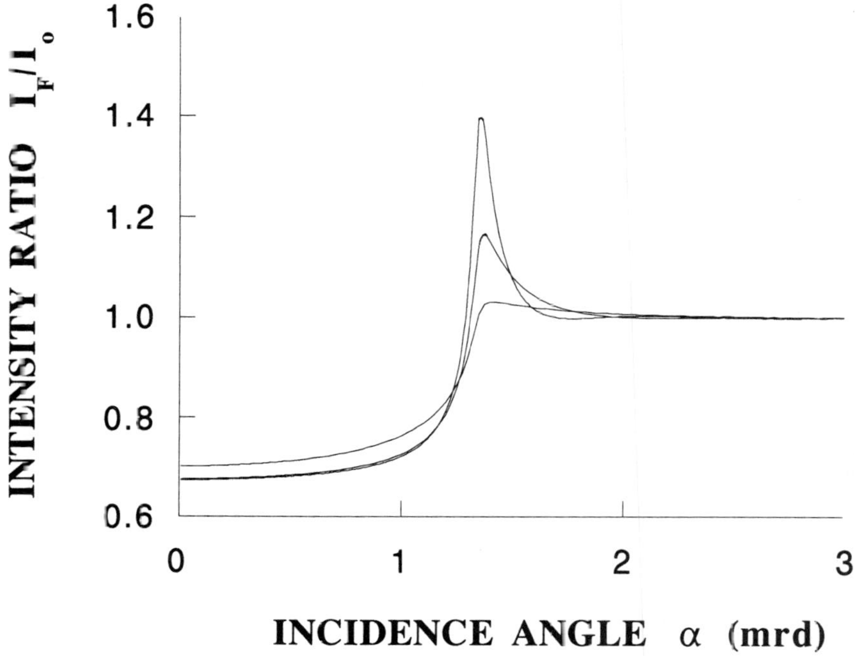

FIG. 4 Radio between the fluorescence intensities emitted under the same geometrical conditions by two different interfaces. The first interface is between vapor and pure bromocyclohexane. The second interface is between vapor and a polymer solution of polydimethylsiloxane in bromocyclohexane, which forms a dense adsorbed layer of thickness a and polymer fraction $\phi_s = 0.40$. The three curves result from calculations and correspond to three different thicknesses of 10, 20 and 30 nm respectively. As the layer thickness get larger, a characteristic peak appears around the critical angle for total reflection.

Although we have so far only discussed the case where there is an excess of polymer chains near the interface, the reciprocal case could be studied as well by the same technique. However, this would require labeling the polymer rather than the solvent. Finally, it is worth pointing out that the x-ray fluorescence method is sufficiently sensitive to be performed using low-power x-ray tubes. The use of high-intensity synchrotron sources is necessary only for adsorption kinetics studies for which long time exposures are unpractical. For instance, a narrowly collimated x-ray beam, from a 1 kW molybdenum tube, by passing through a 6 μm slit, and impinging on the liquid surface with an incidence angle of 1 mrd, gives a signal of 1000 counts/s on a 2-in Ge energy dispersive detector placed directly over the illuminated area (1×2 cm^2). By comparison, the noise background is 100 counts/s, so the main experimental difficulty is therefore not the signal-to-noise ratio but rather to maintain mechanical stability of the various elements so that the unavoidable beam angular wandering stays less than $+/-0.02$ mrd.

In the polymer case, the thickness of the adsorption layer is generally comparable to the chain radius of gyration. For molecular weight higher than 5×10^5 daltons, it is therefore 10–20 times larger than the minimum penetration distance of the evanescent wave. Complete concentration profiles can thus be extracted. There is, however, another case reported in the literature for which the absorbed layer is much thinner and where detailed information can still be obtained. This has been demonstrated recently in studies of the diffuse double layer for the head groups of ionizable surfactants spread at the air-water interface [39, 40]. In Ref. 39 stearic acid was deposited on the surface of a dilute aqueous solution containing 10^{-3} mole L^{-1} of $MnCl_2$. A strong Mn-K_α fluorescence signal, quite incompatible with a homogeneous distribution of the manganese ions, was observed around the critical angle for total reflection. The stearic monolayer caused a dramatic segregation of the metal ions around the carboxylic polar heads, from the liquid bulk to the liquid-gas interface. From the fit, it was possible to determine that the ratio of head group to manganese ion was 0.6 ± 0.2. Complementary surface EXAFS experiments were able to show that there was no long-range in-plane ordering of the metal even at this high coverage. In Ref. 40, x-ray fluorescence and x-ray reflectivity were simultaneously used to study the conformation of behenic acid chains labeled with manganese counterions, at the air-liquid interface. The results were approximately the same as in Ref. 39, except that the measured reflectivity curves allowed one to derive the complete electron density profile across the interface and therefore to compute exactly the electromagnetic field in the evanescent region. The location of the manganese ions is thus obtained with an unprecedented accuracy of better than 1 nm.

III. FLUORESCENCE PHOTOBLEACHING

Fluorescence recovery after photobleaching (FRAP) has become a standard method to study molecular surface diffusion. One reason for its widespread use is its conceptual simplicity and versatility. However, there are some necessary precautions to insure meaningful results.

The basic principle is to use a two-step sequence of light illumination onto the sample. In the first step an intense light beam is used to bleach the fluorescent molecules irreversibly. The region of illumination can be either a narrow spot, e.g., the focal point of a converging lens [41], or a periodic set of stripes [42]. Following this writing pulse, a much weaker light beam at the same wavelength, and with the exact same geometry, is used to probe locally the sample fluorescence. Immediately after photobleaching, the signal is almost zero, but it gradually increases again as unbleached molecules coming from the nonilluminated regions diffuse by Brownian motion into the bleached spots. If the geometry of the illuminated region is well known, and has a size L, the characteristic recovery time will be $\tau \propto L^2D^{-1}$ where D is the self-diffusion coefficient of the fluorescent molecules. More precisely, if the excitation beam has a rectangular intensity profile of width L, the time to recover half of the fluorescence intensity is $\tau_{1/2} = 0.88\ \tau_D$ where $\tau_D = L^2/4D$ for two-dimensional diffusion [41]. It is also possible to illuminate the sample with two crossed laser beams. This produces a well-defined sinusoidal fringe pattern, the period of which can be adjusted by changing the angle θ under which the two laser beams cross [43]. The fluorescence signal is then read with a second identical fringe pattern, but of much lower intensity. This method, although more complex, has several advantages over spot photobleaching recovery. First, the recovery law is purely exponential, in this case with a time constant $\tau = a^2/D$, where $a = 2\pi/P$ and P is the period of the stripe pattern. Second, Brownian diffusion coefficient and flow velocities can be measured independently. Third, the separation between the mobile and the immobile fluorophores can be achieved by spatially modulating the position of the fringe pattern and by using a lock-in detection [44]. Finally, the interfringe spacing can be adjusted continuously over a wide range of distances (0.5–50 μm).

Fluorescence photobleaching techniques have been applied to the measurement of translational diffusion over a large variety of liquid surfaces: spread monolayers at the air-water interface [45], monolayers deposited on alkylated glass slides [46], simple cell membranes [47], and bilayers and closed vesicles [48]. Depending on the experimental system, an enormous range of D values from 10^{-5} to 10^{-13} cm^2 s^{-1} has been covered in the literature. The main reason for this unusual versatility is that the distance over which diffusion takes place can always be adjusted so that the diffusing time does not become prohibitively long. On the other hand, the distance is still large compared to molecular dimensions, and the measured mobility is therefore truly macroscopic in contrast with other techniques like excimer [49] or spin label [50] probes that yield values of the lateral mobility on very local scales ($\cong$10 nm). Spot sizes of 1 μm for instance have been used to probe the Brownian diffusion at the surface of erythrocyte ghost membranes or within the relatively flat regions separating the spicules of the sea-urchinlike, echinocyte red blood cells [51]. In the coexistence region of Langmuir monolayers, it has similarly been possible to separate the very slow diffusion in the solidlike regions from the much faster diffusion in the liquid expanded phase [11].

The diffusion is, of course, quite sensitive to the physical state of the surrounding matrix. For instance, the measured diffusion coefficient can vary by two orders of magnitude between the gel to liquid-crystal phase transition in both single bilayers and multilamellar liquid crystals of various phospholipids

[48] or by more than three orders of magnitude between the liquid expanded and liquid condensed state of Langmuir monolayers [45]. In this latter case, this is quite in agreement with viscosity measurements obtained either from surface dynamic light scattering or from mechanical torsion pendulum [52]. One difficulty peculiar to the air-liquid interface is to get rid of spurious flow processes: the monolayer must be protected from air currents by a protective glass cover, and the bleaching laser beam intensity has to be kept to a minimum to avoid excessive heating and the start-up of correction currents. It is advisable to test various bleaching durations and to check systematically whether the measured fluorescence recovery time after photobleaching scales as the square of the diffusion length. With all these precautions taken, it is generally agreed that $D \sim 10^{-7}$ cm^2 s^{-1} in the liquid expanded phase of a di-myristoyl phosphatidyl choline (DMPC) monolayer and 10^{-10} cm^2 s^{-1} in the liquid condensed or solid phase. Similar values are measured when the same phospholipid is used to form the bilayer membrane of a vesicle. In an interesting experiment where DMPC was mixed with a polymerizable lipid, it was shown that the diffusion of the probe was slowed down only by a factor of five upon polymerization [53]. This indicates that this particular two-dimensional polymer network was not continuous, contrary to expectation, and that cross-linking was not extending beyond 10 nm.

It is fortunate that in all these fluorescence photobleaching experiments the measured diffusion coefficients do not seem to depend much on the exact nature of the diffusing fluorophore. A comparative study between externally added probes and fluorophores that can be added metabolically in the plasma membrane yield differences of no more than a factor of two [54]. More quantitatively, it is generally admitted in the literature that the diffusion coefficient of lipid probes is roughly inversely proportional to the square root of their molecular weight [55]. On the other hand, the lateral mobility of large macromolecules like membrane proteins is only a logarithmic function of the fraction of the diffusing species that penetrates into the bilayer [56].

IV. CONCLUSIONS

We have given here only a few examples of the capabilities of fluorescence techniques for the investigation of fluid interfaces. Although we have confined the discussion to the liquid-vapor interface, it should be clear that similar experiments can also be performed at solid-liquid and liquid-liquid interfaces. For instance, the so-called total internal reflection fluorescence (TIRF) is routinely used by biologists to study protein adsorption from solution onto various solid substrates [57]. In this case, one works in the visible and near-UV optical range, taking advantage of the larger refractive index of the solid substrate to generate the evanescent wave on the solution side. It only requires the protein to be fluorescent (or to be labeled with a fluorescent marker). These experiments are generally performed at fixed incidence angle and one therefore measures essentially the total fluorescence emitted by the protein molecules present in the evanescent wave region. If the quantity of interest is not so much the absorbed amount but the profile, then the experiments are performed at variable angles. Some results exist on the adsorption and depletion layers of flexible polymers in solution [58]. The same technique is, of course, readily applicable to liquid-liquid interfaces and should be of great interest for understanding emulsions and their stabilization by adsorption of amphiphile molecules (short-chain surfactants, block copolymers).

Fluorescence photobleaching recovery measurements of the lateral diffusion of molecules adsorbed on solid substrates are now starting to appear in the literature. Depending on the particular system, they have already revealed several interesting features: (1) existence of an immobile fraction [59]; (2) diffusion by hopping between weakly binding surface sites, with therefore a marked temperature dependence [59]; (3)viscous molecular coupling with the substrate and lubrication by a molecularly thick water film [46]; (4) surface-induced phase transitions in the adsorbed layer [46, 60]. In the above, we have exclusively discussed translational diffusion. Fluorescence, of course, also lends itself to rotational diffusion measurements, via fluorescence depolarization. However, only rotational times shorter than the natural fluorescence lifetime can be measured, and this limits the studies to mostly local motions.

To conclude, the unique sensitivity of the fluorescence methods lend themselves to the study of a wide range of interfacial systems, and there is no doubt that many more applications are still to come.

ACKNOWLEDGMENTS

I am grateful to my colleagues, S. Akamatsu, S. Barton, M. Kent, and E. To, for illuminating discussions and invaluable help while preparing this chapter.

REFERENCES

1. A. Pockels, *Nature 43:* 437 (1981).
2. I. Langmuir, *J. Am. Chem. Soc. 39:* 1848 (1917).
3. G. L. Gaines, *Insoluble Monolayers of Liquid-Gas Interfaces,* Interscience, John Wiley, New York, 1966.
4. K. Kjaer, J. Als-Nielsen, C. A. Helm, L. A. Laxhuber, and H. Möhwald, *Phys. Rev. Lett. 58:* 2224 (1987).
5. N. R. Pallas and B. A. Pethica, *Langmuir 1:* 509 (1985). N. R. Pallas and B. A. Pethica, *J. Chem. Soc. Faraday Trans I, 83:* 585 (1987).
6. G. W. Castellan, *Physical Chemistry,* 3rd ed., Addison-Wesley, 1983, p. 299.

7. M. Lösche and H. Möhwald, *Rev. Scient. Instr. 55:* 1968 (1984).
8. B. G. Moore, C. M. Knobler, S. Akamatsu, and F. Rondelez, *J. Phys. Chem. 94:* 4588 (1990).
9. B. G. Moore, *Ph.D. thesis,* University of California, 1989.
10. A. Fischer, M. Lösche, H. Möhwald, and E. Sackmann, *J. Phys. Lett. 45:* 785 (1984).
11. H. M. McConnell, L. K. Tamm, and R. M. Weis, *Proc. Natl. Acad. Sci. USA 81:* 3249 (1984).
12. D. J. Keller, H. M. McConnell, and V. T. Moy, *J. Phys. Chem. 90:* 2311 (1986).
13. D. Andelman, F. Brochard, and J. F. Joanny, *J. Chem. Phys. 86:* 3673 (1987).
14. H. M. McConnell, *Proc. Natl. Acad. Sci.* (*USA*) *86:* 3452 (1989).
15. W. M. Heckl and H. Möhwald, *Bes. Bunsenges. Phys. Chem. 90:* 1159 (1986).
16. E. To and F. Rondelez, *Compt. Rend. Acad. Sci.* (Paris), Series II, *251:* (1991).
17. A. Miller, W. Knoll, and H. Möhwald, *Phys. Rev. Lett. 56:* 2633 (1986).
18. K. A. Suresh, J. Nittmann, and F. Rondelez, *Prog. Colloid Polym. Sci. 79:* 184 (1989).
19. C. M. Knobler, *Science 249:* 870 (1990).
20. S. Akamatsu, O. Bouloussa, K. To, and F. Rondelez, submitted for publication, 1991.
21. A. Miller and H. Möhwald, *J. Chem. Phys. 86:* 4258 (1987).
22. R. Weis and H. M. McConnell, *J. Phys. Chem. 89:* 4453 (1985).
23. S. Akamatsu and F. Rondelez, *J. Phys. Chem.* (Paris) to appear, 1991.
24. R. M. Weiss and H. M. McConnell, *Nature* (*London*) *310:* 47 (1984).
25. R. M. Weiss and H. M. McConnell, *J. Phys. Chem. 89:* 4453 (1985).
26. H. Bercegol, F. Gallet, D. Langevin, and J. Meunier, *J. Phys.* (*Paris*) *89:* 2277 (1989).
27. P. Walstra "Principles of Foam Formation and Stability," in *Foams: Physics, Chemistry and Structure* (A. J. Wilson, ed.), Springer-Verlag, Berlin, 1990, pp. 1–16.
28. B. G. Moore, C. M. Knobler, D. Broseta, and F. Rondelez, *J. Chem. Soc. Faraday Trans II 82:* 1753 (1968).
29. K. J. Stine, S. A. Rauseo, B. G. Moore, J. A. Wise, and C. M. Knobler, *Phys. Rev. A41:* 6884 (1990).
30. B. Berge, A. J. Simon, and A. Libchaber, *Phys. Rev. A41:* 6893 (1990).
31. R. S. Becker, J. A. Golovchenko, and J. R. Patel, *Phys. Rev. Lett. 50:* 153 (1983).
32. A. H. Compton, *Phil. Mag. 45:* 1121 (1923).
33. See also Chap. 18 by Schlossman, Pershan, and Schwartz, in this book.
34. J. M. Bloch, M. Samsone, F. Rondelez, D. G. Pfeiffer, P. Pincus, M. W. Kim, and P. M. Eisenberger, *Phys. Rev. Lett. 54:* 1039 (1985).
35. A. Braslau, M. Deutsch, P. S. Pershan, A. H. Weiss, J. Als-Nielsen, and J. Bohr, *Phys. Rev. Lett. 54:* 114 (1985).
36. S. Barton, L. Bosio, R. Cortes, and F. Rondelez, submitted for publication, 1991. M. Kent, L. Bosio, and F. Rondelez, submitted for publication, 1991.
37. J. Brandrup and E. H. Immergut, *Polymer Handbook,* John Wiley, New York, 1975.
38. H. Nakanishi and P. A. Pincus, *J. Chem. Phys. 79:* 997 (1983).

39. J. M. Bloch, W. B. Yun, X. Yang, M. Ramanathan, P. A. Montano, and C. Capasso, *Phys. Rev. Lett. 61:* 2941 (1988).
40. J. Daillant, L. Bosio, J. J. Benattar, and C. Blot, to appear in Langmuir (1991).
41. D. Axelrod, D. E. Koppel, J. Schlessinger, E. Elson, and W. W. Webb, *Biophys. J. 16:* 1055 (1976).
42. B. A. Smith and H. M. McConnell, *Proc. Natl. Acad. Sci.* (*USA*) *75:* 2759 (1978).
43. This point was demonstrated earlier in the forced rayleigh scattering technique; see H. Hervet, W. Urbach, and F. Rondelez, *J. Chem. Phys. 68:* 2725 (1978).
44. J. Davoust, P. F. Devaux, and L. Leger, *Embo. J. 1:* 1233 (1982).
45. R. Peters, and K. Beck, *Proc. Natl. Acad. Sci. USA 80:* 7183 (1983).
46. R. Merkel, E. Sackmann, and E. Evans, *J. Phys.* (*Paris*) *50:* 1535 (1989).
47. R. Peters, *Cell. Biol. Intern. Reports 5:* 733 (1981).
48. P. F. Fahey and W. W. Webb, *Biochemistry, 17:* 3046 (1978).
49. R. A. Badley, "Fluorescent Probing of Dynamic and Molecular Organization of Biological Membranes," in *Modern Fluorescence Spectroscopy* (A. Wehry, ed.), Vol. 2, 1976, pp. 91–168.
50. R. D. Kornberg and H. M. McConnell, *Proc. Natl. Acad. Sci.* (*USA*) *68:* 2564 (1971).
51. H. G. Kapitza and E. Sackmann, *Biochim. Biophys. Acta 595:* 56 (1980).
52. P. J. Winch and J. C. Earnshaw, *J. Phys. Condensed Matter 2:* 8699 (1990).
53. H. Gaub, E. Sackmann, R. Büschl, and H. Ringsdorf, *Biophys. J. 45:* 725 (1984).
54. L. Dupou, A. Lopez, and J. F. Tocanne, *Eur. J. Biochem. 171:* 669 (1988).
55. E. Sackmann, J. Engelhardt, K. Fricke, and H. Gaub, *Coll. and Surf. 10:* 321 (1984).
56. P. G. Saffman and M. Delbruck, *Proc. Natl. Acad. Sci.* (*USA*) *72:* 3111 (1975).
57. W. M. Reichert, "Evanescent Detection of Adsorbed Films: Assessment of Optical Considerations for Absorbance and Fluorescence Spectroscopy at the Crystal/Solution and Polymer/Solution Interfaces," in *Crit. Rev. Biocompat. 5:* 173–205 (1989).
58. F. Rondelez, D. Ausserré, and H. Hervet, *Ann. Rev. Phys. Chem. 38:* 317 (1987).
59. D. W. Bjarneson and N. O. Petersen, *J. Am. Chem. Soc. 112:* 988 (1990).
60. M. Seul and H. M. McConnell, *J. Phys.* (*Paris*) *47:* 1587 (1986).

IV

Solid Surfaces

20

Light Scattering by Solid Surfaces and Surface Characterization

DANIEL MAYSTRE Laboratoire d'Optique Electromagnétique,
University of Aix-Marseille 3, Marseille, France

I. INTRODUCTION

This chapter deals with the characterization of solid surfaces, in particular using optical scattering. The characterization of metallic and dielectric surfaces is becoming increasingly important, in particular for its applications in industry and technology. This may be partly explained by the tremendous development of real-time process controls or quality tests. There is nowadays a very strong demand for solid surface characterization in optics (lens surfaces, telescope mirrors), microelectronics (control of silicon plates before and after etching), botanics and biology (membranes), mechanics (piston axes, ball-bearing surfaces), and so on. It is worth noting that sometimes the roughness may be useful, for example microasperities may multiply the second harmonic generation in nonlinear optics by a factor greater than 1000 [1].

The aim of this chapter is to present the possibilities of nondestructive, rapid, and precise characterization of solid surfaces using light scattering. These characterization means are compared with more classical ones. It is very important to notice that, from a fundamental point of view, characterizations of solid and liquid rough surfaces do not differ from each other. In practice, solid rough surfaces may present asperities with very small widths (less than 1 μm) and large slopes (sometimes larger than 1), a kind of roughness rarely countered on liquid surfaces. An accurate theoretical tool for dealing with scattering from solid rough surfaces must include the so-called resonance domain, where the dimensions of the asperities have the same order of magnitude as the incident wavelength. The investigation of the scattering properties of liquid surfaces can be handled with simpler theoretical means.

At the present time, a general surface characterization technique for solid surfaces does not exist, and we cannot expect that such a tool will appear in the future. For a given surface, the choice of the best-adapted technique depends mainly on the material, on the statistical parameters of the roughness, and on the experimental conditions in which the characterization has to be performed. As a consequence, a rough prior knowledge of the statistical dimensions of the asperities is vital for chosing the appropriate tool. Some techniques are very efficient when the width of the asperities is of the order of 100 μm but become catastrophic as this width approaches 1 μm. The depth and the slope of the grooves also have a vital importance. When a surface is formed by multiscale modulations, it may be necessary to use several methods. As regards the material, we shall assume here that it is metallic, even though the major part of the techniques presented here could be used for dielectric surfaces as well.

First, it will be shown that the expression "rough surface" can designate a wide variety of structures. We shall classify these structures in two categories: deterministic and random rough surfaces. Then a nonexhaustive overview of modern characterization tools will be made. A particular emphasis will be given to the insufficiency of the usual means when the width of the asperities becomes smaller than a few μm. Finally, an extension of the actual characterization means, based on the use of light scattering and electromagnetic theory, will be envisaged.

II. THE TWO KINDS OF ROUGH SURFACES

A. Deterministic Rough Surfaces

A deterministic rough surface is a rough surface in the common sense. Such a surface is defined by a particular profile, which for instance may be given by its mathematical representation in Cartesian coordinates $z = f(x, y)$. In practice, it can be, for example, a silicon plate on the surface of which a V-shaped groove has been traced. Here, the problem of characterization is to find a function representing the shape of the surface profile.

B. Random Rough Surfaces

The notion of a random rough surface is more subtle and unfortunately causes much confusion. In order to give first an intuitive idea of a random rough surface, let us notice that the manufacturer of a telescope mirror or of a ball bearing does not want to know the precise microtopography of the surface of such an object, which is very complicated. In general, he prefers to know a few statistical parameters such as the mean height or the mean width of the asperities. These simple parameters are much more appropriate than an exact Cartesian representation of the surface profile if one wants to know the optical or mechanical quality of the parts. Thus one can consider that a random rough surface is defined by statistical data. The precise definition of these statistical data will be given further, but first, another and quite complementary definition of this kind of surface must be given. A random rough surface can also be defined by an infinite (or in practice very large) set of deterministic rough surfaces generated by the same experimental or numerical process. Each of these surfaces, called a sample or realization, has a particular deterministic profile but all these surfaces have the same statistical parameters. One can consider for instance a large number of metallic plane mirrors made with the same polish process (Fig. 1).

From the infinite set of deterministic rough surfaces, it is easy to deduce the statistical parameters. For example, in Fig. 1, the RMS (root mean square), which can be considered as the mean height of the asperities, is defined, for given x and y, by

$$h_{x,y} = \sqrt{\langle(z_i(x, y) - \langle z_i(x, y)\rangle_{x,y})^2\rangle_i}$$

where $z_i(x, y)$ is the ordinate on the mirror M_i at the point of coordinates x and y, $\langle\ \rangle_{x,y}$ its average on the mirror (i.e., the ordinate of the average plane of M_i), and $\langle\ \rangle_i$ the mean value over all the samples M_i. When all the statistical parameters

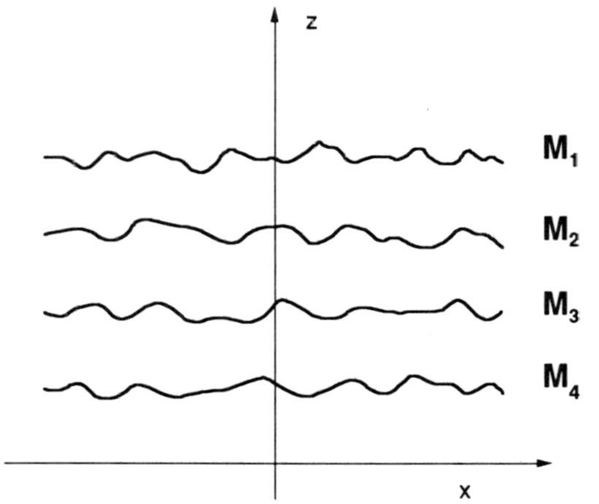

FIG. 1 Deterministic samples of a randomly rough surface.

of the surface are independent of x and y, the surface is stationary and $h_{x,y}$ becomes a parameter h characterizing the mean height at any point of the surface. For a stationary surface, the correlation function $C(u, v)$ is defined by

$$C(u, v) = \langle (z_i(x, y) - \langle z_i(x, y) \rangle_{x,y}) \times (z_i(x + u, y + v) - \langle z_i(x, y) \rangle_{x,y}) \rangle_i$$

and for one-dimensional random rough surfaces ($z = f(x)$) or when the correlation function has a symmetry of revolution, the correlation length l (which is linked to the mean width of the asperities) is defined as the width at $1/e$ of the correlation function. Another important statistical function is the probability density $p(\xi)$ defined by

$$p(\xi) = \frac{P_i(\xi < z_i(x, y) - \langle z_i(x, y) \rangle_{x,y} < \xi + d\xi)}{d\xi}$$

This is the probability (calculated in the set of samples for x and y fixed) that the distance between the surface and its mean plane is between ξ and $\xi + d\xi$.

A random rough surface is ergodic when the mean values $\langle \, \rangle_i$ in the set of samples can be replaced by the mean values $\langle \, \rangle_{x,y}$ in x and y for an arbitrary sample, a definition that assumes that the surface extends to infinity. This remark is at the origin of a widespread confusion. Many people classify as a random rough surface one sample of the actual random rough surface, which in fact is a deterministic rough surface. The interested reader may find in Ref. 2 a more extensive description of the statistical parameters of a random rough surface.

It is important to notice that the major part of the experimental and theoretical studies of random rough surfaces assume that the correlation function and the density of probability are Gaussian. We shall deal with this kind of surface in the following, even though the various techniques are not restricted to this important case. In practice, the surfaces are in general more complicated and, for instance, it is very commonplace to find the superposition of a roughness formed by microasperities and a waviness of much larger correlation length.

III. THE TECHNIQUES OF CHARACTERIZATION

A. The Mechanical Profilometer

The mechanical profilometer measures the topography of a deterministic surface by moving a fine diamond stylus parallel to this surface so that the stylus contacts the surface very lightly. This stylus is coupled to an electromechanical transducer whose output is amplified and may be stored in a computer and then analyzed. Thus the work of this instrument can be compared to that of the sensor head of a phonograph pickup. The head may also be set in order to scan a series of closely spaced cross-sections in order to obtain a pictorial representation of the surface, but this operation requires a very long time.

The range in height sensitivity of this instrument is between 25 and 250 Å with a horizontal resolution between 1 and 10 μm, provided the slope of the asperities is small. Even though the stylus instrument is considered to be the standard technique, it presents many shortcomings. The stylus is fragile, slow, and possibly destructive. It cannot be used for nondestructive evaluations, inspection of large surfaces, on-line surface measurements, or in-process inspection during manufacture.

A comparison between topography measurements made with a mechanical profilometer and optical means described in the following can be found in Ref. 3.

B. The Optical Microscope

Many techniques of surface characterization use an optical microscope. One of the oldest is to cut the surface and then to observe the section with this microscope. The nondestructive technique of light sectioning [4] uses a razor-blade-shaped beam of light projected onto the measured surface at an angle of about 45°, producing an optical section of the surface in the microscope whose axis is placed perpendicular to the plane of the beam. The profile of the surface is measured from a micrograph of the optically sectioned surface after calibration. This method is rapid, noncontacting, and nondestructive, but it is restricted to relatively rough surfaces (more than 1 μm valley-to-peak).

The stereoscopic technique, using two micrographs of the surface realized with slightly different incidences (close to the normal incidence), permits a reconstitution of the profile, but a quantitative estimation of the roughness is delicate. Among the qualitative methods, we can also classify the interference microscope developed by Nomarski [5]. Finally, let us mention the FECO developed by J. Bennett [6] mainly for the study of optical mirrors, thin films, and lenses. This technique based on the observation of chromatic fringes is restricted to shallow roughness.

C. The Electron Microscope

Very often, the direct observation of the profile using an electron microscope is impossible and, at least for one-dimensional surfaces like diffraction gratings, one must cut the surface and observe the cross-section [7] or make a carbon replica by contact.

The scanning electron microscope may lead to many problems in interpreting edges and bright and dark areas as topographic information, and thus quantitative measurements cannot be envisaged without using stereo pairs of micrographs. In practice, these techniques cannot provide much precision once the width of the asperities approaches one micrometer.

D. The Scanning Tunneling Microscope

The scanning tunneling microscope [8] provides a three-dimensional picture of the surface. The principle is the following: a tip is brought close enough to the surface that, at a convenient voltage, the tunneling current is measurable. The tip is scanned over the surface and the current between it and the surface is sensed. Two modes may be envisaged. In the constant current mode, a feedback network changes the altitude of the tip in order to keep the current constant. Since the current varies exponentially with the gap between the tip and the surface, this gap remains nearly constant. Thus a map of the tip altitude gives directly the surface topography. In the constant height mode, the tip is scanned across the surface at nearly constant altitude and constant voltage, the current being monitored.

The lateral resolution of this instrument can reach some angstroms (separation of two carbon atoms on graphite), and the vertical resolution can be sufficient to resolve 0.02 Å in surface characterization.

E. The Optical Profilometer

The optical profilometer technique [9, 10] is the equivalent in the optical domain of the mechanical profilometer. A beam of white or monochromatic light is focused on the mean plane of the surface. By moving the surface horizontally, the defect of focusing due to the roughness is evaluated by observation of the reflected beam, assuming that the laws of geometrical optics are satisfied. Because of this approximation, and also taking into account the diameter of the light spot (some μm), this method is restricted to small slope defects with large widths.

F. The Optical Interferometer

In the last two decades, the optical interferometer technique has become very popular because of its simplicity, its resolution, and its low cost. Two examples of this kind of instrument are the FECO (fringes of equal chromatic order, see also Sec. III.B) developed by J. M. Bennett (U.S. Naval Weapons Center, China Lake, CA 93555) and the TOPO instruments proposed by the WYKO Corporation (2990 East Fort Lowell, Tucson, AZ 85716). For monochromatic light, its principle is given in Fig. 2. The monochromatic light emitted by the source S is focused by the lens L and reflected by S′ at a point placed at the focus of the microscope objective O. The interference figure between the highly reflecting plate R and the rough surface T is observed and recorded in the plane P which is the conjugate of T through O, by image analysis, then treated by a computer using the laws of geometrical optics. It is possible to reconstitute the deterministic profile of the surface and to deduce its statistical parameters. Other kinds of optical interferometers are now in use, some being based on the Michelson interferometer, others using polychromatic light. One of the rare shortcomings of this technique is the problem caused by the interpretation of the interference figure as the width of the asperities approaches 1 μm. In these conditions, it becomes very dangerous to believe that the laws of reflection in geometrical optics can apply. Unfortunately, the instrument still provides a profile topography in these conditions, a fact that may lead a naive user to serious errors.

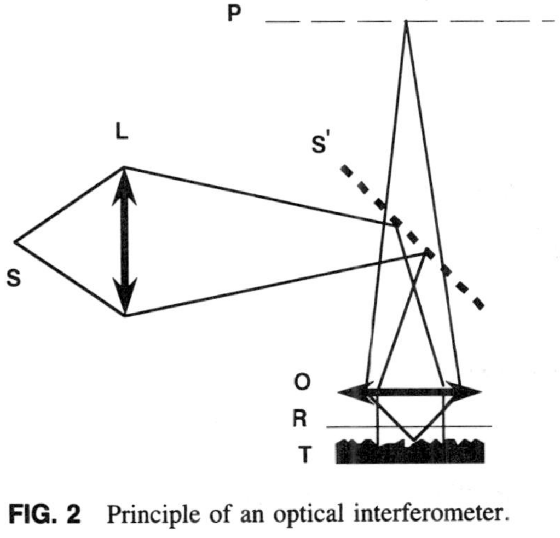

FIG. 2 Principle of an optical interferometer.

G. Scattered Field Analysis

Scattered field analysis is probably the only technique potentially capable of characterizing in a precise, nondestructive, and rapid manner the surface deformations less than 1 or 2 μm in dimension. Furthermore, it can be used for on-line inspections.

Figure 3 gives the principle of the method. The rough surface is illuminated by a laser beam. As the existence of a scattered field is a direct consequence of the roughness, one can believe that an appropriate analysis of the scattering pattern should permit a good estimation of the surface profile, provided the scattering phenomenon can be accurately simulated on a computer. In principle, this method is very general and elegant. Moreover, for random rough surfaces, *it should permit a direct estimation of the statistical parameters of the surface from the statistical parameters of the speckle,* which avoids an intermediate calculation of the deterministic profile of the sample. These remarks indicate why this method of surface characterization is considered as one of the techniques of the future, at least for microroughness.

However, the implementation of this technique requires the solution of a tremendously difficult theoretical problem: deducing the surface profile from the scattering pattern at infinity is *a theoretical and numerical problem that has not yet been solved in a precise and general way*. This explains the great interest devoted in recent years to the numerical simulation of scattering phenomena. In our opinion, it is this method that must be developed now by taking advantage of the exponential development of computer capabilities. This is the reason why we shall present, in the next section, the present limits of the numerical treatment of scattering problems and the first successes of the characterization from scattering analysis.

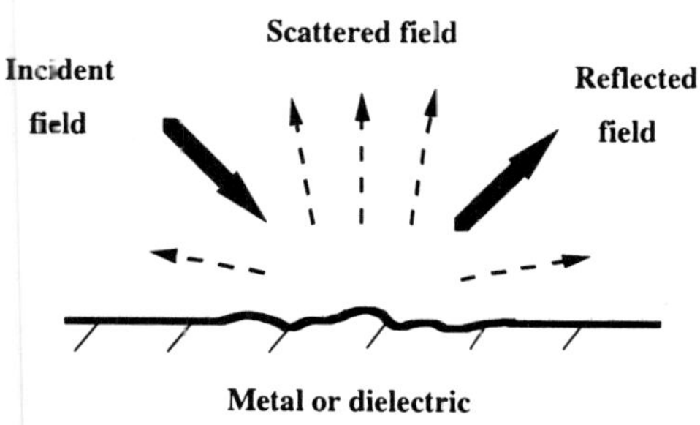

FIG. 3 Measurement of surface microtopography using scattering phenomena.

IV. NUMERICAL SIMULATION OF SCATTERING PHENOMENA

A. Generalities

The use of the laws of optics and electromagnetism for solving scattering problems is an old research field in which Fresnel, Fraunhoffer, and Rayleigh achieved pioneering contributions. Here we are not concerned with the classical optical theories (for instance the famous theory of speckle). Indeed, these theories are accurate and valuable when the sizes of the surface defects are much greater than the light wavelength, but they do not take into account the vector character of the light, so they cannot in a realistic manner handle scattering phenomena as the size of these defects approaches 1 μm. In these conditions (the so-called resonance domain), scattering phenomena strongly depend on the polarization of the incident light; thus the only realistic way of solving the scattering problem is to use the fundamental laws of electromagnetism. Unfortunately, we are then faced with the solution of boundary value problems that require sophisticated theoretical treatments and powerful computers. This explains why, except for some simple shapes of scattering objects where the mathematical problem can be solved in closed form, electromagnetic optics was not developed until the 1960s. Previously some attempts were made in this direction, but without powerful computers it was necessary to introduce approximations in the theory in order to obtain results in closed form. One of the most remarkable of these attempts was the Kirchhoff approximation, which is still used by many people today. In the following, we shall present successively the rigorous boundary value problem, the Kirchhoff approximation, and the rigorous theory.

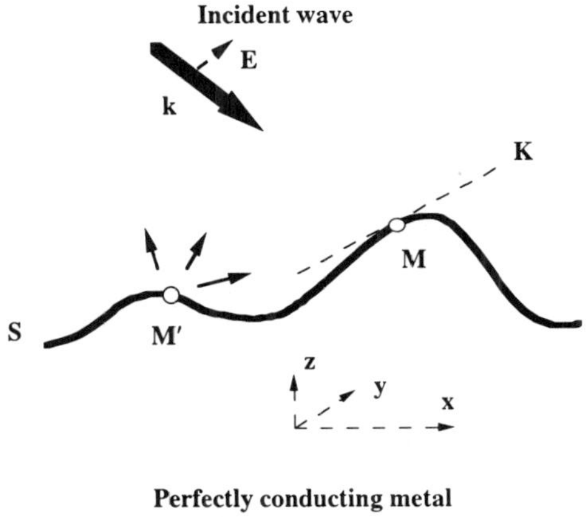

FIG. 4 The Kirchhoff approximation, principle and criticism.

B. The Boundary Value Problem

For the sake of clarity, the boundary value problem is presented in a very simple case. Let us consider in Fig. 4 a one-dimensional rough surface S ($z = f(x)$), considered as perfectly conducting (or in other words perfectly reflecting), illuminated by a homogeneous monochromatic plane wave having the electric field $\mathbf{E}^i$ parallel to the y axis ($\mathbf{E}^i = E^i \cdot \mathbf{u}_y$, with $\mathbf{u}_y$ being the unit vector of the y axis).

For simplicity, we assume here an intuitive property (but unfortunately physical intuition is not mathematical proof) that can be demonstrated from theorems of existence and uniqueness: the total electric field **E** remains parallel to the y axis ($E = E \cdot \mathbf{u}_y$). The scattered field is classically defined as the difference $E^d = E - E^i$ between the total and the incident field. From Maxwell equations and boundary conditions, it can be established easily that E^d satisfies the following boundary value problem in the domain located above S:

$$\nabla^2 E^d + k^2 E^d = 0 \qquad \text{above S}$$

$$E^d = -E^i \qquad \text{on S}$$

$$E^d \text{ satisfies a radiation condition when } z \to \infty$$

with $k = 2\pi/\lambda$, λ being the wavelength of the light. The first equation is the Helmholtz equation and is obtained by using Maxwell equations; the second is the boundary condition on the perfectly conducting material, which states that the tangential component $E^d + E^i$ vanishes on S; the third statement, which is intuitive, assumes that the scattered field can be described by a superposition of outgoing waves when $z \to \infty$. For the sake of simplicity, we shall assume the existence and uniqueness of the boundary value problem.

C. The Kirchhoff Approximation

The basis of this theory is to notice that the scattering process may be split into two steps:

Incident wave ⇒ Surface current ⇒ Scattered field

The second step means that the scattered field is nothing but the field generated by the surface current density on the surface. Indeed, it can be shown mathematically that the knowledge of the surface current density enables the calculation of the scattered field in a very simple way (in practice by performing the numerical calculation of integrals on the profile).

In the Kirchhoff approximation, it is proposed to achieve the first step with no more difficulty. The surface current density generated at the point M of S by the incident wave is assumed to be the surface current density that would exist at the point M if S were replaced by a perfectly conducting plane mirror K tangent to S at the point M (Fig. 4). With this assumption, the calculation of the surface current density at any point of the surface may be achieved in closed form, and the problem of scattering becomes very easy to solve. This of course explains the popularity of this method.

This hypothesis fails because the surface current density in M is generated not only by the incident wave but also by the field reradiated toward M by the surface densities at other points M′ of S. Thus there does exist a cause-effect relationship between the incident field and the surface current density, but the process includes a very complicated coupling phenomenon between the surface currents at the different points of the surface. Roughly speaking, the Kirchhoff approximation is comparable to the assumption of neglecting the interactions between the various skittles in a bowling game!

Obviously, the Kirchhoff approximation should not be used when the dimensions of the asperities have an order of magnitude equal to or less than that of the wavelength λ. On the other hand, this approximation may give accurate results when the wavelength is very small compared to the dimension of the asperities, though this is not a general rule, because of the phenomena of multiscattering. The most serious problem of this theory is that its precise limits of validity are not well known.

D. The Rigorous Theory

It is not our purpose to describe here the rigorous theories of scattering. The integral theory rigorously reduces the boundary value problem to the solution of an integral equation. The various authors give descriptions of this theory based on the theory of distributions, or on the theory of potentials, or sometimes on the theory of Green's function, but from a mathematical point of view these presentations are equivalent. Coming back to the particular scattering problem of the perfecting conducting one-dimensional rough surface, the unknown function of the integral equation is the surface current density, and this equation may be solved on the computer by using a moment method, or a Gallerkin method, or a finite element method (depending on the basis on which the equation and the unknown function are projected). The interested reader may consult Ref. 11 for a review in the particular case of a periodic surface (diffraction grating) or Refs. 12, 13, and 14 for a one-dimensional, arbitrarily shaped rough surface. For numerical reasons, the theoretical studies are practically restricted to one-dimensional rough surfaces. At present, it can be considered that it is a possible to compute with a precision of about 1% the diffraction pattern of an arbitrarily shaped one-dimensional rough surface made with a dielectric or metallic material and illuminated by a beam of arbitrary width and polarization [15], even though numerical problems caused by the propagation of surface plasmon waves still exist when the material is a highly (but not perfectly) conducting metal like gold, silver, or aluminum.

For two-dimensional rough surfaces, some attempts at solving the problem of scattering on powerful computers have been made recently. However, because of huge difficulties in matrix storage and matrix inversion when the size of the matrix becomes very large, the results are restricted to shallow surfaces.

V. THE TECHNIQUE OF SCATTERING PATTERN ANALYSIS

To our knowledge, the technique of scattering pattern analysis was used for the first time in the National Institute of Standards and Technology of Gaithersburg (MD, USA). The purpose of the work [16] was to use a laser and detectors at 10.53 μm in order to determine the exact parameters of a V-shaped groove from the scattering pattern. The V-shaped groove was drawn by a diamond stylus on a copper plate, and the width and the height of the groove were close to 3 and 1 wavelengths respectively.

With this aim, the authors used the computer code described in Ref. 12 and achieved a catalogue giving the evolution of the scattering pattern by varying the two parameters (width and height) of the groove. Their conclusion was that the comparison of the experimental scattering pattern with the scattering patterns of the catalogue allowed them, in general, to control the shape of the groove with a good precision.

At the same time, the same institution elaborated DALLAS (detector array for laser light angular scattering). This instrument is devoted to the characterization of a random rough surface from the angular distribution of the scattering pattern [17]. It consists (Fig. 5) of a 5 mW He-Ne laser illuminating the rough surface with an incidence that may be varied by a rotating assembly of two mirrors. The detection system consists of an array of 87 detectors spaced 2° apart in a semicircular yoke of diameter 164 mm, centered on the illumination spot of the surface (about 2 $\times$ 3 mm). The yoke can be rotated about one axis so that the detectors can sample practically the entire hemisphere of radiation scattered by the surface. The detectors (lens, optical fiber, PIN Si photodiode) transmit the output voltage signals to a 100-channel scanner, and the signals are digitized and stored in a desktop microcomputer. The computer compares these signals with the theoretical predictions based on a simulation of the scattering phenomena.

For the first trial, the apparatus was checked by comparing an experimental scattering pattern to a theoretical pattern of the surface after measurement of the topography using a mechanical profilometer. Even though the comparison did not result in a complete failure, the importance of the discrepancies is such that, in our opinion, the implications of this work are not immediately promising.

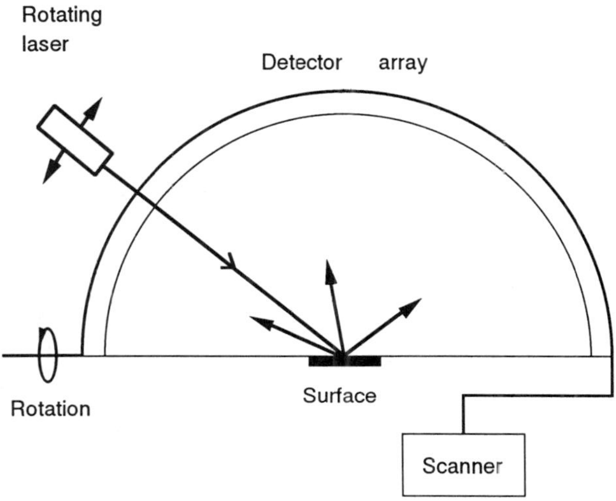

FIG. 5 Scheme of DALLAS.

This is not surprising. Indeed, because the actual surface was two-dimensional, the theory used for this work was that deduced from the Kirchhoff approximation. Since the mean width of the asperities was of the order of one wavelength, it is not surprising to find significant discrepancies between theory and experiment. Thus the pessimistic evaluation of this method could be revised by a better choice of theoretical prediction. Of course, this remark should not be considered as a criticism of this pioneering work in the domain of surface characterization, achieved by the National Institute of Standards and Technology. It must be acknowledged that the state of the art in the theoretical simulation of scattering from two-dimensional rough surfaces is poor; but we shall explain in the next section why certain recent progress in this domain lead us to a reasonable optimism.

VI. PERSPECTIVES OF SURFACE CHARACTERIZATION FROM SCATTERING PATTERN ANALYSIS

A numerical proof of some optical phenomena already known from experimental measurements has been given recently. Figure 6, drawn from Ref. 18, presents one of these phenomena. It shows a rigorous computation of the mean value (on a large number of samples) of the intensity scattered by a random rough surface versus the sine of the scattering angle when the surface is illuminated by a Gaussian beam. The columns show, from left to right, the mean values of the

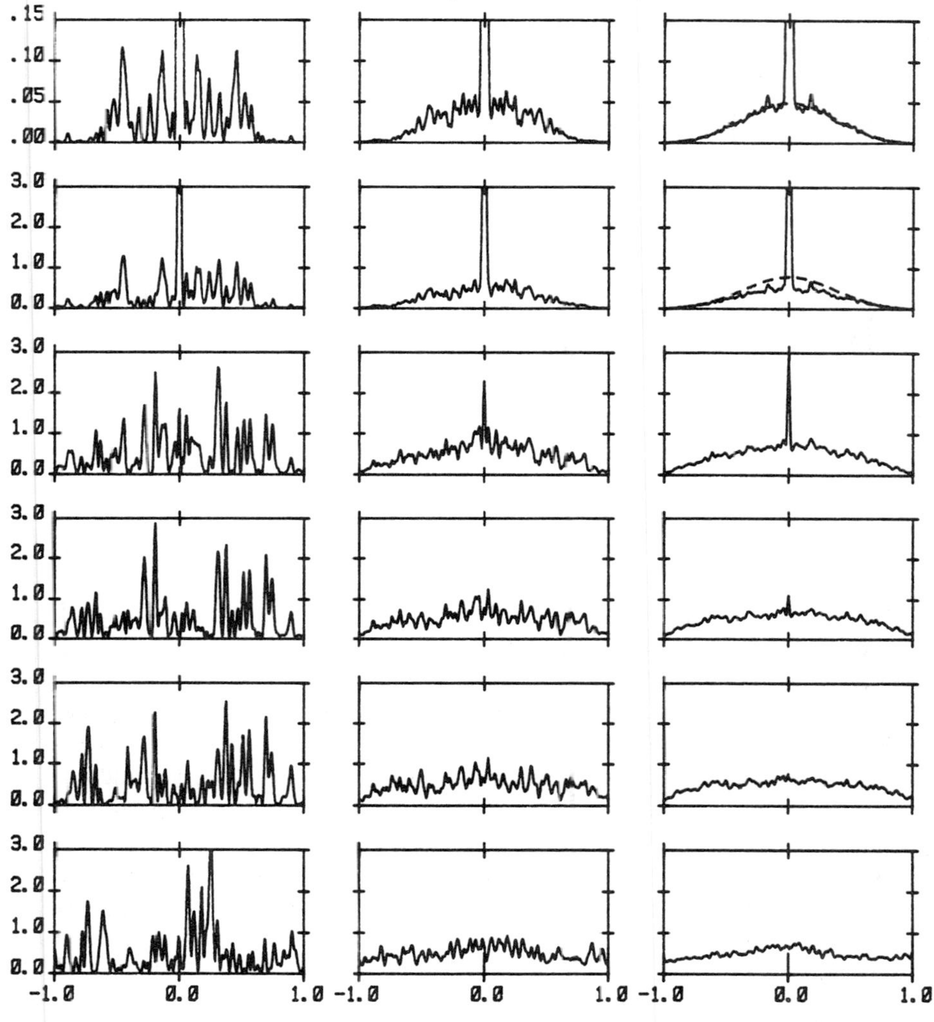

FIG. 6 Rigorous theoretical scattering diagram of a randomly rough surface of correlation length 0.5642 μm, illuminated in normal incidence by a Gaussian beam of wavelength 1 μm and width 50 μm. The abscissa represents the sine of the scattering angle. (With permission of *Optical Engineering*.)

intensities scattered by 1, 10, and 100 samples respectively. From top to bottom, the RMS takes the values 0.0177, 0.0707, 0.1414, 0.177, 0.212, and 0.354 μm.

This figure shows that, when the number of samples is increased, the mean intensity curve tends to a very smooth limit, except in the vicinity of the normal direction, where for the smallest values of the RMS a narrow peak is present. This peak may be explained easily. When the surface is perfectly planar, the scattered field reduces to the very narrow specularly reflected beam and the speckle does not exist. As the RMS is increased, an increasing part of the energy contained in the specularly reflected beam goes to the speckle and finally, above a critical value of the RMS, the speckle is completely developed, which means that the peak of the specularly reflected beam has completely disappeared.

This phenomenon may give a very simple means for characterizing random rough surfaces. Figure 7, derived from Fig. 6, shows the total integrated intensity $T(h)$ of the speckle (in other words, the difference between the total scattered intensity and the specularly reflected beam intensity), h being the RMS. This function may be represented accurately in Gaussian form

$$T(h) \cong 1 - e^{-h^2/c^2}$$

where c is a real positive constant.

It has been verified numerically that in most cases the above expression gives a precision of about 1% on the integrated intensity of the speckle. So a precise measurement of T should permit a precise evaluation of the RMS, provided that this RMS is smaller than about 0.15 μm (a value for which T becomes very close to its asymptotic value of unity) and that the constant c of this expression is known. This phenomenon is known by many experiments, but since they do not know the constant c, they generally estimate it from a perturbation theory or from the Kirchhoff approximation. In fact, we have found numerically that this kind of theory was unable to provide a good precision on the constant c. As a consequence, only two possibilities can be envisaged, an experimental

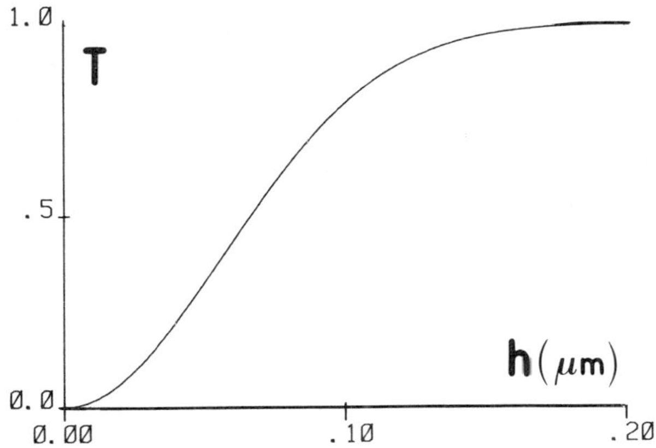

FIG. 7 Total intensity of the speckle (unity = incident beam intensity) versus the RMS. (With permission of Optical Engineering.)

calibration or a rigorous numerical calculation. In Fig. 7, the random rough surface is one-dimensional and, as mentioned before, the rigorous computation of the field scattered by two-dimensional randomly rough surfaces is much more difficult. However, it is worth noting that the determination of the constant c of the Gaussian curve of Fig. 7 does not require the computation of the intensity scattered by deep rough surfaces. In practice, an RMS of 0.05 μm is sufficient for that. If the perturbation theories cannot accurately treat this RMS, this is not so for the recently developed numerical codes for two-dimensional rough surfaces.

Even though the RMS is generally considered the most important parameter of a random rough surface, many experimenters emphasize the vital importance of the correlation function, which may be different from a Gaussian. The perturbation theory shows that as long as the RMS is small compared to the wavelength λ, the mean speckle intensity (without the specularly reflected intensity) is the power spectrum of the surface, i.e., the Fourier transform of the correlation function. Thus in Fig. 6, the top right curve may permit, through a simple integral, the determination of the correlation function, which in that case is Gaussian. When the RMS is increased, obviously the Gaussian shape of the mean intensity disappears and is replaced by a uniform intensity, a fact that shows that the simple rule mentioned above, linking the mean intensity with the spectral density of the surface, does not hold. In order to replace it, we cannot avoid the rigorous numerical simulation, in order to deduce the power spectrum from the mean intensity; but of course then we are faced with an inverse scattering problem.

VII. CONCLUSION

It has been shown that the surface characterization techniques, which include numerous mechanical and optical tools, are inadequate in the field of rapid, nondestructive, and on-line methods. In this case, the analysis of the scattering intensity should be the most powerful and elegant tool. The largest obstacle in the implementation of this method is the tremendous difficulty of handling the precise numerical simulation of scattering by two-dimensional rough surfaces. However, some research groups have made significant progress in this direction recently; and with the development of computer technologies one may hope that this problem will be solved in the next few years.

REFERENCES

1. D. Maystre, M. Neviere, R. Reinisch, and J. L. Coutaz, *J. Opt. Soc. Am. B 5:* 338 (1988).

2. P. Beckmann and A. Spizzichino in *The Scattering of Electromagnetic Waves from Rough Surfaces,* Pergamon Press, Oxford, 1963, pp. 185–192.
3. E. L. Church, T. V. Vorburger, and J. C. Wyant, *Optical Engineering 24:* 388 (1985).
4. G. Schmaltz in *Technische Oberflächenkunde,* Springer-Verlag, Berlin, 1936.
5. R. Hoffman and L. Gross, *J. Microscopy 91:* 149 (1970).
6. H. Bennett and J. Bennett in *Physics of Thin Films* (G. Haas, ed.), Academic Press, New York, 1967.
7. A. Roger and M. Breidne, *Optics Commun. 35:* 299 (1980).
8. G. Binnig, H. Rohrer, Ch. Gerber and E. Weibeil, *Phys. Rev. Lett. 49:* 57 (1982).
9. J. Simon, *Applied Optics 9:* 2337 (1970).
10. C. Gorecki, G. Tribillon, and J. Mignot, *J. Optics 14:* 19 (1983).
11. D. Maystre in *Progress in Optics, Vol. 21* (E. Wolf, ed.), North Holland, Amsterdam, 1984, pp. 1–67.
12. D. Maystre, *IEEE Trans. Antennas and Propagation 31:* 883 (1983).
13. A. A. Maradudin, E. R. Mendez, and T. Michel, *Opt. Lett. 14:* 151 (1989).
14. M. Nieto Vesperinas and J. M. Soto Crespo, *Optics Commun. 75:* 215 (1990).
15. M. Saillard and D. Maystre, *J. Opt. Soc. Am. A 7:* 982 (1990).
16. G. S. White and J. Marchiando, *Applied Optics 22:* 2308 (1983).
17. T. V. Vorburger, E. C. Teague, F. E. Scire, M. J. McLay, and D. E. Gilsinn, *J. Res. Nat. Bureau of Standards 89:* 3 (1984).
18. J. P. Rossi and D. Maystre, *Opt. Engineering 25:* 613 (1986).

Symbols

$\mathbf{u}$	surface displacement
ζ	surface vertical displacement
$\mathbf{r}$	position in the surface plane
$\mathbf{q}$	conjugate variable in the Fourier space
ζ_q	Fourier component of ζ having the wavevector $\mathbf{q}$
ξ_q	Fourier component of the horizontal surface displacement parallel to $\mathbf{q}$
$G(\mathbf{q}, t)$	space Fourier transform of the correlation function $\langle\zeta(\mathbf{r}, t)\zeta(\mathbf{r}', t')\rangle$
$P(\mathbf{q}, \omega)$	space-time Fourier transform of the correlation function $\langle\zeta(\mathbf{r}, t)\ \zeta(\mathbf{r}', t')\rangle$
$\chi(\mathbf{q}, \omega)$	response function to an external pressure p_{ext}: $\chi(\mathbf{q}, \omega) = \zeta_q(\omega)/p_{ext}$
γ	interfacial tension
g	gravity constant
T	absolute temperature
k_B	Boltzmann constant
ρ_1	density of the upper fluid
ρ_2	density of the lower fluid
l_c	capillary length $(\gamma/(\rho_2 - \rho_1)g)^{1/2}$
η_1	viscosity of the upper fluid
η_2	viscosity of the lower fluid
n_1	refractive index of the upper fluid
n_2	refractive index of the lower fluid
h	monolayer thickness
Π	monolayer surface pressure

γ_0	interfacial tension without the monolayer ($\gamma = \gamma_0 - \Pi$)
ε	surface compression elasticity
κ	surface compression viscosity
s	surface shear elasticity
η_s	surface shear viscosity
μ	surface transverse viscosity
$\tilde{\gamma}$	complex surface tension ($\tilde{\gamma} = \gamma + i\omega\mu$)
$\tilde{\varepsilon}$	complex surface compression elastic modulus ($\tilde{\varepsilon} = \varepsilon + i\omega\kappa$)
$\tilde{s}$	complex surface shear elastic modulus ($\tilde{s} = s + i\omega\kappa$)
K	bending elastic modulus
ξ_K	persistence length
Σ	area per molecule at the surface
Γ	surface concentration of an adsorbed species
c	bulk concentration of this species
$\mathbf{n}$	unit vector parallel to the mean molecular orientation in liquid cristalline phases
ξ	correlation length
$\mathbf{v}$	hydrodynamic velocity
p	pressure
m	inverse penetration depth of surface waves: $m = q(1 - i\omega\rho/\eta q^2)^{1/2}$
σ_{ij}	strain tensor
$\mathbf{F}$	surface force
F	free energy
f	free energy per unit volume
$\mathcal{F}$	free energy per unit area
A	surface area
$\Delta = 0$	dispersion equation of surface waves
S	reduced frequency: $S = i\omega\tau_0$
τ_0	characteristic viscous time: $\tau_0 = \rho/(2\eta q^2)$
y	reduced capillary number: $y = \gamma\rho/(4\eta^2 q)$
ω_c	propagation frequency of the capillary waves
$\Delta\omega_d$	inverse damping time of the capillary waves
$\mathbf{E}$	electric field created by the electromagnetic radiation
$\mathbf{P}$	electric polarization created by the electromagnetic radiation
$\mathbf{E}_i$	incident electric field
$\mathbf{E}_r$	reflected electric field
$\mathbf{E}_t$	transmitted electric field
$\mathbf{E}_s$	scattered electric field
$\mathbf{E}_{lo}$	local oscillator electric field
k_0	modulus of the wavevector of the electromagnetic radiation in vacuum

k	modulus of the wavevector of the incident electromagnetic radiation; $k = n_i k_0$
$\mathbf{k}_i$	incident wavevector
$\mathbf{k}_r$	reflected wavevector
$\mathbf{k}_t$	transmitted wavevector
$\mathbf{k}_d$	scattered wavevector around refracted beam
$\mathbf{k}$	scattered wavevector around reflected beam
ε_0	electrical permittivity in vacuum
μ_0	magnetic permittivity in vacuum
ω_0	frequency of the incident electromagnetic radiation
λ	wavelength of the incident radiation
$\mathbf{A}$	amplitudes of the electric fields
$\mathbf{s}$	unit vectors along the light beams
θ_i	angle between the incident beam and the normal to the interface
θ_t	angle between the transmitted beam and the normal to the interface
θ	angle between the beam scattered around reflection and the normal to the interface
θ_d	angle between the beam scattered around transmission and the normal to the interface
$\Delta\theta$	angle between the incident and the scattered beams
θ_{lo}	angle between the beam scattered by the local oscillator and the normal to the interface
$\delta\theta_o$	angular beam diameter due to diffraction
ϕ	angle between the planes of incidence and of scattering
n_i	refractive index for the incident beam
n_t	refractive index for the transmitted beam
n	relative refractive index n_t/n_i
θ_1	limit angle: $\sin \theta_1 = n$
R	reflection coefficient
I	scattered light intensity
$dI/d\Omega$	scattered light intensity per unit solid angle
I_0	incident light intensity
ω_m	peak frequency of the power spectrum of the scattered light
$\Delta\omega$	half-width of the power spectrum
$\Delta\omega_I$	instrumental half-width of the power spectrum
l	size of the illuminated area
d	distance between the pinhole and the reflected spot in the focusing plane
D	distance between the surface and the reflected spot in the focusing plane

Index